Sustainability Sciences in Asia and Africa

Agroecosystem Dynamics and Sustainable Practices

Sustainability as a scientific endeavor requires broad understanding of the interconnections in our global environment. It is therefore positioned at the crossroads of research in natural sciences, technology and human sciences. This book series supports the global efforts by providing timely coverage of the progress, opportunities, and challenges towards achieving sustainable development goals (SDGs) in Asia and Africa. It fosters the research in transdisciplinary academic fields spanning across agriculture and global food security, health and diseases, access to clean water and affordable energy, climate resilience and effects of global climate change on biodiversity, water and land. This series is planned to narrate the stories of scientific development and progression towards SDGs from the regions of Asia and Africa. The overall aim of the series is to provide a comprehensive publication platform and act as a knowledge engine in the growth of sustainability sciences with a special focus on developing nations. The series constitute of four sub-series each focusing on a specific theme closely linked to a set of SDGs. Each sub-series is leaders and experts in agricultural sciences, environmental sciences, biomedicine and medical sciences respectively. Books in this series are useful for environmentalists, biotechnologists, plant scientists, medical practitioners, conservation biologists, policy makers, NGOs and scientific bodies working in the areas of sustainability sciences in the developing regions.

This book series publishes research associated with the prevailing opportunities and challenges faced by agroecosystems in Asia and Africa. It also explores the future scope of agroecological techniques and offers guidance on sustainable management of these systems. The series discusses important areas such as sustainable agricultural production, soil health, water resource management, biodiversity conservation, animal health and productivity, insect management strategies, genome editing and climate change mitigation. Further, it includes the challenges faced by African agriculture in arid and semi-arid regions, particularly in response to climate change. It highlights issues such as water shortages, soil degradation, desertification, and the need for sustainable farming techniques.

Sustainable agroecosystem maintenance entails strategies that improve overall environmental health and economic viability while limiting negative environmental impacts. The series takes a multidisciplinary approach to investigate creative ways for boosting sustainable agriculture, increasing food security, and improving lives for people in global south. This book series acts as a global publishing platform to collate and disseminate knowledge on fostering resilient and productive agroecosystems, thus enhancing accessibility, and fostering informed decision-making.

Abd El-Hafeez Zohry • Samiha Ouda
Editors

Resilient Agroecosystems

Innovations in Cropping Systems and Climate Change Mitigation

Springer

Editors
Abd El-Hafeez Zohry
Crops Intensification Research Department,
Field Crops Research Institute
Agricultural Research Center
Giza, Egypt

Samiha Ouda
Water Requirements and Field Irrigation
Research Department, Soils, Water and
Environment Research Institute
Agricultural Research Center
Giza, Egypt

ISSN 2730-6771 ISSN 2730-678X (electronic)
Sustainability Sciences in Asia and Africa
ISSN 3059-247X ISSN 3059-2488 (electronic)
Agroecosystem Dynamics and Sustainable Practices
ISBN 978-981-96-9629-1 ISBN 978-981-96-9630-7 (eBook)
https://doi.org/10.1007/978-981-96-9630-7

This Springer imprint is published by the registered company Springer Nature Singapore Pte Ltd.
The registered company address is: 152 Beach Road, #21-01/04 Gateway East, Singapore 189721, Singapore

Preface

Degradation is adversely affecting our agroecosystem, hence impacting food security. These realities prompted the creation of this book. It is crucial to conserve our agroecosystem to maintain its vital functions and avert damage. There is a necessity to implement innovations that will enhance our agricultural systems, while concurrently strengthening the resilience of Egypt's agroecosystem and mitigating its deprivation. This book aims to highlight the significance of innovations in agricultural systems and climate change mitigation for Egypt and other scholars in the discipline. It integrates innovations that enhance the agroecosystem's resistance to environmental stressors and provides a range of sustainable agroecological systems by prioritizing the use of diverse resources in agricultural output. This book may be advantageous for graduate students, researchers, policymakers, and extension workers.

Finally, we would like to extend our heartfelt appreciation to the Springer team, with a special mention to Sanchi Bhimrajka, our dedicated Publishing Editor. Your unwavering support and expert guidance have been genuinely invaluable throughout the publishing journey.

We are deeply grateful to the Series Editors, Prof. Kamel A. Abd-Elsalam from the Agricultural Research Center in Giza, Egypt, and Prof. Olubukola O. Babalola, Vice-President of The World Academy of Sciences (TWAS) and a distinguished Research Director at North-West University, South Africa. Your visionary contributions and exceptional expertise have significantly improved the Agroecosystem Dynamics and Sustainable Practices series, solidifying it as a significant asset in the field.

Additionally, we acknowledge Prof. Adel Abdel Azeem, President of the Agricultural Research Center in Egypt, and the series editorial board, for their unwavering commitment to the advancement of knowledge and the promotion of development in the field of agroecosystem dynamics and sustainable practices. Your unwavering dedication has been instrumental in the successful completion of this substantial undertaking.

Giza, Egypt
March 2025

Abd El-Hafeez Zohry
Samiha Ouda

Contents

Introduction

1

Abd El-Hafeez Zohry

Abstract

This book examines advancements in agricultural systems and climate change mitigation, emphasizing their importance for Egypt and other nations. It offers various designs of sustainable agroecological systems by emphasizing the use of diverse resources in agricultural production and presents innovations that improve the agroecosystem's resilience to environmental stresses. Thus, the book intends to encompass five main divisions, each comprising many chapters, in addition to a chapter dedicated to conclusions and recommendations. The five primary divisions encompass innovative cropping systems for resilience, augment the soybean intercropping system with maize, untraditional approaches to maximize agricultural production, climate change impacts and mitigation, and machine learning use in crop production. This book's expertise is beneficial for graduate students, researchers, policymakers, and extension workers.

The objective of the 2030 Agenda for Sustainable Development, ratified by all United Nations member states in 2015, is to attain peace and prosperity. It established 17 global Sustainable Development Goals (SDGs) to achieve this objective. The SDGs emphasize the interrelations among the environmental, social, and economic dimensions of sustainable development. Goal 15 is focused on "protecting, restoring, and promoting the sustainable use of terrestrial ecosystems, managing forests sustainably, combating desertification, halting and reversing land degradation, and preventing biodiversity loss." Sustainability, central to the Sustainable Development Goals (SDGs), necessitates the judicious use of soil and water resources as a critical milestone for preserving our agroecosystem for future generations and enhancing its resilience against climate change to ensure food production.

A. E.-H. Zohry (✉)
Crop Intensification Research Department, Field Crops Research Institute, Agricultural Research Center, Giza, Egypt
e-mail: azohry1954@arc.sci.eg

A. E.-H. Zohry, S. Ouda (eds.), *Resilient Agroecosystems*, Sustainability Sciences in Asia and Africa, https://doi.org/10.1007/978-981-96-9630-7_1

Agricultural systems are fundamentally linked to the environment. Human land use practices have negatively impacted natural resources, including freshwater, and arable soil, while also exacerbating climate change effects, such as rising temperatures, variable precipitation patterns, and extreme weather events. The difficulties of salinity and drought are concurrently intensifying. These conditions make agriculture uncertain in several areas, particularly for small and marginal farmers. The term agroecosystem includes plant populations, their genetic resources, and the environmental conditions they depend on, such as soil, hydrology, climate patterns, fauna, and environmental management practices informed by indigenous technological knowledge.

This book emphasizes the importance of innovations in agricultural systems and climate change mitigation for Egypt and other countries. Thus, the book intends to encompass five primary divisions, each comprising many chapters, in addition to an introduction (Chap. 1) and a chapter dedicated to findings and recommendations (Chap. 18).

Part I, titled "Innovative Cropping Systems for Resilience," comprised two chapters. Chapter 2 titled "Agroecological Intensification of Legume-Cereal Intercropping Systems: The 4C Approach," examined the notion of agroecological intensification in legume-cereal intercropping systems as an innovative method that enhances the resilience of the agroecosystem. The 4C framework categorizes interspecific plant-plant interactions into four principles: competition, complementarity, collaboration, and compensation, which were comprehensively examined. The ratios of land and not to water equivalents were analyzed using examples from Egypt to assess the partial relationships among species within intercropping systems.

Chapter 3, titled "Innovative Cropping Systems to Maximize the Use of Available Natural Resources," underscores the significance of innovative cropping systems in increasing land productivity and use efficiency, thus promoting sustainability. The discussion focused on using intensive cropping systems to mitigate shading stress and natural insect issues, utilizing results from prior field trials undertaken in Egypt.

Part II, titled "Augment the Soybean Intercropping System with Maize," comprises three chapters. Due to its significance as a major global legume crop, the enhancement of soybean intercropping with maize for light interception, along with the contribution of advanced breeding to develop resistant varieties against shading stress and insect infestations, was also addressed.

Chapter 4, titled "Enhancing the Efficiency of Light Interception in Soybean and Maize Intercropping System," examines the correlation between light interception and soybean productivity within an intercropping system with maize. This chapter discusses methods to enhance soybean's resistance to shading stress in its intercropping system with maize. The chapter examined various issues concerning the interplay between light intensity and the soybean rhizosphere within its intercropping system with maize, the influence of soybean plant height and maize plant spacing on light interception and subsequent productivity, and the impact of diverse planting arrangements of soybean and maize on light use efficiency and relative photosynthetic potential.

Chapter 5 presents an intriguing discourse on the "Mitigation of Shading Stress on Soybean Varieties Through Intercropping with Maize." The chapter analyzed the role of "land equivalent ratio" in determining compatibility and competitiveness within the intercropping system of soybean and maize. Additionally, the function of the "system production index" in differentiating between complementary and complimentary within the intercropping system of soybean and maize was addressed. The chapter also addressed the parameters for implementing a successful soybean and maize intercropping system, namely, the selection of shade-tolerant soybean cultivars and the optimal plant population densities and spatial arrangements within the intercropping system. Finally, the capacity of five examined Egyptian soybean cultivars to endure shading stress within an intercropping system with maize was comprehensively analyzed in prior research conducted in Egypt.

Chapter 6 introduced a shift in the discourse around breeding procedures. The title of the chapter is "Advanced Soybean Breeding Techniques to Alleviate Biotic and Abiotic Stresses Under Intercropping." The principal elements of soybean breeding, encompassing genetic breeding and molecular techniques, were discussed. Additionally, the effects of intercropping soybean with maize on mitigating soybean mosaic virus damage and insect infestations were examined. The impact of cropping systems on gene expression as a potential technique for disease management in soybean fields was also addressed. Genetic investigation of certain Egyptian soybean cultivars for their susceptibility to soybean mosaic virus and insect infestations under intercropping conditions was also presented.

Part III, titled "Untraditional Approaches to Maximize Agricultural Production," presents unconventional strategies aimed at enhancing agricultural output sustainably. Chapter 7, titled "Opportunities and Challenges of Greenhouses and Soilless Agricultural Systems in Combating Food Insecurity," examined the cultivation techniques of greenhouses and soilless agricultural systems, necessitated by water scarcity, especially in arid and semi-arid regions. The chapter examined the efficacy of these potential strategies in tackling food production difficulties and addressed the obstacles to their implementation. Additionally, an evaluation of prior scientific endeavors aimed at innovating technology to address barriers to their general acceptance and to attain resilient and sustainable agroecosystems for future generations was also presented.

Chapter 8, titled "Using Fishponds Wastewater to Enhance Agricultural Crop Productivity and Sustainability," discusses the advantages of employing fishpond wastewater to sustainably improve crop output through nutrient recycling and waste reduction. The chapter also detailed the precautions that tackle issues such as nutrient imbalance and regulatory compliance. A thorough evaluation of existing methods, technology, and research in aquaculture systems was conducted, emphasizing the enhancement of production efficiency, the promotion of environmental sustainability, and the assurance of economic viability. This review integrates ideas from current studies to provide a valuable resource for researchers, policymakers, and stakeholders, offering suggestions to enhance sustainable aquaculture methods.

Part IV, titled "Climate Change Impacts and Mitigation," was delineated throughout six chapters. Chapter 9, titled "Adaptation Techniques to Irrigation Water

Shortages Due to Climate Change," discusses potential solutions to alleviate the impacts of climate change on water availability. These solutions include enhancing water storage and management, improving drought monitoring and early warning systems, adopting climate-smart agricultural practices and water reuse, protecting natural buffers, promoting rainwater harvesting, and increasing irrigation water productivity. Implementing these solutions requires a combination of technological advancements, governmental initiatives, and community-based methods to enhance water security and build resilience against the impacts of climate change on water supply.

Chapter 10, titled "Climate Change Effects on Sustainable Field Crops Production in Egypt: Challenges and Prospective," examined the impact of climate change on agricultural crops in Egypt, highlighting its effects on water resources and crop yield. The initiatives from the National Egyptian Strategic Plan for climate change mitigation and adaptation were also analyzed. Additionally, the review encompassed biotechnological knowledges, including genetic engineering and other molecular breeding techniques with traditional breeding, which are crucial for cultivating robust crops that can withstand the biotic and abiotic difficulties posed by climate change.

Chapter 11, titled "AquaCrop Model for Simulating Maize Productivity with Varying Planting Dates Under Deficit Irrigation in the Context of Climate Change," serves as a case study examining the impact of climate change on maize productivity in Egypt, where maize is a crucial summer cereal crop for food and feed purposes. The effects of three planting dates and three irrigation levels on maize grain yield, biological yield, and water productivity were simulated using the AquaCrop version 7.1 model using data from a 2-year field experiment. Subsequent to model validation, data from the CMIP6 mean projection for the SSP5-8.5 future scenario spanning two-time intervals, namely, 2020–2039 and 2040–2059, were utilized to execute the model and assess projected losses in maize grain and biological yields under the specified future scenario.

Chapter 12, "Legume-Based Intercropping Systems Mitigate Climate Change," emphasizes the crucial importance of legume-based intercropping systems in bolstering resilience to climate change. Legume-based intercropping systems offer numerous advantages that serve as an advanced agroecosystem strategy to alleviate climate change impacts, hence improving agricultural sustainability and soil security in various agroecosystems.

Chapters 13 and 14 concentrated on the soil of the Nile Delta in Egypt. Chapter 13 titled "Evaluating Soil Degradation Processes and Their Exacerbation by Climate Change in the Nile Delta of Egypt," a crucial area for the nation's agricultural output, where climate change accelerates various degradation phenomena, including erosion, waterlogging, land subsidence, salinization, and depletion of soil organic matter. The issues associated with soil degradation in the Nile Delta were analyzed, particularly the impact of escalating salinization and land subsidence, notably in coastal regions, as a consequence of rising sea levels induced by global warming. Additional factors influencing land degradation were also addressed, specifically drought, depletion of soil organic matter, soil compaction, and desertification. An analysis of long-term temperature and precipitation in the Nile Delta governorates was conducted to assess historical trends and predict their future impact on land degradation. A range of adaptive techniques were also suggested.

Chapter 14, titled "Impact of Climate Change Drivers on Soil Microbial Communities and Nutrient Cycling in Agricultural Soils in the Nile Delta of Egypt," examines how climate change factors influence microbial diversity and functionality. The chapter underscores the role of soil microbial populations in mitigating climate change through nitrogen fixation, carbon sequestration, and biological control of plant diseases, with a focus on their impact on nutrient cycle and greenhouse gas emissions. Additionally, effective strategies for enhancing soil resilience were analyzed. This chapter emphasizes the importance of integrating microbial-based solutions into agricultural practices to mitigate the adverse effects of climate change on soil ecosystems.

Part V, titled "Machine Learning Use in Crop Production," delineates the utilization of machine learning in crop production over three chapters. Machine learning is becoming integral to agricultural systems, optimizing crop yield and managing water and soil resources effectively. Chapter 15, titled "Climate Change Impacts on Crop Yield Prediction: A Machine Learning Perspective," concentrated on the effects of climate change on crop output. This review specifically examines the primary machine learning methodologies utilized for investigating and assessing the effects of climate change on crop yield prediction.

Chapter 16, titled "Coupling Process-Based Models with Machine Learning for Robust Predictions of Soil, Water, and Crop Dynamics," addresses the enhancement of process-based models by integrating the mechanistic insights from these models with the data-driven strengths of machine learning algorithms, thereby improving forecasting accuracy and reliability in agricultural systems. This chapter illustrates the effectiveness of integrating process-based biophysical models with machine learning to enhance regional simulations and assesses the ramifications of this methodology, particularly concerning crop yields and soil organic carbon dynamics across different agricultural management practices.

Chapter 17, titled "Fusion of Process-Based Models, Machine Learning, and Remote Sensing for Yield Gap Assessment," explores the innovative amalgamation of process-based models, machine learning methodologies, and remote sensing to enhance the accuracy and dependability of yield gap evaluations. This chapter presents a comprehensive technique for yield gap assessments by integrating the benefits of different technologies, resulting in enhanced prediction accuracy and scalability. This amalgamation of sophisticated methodologies not only improves the accuracy of yield predictions but also offers practical insights for policymakers and agricultural stakeholders, facilitating the formulation of targeted, sustainable strategies vital for future food security amid resource limitations and climate change.

Part VI, Chap. 18 encompassing the final chapter delineates the conclusions and recommendations articulated by the writers.

Part I

Innovative Cropping Systems for Resilience

Agroecological Intensification of Legume-Cereal Intercropping Systems: The 4C Approach

2

Abd El-Hafeez Zohry, Samiha Ouda, and Fouad Khalil

Abstract

Agroecological intensification enhances the resilience and sustainability of food production by employing ecological principles to diversify cropping systems and use ecosystem services instead of agrochemicals and fossil fuels. Intercropping systems of legumes and cereal crops offer numerous advantages over monocrop farming, including sustained yields and enhanced quality of cereal seeds. It enhances the efficiency of soil nitrogen resources and biological nitrogen fixation, hence reducing the necessity for chemical nitrogen and phosphorus fertilizers. Additionally, the legume-cereal intercropping system demonstrated an enhancement in water use efficiency. The 4C approach can be utilized to assess the utilization of natural resources in intercropping systems relative to sole crop cultivation. The 4C approach can be employed to evaluate the exploitation of natural resources in intercropping systems compared to monocrop cultivation. The four joint impacts of competition, complementarity, collaboration, and compensation, which classify interspecific plant-plant interactions, are integrated to elucidate the 4C technique. This chapter examined the implementation of legume-cereal intercropping systems, to enhance the sustainable utilization of natural resources through promoting beneficial above- and below-ground interactions. The characterization and performance assessment of legume-cereal intercropping systems were conducted to understand the 4C effects. We examined the interrelations among the 4Cs to ascertain and elucidate the yield advan-

A. E.-H. Zohry (✉)
Crop Intensification Research Department, Field Crops Research Institute, Agricultural Research Center, Giza, Egypt
e-mail: azohry1954@arc.sci.eg

S. Ouda · F. Khalil
Water Requirements and Field Irrigation Research Department, Soils, Water and Environment Research Institute, Agricultural Research Center, Giza, Egypt

A. E.-H. Zohry, S. Ouda (eds.), *Resilient Agroecosystems*, Sustainability Sciences in Asia and Africa, https://doi.org/10.1007/978-981-96-9630-7_2

tages. Utilizing examples from Egypt, we examine the land and water equivalent ratios to evaluate the partial relation across species within the intercropping systems.

Keywords

Competition · Complementarity · Cooperation · Compensation · Land equivalent ratio · Water equivalent ratio

Introduction

The swift rise in global food demand, especially for cereal-based cropping systems, has necessitated agricultural development, resulting in economic and environmental consequences (van Dijk et al. 2021) and requiring sustainable practices (Klöffel et al. 2022). The heightened significance of preserving environmental quality led to the emergence of the concept of "agroecological intensification" (Milder et al. 2019). Agroecological intensification has been advocated to mitigate adverse environmental impacts, increasing interest in different cropping systems (Bommarco et al. 2013). Wezel et al. (2015) characterized agroecological intensification as the enhancement of agricultural performance through the reduction of external inputs and the minimization of environmental impact by incorporating ecological principles into farm system management. Lampkin et al. (2015) defined agroecological intensification as a means to augment the resilience and sustainability of food production.

Agroecological intensification enhances the agricultural system by replacing agrochemicals and fossil fuels with environmental services through the adaption of ecological concepts. Agroecological intensification seeks to optimize productivity through nutrient recycling on the farm, while safeguarding natural resources, maintaining soil structure (Gambart et al. 2020), mitigating land degradation (Jhariya et al. 2021), and enhancing food production with constrained resources (Naazie et al. 2023). Intercropping as a method of crop diversification is a crucial strategy for agroecological intensification (Justes et al. 2021).

Intercropping is an agroecological method involving the cultivation of two or more crops simultaneously on the same field for part of the growing season (Bedoussac et al. 2015). Xu et al. (2020) stated that intercropping is regarded as the "new green revolution" since it enhances land efficiency by utilizing knowledge of species complementarity and facilitates agroecological intensification. Crops in an intercropping system are typically chosen for their complementing species traits, such as variations in height, growth rates, root system configurations, or leaf structure (Wezel et al. 2014). Intercropping often improves soil quality (Bernard and Lux 2017), maximizes solar radiation absorption (Tilman 2020), mitigates diseases, insects, and weeds (Bybee-Finley and Ryan 2018), and promotes natural pest antagonists (Yousefi et al. 2024). Furthermore, intercrops may offer greater output per unit area or enhance farmer profitability compared to sole crops (Dowling et al. 2021).

The challenge of developing agroecosystems that achieve high crop yields with low input levels has heightened interest in the adoption of legume-cereal intercropping systems (Brooker et al. 2015). Legume-cereal intercropping systems confer numerous beneficial agroecological effects: they enhance crop diversification and farm nitrogen self-sufficiency (Tosti et al. 2023), improve soil health (Ditzler et al. 2021), control pests and weeds (Daryanto et al. 2020), and mitigate yield loss while achieving yield stability (Koskey et al. 2022). Legumes form symbiotic relationships with rhizobia in the soil, fix atmospheric nitrogen (Quinones et al. 2022), facilitate nutrient cycling, and improve soil water retention (Mrunalini et al. 2022).

The interaction processes among intercrops, which facilitate resource utilization, elucidate the potential advantages of legume intercropping systems. The 4C approach facilitates a comprehensive investigation of how its various principles enhance the good consequences of intercropping (Justes et al. 2021). The 4C approach categorizes interspecific plant-plant interactions into four principles: competition, complementarity, collaboration, and compensation. Multiple ecological processes occur concurrently and dynamically during the growing season, resulting in these consequences (Justes et al. 2021).

This chapter examined the implementation of agroecological intensification, namely, the legume-cereal intercropping systems, to enhance the sustainable utilization of resources by promoting beneficial above- and below-ground interactions and maximizing the relationships among the diverse species involved. The characterization and performance assessment of legume-cereal intercropping systems were conducted to understand the 4C effects. We examined the interrelations among the 4Cs to ascertain and elucidate the yield advantages. Utilizing examples from Egypt, we examine the land and water equivalent ratios to evaluate the partial relation across species within the intercropping system.

The Principles of the 4C Approach in Intercropping Systems

According to Justes et al. (2021), the relative contribution of the principles of the 4C approach can be utilized to evaluate the yield-related advantages of intercropping. When interspecific competition among intercrops is inferior to intraspecific competition within a sole crop, complementarity effects manifest, leading to productivity enhancements (Brooker et al. 2015). Improving agronomic features in intercropping systems might mitigate the adverse consequences of competition while promoting complementarity, cooperation, and compensation (Kiaer et al. 2022). Therefore, it is essential to pursue the task aimed at optimizing the advantages of intercropping.

Competition

Competition arises when a finite availability of abiotic resources is present in both space and time. If a species can utilize these scarce resources more effectively, it is deemed the superior competitor (Holomuzki et al. 2010). Moreover, if species within the mixture possess identical or comparable resource acquisition capabilities, competition persists throughout their overlapping growing season (Justes et al. 2021). Plants engage in competition when abiotic resources are scarce in both spatial and temporal contexts. A species is deemed the superior competitor if it can more effectively exploit these limited resources (Holomuzki et al. 2010). Moreover, competition continues during their mutual growth period if the intercropped species exhibit similar or equivalent abilities to acquire the available resources (Justes et al. 2021).

In terrestrial competition, plants vie for light, water, and nutrients. Competition can be categorized into two types: above-ground and below-ground competition. Light is a vital abiotic resource for plants throughout their life cycle. The quality, intensity, and spectral composition of light can regulate the spatial structure of certain crops, such as soybeans, whose leaves' photosynthetic capacity is affected by their spatial arrangement (Paradiso and Proietti 2021). As suggested by Black and Ong (2000), intercropping is a three-dimensional optical system with a substantial light-specific surface area, enhancing lateral light interception. Cui et al. (2015) asserted that the increase in dry matter accumulation in intercropping systems results from the multi-tiered nature of the crop group, which may enhance solar radiation and stratified light energy use, along with its efficient application. Additionally, Zhang et al. (2020) reported that increased light interception and utilization efficiency correlate with enhanced productivity in intercropping systems.

Competition for light in intercropping systems arises when shorter plants are intercropped with taller ones. Intercropping shorter C3 plants with taller C4 plants enhances light utilization efficiency. C4 species possess higher saturation point than C3 species, enabling them to exploit greater light intensities by intercropping compared to monocropping (Reynolds et al. 2007). Moreover, the C3 intercrops experience shading in intercropping, resulting in reduced light intensity, which enables them to utilize light more efficiently than monocultures (Harris et al. 1987). Consequently, the detrimental impacts of competition are mitigated. MacLaren et al. (2023) proposed an approach including the intercropping of tall plants with a high specific leaf area alongside short plants with a low specific leaf area, perhaps reducing competition.

In the soybean and maize intercropping system, soybean plants experience shading throughout the concurrent growth period with maize (Hussain et al. 2020). The red-to-far-red light ratio within the soybean canopy is decreased in this intercropping system (Yang et al. 2020). Liu et al. (2008) proposed that modifying soybean planting density is crucial for optimal crop growth, maximizing biomass and grain yield. In Egypt, Rashwan and Zen El-Dein (2017) examined various planting densities of intercropped soybean and maize, concluding that intercropping 50% soybean with 100% maize yielded the highest soybean productivity due to reduced interspecific competition between the two crops. Simultaneously, intercropping 66% soybean with 66% maize produced the maximum maize production, attributed to significant intraspecific competition among soybean plants in this arrangement. Abdel-Wahab and Abdel-Wahab (2020) determined that the soybean cultivar Giza11 exhibited greater tolerance for intercropping with maize at its recommended planting density and yielded the maximum productivity, in comparison to the Giza21 and Giza82 cultivars. Furthermore, Abdel-Wahab and Abd El-Rahman (2016) demonstrated that the canopy architecture of soybean influences light intensity within the canopy, hence affecting its seed yield. They also stated that soybean cultivar Giza22 yielded more than the Giza11 variety. Metwally et al. (2012) indicated that the earlier cultivar possessed narrow leaves, allowing greater solar radiation to infiltrate its canopy, whereas the latter cultivar featured broad leaves, resulting in reduced solar radiation penetration. Consequently, light intensity within the Giza22 canopy exceeded that of Giza11.

A further instance of cultivating tall and short species as intercrops is the maize intercropping system with peanut, which enhances light interception through expanded ground cover (Zhang et al. 2020). Li et al. (2019) showed that maize productivity in intercropping systems with peanuts surpassed that of peanuts, which experienced a 32% reduction relative to their yields under monoculture conditions. They also indicated an enhancement in light transmission ratio under the maize canopy and an increase in post-silking dry matter accumulation, relative to the values recorded under mono-maize cultivation. Hamd-Alla et al. (2023) in Egypt indicated that intercropping 50% of maize planting density with 100% of peanut planting density revealed maize as the more competitive species.

Below-ground competition generally exerts a greater influence on the equilibrium of competing species than above-ground competition (Hassan et al. 2020). Reports indicate that, in a legume-cereal intercropping system, below-ground resource acquisition is significantly linked to light partitioning (Neugschwandtner and Kaul 2014). Cereal crops have greater competitiveness for soil nitrogen compared to legume crops, attributed to their accelerated root development, significantly influencing N_2 fixation, foliar growth, and overall crop development (Corre-Hellou et al. 2006). Ren et al. (2021) discovered that in a soybean-maize intercropping system, maize exhibited greater nitrogen and phosphorus uptake compared to the sole crop, but soybean shown reduced nitrogen and phosphorus uptake relative to the sole crop. These findings indicate that maize benefits from enhanced nutrient uptake when intercropped with soybeans, as opposed to sole cropping, highlighting the major impact of below-ground competition on crop development.

Complementarity

According to Brooker et al. (2021), complementarity is defined as "the effects that occur when niche space (i.e., physical, chemical and biological resources) is used more completely as more species are added in (niche differentiation), or when interactions between neighboring plants have beneficial outcomes for at least one of the neighbors (facilitation)." Therefore, complementarity may arise when diverse species are grown with varying abiotic resource requirements in time and place (Justes et al. 2021). The greater and more steady yield produced from intercropping is mainly attributed to abiotic resource complementarity (Raza et al. 2020), where adequate use of the available resources by intercrops exists due to different spatial

(Schöb et al. 2015), temporal (Yang et al. 2019), and phenological characteristics (Martín-Robles et al. 2019). Complementarity can be attributed to a two-sided interaction: niche differentiation or facilitation (Loreau and Hector 2001). Schöb et al. (2015) stated that the overall performance of intercrops in a mixture is promoted by complementarity effects, causing a more significant net resource usage.

The degree of complementarity is influenced by various factors, including the planting density of intercropped species (Ren et al. 2016) and both intra- and interspecific competition (Yang et al. 2015). Intercropping has been shown to facilitate complementary nitrogen utilization between legumes and cereals by enhancing N_2 fixation and the acquisition of soil nitrogen by cereals (Rodriguez et al. 2020). Moreover, the nitrogen-fixing capacity of legumes is significantly enhanced when intercropped with cereals, resulting in increased soil nitrogen levels and improved cereal yields (Yu et al. 2016).

A prominent example of complementarity is intercropping legumes with cereals, where legumes use atmospheric N_2 through biological nitrogen fixation in symbiosis with soil bacteria (Fustec et al. 2010). In contrast, cereals cannot fix nitrogen; instead, they utilize it from the soil. Thus, legumes' enrichment in soil nitrogen subsequently enhances cereal yield (Ivanov et al. 2012). According to Ren et al. (2017), intercropping soybean with maize improves nitrogen uptake by maize through absorption of the leached nitrogen from the root zone of soybean, in addition to promoting more nitrogen fixation by soybean to be transferred to maize. Furthermore, Ren et al. (2021) reported that soybeans could release protons and mobilize insoluble soil phosphorus as a result of acidification of the rhizosphere through exuding malate and citrate, which could increase the soil phosphorus and increase its absorption by maize. According to Guo et al. (2021), intercropping maize with peanuts increases beneficial soil microbes, improves phosphonate and phosphinate metabolism, and makes phosphorus more accessible for maize. Moreover, intercropping cowpeas with maize enhanced soil health by fixing more N_2 replacing soil nitrogen absorbed by cereals (Boukar et al. 2013).

An additional illustration occurs when the abiotic resource requirements of intercrops diverge, such as when two intercrops exploit light across distinct canopy strata (Mao et al. 2012). Ghanbari et al. (2010) assert that intercropping cowpea with maize is effective because cowpea establishes ground cover early in the season, unlike maize (Haruna et al. 2018), facilitating the cowpea's absorption of photosynthetically active radiation in the lower strata. Additionally, a complementary effect was observed in the soybean and maize intercropping system, where the canopy's light transmission improved and light energy wastage diminished (Ren et al. 2021).

Raza et al. (2024) indicated that the efficiency of an intercropping system can be quantified through an index termed "net biodiversity effect (NE)," which comprises two components: complementarity effect (CE) and selection effect (SE). According to Mahmoud et al. (2022), CE is defined by the total productivity derived from both intercrops, while SE reflects the competitive superiority of one intercrop over another within the intercropping system. An NE value over zero indicates that the yield of intercrops surpasses that of sole crops, suggesting a benefit to adopting an intercropping system (Elhakeem et al. 2019). Raza et al. (2024) demonstrated that intercropping chickpeas with wheat and peas with wheat on a 0.6 m strip width yielded a value of CE higher than the value of SE and resulted in a positive value of NE. A negative NE value was seen due to the SE exceeding the CE values when these systems were established on a 1.2 m strip width. The findings underscore the significance of strip width and spatial configuration in facilitating more complementary and less competitive interactions among intercrops in resource acquisition (Khalid et al. 2023).

Cooperation

Cooperation is characterized as: "a behavior that yields an advantage to another individual (recipient), and is chosen due to its advantageous impact on the recipient" (West et al. 2007). Hinsinger et al. (2011) asserted that "cooperation is attained when the modification of the environment by one species is beneficial to the other(s), i.e., increasing the availability of soil nutrients." Consequently, cooperation is synonymous with facilitation, as noted by Liebmann (1995). The cooperative effect is observed in legume intercropping systems with cereals. Stagnari et al. (2017) reported enhanced soil organic carbon and humus content when legumes were intercropped with cereals. Legume crops can elevate soil organic carbon and nitrogen; hydrogen gas is released during biological nitrogen fixation, promoting bacterial growth in the rhizosphere (Li et al. 2020). Furthermore, Ferguson et al. (2013) illustrated that legumes can more effectively extract calcium and magnesium from the soil than cereals, substituting them with hydrogen, which results in the removal of OH^- ions and an increase in H^+, thereby lowering soil pH. In Egypt, Abbady et al. (2016) reported a 17% increase in soil organic matter under a pea-wheat intercropping system compared to sole wheat cultivation. Zohry et al. (2020) noted an 11% rise in soil organic matter with the same intercropping system relative to sole wheat. Consequently, wheat yields were superior under intercropping compared to monocultivation. Additionally, Ouda et al. (2020) observed an enhancement in maize yield when intercropped with cowpea, attributed to improved soil nitrogen content compared to maize mono-cropping yields.

As explained by Chaoui et al. (2023), cooperation occurred when wheat was intercropped with chickpeas in low-phosphorus soil, resulting in increased inorganic phosphorus availability in the wheat rhizosphere and enhanced utilization efficiency. Moreover, in the soybean-maize intercropping system, there was an increase in soil phosphorus availability, alongside a rise in arbuscular mycorrhizal fungi diversity and macroaggregates within both maize and soybean rhizospheres (Lu et al. 2023), exemplifying another instance of cooperation. Additionally, the cultivation of peanuts in alkaline and calcareous soils, characterized by severe iron deficiency, adversely affects peanut yield and quality. Conversely, intercropping with maize enhances iron availability, thereby improving peanut photosynthetic efficiency and yield (Dai et al. 2019).

Cooperation in intercropping is most pronounced during the seedling phase. The mechanical resistance of the surface soil layer poses a challenge, yet intercrop seedlings can mitigate this issue (Klimek-Kopyra et al. 2013). Additionally, one species can physically support another, thereby reducing lodging (Justes et al. 2021). For instance, Cai et al. (2019) found that incorporating wheat into an intercropping system enhanced its lodging resistance, which subsequently improved photosynthesis and diminished yield losses. Another example is the intercropping of peas with barley, which facilitates the growth of peas by preventing lodging (Podgórska-Lesiak and Sobkowicz 2013). Likewise, the wheat-lentil intercropping system allows lentils to support wheat, thereby reducing mechanical harvest losses in organic farming (Loïc et al. 2018).

Cooperation is also evident when one intercrop enhances the biological control of the pests affecting another intercrop. Intercropping peanuts with maize has been demonstrated to improve pest management in peanuts by providing a habitat for predators that suppress peanut aphids (Ju et al. 2019). Additionally, intercropping maize with peanut and soybean significantly mitigated termite damage and reduced maize grain yield loss compared to monoculture maize (Sekamatte et al. 2003). According to Fernandez-Aparicio et al. (2007), the infestation of faba bean and pea by *Orobanche* was inhibited in an intercropping system with oat, where cereal roots exude allelochemicals to diminish infestation. In Egypt, Hamd-Alla et al. (2014) noted an enhancement in maize yield when intercropped with cowpea, attributable to a decrease in insect pests affecting maize and a reduction in accompanying weed growth (Zohry 2005).

Altering the microclimate around intercrops is an additional benefit of intercropping systems. In intercropping, the canopy's microclimate can mitigate elevated temperatures by diminishing air circulation, thereby decreasing evaporation rates and increasing relative humidity (Shumet et al. 2022). Midmore et al. (1988) indicate that the reduced temperature beneath the maize canopy, compared to the ambient air temperature, promotes the growth of groundnut intercropped with maize. Moreover, Wu et al. (2012) noted that taller maize plants can alter the microclimate of shorter cowpea plants, specifically regarding light and temperature.

Compensation

Döring and Elsalahy (2022) defined compensation as "a process by which an entity (an individual, system, or group of individuals) can regain or restore its normal functionality after losing a part of its structure or several individuals due to disturbance or stress." Compensation frequently occurs in intercropping, where one crop yields less than anticipated while another yields more (Willey 1979). The susceptibility of one intercrop to abiotic and biotic stress compared to another initiates compensation, whereby the failure of the first intercrop is offset by the second intercrop (Justes et al. 2021). Various approaches have been established for evaluating compensation in communities, wherein negative interactions among associates serve as metrics for compensation assessment (Bai et al. 2004). Gonzalez and Loreau (2009) propose an alternative conceptualization of compensation as "the ratio of the relationship between the variance of the aggregate variable (usually the sum of the component populations) and the variances of the individual species populations." In multi-species studies examining the relationship between production and diversity, Adler and Bradford (2002) recommended assessing compensation by excluding species from a species pool. The authors formulated a compensation index by calculating the difference between the observed and predicted biomass, divided by the anticipated biomass loss resulting from extinction.

Franco et al. (2018) indicated that the potential compensatory effect among intercrops can mitigate interspecific resource competition. Chen et al. (2018) observed that shorter-season peas, when relay intercropped in alternating strips with longer-season maize, primarily extracted soil water from the uppermost layers, while maize plants accessed water from the lower layers of the adjacent pea strips. The intercropped maize subsequently compensated for soil water loss from the pea strips post-harvest, fully recovering from growth suppression experienced during the concurrent growth period (Li et al. 2014). Additionally, Zhao et al. (2019) reported that this intercropping system resulted in higher accumulated nitrogen and carbon values in both maize and pea intercrops compared to those in sole crops, attributable to the compensatory effect of pea relay intercropped with maize.

Hu et al. (2016) established a measurement for compensatory effect (CE) utilizing crop growth rate (CGR) for both intercropped and sole main crop. A CE value exceeding 1.0 indicates that the intercropped main crop exhibited compensatory growth through enhanced dry biomass accumulation, whereas a value below 1.0 suggests the contrary:

$$\mathrm{CGR} = \left(W2 - W1\right) / \left(T2 - T1\right) \tag{2.1}$$

$$\mathrm{CE} = \mathrm{CGR}_{\mathrm{int}} / \mathrm{CGR}_{\mathrm{sol}} \tag{2.2}$$

where $W1$ and $W2$ are the biomass of the main crop in two growth stages: mid vegetative and maturity, respectively; $T1$ and $T2$ are days from planting to sampling, respectively; and $\mathrm{CGR}_{\mathrm{int}}$ and $\mathrm{CGR}_{\mathrm{sol}}$ are the growth rate of intercropped and sole crop, respectively.

Intercropping cowpeas with maize achieves a compensatory soil water effect, as noted by Ouda et al. (2020) in Egypt, where irrigation was administered to the system in the same quantity as for sole maize cultivation. Consequently, cowpea plants extract water from the upper soil layers, while maize plants absorb water from deeper layers. Following the harvest of cowpeas, the intercropped maize continued to utilize soil water that would have been consumed by cowpeas, resulting in a cowpea yield comparable to that of its sole crop and a maize yield exceeding that of its sole crop. Similarly, Hamd-Alla et al. (2023) intercropped peanuts (at 100% of the recommended plant population) with maize (at 25% of the recommended plant population), achieving a peanut yield equivalent to 96% of that from the sole crop and a maize yield surpassing its corresponding planting density by 42%, thereby underscoring the significance of the compensatory effect.

Table 2.1 Comparative total income value from intercropping systems and the value of the sole main crop

Intercropping systems	Income ($/ha)				References
	Intercropped crops				
	Secondary crop	Main crop	Total	Sole main crop	
Soybean/maize	578	1748	2326	2166	Abdel-Wahab and Abd El-Rahman (2016)
Cowpea/maize	323	1371	1694	1555	El-Ghobashi et al. (2020)
Maize/peanut	944	2265	3209	2430	Hamd-Alla et al. (2023)
Pea/wheat	1020	2801	3821	3169	Abd-Rabboh and Koriem (2021)
Faba bean/wheat	456	2064	2520	2334	Abdel-Wahab and El Manzlawy (2016)

Calculated by the authors

Willey (1979) posits that the compensatory effect of intercropping manifests when one intercrop yields less while the other yields more. Consequently, we compared the reported total income figures (yield multiplied by price for each crop) from various intercropping systems of legumes with cereals against the income from the sole main crop (Table 2.1). The table clearly indicated that the income from a sole main crop surpasses that of an intercropped main crop. However, when the income from the secondary crop was factored in, the overall income from the intercropping system exceeded that of the sole main crop, thereby underscoring the compensatory effect of intercropping.

Utilizing the 4C Approach to Assess Intercrops Performance

In a natural plant community, none of the 4C can exist autonomously. For example, complementarity does not preclude concurrent competition; it merely indicates that adequate resources for one intercrop can still be obtained despite resource appropriation by another intercrop (Justes et al. 2021). The predominant effects of 4C occur during the growing season and are notably assessed through the land equivalent ratio (LER) and the partial land equivalent ratio (pLER) of each intercrop within the system. LER quantifies the overall net impact on the total biomass of both intercrops in comparison to the biomass generated by sole crops (Yu et al. 2015). It serves as an index for evaluating the efficiency of land utilization within the intercropping system, as articulated by Rao and Willey (1980):

$$\text{LER} = \text{LER}_{\text{A}} + \text{LER}_{\text{B}} = \left(\text{Y}_{\text{int,A}} / \text{Y}_{\text{mono,A}}\right) + \left(\text{Y}_{\text{int,B}} / \text{Y}_{\text{mono,B}}\right) \tag{2.3}$$

where LER_{A}, $\text{Y}_{\text{int, A,}}$ and Y_{mono} are partial land equivalent ratios of crop A, the yield of intercrop A, and sole A, respectively. LER_{B}, $\text{Y}_{\text{int,B}}$, and $\text{Y}_{\text{mono,B}}$ are partial land equivalent ratio of crop B, of yield of intercrop B, and the yield of sole B, respectively. A

LER value exceeding 1 indicates superior land utilization efficiency for the intercropping system in comparison to sole cropping.

The water equivalent ratio (WER) is another statistic that correlates the yield of intercrops to their water utilization. Mao et al. (2012) defined WER as the volume of water necessary for a single crop to produce an equivalent yield to that of one unit of water used for intercrops, as follows:

$$\begin{aligned} \mathrm{WER} &= \left[\left(Y_{\mathrm{int,A}} / \mathrm{WU}_{\mathrm{int}} \right) / \left(Y_{\mathrm{mono,A}} / \mathrm{WU}_{\mathrm{mono,A}} \right) \right] \\ &+ \left[\left(Y_{\mathrm{int,B}} / \mathrm{WU}_{\mathrm{int}} \right) / \left(Y_{\mathrm{mono,B}} / \mathrm{WU}_{\mathrm{mono,B}} \right) \right] \end{aligned} \tag{2.4}$$

where $Y_{int,A}$, WU_{int}, $Y_{mono,A}$, and $WU_{mono,A}$ are yield of intercropped crop A, water use by the intercropping system, yield of monocrop A, and water use by monocrop A, respectively. $Y_{int,B}$, $Y_{mono,B}$, and $WU_{mono,B}$ are yield of intercropped crop B, yield of monocrop B, and water use by monocrop B, respectively. Similar to LER, a WER value exceeding 1 indicates superior productivity per unit of water compared to monoculture.

Impact of the 4C Method on the Final Yield of Intercrops

Abdel-Wahab and El Manzlawy (2016) intercropped faba bean (50% of its plant density) with wheat under different plant densities, namely, 100%, 83.3%, and 67%. They found that intercropped wheat plants using 67% density produced lower yield than its value under 100% density by only 7%, although the density was lower by 33%, which implied the complementary impact of faba bean in fixing atmospheric N_2 (Ivanov et al. 2012); thus, an increase in wheat productivity was observed. Cooperation effect was also observed under this intercropping system, where an increase in 100-grain weight in wheat under the lowest density was reported, which was higher than its value under sole wheat planting (high intraspecific competition exists) due to the increasing availability of soil phosphorus under intercropping with faba bean as reported by Kang et al. (2014) who stated that phosphorus is a vital macronutrient affecting wheat grain weight and yield. Furthermore, they found that the highest total LER was found under 50% faba bean intercropped with 67% wheat, namely, 1.32, which resulted in the highest pLER for faba bean (0.43) and the lowest pLER for wheat (0.89). Both values were higher than the percentage of planting density, and that reflected the compensatory effect of this intercropping system. The highest wheat yield was found under the highest plant density, where the wheat yield was lower than the obtained value of monoculture by only 3%, reflecting the interspecific competition between faba bean and wheat under intercropping.

El-Shamy et al. (2015) executed two intercropping configurations of soybean and maize, with soybean seeds sown on either side of a 70 cm ridge. Conversely, two maize intercropping patterns were established: the first involved planting one maize plant per hill at a spacing of 30 cm (1P/30 cm), while the second involved two plants per hill spaced at 60 cm (2P/60 cm). These configurations achieved 50% of

Fig. 2.1 Soybean intercropping system with maize

the recommended planting densities for both maize and soybean (Fig. 2.1). They also planted sole maize and soybeans, implemented under the same treatments. Their results indicated a high yield of intercropped maize under 2P/60 cm due to low interspecific competition between soybean and maize (short versus tall plants) and low intraspecific competition between maize plants. It was also higher than the yield obtained by sole maize due to high intraspecific competition between the plants. The highest intercropped soybean yield was obtained under maize intercropping pattern 2P/60 cm. The high yield of maize is attributed to the complementary effect of fixing atmospheric N_2 by soybean, which contributed to the increase in maize yield.

They also indicated that the highest percentage of light intensity was found for the 2P/60 cm pattern for both intercropping patterns when measured in the middle and bottom of the canopy, where low competition exists for light and intercropped maize benefited from higher light intensity than in sole cropping.

Furthermore, the authors found that the total rhizobia count was higher in both intercropping patterns than the value found under sole maize. Thus, soybeans promoted the complementary effect in its intercropping system with maize.

The authors of this study did not compute LER; consequently, we utilized their data to determine the partial and total LER (Table 2.2). The maximum pLER for soybean was recorded at 2P/60 cm, with a value of 0.61. This figure was deemed low due to the 100% planting density of soybean, indicating significant shading from maize plants adversely affecting soybean yield. Conversely, the highest pLER for maize was noted at 1P/30 cm, specifically 1.06, reflecting the substantial

Table 2.2 The values of partial LER and total LER of soybean intercropped with maize under two intercropping patterns

Intercropping pattern	$pLER_{soybean}$	$pLER_{maize}$	LER_{total}
1P/30 cm	0.55	1.06	1.61
2P/60 cm	0.61	1.04	1.65

Prepared by the authors

positive impact of this system on maize yield. Nonetheless, the peak LER_{total} was observed at 2P/60 cm, amounting to 1.65 (Table 2.2).

The Impact of the 4C Approach on Water Utilization by Intercrops

Intercropping systems confer multiple advantages regarding water utilization, such as reduced runoff, improved use of available soil moisture, increased overall yield, and enhanced productivity per unit of applied irrigation water, thereby augmenting water use efficiency (Blum 2011; Yang et al. 2011; Fan et al. 2006; Rahman et al. 2017; Chen et al. 2018).

Fahl clover, a variety of Egyptian clover, was intercropped in three recommended planting densities, i.e., 15%, 25%, and 35%, with 100% of wheat recommended planting density using three cultivation techniques, i.e., in basins (control), furrows, and raised beds under surface irrigation system, to determine which of the three techniques result in maximizing the use efficiency of land and water (Ouda and Zohry 2024). In the first season, they found that the highest yield of intercropped wheat was obtained under 15% planting density of fahl clover on a raised bed, which was lower than the obtained yield for sole wheat. However, in the second season, the intercropped yield of wheat was higher than the sole wheat yield under all cultivation techniques, which is attributed to the complementary effect of fahl clover in improving soil nitrogen and cooperation effect in mobilization of phosphorus in the soil, with a significant agroecological facility provided by legumes (Mat Hassan et al. 2012). They also reported that the highest value of soil nitrogen content and organic matter was found when 35% of fahl clover was intercropped with wheat, especially in the second growing season. It has been reported that a significant amount of nitrogen in the soil after legume crop harvest is available for subsequent crops, which supplies some amount of organic matter that enriches soil fertility (Banik et al. 2018). This also highlighted the complementary effect that positively affected the dry matter production and was reflected by the high value of WER found under cultivation on raised beds. The highest value of pWER for fahl clover, 0.45, was obtained when 35% fahl clover was intercropped with wheat. The highest value of wheat pWER, 1.00, was obtained when 15% fahl clover was intercropped with wheat. Regarding total WER, the highest value, 1.26, was obtained when 25% fahl clover was intercropped with wheat.

The cowpea and maize intercropping system exemplifies the 4C approach. While there is minimal competition for light and soil nutrients, a complementary effect is observed, resulting in enhanced soil nitrogen and organic matter. This system also

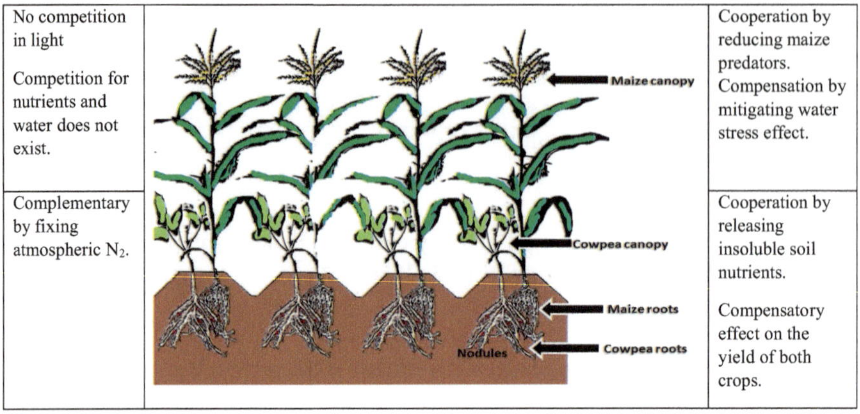

Fig. 2.2 The 4C approach explains the system of cowpea intercropping with maize

promotes the release of insoluble soil nutrients, such as phosphorus. Furthermore, cowpea attracts beneficial insects, including *C. septempunctata*, *C. undecimpunctata*, and *Chrysopa pallens*, which prey on aphids that infest maize tassels and lead to significant yield losses (Hamd-Alla et al. 2014). Cowpea attracts fall armyworms that infest maize during midseason, resulting in significant yield losses. Moreover, it offers a compensatory effect by establishing ground cover early in the season, thereby minimizing water loss through evaporation under conditions of water stress. Figure 2.2 illustrates the 4C benefits of the cowpea intercropping system with maize.

Taha and Kasem (2022) intercropped 50% of cowpea planting density with 100% of maize planting density under three irrigation treatments (required irrigation (100% RI, 80% RI, and 60% RI). They obtained the highest values of intercropped maize yields under the application of 100% RI amounts, which was higher than its value under sole cultivation. Furthermore, when 80% RI was applied to the system, the obtained intercropped maize yield was also higher than that of sole cultivation. This result is attributed to the intercropped cowpea's complementary and cooperation effects on helping maize withstand low water application. The yield of intercropped cowpea was lower than its value under sole cultivation and 100% and 80% RI due to higher interspecific competition between cowpea and maize plants. The authors also calculated WER, finding that the pWER for cowpea and the pWER for maize were the highest under low irrigation application, 0.36 and 0.90, respectively, with a total value of 1.26. These results implied the highly efficient use of water by this intercropping system.

Conclusion

The various advantageous agroecological effects of legume-cereal intercropping systems assist in addressing the challenge of creating agroecosystems that attain high crop yields with minimal input levels. The interactions among intercrops enhance resource utilization and clarify the potential benefits of legume intercropping systems. The 4C approach enables a thorough examination of how its distinct principles augment the beneficial effects of intercropping by classifying interspecific plant-plant interactions into four categories: competition, complementarity, collaboration, and compensation.

Various practices can be implemented to mitigate the detrimental effects of competition for light between intercropped shorter and taller plants, specifically intercropping shorter C3 plants (which experience shading, thereby reducing light intensity and enhancing light use efficiency) with taller C4 plants (which exploit higher light intensities to improve light utilization efficiency). Intercropping tall plants with a high specific leaf area alongside short plants with a low specific leaf area helps mitigate competition. In the intercropping system of soybean (C3) and maize (C4), adjusting soybean planting density is essential for optimal crop development, maximizing biomass and grain production. Furthermore, the canopy architecture of soybean can affect light intensity within the canopy, hence impacting seed yield. A further instance is the intercropping of maize with peanuts, wherein light interception is improved via increased ground cover. The acquisition of belowground resources is closely associated with light partitioning in a legume-cereal intercropping system. In a soybean-maize intercropping system, maize demonstrated enhanced nitrogen and phosphate absorption, whereas soybean revealed diminished nitrogen and phosphorus uptake compared to sole cropping.

The extent of complementarity is affected by the planting density of intercropped species and both intra- and interspecific competition. Intercropping with legumes enhances complementary nitrogen usage between legumes and cereals by improving N_2 fixation and soil nitrogen acquisition by cereals, leading to elevated soil nitrogen levels and enhanced cereal yields. Intercropping soybean with maize enhances nitrogen absorption by maize through the uptake of leached nitrogen from the soybean root zone, while also facilitating increased nitrogen fixation by soybean for transfer to maize. Additionally, soybeans may release protons and facilitate the mobilization of insoluble soil phosphorus due to the acidity of the rhizosphere through the exudation of malate and citrate, potentially enhancing soil phosphorus availability and its absorption by maize. Moreover, a synergistic impact was noted in the soybean and maize intercropping system, wherein the canopy's light transmission enhanced and light energy loss reduced. Another instance is the intercropping of maize with peanuts, which enhances beneficial soil bacteria, optimizes phosphonate and phosphinate metabolism, and increases phosphate availability for maize. In a cowpea-maize intercropping system, cowpea rapidly forms ground cover at the season's onset, enhancing its absorption of photosynthetically active radiation in lower layers.

An increase in soil organic carbon and humus content was found in legume-cereal intercropping systems due to cooperative effects. Leguminous crops can enhance soil organic carbon and nitrogen levels, as well as facilitate the production of hydrogen gas during biological nitrogen fixation, hence increasing bacterial proliferation in the rhizosphere. Moreover, soil organic matter was documented to rise by 17% in a pea intercropping system with wheat. Furthermore, the augmentation of soil nitrogen levels increases maize output when intercropped with cowpea.

The soybean and maize intercropping system demonstrated an increase in arbuscular mycorrhizal fungi diversity and macroaggregates in the rhizospheres of both maize and soybean, accompanied by enhanced soil phosphorus availability. This system also significantly mitigates termite damage and reduces maize grain yield loss. Moreover, intercropping maize with peanuts in alkaline and calcareous soils mitigates acute iron shortage, enhances its availability, and improves the photosynthetic efficiency and production of peanuts. It enhances pest management in peanuts by creating a habitat for predators that control peanut aphids. Intercropping peas with barley promotes pea growth by mitigating lodging. The wheat-lentil intercropping system enables lentils to assist wheat, therefore minimizing mechanical harvest losses in organic agriculture. Modifying the microclimate surrounding intercrops is an added advantage of intercropping systems. Intercropping can alter the microclimate of the canopy, reducing increased temperatures by limiting air circulation, which in turn lowers evaporation rates and enhances relative humidity. The reduced temperature under the maize canopy, relative to the ambient air temperature, facilitates the growth of groundnut intercropped with maize. Furthermore, the taller maize plants can modify the microenvironment of the shorter cowpea plants, particularly in terms of light and temperature.

The possible compensatory effect among intercrops can alleviate interspecific resource competition, wherein shorter-season peas, intercropped in alternating strips with longer-season maize, primarily extract soil water from the uppermost layers, while maize plants access water from the deeper layers of the adjacent pea strips. The intercropped maize subsequently mitigated soil water depletion from the pea strips after harvest, completely recuperating from the growth inhibition encountered during the overlapping growth phase. This intercropping strategy yielded greater accumulated nitrogen and carbon levels in both maize and pea intercrops relative to sole crops. A compensatory soil water effect is attained in a cowpea-maize intercropping system when irrigation is applied in the same volume as for sole maize. Thus, cowpea plants extract moisture from the higher soil strata, but maize plants draw water from deeper levels. After the cowpea harvest, the intercropped maize persisted in utilizing soil moisture that would have been absorbed by the cowpeas, leading to a cowpea yield equivalent to that of its monoculture and a maize yield surpassing that of its monoculture.

In a natural plant community, none of the 4C can exist independently. The primary impacts of 4C manifest throughout the growing season and are mostly evaluated using the land equivalent ratio and water equivalent ratio. Instances of intercropping systems from Egypt were showcased to illustrate the presence of the 4C components in the faba bean intercropping system with wheat, and soybean intercropped with maize, along with their impact on land equivalent ratio. Likewise, further instances of intercropping systems and their impacts on water equivalent ratio were demonstrated, specifically fahl clover intercropped with wheat and cowpea intercropped with maize.

Innovative research is necessary to provide the criteria for identifying the optimal species and cultivar combinations that would enhance niche complementarity for the acquisition of abiotic resources. Simulation models may be employed to evaluate multiple combinations of species and cultivars across diverse pedoclimatic conditions.

References

Abbady K, El-Maaz E, Ahmed H, Zohry AA (2016) Carbon sequestration as a function of intercropping management practice and different nitrogenous fertilizer types. J Soil Sci Agric Eng 7(8):565–586

Abd-Rabboh AMK, Koriem MHM (2021) Increasing unit area productivity and net income by intercropping peas with some wheat cultivars under different levels of nitrogen fertilizer. J Adv Agric Res 27(2):260–276

Abdel-Wahab TI, Abd El-Rahman RA (2016) Response of some soybean cultivars to low light intensity under different intercropping patterns with maize. Int J Appl Agric Sci 2(2):21–31. https://doi.org/10.11648/j.ijaas.20160202.11

Abdel-Wahab SI, Abdel-Wahab EI (2020) Competitive and facilitative effects of intercropping some soybean varieties with corn under different soybean plant densities. Plant Arch 20(S2):1631–1639

Abdel-Wahab TI, El Manzlawy AM (2016) Yield and quality of intercropped wheat with faba bean under different wheat plant densities and slow—release nitrogen fertilizer rates in sandy soil. Am J Exp Agric 11(6):1–22

Adler PB, Bradford JB (2002) Compensation: an alternative method for analyzing diversity-productivity experiments. Oikos 96:411–420

Bai Y, Han X, Wu J, Chen Z, Li L (2004) Ecosystem stability and compensatory effects in the Inner Mongolia grassland. Nature 431:181–184

Banik D, Upadhyay H, Devi BP, Darvhankar M (2018) Effect of different irrigation system on growth and development of wheat—chickpea intercropping. Bull Environ Pharmacol Life Sci 8(1):32–34

Bedoussac L, Journet E-P, Hauggaard-Nielsen H, Naudin C, Corre-Hellou G, Jensen ES, Prieur L, Justes E (2015) Ecological principles underlying the increase of productivity achieved by cereal-grain legume intercrops in organic farming: a review. Agron Sustain Dev 35:911–935. https://doi.org/10.1007/s13593-014-0277-7

Bernard B, Lux A (2017) How to feed the world sustainably: an overview of the discourse on agroecology and sustainable intensification. Reg Environ Chang 17:1279–1290

Black C, Ong C (2000) Utilization of light and water in tropical agriculture. Agric For Meteorol 104:25–47

Blum A (2011) Plant breeding for water-limited environment. Springer, New York. ISBN 978-1-4419-7490-7

Bommarco R, Kleijn D, Potts SG (2013) Ecological intensification: harnessing ecosystem services for food security. Trends Ecol Evol 28(2013):230–238. https://doi.org/10.1016/j.tree.2012.10.012

Boukar O, Bhattacharjee R, Fatokun C, Kumar PL, Gueye B (2013) Cowpea. In: Singh M, Upadhyaya HD, Bisht IS (eds) Genetic and genomic resources of grain legume improvement. Elsevier, pp 137–155

Brooker RW, Bennett AE, Cong W-F, Daniell TJ, George TS, Hallett PD, Hawes C, Iannetta PPM, Jones H, Karley A et al (2015) Improving intercropping: a synthesis of research in agronomy, plant physiology and ecology. New Phytol 206:107–117. https://doi.org/10.1111/nph.13132

Brooker R, George T, Homulle Z et al (2021) Facilitation and biodiversity ecosystem function (BEF) relationships in crop production systems and their role in sustainable farming. J Ecol 109(5):2054–2067. https://doi.org/10.1111/1365-2745.13592
Bybee-Finley K, Ryan MR (2018) Advancing intercropping research and practices in industrialized agricultural landscapes. Agriculture 8:80
Cai T, Peng D, Wang R, Jia X, Qiao D, Liu T, Jia Z, Wang Z, Ren X (2019) Can intercropping or mixed cropping of two genotypes enhance wheat lodging resistance? Field Crop Res 239:10–18. https://doi.org/10.1016/j.fcr.2019.05.009
Chaoui R, Boudsocq S, Taschen E, Sentenac H, Farissi M, Lazali M (2023) Intercropping durum wheat and chickpea increases nutrient availability and use efficiency under low phosphorus soils. J Plant Nutr 46(17):4125–4139. https://doi.org/10.1080/01904167.2023.2221677
Chen G, Kong X, Gan Y et al (2018) Enhancing the systems productivity and water use efficiency through coordinated soil water sharing and compensation in strip-intercropping. Sci Rep 8:10494. https://doi.org/10.1038/s41598-018-28612-6
Corre-Hellou G, Fustec J, Crozat Y (2006) Interspecific competition for soil N and its interaction with N2 fixation, leaf expansion and crop growth in pea-barley intercrops. Plant Soil 282:195–208. https://doi.org/10.1007/s11104-005-5777-4
Cui L, Yang WY, Huang N, Liu J, Wang YL, Wang XH, Liu Y, Yan S (2015) Effects of maize plant types on dry matter accumulation characteristics and yield of soybean in maize soybean intercropping systems. Chin J Appl Ecol 26:2414–2420
Dai J, Qiu W, Wang N, Wang T, Nakanishi H, Zuo Y (2019) From leguminosae/gramineae intercropping systems to see benefits of intercropping on iron nutrition. Front Plant Sci 10:605. https://doi.org/10.3389/fpls.2019.00605
Daryanto S, Fu B, Zhao W, Wang S, Jacinthe P, Wang L (2020) Ecosystem service provision of grain legume and cereal intercropping in Africa. Agric Syst 178:102761. https://doi.org/10.1016/j.agsy.2019.102761
Ditzler L, van Apeldoorn DF, Pellegrini F, Antichi D, Bàrberi P, Rossing WAH (2021) Current research on the ecosystem service potential of legume inclusive cropping systems in Europe: a review. Agron Sustain Dev 41(2):26. https://doi.org/10.1007/s13593-021-00678-z
Döring TF, Elsalahy H (2022) Quantifying compensation in crop mixtures and monocultures. Eur J Agron 132:126408. https://doi.org/10.1016/j.eja.2021.126408
Dowling A, Sadras VO, Roberts P, Doolette A, Zhou Y, Denton MD (2021) Legume-oilseed intercropping in mechanized broadacre agriculture: a review. Field Crop Res 2021:260
El-Ghobashy YE, El-Mehy AA, El-Douby KA (2020) Influence of intercropping cowpea with some maize hybrids and N nano-mineral fertilization on productivity in salinity soil. Egypt J Agron 42(1):63–78
El-Shamy MA, Abdel-Wahab TI, Abdel-Wahab SI, Ragheb SB (2015) Advantages of intercropping soybean with maize under two maize plant distributions and three mineral nitrogen fertilizer rates. Adv Biosci Bioeng 3(4):30–48. https://doi.org/10.11648/j.abb.20150304.11
Elhakeem A, van der Werf W, Ajal J, Luc D, Claus S, Vico RA, Bastiaans L (2019) Cover crop mixtures result in a positive net biodiversity effect irrespective of seeding configuration. Agric Ecosyst Environ 285:106627
Fan F, Zhang F, Song Y, Sun J, Bao X, Guo T, Li L (2006) Nitrogen fixation of faba bean (*Vicia faba* L.) interacting with a non-legume in two contrasting intercropping systems. Plant Soil 283(1):275–286
Ferguson BJ, Lin MH, Gresshoff PM (2013) Regulation of legume nodulation by acidic growth conditions. Plant Signal Behav 8(3):e23426
Fernandez-Aparicio M, Sillero JC, Rubiales D (2007) Intercropping with cereals reduces infection by *Orobanche crenata* in legumes. Crop Prot 26(8):1166–1172. https://doi.org/10.1016/j.cropro.2006.10.012
Franco JG, King SR, Volder A (2018) Component crop physiology and water use efficiency in response to intercropping. Eur J Agron 93:27–39. https://doi.org/10.1016/j.eja.2017.11.005
Fustec J, Lesuffleur F, Mahieu S et al (2010) Nitrogen rhizodeposition of legumes. A review. Agron Sustain Dev 30:57–66. https://doi.org/10.1051/agro/2009003

Gambart C, Swennen R, Blomme G, Groot JCJ, Remans R, Ocimati W (2020) Impact and opportunities of agroecological intensification strategies on farm performance: a case study of banana-based systems in central and South-Western Uganda. Front Sustain Food Syst 4:87. https://doi.org/10.3389/fsufs.2020.00087

Ghanbari A, Dahmardeh M, Siahsar BA, Ramroudi M (2010) Effect of maize (*Zea mays* L.)—cowpea (*Vigna unguiculata* L.) intercropping on light distribution, soil temperature and soil moisture in arid environment. J Food Agric Environ 8(1):102–108

Gonzalez A, Loreau M (2009) The causes and consequences of compensatory dynamics in ecological communities. Annu Rev Ecol Evol Syst 40:393–414

Guo F, Wang M, Si T, Wang Y, Zhao H, Zhang X, Yu X, Wan S, Zou X (2021) Maize-peanut intercropping led to an optimization of soil from the perspective of soil microorganism. Arch Agron Soil Sci 67:1986–1999

Hamd-Alla WA, Shalaby EM, Dawood RA, Zohry AA (2014) Effect of cowpea (*Vigna sinensis* L.) with maize (*Zea mays* L.) intercropping on yield and its components. Int Scholar Sci Res Innov 8(11):1170–1176

Hamd-Alla W, Shehata MAK, Leilah AAA, Darwesh RKH, Hefzy M (2023) Impact of irrigation regimes on productivity and profitability of maize + peanut intercropping system in Upper Egypt. Eur J Biol Res 13(4):218–231. https://doi.org/10.5281/zenodo.10413383

Harris D, Natarajan M, Willey RW (1987) Physiological basis for yield advantage in a sorghum/groundnut intercrop exposed to drought. 1. Dry-matter production, yield, and light interception. Field Crop Res 17(3–4):259–272

Haruna A, Kombiok JM, Mohamed AM, Sarkodie-Addo J, Larbi A, Rahman NA (2018) Profitability of cowpea intercropped with maize in West Africa Guinea savanna. J Agric Sci 10:185. https://doi.org/10.5539/jas.v10n11p185

Hassan A, Kodoh J, Kamu A, Majius R, Yunus N (2020) Root interactions on sole crop and intercrop give different effects of competition. IOP Conf Ser Earth Environ Sci 449:1–5. https://doi.org/10.1088/1755-1315/449/1/012012

Hinsinger P, Betencourt E, Bernard L, Brauman A, Plassard C, Shen J, Tang X, Zhang F (2011) P for two, sharing a scarce resource: soil phosphorus acquisition in the rhizosphere of intercropped species. Plant Physiol 156(3):1078–1086

Holomuzki JR, Feminella JW, Power ME (2010) Biotic interactions in freshwater benthic habitats. J N Am Benthol Soc 29:220–244

Hu FL, Gan YT, Chai Q, Feng F, Zhao C, Yu AZ, Mu Y, Zhang Y (2016) Boosting system productivity through the improved coordination of interspecific competition in maize/pea strip intercropping. Field Crop Res 198:50–60. https://doi.org/10.1016/j.fcr.2016.08.022

Hussain S, Liu T, Iqbal N, Brestic M, Pang T, Mumtaz M et al (2020) Effects of lignin, cellulose, hemicellulose, sucrose and monosaccharide carbohydrates on soybean physical stem strength and yield in intercropping. Photochem Photobiol Sci 19:462–472. https://doi.org/10.1039/cspp00369j

Ivanov S, Fedorova EE, Limpens E, De Mita S, Genre A, Bonfante P et al (2012) Rhizobium–legume symbiosis shares an exocytotic pathway required for arbuscule formation. Proc Natl Acad Sci 109:8316–8321. https://doi.org/10.1073/pnas.1200407109

Jhariya MK, Meena RS, Banerjee A (2021) Ecological intensification of natural resources for sustainable agriculture. Springer Singapore. https://doi.org/10.1007/978-981-33-4203-3_1

Ju Q, Ouyang F, Gu S, Qiao F, Yang Q, Qu M, Ge F (2019) Strip intercropping peanut with maize for peanut aphid biological control and yield enhancement. Agric Ecosyst Environ 286:106682. https://doi.org/10.1016/j.agee.2019.106682

Justes E, Bedoussac L, Dordas C et al (2021) The 4C approach as a way to understand species interactions determining intercropping productivity. Front Agric Sci Eng. https://doi.org/10.15302/J-FASE-2021414

Kang L, Shan-chao Y, Shi-qing L (2014) Effects of phosphorus application in different soil layers on root growth, yield, and water-use efficiency of winter wheat grown under semi-arid conditions. J Integr Agric 13(9):2028–2039. https://doi.org/10.1016/S2095-3119(14)60751-6

Khalid MHB, Cui L, Abbas G, Raza MA, Anwar A, Ahmed Z, Waheed A, Saeed A, Ahmed W, Babar MJ (2023) Effect of row spacing under maize-soybean relay intercropping system on yield, competition, and economic returns. Turk J Agric For 47:390–401

Kiaer LP, Weedon OD, Bedoussac L, Bickler C, Finckh MR, Haug B, Iannetta PPM, Raaphorst-Travaille G, Weih M, Karley AJ (2022) Supply chain perspectives on breeding for legume–cereal intercrops. Front Plant Sci 13:844635

Klimek-Kopyra A, Zaja T, Rebilas K (2013) A mathematical model for the evaluation of cooperation and competition effects in intercrops. Eur J Agron 51:9–17. https://doi.org/10.1016/j.eja.2013.06.002

Klöffel T, Young EH, Borchard N, Vallotton JD, Nurmi E, Shurpali NJ, Urbano-Tenorio F, Liu X, Young GHF, Unc A (2022) The challenges fraught opportunity of agriculture expansion into boreal and Arctic regions. Agric Syst 203:103507. https://doi.org/10.1016/j.agsy.2022.103507

Koskey G, Leoni F, Carlesi S, Avio L, Bàrberi P (2022) Exploiting plant functional diversity in durum wheat–lentil relay intercropping to stabilize crop yields under contrasting climatic conditions. Agronomy 12(1):210. https://doi.org/10.3390/agronomy12010210

Lampkin NH, Pearce BD, Leake AR, Creissen H, Gerrard CL, Girling R, Lloyd S, Padel S, Smith J, Smith LG, Vieweger A, Wolfe MS (2015) The role of agroecology in intensification. http://publications.naturalengland.org.uk/file/6602354724962304

Li L, Tilman D, Lambers H, Zhang FS (2014) Plant diversity and overyielding: insights from belowground facilitation of intercropping in agriculture. New Phytol 203:63–69. https://doi.org/10.1111/nph.12778

Li Y-H, Shi D-Y, Li G-H, Zhao B, Zhang J-W, Liu P, Ren B, Dong S-T (2019) Maize/peanut intercropping increases photosynthetic characteristics, 13C-photosynthate distribution, and grain yield of summer maize. J Integr Agric 18(10):2219–2229

Li C, Hoffland E, Kuyper TW, Yu Y, Li H, Zhang C, Zhang F, van der Werf W (2020) Yield gain, complementarity and competitive dominance in intercropping in China: a meta-analysis of drivers of yield gain using additive partitioning. Eur J Agron 113:125987

Liebmann M (1995) Polyculture cropping systems. In: Altieri MA (ed) Agroecology: the science of sustainable agriculture. Westview Press, pp 205–218

Liu XB, Jin J, Wang GH, Herbert SJ (2008) Soybean yield physiology and development of high-yielding practices in Northeast China. Field Crop Res 105:157–171

Loïc V, Laurent B, Etienne-Pascal J, Eric J (2018) Yield gap analysis extended to marketable yield reveals the profitability of organic lentil–spring wheat intercrops. Agron Sustain Dev 38(4):39

Loreau M, Hector A (2001) Partitioning selection and complementarity in biodiversity experiments. Nature 412:72–76. https://doi.org/10.1038/35083573

Lu M, Zhao J, Lu Z et al (2023) Maize–soybean intercropping increases soil nutrient availability and aggregate stability. Plant Soil. https://doi.org/10.1007/s11104-023-06282-2

MacLaren C, Waswa W, Aliyu KT, Claessens L, Mead A, Schob C, Vanlauwe B, Storkey J (2023) Predicting intercrop competition, facilitation, and productivity from simple functional traits. Field Crop Res 297:108926

Mahmoud R, Casadebaig P, Hilgert N, Alletto L, Freschet GT, de Mazancourt C, Gaudio N (2022) Species choice and N fertilization influence yield gains through complementarity and selection effects in cereal-legume intercrops. Agron Sustain Dev 42:12

Mao LL, Zhang LZ, Li WQ, van der Werf W, Sun JH, Spiertz HLL, Li L (2012) Yield advantage and water saving in maize/pea intercrop. Field Crop Res 138:11–20

Martín-Robles N, Morente-López J, Freschet GT et al (2019) Root traits of herbaceous crops: pre-adaptation to cultivation or evolution under domestication? Funct Ecol 33:273–285. https://doi.org/10.1111/1365-2435.13231

Mat Hassan H, Marschner P, McNeill A, Tang C (2012) Grain legume pre-crops and their residues affect the growth, P uptake and size of P pools in the rhizosphere of the following wheat. Biol Fertil Soils 48:775–785. https://doi.org/10.1007/s00374-012-0671-8

Metwally AA, Shafik MM, El-Habbak KE, Abdel-Wahab SI (2012) Yield and soybean characters under some intercropping patterns with corn. Soybean Res 10:24–42

Midmore DJ, Roca J, Berrios D (1988) Potato solanum spp. in the hot tropics. IV. Intercropping with maize and the influence of shade on potato microenvironment and crop growth. Field Crop Res 18(2–3):141–215. https://doi.org/10.1016/0378-4290(88)90005-6

Milder JC, Garbach K, DeClerck FAJ, Driscoll L, Montenegro M (2019) An assessment of the multi-functionality of agroecological intensification. Gates Open Res 3:279. https://doi.org/10.21955/gatesopenres.1115387.1

Mrunalini K, Behera B, Chandana C, Patnaik GP, Modi RU, Saraswat A, Rathi N, Kumar N (2022) Legumes to reduce ecological footprints for climate-smart cropping systems. In: Advances in legumes for sustainable intensification. Academic, pp 403–420. https://doi.org/10.1016/B978-0-323-85797-0.00032-X

Naazie GK, Dakyaga F, Derbile EK (2023) Agro-ecological intensification for climate change adaptation: Tales on soil and water management practices of smallholder farmers in rural Ghana. Discov Sustain 4(1):27. https://doi.org/10.1007/s43621-023-00142-w

Neugschwandtner RW, Kaul H-P (2014) Sowing ratio and N fertilization affect yield and yield species of oat and pea in intercrops. Field Crop Res 155:159–163. https://doi.org/10.1016/j.fcr.2013.09.010

Ouda S, Zohry A (2024) Increasing land and water use efficiencies of wheat: case study of Egypt. In: Integration of legume crops with cereal crops under changing climate: sustainably increases food production. Springer. ISBN: 978-3-031-68101-1

Ouda S, Zohry A, Abdel-Wahab S (2020) Sustainable intensive cropping to reduce irrigation-induced erosion: II. Changing cropping sequence under sprinkler irrigation practice. Mor J Agric Sci 1(1):26–33

Paradiso R, Proietti S (2021) Light-quality manipulation to control plant growth and photomorphogenesis in greenhouse horticulture: the state of the art and the opportunities of modern LED systems. J Plant Growth Regul 15:1–39. https://doi.org/10.1007/s00344-021-10337-y

Podgórska-Lesiak M, Sobkowicz P (2013) Prevention of pea lodging by intercropping barley with peas at different nitrogen fertilization levels. Field Crop Res 149:95–104. https://doi.org/10.1016/j.fcr.2013.04.023

Quinones M, Lucas M, Pueyo J (2022) Adaptive mechanisms make lupin a choice crop for acidic soils affected by aluminum toxicity. Front Plant Sci 12:810692. https://doi.org/10.3389/fpls.2021.810692

Rahman T, Liu X, Hussain S, Ahmed S, Chen G, Yang F (2017) Water use efficiency and evapotranspiration in maize-soybean relay strip intercrop systems as affected by planting geometries. PLoS One 12(6):e0178332. https://doi.org/10.1371/journal.pone.0178332

Rao MR, Willey RW (1980) Evaluation of yield stability in intercropping: studies on sorghum/pigconpea. Exp Agric 16(2):105–116

Rashwan EA, Zen El-Dein AA (2017) Effect of two patterns of intercropping soybean with maize on yield and its components under different nitrogen fertilizer levels. Egypt J Agron 39(3):449–466

Raza MA, Feng LY, van derWerf W et al (2020) Optimum strip width increases dry matter, nutrient accumulation, and seed yield of intercrops under the relay intercropping system. Food Energy Secur. https://doi.org/10.1002/fes3.199

Raza MA, Ud Din AM, Shah GA, Zhiqi W, Feng LY, Gul H, Yasin HS, Ur Rahman MS, Juan C, Liang X, Rehman R, Al Garawi AM, van der Werf W, Qin R, Xin L, Bin Khalid MH, Zhongming M (2024) Legume choice and planting configuration influence intercrop nutrient and yield gains through complementarity and selection effects in legume-based wheat intercropping systems. Agric Syst 220:104081. https://doi.org/10.1016/j.agsy.2024.104081

Ren BZ, Cui HY, Camberato JJ, Dong ST, Liu P, Zhao B, Zhang JW (2016) Effects of shading on the photosynthetic characteristics and mesophyll cell ultrastructure of summer maize. Sci Nat 103(7–8):67. https://doi.org/10.1007/s00114-016-1392-x

Ren Y, Wang X, Zhang S, Palta J, Chen Y (2017) Influence of spatial arrangement in maize–soybean intercropping on root growth and water use efficiency. Plant Soil 415:131–144. https://doi.org/10.1007/s11104-016-3143-3

Ren Y, Zhang L, Yan M, Zhang Y, Chen Y, Palta JA, Zhang S (2021) Effect of sowing proportion on above- and below-ground competition in maize–soybean intercrops. Sci Rep 11:15760. https://doi.org/10.1038/s41598-021-95242-w

Reynolds PE, Simpson JA, Thevathasan NV, Gordon AM (2007) Effects of tree competition on corn and soybean photosynthesis, growth, and yield in a temperate tree-based agroforestry intercropping system in southern Ontario, Canada. Ecol Eng 29(4):362–371

Rodriguez C, Carlsson G, Englund J-E, Flöhr A, Pelzer E, Jeuffroy M-H, Makowski D, Jensen ES (2020) Grain legume-cereal intercropping enhances the use of soil-derived and biologically fixed nitrogen in temperate agroecosystems. A meta-analysis. Eur J Agron 118:126077. https://doi.org/10.1016/j.eja.2020.126077

Schöb C, Kerle S, Karley A et al (2015) Intra-specific genetic diversity and composition modify species-level diversity productivity relationships. New Phytol 205:720–730. https://doi.org/10.1111/nph.13043

Sekamatte BM, Ogenga-Latigo M, Russell-Smith A (2003) Effects of maize–legume intercrops on termite damage to maize, activity of predatory ants and maize yields in Uganda. Crop Prot 22(1):87–93. https://doi.org/10.1016/S0261-2194(02)00115

Shumet ST, Ayalew T, Roro AG, Beshir HM (2022) Intercropping and rhizobium inoculation affected microclimate and performance of common bean (*Phaseolus vulgaris* L.) varieties. Scientifica (Cairo) 6:3471912. https://doi.org/10.1155/2022/3471912. PMID: 36523483

Stagnari F, Maggio A, Galieni A, Pisante M (2017) Multiple benefits of legumes for agriculture sustainability: an overview. Chem Biol Technol Agric 4:1–13. https://doi.org/10.1186/s40538-016-0085-1

Taha A, Kasem E (2022) Water management of maize-cowpea intercropping system under surface irrigation. Mor J Agri Sci 3(1):29–38

Tilman D (2020) Benefits of intensive agricultural intercropping. Nat Plants 6:604–605

Tosti G, Falcinelli B, Guiducci M (2023) Lentil–cereal intercropping in a Mediterranean area: yield, pests and weeds. Agron J 115:2570–2578

Van Dijk M, Morley T, Rau ML, Saghai Y (2021) A meta-analysis of projected global food demand and population at risk of hunger for the period 2010–2050. Nat Food 2(7):494–501. https://doi.org/10.1038/s43016-021-00322-9

West SA, Griffin AS, Gardner A (2007) Social semantics: altruism, cooperation, mutualism, strong reciprocity and group selection. J Evol Biol 20:415–432

Wezel A, Casagrande M, Celette F, Vian J-F, Ferrer A, Peigné J (2014) Agroecological practices for sustainable agriculture. Agron Sustain Dev 34:1–20. https://doi.org/10.1007/s13593-013-0180-7

Wezel A, Soboksa G, McClelland S, Delespesse F, Boissau A (2015) The blurred boundaries of ecological, sustainable, and agroecological intensification: a review. Agron Sustain Dev 35(4):1283–1295. https://doi.org/10.1007/s13593-015-0333-y

Willey RW (1979) Intercropping: its importance and research needs. Part-2. Agronomy and research approaches. Field Crop Abstr 32:73–81

Wu W, Fan L, Li MF, Liu HB, Li YQ (2012) Sensitivity analysis of crop growth models to multi-temporal scale solar radiation. Trans Chin Soc Agric Eng 28:123–128

Xu Z, Li C, Zhang C, Yu Y, van der Werf W, Zhang F (2020) Intercropping maize and soybean increase efficiency of land and fertilizer nitrogen use: a meta-analysis. Field Crop Res 246:107661

Yang C, Huang G, Chai Q, Luo Z (2011) Water use and yield of wheat/maize intercropping under alternate irrigation in the oasis field of Northwest China. Field Crop Res 124:426–432

Yang F, Wang X, Liao D, Lu F, Gao R, Liu W, Yang W (2015) Yield response to different planting geometries in maize-soybean relay strip intercropping systems. Agron J 107:296–304. https://doi.org/10.2134/agronj14.0263

Yang T, Zhu Y, Duan ZP et al (2019) Root distribution and productivity in a poplar tree + alfalfa silvopastoral system in Northwest China's Xinjiang Province. Agrofor Syst. https://doi.org/10.1007/s10457-019-00466-5

Yang F, Liu Q, Cheng Y, Feng L, Wu X, Fan Y et al (2020) Low red/far-red ratio as a signal promotes carbon assimilation of soybean seedlings by increasing the photosynthetic capacity. BMC Plant Biol 20:148. https://doi.org/10.1186/s12870-020-02352-0

Yousefi M, Marja R, Barmettler E et al (2024) The effectiveness of intercropping and agri-environmental schemes on ecosystem service of biological pest control: a meta-analysis. Agron Sustain Dev 44:15. https://doi.org/10.1007/s13593-024-00947-7

Yu Y, Stomph TJ, Makowski D, van der Werf W (2015) Temporal niche differentiation increases the land equivalent ratio of annual intercrops: a meta-analysis. Field Crop Res 184:133–144

Yu Y, Stomph T-J, Makowski D, Zhang L, van der Werf W (2016) A meta-analysis of relative crop yields in cereal/legume mixtures suggests options for management. Field Crop Res 198:269–279

Zhang D, Sun Z, Feng L, Bai W, Yang N, Zhang Z, Du G, Feng C, Cai Q, Wang Q, Zhang Y, Wang R, Arshad A, Hao X, Sun M, Gao Z, Zhang L (2020) Maize plant density affects yield, growth and source-sink relationship of crops in maize/peanut intercropping. Field Crop Res 257:107926

Zhao Y, Fan Z, Hu F, Yin W, Zhao C, Yu A, Chai Q (2019) Source-to-sink translocation of carbon and nitrogen is regulated by fertilization and plant population in maize-pea intercropping. Front Plant Sci 10:891. https://doi.org/10.3389/fpls.2019.00891

Zohry AA (2005) Effect of preceding winter crops and intercropping on yield, yield components and associated weeds in maize. Ann Agric Sci Moshtohor 43(1):139–148

Zohry A, Ouda S, Abdel-Wahab T (2020) Sustainable intensive cropping to reduce irrigation-induced erosion: intercropping systems under surface irrigation practice. Mor J Agric Sci 1(2):63–71

3 Innovative Cropping Systems to Maximize the Use of Available Natural Resources

Sherif Ibrahim Abdel-Wahab, Tamer Ibrahim Abdel-Wahab, and Eman Ibrahim Abdel-Wahab

Abstract

Innovative cropping systems can optimize land productivity and efficiency, thereby enhancing sustainability. Intercropping and crop sequencing are two strategies that can help in maintaining soil fertility and health. In this chapter, the benefits of implementing intensive cropping systems, such as changing cropping sequence and intercropping systems including legume crops, intercropping systems that can reduce shading stress, as well as the natural pests, were presented using results of field experiments previously conducted under the Egyptian conditions. Furthermore, evolutionary breeding can be used to develop crop varieties that are better adapted to changing environmental conditions, which can play a crucial role in crop intensification with enhanced resistance to environmental stresses. These types of practices can help farmers to improve yields, maximize the use of agricultural resources, prevent its deterioration, and increase food security in Egypt.

Keywords

Intercropping systems · Changing cropping sequence · Cropping index · Land equivalent ratio · Shading stress

S. I. Abdel-Wahab (✉) · T. I. Abdel-Wahab
Crop Intensification Research Department, Field Crops Research Institute, Agricultural Research Center (ARC), Giza, Egypt
e-mail: sherif.abdel-wahab@arc.sci.eg

E. I. Abdel-Wahab
Food Legumes Research Department, Field Crops Research Institute, Agricultural Research Center (ARC), Giza, Egypt

A. E.-H. Zohry, S. Ouda (eds.), *Resilient Agroecosystems*, Sustainability Sciences in Asia and Africa, https://doi.org/10.1007/978-981-96-9630-7_3

Introduction

Egypt's agricultural sector benefits from the fertile soil along the Nile River and abundant sunlight, allowing for the cultivation of a variety of field and vegetable crops such as wheat (*Triticum aestivum* L.), Egyptian clover (*Trifolium alexandrinum* L.), faba bean (*Vicia faba* L.), peanut (*Arachis hypogaea* L), maize (*Zea mays* L.), soybean (*Glycine max* L. [Merrill]), fenugreek (*Trigonella foenum-graecum* L.), flax (*Linum usitatissimum* L.), sunflower (*Helianthus annuus* L.), garlic (*Allium sativum* L.), radish (*Raphanus sativus* L.), and cowpea (*Vigna unguiculata* L.), which are all vital to the country's economy. Yet, the agricultural sector faces challenges such as water scarcity, soil salinity, and limited arable land (Ouda and Zohry 2018). The limited arable land in Egypt poses a major challenge for agricultural production, as only a small portion of the country's land is suitable for farming. Addressing these issues requires the implementation of sustainable agricultural practices and investments in technological interventions to maintain high crop productivity and ensure food security to meet the demands of the increasing population. Investing in research on intensive cropping systems could help address these issues. Intensive cropping systems are characterized by growing several crops on the same plot of land in a single year to increase agricultural production (Waha et al. 2020). These systems need appropriate management of available agricultural resources to sustain high yields. Promoting environmentally friendly farming practices may also boost Egypt's agriculture industry.

Intercropping and changing cropping sequence are two strategies that can help in maintaining soil fertility and health. These practices can enhance the overall resilience of the agricultural system and promote biodiversity. Intercropping involves planting multiple crops in the same field simultaneously, which can enhance nutrient cycling and reduce pests and diseases (Toker et al. 2024). Furthermore, intercropping is considered one of the most intensive cropping techniques, in which insect pressure is lowered as well as diversifying plant species and land use efficiency are optimized, contributing to better soil health (Barbieri et al. 2019). Implementing intercropping systems often achieves yield advantages through coordinated interspecies interactions between the intercrops for both aboveground and belowground resources (Li et al. 2006). Niche differentiation involves competitive species utilizing resources differently in time or space, reducing inter-specific competition, and promoting species coexistence and resource complementarity (Loreau 2000).

Various measures of the efficiency of intercropping systems have been utilized for evaluation purposes (Hiebesch and McCollum 1987). The most important one is the land equivalent ratio (LER), which is a key focus in intercropping research since the 1970s, when first initiated by Willey and Osiru (1972). LER is a measure of biological efficiency based on the total yield of the intercrops and the area of sole crops needed to achieve the same yields (Willey 1979). LER values above one indicates that intercropping provides a land use advantage over sole planting. This advantage is typically seen when there is strong complementarity for nutrient and light resources between species, outweighing competition effects (Wang et al. 2024).

Changing cropping sequence, on the other hand, includes the insertion of legume crops in the system, which is considered as one of the effective tools for nutrient recycling and nitrogen fixation, resulting in the acceleration of the microbial activity and availability of more nutrients (Pokhrel and Pokhrel 2013). Changing cropping sequence involves studying the residual effects of preceding crops on successive crops to maintain soil fertility and to prevent soil degradation (Negash et al. 2018). Considerable amounts of residue after harvesting legume crops with narrower C/N ratio are left in the soil, which upon decomposition improves the physical condition and fertility of the soil (El-Sodany and Abou-Elela 2010).

Ecological approach in crop production focuses on sustainable farming practices that work in harmony with natural ecosystems (Ranjan 2024). Agroecology relies on sustainable farming practices that prioritize biodiversity, soil health, and organic pest management (Yadav et al. 2021). Crop intensification is a key aspect of agroecology, which integrates modern technology with traditional knowledge to enhance agricultural productivity while minimizing environmental impact (Çakmakçı et al. 2023). It enables the efficient use of natural resources like water and nutrients, as well as promoting sustainable agricultural practices (Metwally et al. 2019). In the meantime, it can enhance the livelihood of small-scale farmers by boosting crop yields and diversifying sources of income. Thus, agroecology-based crop intensification can promote sustainable farming practices and enhances food production (Ranjan 2024). By implementing techniques such as including legume crops in the crop sequence and intercropping systems, soil health can be improved. Furthermore, implementing intercropping system can reduce the natural pests and help farmers to improve yields without relying on synthetic inputs (Çakmakçı et al. 2023). Additionally, agroecology fosters knowledge-sharing and community engagement among farmers, contributing to the long-term sustainability of food production (Rossi 2019).

In this chapter, the advantages of implementing intensive cropping systems, such as changing cropping sequence and intercropping systems including legume crops, intercropping system that can reduce shading stress, as well as reducing the natural pests, were presented using the results of field experiments previously conducted under the Egyptian conditions. These types of practices can help farmers to improve yields, maximize the use of agricultural resources, prevent its deterioration, and increase food security in Egypt.

The Problems of the Egyptian Agriculture

The agricultural lands of Egypt are divided into two main categories: the old lands, which is located in the Nile Valley and Delta Regions and the new lands located mainly on both the east and west fringes of the Nile Delta and Valley and scattered over various areas in the country. The old lands cover a total area of 2.38 million hectares or about 68% of the total cultivated area (Zohry and Ouda 2020). The old lands are characterized by being alluvial soils (clay to loamy), in which the available phosphorus is generally moderate, and the available potassium (soluble and exchangeable) is high in most of Egyptian alluvial soils (FAO 2005). The main source for irrigation in the old lands is the Nile River, and the prevailing irrigation system is surface irrigation.

Furthermore, the area of the new lands covers approximately 1.01 million hectares. The sandy and calcareous soils are the prevailing soil types in these areas. According to Hegazi et al. (2005), the micronutrients in this type of soil are above the critical limits, and available phosphorus, potassium, and micronutrient levels are impartially low. Some of these areas are recently reclaimed and became productive lands. Zohry and Ouda (2020) stated that the newly reclaimed lands are considered as an opportunity for elevating agricultural production to enhance the Egyptian food security. Groundwater is the main source for irrigation water in the newly reclaimed lands, and the prevailing irrigation systems are sprinkler and drip irrigation.

Egypt is confronted by many agricultural challenges, including limited arable land and rising water deficiency conditions. These problems have negatively affected farmers' livelihoods and the country's food security, leading to decreased crop yields and overall production (Ouda and Zohry 2022). This constraint has resulted in a heavy dependence on irrigation and chemicals to enhance crop yields. Additionally, pest pressure worsens the difficulties encountered by Egyptian farmers, necessitating the deployment of extra cost resources and technologies to safeguard crops. The common biotic predators spreading in Egypt are aphids (*Aphis gossypii*), cotton leafworms (*Spodoptera littoralis*), broomrape (*Orobanche crenata* Forsk.), jassids (*Amrasca biguttula*), and whiteflies (*Bemisia tabaci*) that can significantly damage various crops (Abd El-Hady et al. 2020; Abdel-Wahab et al. 2021; Gad and Abdel-Aziz 2024).

Moreover, farmers face challenges in efficiently marketing their products and expanding their market reach when they depend on a single crop, especially in the midst of market price fluctuations and limited transportation and storage facilities. Therefore, crop intensification can offer farmers more opportunities to reach various customer segments and mitigate risks associated with market fluctuations.

Intensive Cropping Using Innovative Cropping Systems

The development of cropping systems and sustainable agriculture involves a complex and intricate relationship (Rodriguez et al. 2021). Achieving long-term success in food production requires balancing social equity, environmental impact, and economic viability (Bisht et al. 2020). Establishing appropriate cropping systems is crucial for promoting biodiversity, soil health, and resilience to environmental stressors. Farmers can enhance yields and preserve natural resources for future generations by embracing sustainable practices. Regenerative agriculture approaches that prioritize soil health and minimize chemical inputs are essential for farmers to adopt (Kabenomuhangi 2024). Additionally, supporting research and education on sustainable farming systems can stimulate innovation and the adoption of environmentally friendly practices. These practices can promote sustainable agriculture by transforming traditional cropping systems into innovative ones. The cultivation of legume crops and intercropping techniques are commonly used in innovative cropping systems to enhance soil health and reduce the need for chemical inputs. According to Ouda and Zohry (2024), including legume crops in a cereal-based cropping system is a practical way to boost the production of cereals.

Replacing Fallow with Short Season Legume Crop to Improve Wheat Yield

An example of innovative cropping system with potential high yield output was presented in a previously published field study by Abdel-Galil et al. (2015). In a 2-year field experiment conducted at Ismailia Governorate, East Egypt, the impact of changing traditional cropping system by including fahl clover (a variety of Egyptian clover and an early winter crop) between summer peanut and winter wheat (innovative cropping system), instead of cultivating peanut followed by wheat (conventional cropping system), on wheat yield was studied. The innovative cropping system increased the yield of wheat and peanut pods while also producing a significant amount of fahl clover cuttings, which provide an additional fodder amount for livestock.

Figure 3.1 illustrates the traditional cropping system and changing of cropping system by including fahl clover (innovative cropping systems).

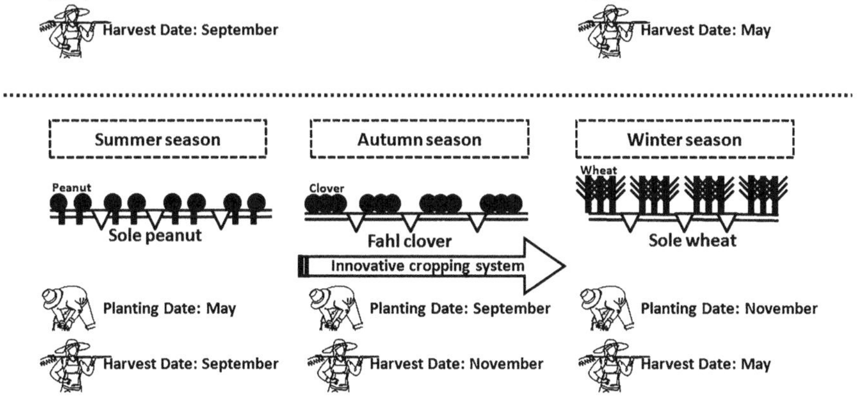

Fig. 3.1 Innovative versus traditional cropping systems of summer peanut and successive winter wheat

In comparison to the traditional cropping system, the implementation of the innovation cropping system by planting fahl clover during the transition period as an early winter crop enhanced wheat productivity by 3% averaged over both seasons (Table 3.1). Due to the short transition time, planting fahl clover resulted in a quantity of accessible N and in attaining the slight improvement in wheat yield. Additionally, the presence of fahl clover may be contributed to improved soil health and reduced weed pressure in the field. These benefits further supported the increase in wheat production observed with the innovative cropping system.

Furthermore, peanuts have high N-fixing capacity, resulting in a positive residual impact on soil N content. According to Dinh et al. (2013), there is a strong association between fixed N, biomass output, and the N concentration in the soil which allows the highest assimilate partitioning, which increased the number of pods produced per hectare (Gao et al. 2023). The nodulation on the roots of peanut caused a considerable quantity of N fixation as a consequence (Koolachart et al. 2013). This effect was further enhanced by growing fahl clover, leading to the improvement in wheat yield. Thus, growing fahl clover during the transition period, between peanut harvest and wheat cultivation, had growth-promoting effects on the soil's chemical and biological properties, enhancing nutrient absorption and availability in the soil.

According to Waha (2020), the cropping index is defined as the ratio of the total planting area to the arable land area. It indicates the frequency of crop planting on the same arable land within a year. Accordingly, in this studied crop sequences, the cropping index in the traditional cropping systems was 2.0, while it increased to reach 3.0 in the innovative cropping systems (Table 3.1). This suggests that innovative cropping systems can achieve higher yields per unit area compared to traditional cropping systems.

Accordingly, this innovative approach showcases the potential for sustainable agricultural practices that enhance productivity and resource utilization. By intensifying field crop production, the cropping system not only increases output but also enhances ecological balance. Thus, it could serve as a blueprint for other farmers seeking to enhance efficiency and sustainability in their practices. Furthermore,

Table 3.1 Innovative versus traditional cropping system effects on economic yield of peanut, fahl berseem, and wheat, as well as cropping index

		Economic yield (ton/ha)		
Cropping system	Crop	First season	Second season	Cropping index
Traditional system	Peanut	2.52	2.68	1.0
	Wheat	4.85	5.09	1.0
Total		–	–	2.0
Innovative system	Peanut	2.52	2.68	1.0
	Fahl clover	22.70	23.50	1.0
	Wheat	5.00	5.24	1.0
Total		–	–	3.0
F test 0.05 wheat yield		*(0.12)	*(0.10)	–

Prepared by the authors from Abdel-Galil et al. (2015)

improved soil health and nutrients availability may be the cause of the beneficial effect of the innovative cropping system on productivity of wheat.

Enhancing Maize Yield by Intercropping with Soybean Under Different Crop Sequences

Another example of changing cropping sequence by implementing an intensive cropping system was previously published by Lamlom et al. (2015). A 2-year field experiment was conducted at Beni-Suef governorate, Middle Egypt, to investigate the residual effects of three preceding winter crops, namely, Egyptian clover, sugar beet, and wheat (winter crops), on the yield of maize (summer crops) as the traditional cropping system and compare it with an innovative cropping system, in which soybean was intercropped with maize. The intercropping system involved two maize ridges alternating with two soybean ridges (2:2). Maize plants under intercropping system have a planting density of 75% of that of sole maize plants, while soybean plants under intercropping system have a plant density of 50% of that of sole soybean plants. Figure 3.2 illustrates the traditional and the innovative cropping systems comprising successive summer maize and three winter crops.

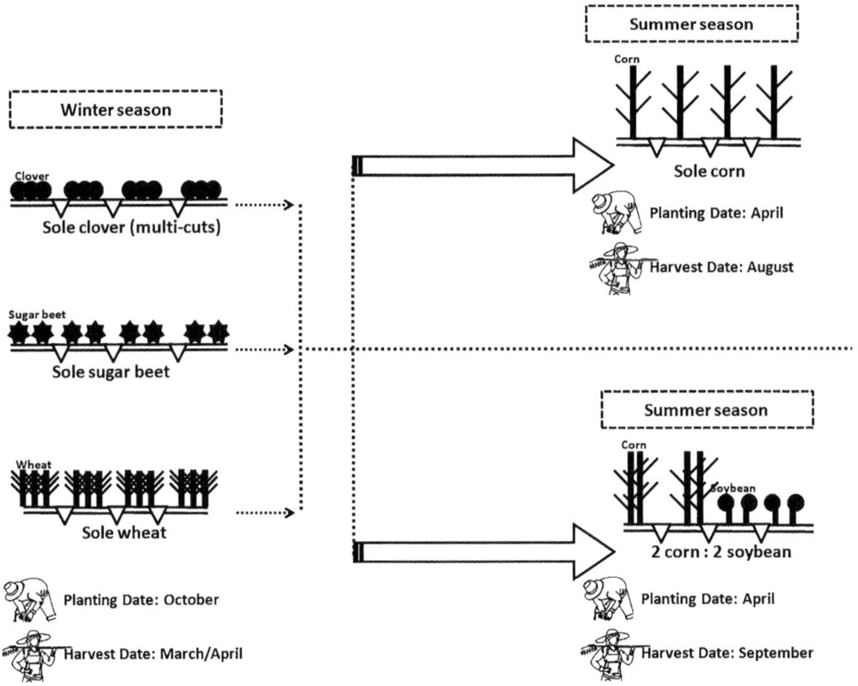

Fig. 3.2 Innovative versus traditional cropping systems of three winter crops and successive summer maize

Accordingly, the traditional cropping system includes planting maize after the harvest of Egyptian clover, rose maize yield by 4% and 3%, compared to maize yield followed sugar beet and wheat, respectively averaged over the two growing seasons (Table 3.2). This result is attributed to the role that Egyptian clover, as a legume crop, plays in enhancing the yield of these crops. It has been reported that legume enhances N_2 fixation and the acquisition of soil nitrogen by the following cereals (Rodriguez et al. 2020). Furthermore, Egyptian clover produced growth-promoting substances that affect the chemical and biological characteristics of the soil, which improved the soil's capacity to absorb nutrients and increase its availability (Fetyan et al. 2015).

It has been reported that the breakdown of Egyptian clover residues may help soybean plants have better soil structure and nutrients availability (Lamlom et al. 2015). Consequently, the nutrient supply and improved soil conditions for the growth of roots and shoots of the soybean crop led to improved maize growth and development following Egyptian clover cutting (Aita and Giacomini 2003). Because leguminous plants fix atmospheric N_2 and reduce the C/N ratio, residues decompose more quickly, which releases stored N and other nutrients (Marcelo et al. 2009).

Furthermore, the innovative cropping system, which includes intercropping soybeans with maize after the harvest of Egyptian clover, increased maize yield by 5% and 11% compared to the yield of maize obtained from sugar beet and wheat followed by soybean intercropped with maize, respectively averaged over the two growing seasons (Table 3.3). Although the intercropped maize plants represented 75% of its sole planting density, it produced 89%, 87%, and 93% of its yield under the traditional cropping system when Egyptian clover, sugar beet, and wheat preceding it averaged over the two growing seasons (Table 3.2 versus Table 3.3). These findings illustrate the possible advantages of implementing innovative cropping system to increase productivity and sustainability. Implementing intercropping systems

Table 3.2 Effect of traditional cropping systems on maize grain yield and cropping index

Cropping system	Crop	Economic yield (ton/ha)		Cropping index	
		First season	Second season	First season	Second season
Traditional cropping system Egyptian clover/maize	Egyptian clover Maize	60.92 6.36	61.35 6. 47	1.00 1.00	1.00 1.00
Total		–	–	2.00	2.00
Sugar beet/maize	Sugar beet Maize	62.64 6.11	63.28 6.29	1.00 1.00	1.00 1.00
Total		–	–	2.00	2.00
Wheat/maize	Wheat Maize	5.50 5.99	5.72 6.19	1.00 1.00	1.00 1.00
Total		–	–	2.00	2.00
Significance for maize grain yield					
L.S.D. 0.05		0.12	0.18	–	–

Prepared by the authors from Lamlom et al. (2015)

Table 3.3 Effect of innovative cropping systems on maize grain yield, LER, and cropping index

Innovative cropping system	Crop	Economic yield/ha (ton) First season	Economic yield/ha (ton) Second season	Relative yield First season	Relative yield Second season
Egyptian clover/soybean + maize	Egyptian clover Soybean Maize	60.92 1.64 5.60	61.35 1.82 5.76	1.00 0.61 0.92	1.00 0.61 0.93
LER		–	–	1.53	1.54
Cropping index		–	–	2.53	2.54
Sugar beet/soybean + maize	Sugar beet Soybean Maize	62.64 1.52 5.28	63.28 1.88 5.46	1.00 0.56 0.88	1.00 0.63 0.88
LER		–	–	1.44	1.51
Cropping index		–	–	2.44	2.51
Wheat/soybean + maize	Wheat Soybean Maize	5.50 1.55 4.93	5.72 1.71 5.15	1.00 0.57 0.82	1.00 0.58 0.83
LER		–	–	1.39	1.41
Cropping index		–	–	2.39	2.41
L.S.D. 0.05		0.10	0.16	–	–

Sole soybean yield = 2.70 and 2.96 ton/ha in the first and second seasons, respectively
Prepared by the authors from Lamlom et al. (2015)

can help farmers to improve soil fertility and reduce the need for synthetic fertilizers. Not only does it benefit crop yields, but it also promotes environmental conservation in the long run.

The findings in Table 3.3 also show that the innovative cropping system, which involved intercropping soybeans with maize after harvesting clover, sugar beet, and wheat, attained values of LER reaching 1.54, 1.48, and 1.40, respectively averaged over the two growing seasons. Furthermore, the above three innovative cropping systems increased the value of cropping index by 21%, 20%, and 17%, respectively averaged over the two growing seasons, which suggest that, in contrast to traditional cropping systems, innovative cropping systems can optimize land productivity and efficiency.

Dynamic Cropping Systems Mitigate Shading Stress

A dynamic cropping system is where crop sequencing decisions are made on an annual basis; thus, it is ascribed by a management approach to improve the adaptability of cropping practices to environmental stressors (Tanaka et al. 2007). A dynamic cropping system can enhance crop resilience and productivity, in which sustainable agricultural practices are promoted and improve the overall crop performance (Liebig et al. 2007). By integrating ecological principles into farming practices, farmers can improve soil health, reduce reliance on synthetic inputs, and enhance biodiversity on their farms (Wagh et al. 2024). This strategy also promotes efficient resource utilization, such as sunlight and nutrients, leading to more resilient and productive farming systems (Çakmakçı et al. 2023).

Enhancing Light Interception in Faba Bean Intercropping System with Wheat

Enhancing light interception in faba bean intercropping system with wheat and increase the economic yields of both crops was studied in a 2-year field experiment conducted in El-Boustan region, El-Behira Governorate, north Egypt, by Abdel-Wahab and El Manzlawy (2016a). The intercropped faba bean planting density was two rows (50% of its recommended planting density). The intercropped wheat planting density was six, five, and four rows (100%, 88.3%, and 67% of its recommended planting density). Their results presented in Table 3.4 indicated that reducing the number of intercropped wheat rows from six to four resulted in an increase in faba bean seed yield by 54% averaged over the two growing seasons, in which competition between faba bean and wheat plants for light and other resources was reduced. This result emphasizes the advantage of optimizing planting density for enhanced productivity.

Furthermore, reducing the number of intercropped wheat rows from six to four resulted in lower wheat yield by 8%. However, although 88.3% and 67% of wheat density was intercropped with faba bean, wheat yield attained 89% and 86% of the yield attained from sole wheat planting. This result emphasizes the role of faba bean in enhancing soil nutrients, which positively reflected on wheat yield. Furthermore, reducing the number of wheat rows from six to four increased the LER values from 1.22 to 1.30, averaged over the two growing seasons. This implies that adjusting wheat row density can enhance land utilization efficiency and boost crop

Table 3.4 Economic yield of faba bean and wheat and LER values under their intercropping systems

	Economic yield (ton/ha)					
	Faba bean		Wheat		LER	
Cropping system	First season	Second season	First season	Second season	First season	Second season
2 faba bean + 6 wheat	0.57	0.81	5.15	5.94	1.14	1.29
2 faba bean + 5 wheat	0.81	0.97	5.04	5.62	1.22	1.30
2 faba bean + 4 wheat	0.85	1.28	4.97	5.27	1.23	1.37
$LSD_{0.05}$	0.26	0.41	0.12	0.24	0.09	0.07
Sole planting	2.29	2.45	5.76	6.18	1.00	1.00

Prepared by the authors from Abdel-Wahab and El Manzlawy (2016a)

productivity. These results highlight the importance of optimizing wheat plant density for maximizing crop yields under intercropping conditions.

The results of this study also showed that light intensity in the middle of the wheat canopy was increased when the number of wheat rows was reduced from six to four rows. The lower wheat planting density resulted in reduced competition for sunlight, allowing more light to reach the middle of wheat canopy boosting light intensity.

Furthermore, light intensity in the middle of wheat canopy showed a significant negative correlation with wheat grain yield. On the contrary, light intensity in the middle of wheat canopy showed a significant positive correlation with both faba bean seed yield and LER, as per the results of simple phenotypic correlation. These findings underscore the importance of considering crop-specific responses to environmental factors in agricultural management practices (Table 3.5).

These results suggest that light availability may have varying effects on the productivity of wheat and faba bean crops under their intercropping systems. Wheat grain yield was significantly negatively correlated with both faba bean seed yield and LER, while faba bean seed yield was significantly positively correlated with LER, which imply potential competition between wheat and faba bean crops for light, resulting in a negative impact on wheat grain yield (Table 3.5). Cui et al. (2015) asserted that the increase in dry matter accumulation in intercropping systems results from the multi-tiered nature of the crop group, which may enhance solar radiation and stratified light energy use, along with its efficient application. Additionally, Zhang et al. (2020) reported that increased light interception and utilization efficiency correlated with enhanced productivity in intercropping systems. Farmers should take these crop-specific responses into account when planning intercropping strategies to enhance overall productivity.

Enhancing Light Interception in Sunflower Intercropping System with Soybean

Enhancing light interception in sunflower intercropping system with soybean was implemented in a 2-year field experiment conducted in the El-Boustan area, El-Behira Governorate, north Egypt, by Abdel-Wahab and El Manzlawy (2016b) to

Table 3.5 Phenotypic simple correlation coefficients between light intensity in the middle of wheat canopy and economic yields of faba bean and wheat, as well as LER—combined data across the two seasons

Traits	Grain yield of wheat	Seed yield of faba bean	LER
Light intensity at the middle of wheat plants	−0.689*	0.788*	0.693*
Grain yield of wheat	0	−0.747*	−0.888**
Seed yield of faba bean		0	0.867**
LER			0

study the effect of planting density of sunflower on light interception in its intercropping system with soybean on final yield of both crops. The study included three sunflower plant densities (recommended planting density, 75% of recommended planting density, and 50% of recommended planting density) intercropped with 50% of the recommended planting density of soybean.

Their results showed that reducing sunflower plant density from 100% to 50% led to a 65% increase in soybean seed yield, and a decrease in sunflower yield by 40% averaged over the two growing seasons (Table 3.6). This suggests that competition from sunflower plants has a significant negative impact on soybean seed yield. Furthermore, the highest value of LER was found under 100% sunflower plant density intercropped with 50% soybean planting density, namely, 1.21, averaged over the two growing seasons (Table 3.6). This suggests that higher sunflower plant density can significantly improve LER in both seasons, indicating a more efficient use of resources and potentially higher yields. Farmers may consider adjusting planting densities to optimize crop yields and overall farm productivity under intercropping conditions.

Moreover, the results in Table 3.6 also revealed that reducing sunflower plant density from 100% to 50% resulted in an increase in light intensity in the middle of soybean canopy. This increase in light intensity is expected to positively influence the growth and development of soybean plants, potentially leading to increased

Table 3.6 The economic yields of soybean and sunflower under its intercropping systems and LER values

Cropping system	Economic yield/ha (ton)				LER	
	Soybean		Sunflower			
	First season	Second season	First season	Second season	First season	Second season
50% soybean + 100% sunflower	0.78	0.95	2.83	3.19	1.21	1.20
50% soybean + 75% sunflower	1.04	1.26	2.41	2.62	1.15	1.13
50% soybean + 50% sunflower	1.32	1.54	1.71	1.92	1.01	1.01
L.S.D. 0.05	0.14	0.19	0.34	0.47	0.12	0.15
Sole planting	3.02	3.34	2.96	3.46	1.00	1.00
Cropping system	**Light intensity in the middle of soybean plants (lux)**					
	First season		**Second season**			
50% soybean + 100% sunflower	90.33		101.00			
50% soybean + 75% sunflower	107.33		121.66			
50% soybean + 50% sunflower	141.66		158.66			
L.S.D. 0.05	19.55		21.66			
Sole planting	174.00		198.00			

Prepared by the authors from Abdel-Wahab and El Manzlawy (2016b)

Table 3.7 Phenotypic simple correlation coefficients between light intensity in the middle of soybean canopy and economic yields of sunflower and soybean, as well as LER—combined data across the two seasons

Traits	Seed yield of soybean	Seed yield of sunflower	LER
Light intensity at the middle of soybean plants	0.909**	−0.639*	−0.727*
Seed yield of soybean	0	−0.673*	−0.933**
Seed yield of sunflower		0	0.703*
LER			0

yields. Additionally, light intensity in the middle of soybean canopy showed a significant positive correlation with soybean seed yield and a significant negative correlation with both sunflower seed yield and LER, as per the results of simple phenotypic correlation (Table 3.7). These findings indicate that soybean plants could benefit from higher light intensity levels, leading to an increase in seed yield. Liu et al. (2008) proposed that modifying soybean planting density is crucial for optimal crop growth, maximizing biomass and grain yield. Conversely, sunflower plants and LER may be adversely affected by higher light intensity in the middle of soybean canopy, suggesting a potential trade-off in optimizing light exposure for different crop species.

The results also discovered that soybean seed yield was significantly negatively correlated with both sunflower seed yield and LER, while sunflower seed yield was significantly positively correlated with LER. These results indicate that there could be competition between soybean and sunflower for resources, resulting in reduced seed yields when they are grown together. Furthermore, the existing strong correlation between sunflower seed yield and LER suggests that sunflower may be more effective in resource utilization compared to soybean (Table 3.7).

Dynamic Cropping Systems Combat Insect and Weed Infection

Insects

Insects like whiteflies, cotton leaf worms, jassids, and aphids feed on plant tissues, causing damage and compromising crop health, and can greatly decrease crop yields (Ahmad et al. 2011). Integrated pest management strategies, such as intercropping cowpea with maize to disrupt pest life cycles and using insecticides as a last resort, can help control these pests and minimize their impact on agricultural production (Abdel-Wahab et al. 2016). By implementing these approaches, farmers can effectively manage insect pests while mitigating the negative effects on crop yields.

In a 2-year field study conducted by Abdel-Wahab et al. (2021) at Giza in Middle Egypt, the effectiveness of intercropping cowpea with maize on pest control was assessed. The study focused on evaluating the presence of whiteflies, cotton leaf

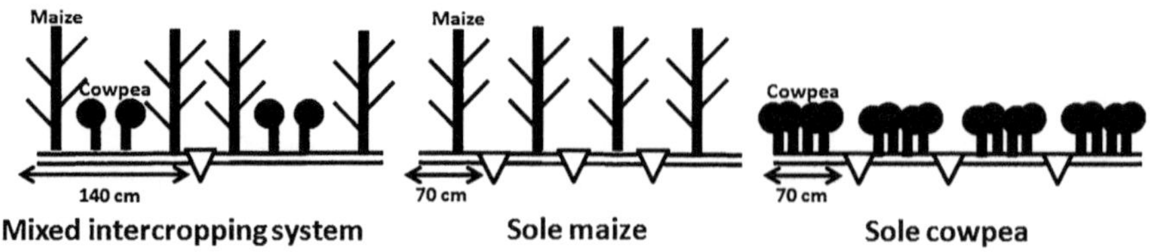

Fig. 3.3 Intercropping cowpea with maize and sole plantings of both crops. (Prepared by the authors from Abdel-Wahab et al. 2021)

worms, jassids, and aphids as part of a sustainable pest management strategy. The plant density in this system was 50% cowpea and 100% maize (Fig. 3.3).

Their results indicate that whitefly assemblages on maize plants were lower, namely, 14.40 in intercropped maize, compared to 15.0 in sole maize. Conversely, whitefly assemblages on cowpea plants decreased from 14.40 in intercropped cowpea to 13.95 in sole cowpea. Furthermore, cotton leaf worm assemblages on maize plants increased from 9.80 in intercropped maize to 10.95 in sole maize. On the contrary, the assemblages of cotton leaf worms on cowpea plants decreased from 9.80 in intercropped to 9.44 in sole cowpea.

The results of this study suggest that cowpea plants are more susceptible to cotton leaf worm infestation when grown alongside maize, as the insects are more attracted to cowpea plants than the maize plants. This could have negative implications for crop productivity and overall plant growth in intercropping system.

The results also indicated that jassid assemblages on maize and cowpea plants increased from 13.80 in intercropped maize to 15.15 in sole maize and 15.05 in sole cowpea. These findings suggest that maize or cowpea plants are more vulnerable to jassid infestation when grown alone compared to when grown together in intercropping.

The presence of other plant species may act as a deterrent, preventing jassids from concentrating on a single crop. This highlights the potential effectiveness of intercropping as a strategy for managing jassid assemblages in agricultural fields.

Moreover, these results indicate that there was no significant difference in aphids' assemblages on maize plants between intercropping and sole planting. However, aphids' assemblages decreased from 15.25 in intercropping to 13.60 in sole cowpea.

These findings suggest that cowpea plants may be more vulnerable to aphid infestation when grown alongside maize, possibly due to competition for resources between the two plant species. This could be due to variations in plant chemistry or attractiveness to aphids, resulting in a higher number of aphids on cowpea plants in intercropping systems compared to sole planting.

Broomrape (*Orobanche crenata* Forsk.)

Broomrape is a harmful weed that poses a threat to various field crops, particularly legumes, and has a wide range of hosts from different plant families (Negewo et al.

2022). It is an obligate parasite that relies entirely on its host plants for water and nutrients. Broomrape parasitizes the roots of its host and comes in various types distinguished by shape, colour, and size (El-wahab et al. 2024). The damage caused by broomrape infection to crops is significant, in which the weed absorbs water and nutrients from the host plant roots, resulting in wilting and drying of the vegetation, leading to reduced crop yield and quality (Habimana et al. 2014). This leads to substantial losses for farmers and the economy. Broomrape seeds attach to the roots of host crops under suitable conditions of moisture and temperature (López-Granados and García-Torres 1996). A small tube-like structure emerges from the seed, attaching to the crop root and penetrating it with enzymes. This attachment leads to the formation of tubers and roots, eventually developing into a flower spike above the soil surface (Foy et al. 1989). Capsules containing seeds are formed, which disperse with the wind, spreading the infection (Yoneyama et al. 2010). The challenge in eradicating broomrape lies in its prolific seed production, with a single weed producing thousands of seeds (Qasem 2021). Some types can remain dormant in the soil for up to 15 years (Habimana et al. 2014). Drought conditions exacerbate the damage caused by broomrape, affecting the host plant's photosynthetic efficiency (Ncgcwo et al. 2022).

Previous research by Abdel-Wahab et al. (2021) has shown that intercropping garlic, radish, fenugreek, and flax with faba beans can help in reducing broomrape infestation and improve seed yield of faba beans. Their study was conducted at Giza governorate, in Middle Egypt to assess the impact of intercropping these four crops with faba beans on broomrape occurrence and faba bean yield in heavily infested soil. Two faba bean cultivars (Nubaria 1 and Nubaria 2) were intercropped with the four field crops as treatments, and all of the crops that were evaluated were planted alone. The results indicated that, when faba beans were planted alongside the studied crops in the severely infested soil with broomrape, the interstitial spaces of the faba bean root were completely close (Fig. 3.4). This enhances the xylem and phloem transport vessels, preventing the transfer of water and photosynthates from the plant to the parasite. This process effectively starves the broomrape, reducing its ability to reproduce and spread throughout the field. The accumulation of cell wall materials was found to be the main mechanism driving the ability of Nubaria 2 cultivar to tolerate broomrape, with residues from various intercrops affecting vascular tissues. Nubaria 2 cultivar demonstrated greater tolerance to broomrape compared to Nubaria 1 cultivar based on these mechanisms.

Furthermore, intercropping these four crops with faba beans can be an effective strategy for controlling broomrape infestations in infected fields as a result of formation of completely several compounds, namely, formic acid from garlic, salicylic acid from radish, citric and ascorbic acids from flax, and oxalic acid from fenugreek. These compounds were identified as the primary sources of soil organic acids produced in the rhizosphere of intercropped faba beans (Abdel-Wahab and Abdel-Wahab 2021). The presence of these organic acids in the rhizosphere significantly inhibited broomrape growth, indicating that intercropping with faba beans could effectively manage broomrape infestations. This approach hinders broomrape's ability to establish a connection with the host plant, resulting in a reduction of

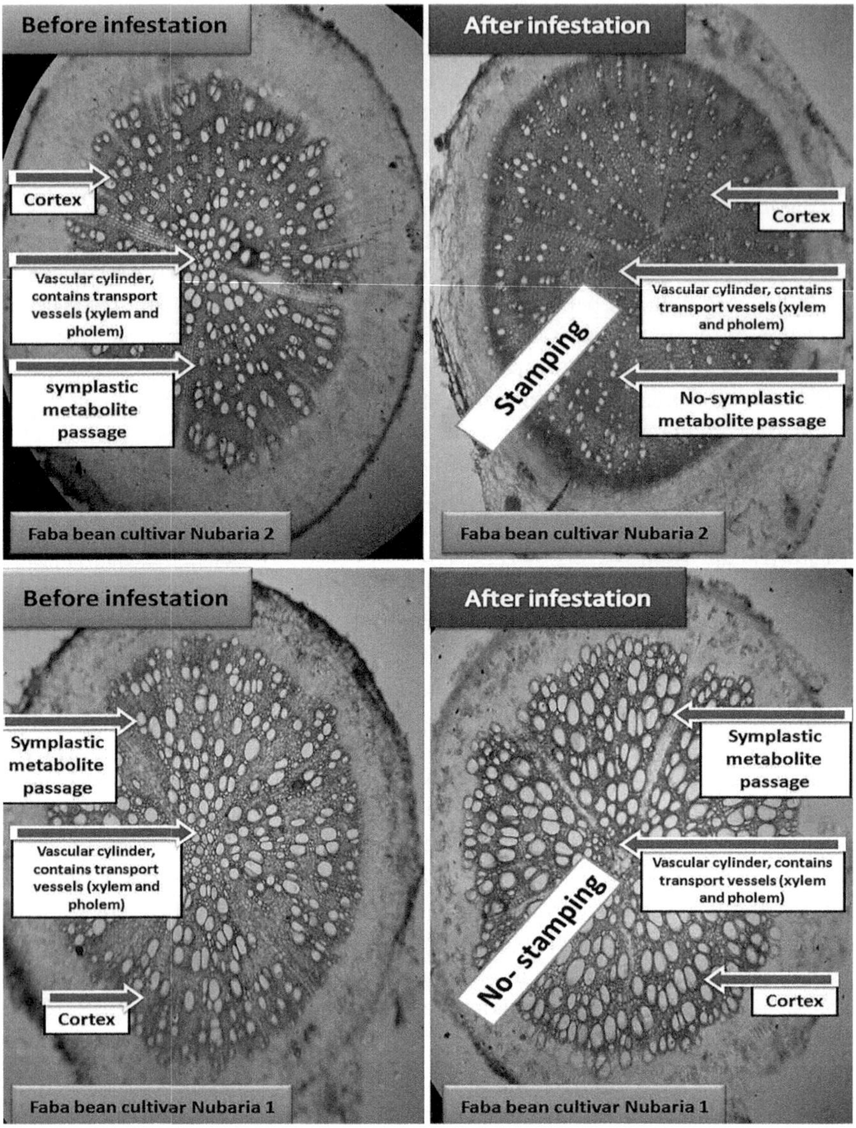

Fig. 3.4 Vascular system of root cells in faba bean cultivars before and after broomrape infestation. (Source: Abdel-Wahab et al. 2021)

broomrape presence in the field. These findings underscore the potential of intercropping as a sustainable strategy for controlling parasitic weeds like broomrape.

Role of Evolutionary Breeding in Enduring Environmental Stressors

Evolutionary breeding aims to develop crop varieties that are better adapted to changing environmental conditions (Ranjan 2024). Evolutionary approach depends on what selective breeding has been utilized for centuries to enhance the genetic traits of crops for human advantage (Venske et al. 2019). This method entails choosing individuals with desired characteristics and mating them to transmit those traits to subsequent generations, leading to more favourable population over time (Limeri 2023). Evolutionary breeding plays a crucial role in crop intensification by producing new varieties that are more resistant to environmental stresses (Metwally et al. 2021). Annicchiarico et al. (2019) have emphasized the potential role of evolutionary breeding as a beneficial tool in intercrop breeding, in which the intercrop species are permitted to co-evolve over a number of generations. This helps in meeting the needs of a growing global population by increasing agricultural productivity and improving nutritional content. Furthermore, by promoting biodiversity in agricultural systems and reducing the reliance on chemical inputs, evolutionary breeding is expected to support sustainable agriculture (Limeri 2023). This strategy ensures long-term food security and helps preserve genetic diversity within crop species.

Conclusions

Innovative cropping systems demonstrated potential for lowering resource inputs which would result in a more ecological friendly agricultural practice. Future perspectives for maximizing available agricultural resources through intensive cropping systems include exploring the potential of precision agriculture technologies to improve nutrient management and monitoring. Additionally, investigating the use of crop sequence with intercropping to increase food production and improve land use efficiency can be beneficial. By incorporating these strategies, farmers can increase productivity and efficiency while promoting sustainability in cropping systems production. Furthermore, crop intensification can be an important tool in controlling insect and weed infestation. Thus, educating farmers on the modern agricultural techniques can enhance their skills and knowledge, leading to higher yields and sustainable farming practices.

References

Abd El-Hady AA, El-Din Shaheen FAH, Saleh AA, Hamid AA (2020) Comparative efficiency of certain products against some piercing sucking pests and their associated predators on cotton plants. J Plant Prot Pathol (Mansoura Univ) 11(2):153–159

Abdel-Galil AM, Abdel-Wahab SI, Abdel-Wahab TI (2015) Effect of some preceded peanut cultivars on wheat yield and agro–economic feasibility under two cropping systems in sandy soil. Sustain Agric Res 4(2):47–56

Abdel-Wahab TI, Abdel-Wahab EI (2021) Impact of intercropping of different crops with two faba bean cultivars on infestation with broomrape. Indian J Agric Res 55(3):245–256

Abdel-Wahab TI, El Manzlawy AM (2016a) Yield and quality of intercropped wheat with faba bean under different wheat plant densities and slow—release nitrogen fertilizer rates in sandy soil. Am J Exp Agric 11(6):1–22

Abdel-Wahab SI, El Manzlawy AM (2016b) Yield and quality of intercropped sunflower with soybean under different sunflower plant spacings and slow—release nitrogen fertilizer rates in sandy soil. Int J Appl Agric Sci 2(3):32–43

Abdel-Wahab SI, El Sayed WM, El Manzlawy AM (2016) Influences of some preceding winter crops and nitrogen fertilizer rates on yield and quality of intercropped maize with cowpea. Am J Exp Agric 11(6):1–19

Abdel-Wahab TI, Abdel-Wahab SI, Shams AS, Taha AM, Saied SM, Adel MM, Hussein HM (2021) Study of productivity and economic evaluation of intercropping cowpea with maize under three irrigation water levels and their response to insect infestation and viral infection. Plant Arch 21(Suppl Issue 1):996–1019

Ahmad S, Veyrat N, Gordon-Weeks R, Zhang Y, Martin J, Smart L, Glauser G, Erb M, Flors V, Frey M et al (2011) Benzoxazinoid metabolites regulate innate immunity against aphids and fungi in maize. Plant Physiol 157:317–327

Aita C, Giacomini SJ (2003) Decomposicao e liberacao de nitrogenio de residuosculturais de plantas de cobertura de solo solteiras e consorciadas. Rev Bras Cienc Solo 27:601–612

Annicchiarico P, Collins RP, De Ron AM, Firmat C, Litrico I, Hauggaard-Nielsen H (2019) Do we need specific breeding for legume-based mixtures. Adv Agron 157:141–215. https://doi.org/10.1016/bs.agron.2019.04.001

Barbieri P, Pellerin S, Seufert V, Nesme T (2019) Changes in crop rotations would impact food production in an organically farmed world. Nat Sustain 2:378–385

Bisht IS, Rana JC, Ahlawat SP (2020) The future of smallholder farming in India: some sustainability considerations. Sustain For 12:3751

Çakmakçı R, Salık MA, Çakmakçı S (2023) Assessment and principles of environmentally sustainable food and agriculture systems. Agriculture 13(5):1073

Cui L, Yang WY, Huang N, Liu J, Wang YL, Wang XH, Liu Y, Yan S (2015) Effects of maize plant types on dry matter accumulation characteristics and yield of soybean in maize soybean intercropping systems. Chin J Appl Ecol 26:2414–2420

Dinh HT, Kaewpradit W, Jogloy S, Vorasoot N, Patanothai A (2013) Biological nitrogen fixation of peanut genotypes with different levels of drought tolerance under mid-season drought. SABRAO J Breed Genet 45(3):491–503

El-Sodany ME, Abou-Elela AM (2010) Effect of preceding winter crops relay cropping, intercropping system and nitrogen fertilizer on soil physical and chemical properties. J Soil Sci Agric Eng 1(3):237–298

El-wahab M, Hassan M, Aboulela M (2024) New hosts and diagnostic characteristics of *Orobanche crenata* (Orobanchaceae) in Egypt. Acta Bot Croat 83(1):32–42

FAO (2005) Fertilizer use by crop in Egypt. Land and Plant Nutrition Management Service, Land and Water Development Division, Rome

Fetyan NAH, Massoud ON, Morsy EM, Khalil HM (2015) Biological evaluation of soil cultivated with Egyptian clover (*Trifolium alexndrinum* L.) though long term trial at Bahtim region, Egypt. Middle East J Appl Sci 5(2):515–525

Foy CL, Jain R, Jacobsohn R (1989) Recent approaches for chemical control of broomrape (Orobanche spp.). Rev Weed Sci 4:123–152

Gad HAA, Abdel-Aziz AAA (2024) Field efficacy of selected insecticides against Spodoptera littoralis, Aphid species and Bemisia tabaci on stevia plants. Middle East J Agric Res 13(2):470–476

Gao H, Zhang C, Zhang L, Guo F, Cross P, Zhang Z, Wan S, Zhang F (2023) Morphological responses in peanut pod development to intercropping and nitrogen application rates. Field Crop Res 302:109101

Habimana S, Nduwumuremyi A, Chinama JD (2014) Management of orobanche in field crops—a review. J Soil Sci Plant Nutr 14(1):43–62

Hegazi AM, Afifi MY, El Shorbagy MA, Elwan AA, El-Demerdashe S (2005) Egyptian national action program to combat desertification. Ministry of Agriculture and Land Reclamation, UNCCD and Desert Research Center

Hiebesch CK, McCollum RE (1987) Area × time equivalency ratio: a method for evaluating the productivity of intercrops. Agron J 79:15–22

Kabenomuhangi R (2024) Regenerative agriculture and soil health: enhancing biodiversity through sustainable farming practice. Int J Res Publ Rev 5(9):3203–3215

Koolachart R, Jogloy S, Vorasoot N, Wongkaew S, Holbrook C, Jongrungklang N, Kesmala T, Patanothai A (2013) Rooting traits of peanut genotypes with different yield responses to terminal drought. Field Crop Res 149:366–378

Lamlom MM, Abdel-Wahab SI, Abdel-Wahab TI, Gendy EK (2015) Residual effects of some preceded winter field crops on productivity of intercropped soybean with three maize cultivars. Am J Biol Sci 3(6):226–242

Li L, Sun J, Zhang F, Guo T, Bao X, Smith FA (2006) Root distribution and interactions between intercropped species. Oecologia 147:280–290

Liebig MA, Tanaka DL, Krupinsky JM, Merrill SD, Hanson JD (2007) Dynamic cropping systems: contributions to improve agroecosystem sustainability. Agron J 99:899–903. https://doi.org/10.2134/agronj2006.0131s

Limeri L (2023) Introductory biology 2. In: Artificial selection. Press Books Publication. https://raider.pressbooks.pub/biology2/chapter/2-artificial-selection/

Liu XB, Jin J, Wang GH, Herbert SJ (2008) Soybean yield physiology and development of high-yielding practices in Northeast China. Field Crop Res 105:157–171

López-Granados F, García-Torres L (1996) Effects of environmental factors on dormancy and germination of crenate broomrape (*Orobanche crenata*). Weed Sci 44:284–289

Loreau M (2000) Biodiversity and ecosystem functioning: recent theoretical advances. Oikos 91:3–17

Marcelo AV, Corã JE, Fernandes C, Martins MR, Jorge RF (2009) Crop sequences in no-tillage system: effects on soil fertility and soybean, maize and rice yield. Rev Bras Ciênc Solo 33:417–428

Metwally AA, Abdel-Wahab TI, Abdel-Wahab SI (2019) Increasing land and water use efficiencies by intercropping summer legumes with corn in Egypt. Agric Biol Res 35(2):6–10

Metwally AA, Safina SA, Abdel-Wahab Eman I, Abdel-Wahab SI, Abdel-Wahab TI (2021) Screening thirty soybean genotypes under solid and intercropping plantings in Egypt. J Crop Sci Biotechnol 24:203–220

Negash F, Mulualem T, Fikirie K (2018) Effect of cropping sequence on agricultural crops: implications for productivity and utilization of natural resources. Adv Crop Sci Tech 6:326

Negewo T, Ahmed S, Tessema T, Tana T (2022) Biological characteristics, impacts, and management of crenate broomrape (*Orobanche crenata*) in faba bean (*Vicia faba*): a review. Front Agron 4:708187

Ouda S, Zohry A (2018) Introductory synopsis of the natural resources involved in food production. In: Crop rotation: an approach to secure future food. Springer. ISBN 978-3-030-05350-5

Ouda S, Zohry A (2022) Water-smart practices to manage water scarcity. In: Climate-Smart agriculture: reducing food insecurity. Springer. ISBN 978-3-030-93110-0

Ouda S, Zohry A (2024) Integration of legume crops with cereal crops under changing climate. Springer

Pokhrel S, Pokhrel S (2013) Legumes crop rotation can improve food and nutrition security in Nepal. Agron J Nepal 3:123–127

Qasem JR (2021) Broomrapes (Orobanche spp.) the challenge and management: a review. Jor J Agric Sci 17(Suppl 3):117–150

Ranjan R (2024) Agroecological approaches to sustainable development. Front Sustain Food Syst 8:1405409

Rodriguez C, Carlsson G, Englund J-E, Flöhr A, Pelzer E, Jeuffroy M-H, Makowski D, Jensen ES (2020) Grain legume-cereal intercropping enhances the use of soil-derived and biologically fixed nitrogen in temperate agroecosystems. A meta-analysis. Eur J Agron 118:126077. https://doi.org/10.1016/j.eja.2020.126077

Rodriguez C, Dimitrova Mårtensson L, Zachrison M, Carlsson G (2021) Sustainability of diversified organic cropping systems—challenges identified by farmer interviews and multi-criteria assessments. Front Agron 3:698968

Rossi A (2019) From co-learning to shared commitment to agroecology. Some insights from initiatives aimed at reintroducing agrobiodiversity. Sustainability 12(18):7766

Tanaka DL, Krupinsky JM, Merrill SD, Liebig MA, Hanson JD (2007) Dynamic cropping systems for sustainable crop production in the Northern Great Plains. Agron J 99(4):904–911

Toker P, Canci H, Turhan I, Isci A, Scherzinger M, Kordrostami M, Yol E (2024) The advantages of intercropping to improve productivity in food and forage production—a review. Plant Prod Sci 27(3):155–169

Venske E, dos Santos RS, Busanello C et al (2019) Bread wheat: a role model for plant domestication and breeding. Hereditas 156:16. https://doi.org/10.1186/s41065-019-0093-9

Wagh P, Singh A, Singh G, Chouhan SH (2024) Agro-ecology: integrating ecology into agricultural systems. In: Fields of tomorrow: advancements in modern agriculture, 1st edn. Astitva Prakashan, pp 1–20

Waha K, Dietrich JP, Portmann FT, Siebert S, Thornton PK, Bondeau A, Herrero M (2020) Multiple cropping systems of the world and the potential for increasing cropping intensity. Glob Environ Chang 64:102131

Wang Y, Han X, Zhao X, Zhang Y, Qi B, Li L (2024) Grain yield and interspecific competition in an oat-common vetch intercropping system at varying sowing density. Front Plant Sci 15:1344110

Willey RW (1979) Intercropping: its importance and research needs. Part-2. Agronomy and research approaches. Field Crop Abstr 32:73–81

Willey RW, Osiru DSO (1972) Studies on mixtures of maize and beans (*Phaseolus vulgaris*) with particular reference to plant population. J Agric Sci (Camb) 79:519–529

Yadav SK, Banerjee A, Jhariya MK, Raj A, Khan N, Meena RS, Kumar S (2021) Agroecology towards environmental sustainability. Sustainable intensification for agroecosystem services and management. Springer, Cham, pp 323–352

Yoneyama K, Awad AA, Xie X, Yoneyama K, Takeuchi Y (2010) Strigolactones as germination stimulants for root parasitic plants. Plant Cell Physiol 51:1095–1103. https://doi.org/10.1093/pcp/pcq055

Zhang D, Sun Z, Feng L, Bai W, Yang N, Zhang Z, Du G, Feng C, Cai Q, Wang Q, Zhang Y, Wang R, Arshad A, Hao X, Sun M, Gao Z, Zhang L (2020) Maize plant density affects yield, growth and source-sink relationship of crops in maize/peanut intercropping. Field Crop Res 257:107926

Zohry A, Ouda S (2020) Egypt faces water deficiency and food insufficiency. In: Deficit irrigation: a remedy for water scarcity. Springer Publishing House. ISBN 978-3-030-35585-2

Part II

Augment the Soybean Intercropping System with Maize

Enhancing the Efficiency of Light Interception in Soybean and Maize Intercropping System

4

Tamer Ibrahim Abdel-Wahab, Sherif Ibrahim Abdel-Wahab, and Eman Ibrahim Abdel-Wahab

Abstract

Intercropping soybean with maize can be a more effective method of utilizing sunlight than growing them individually. The system of intercropping soybean with maize can enhance soil microbial communities by improving soil fertility and nutrient availability due to the light harvesting efficiency compared to sole soybean planting. Studying how soybeans react to various light environments can aid in optimizing agricultural practices for increased yields. This chapter highlights several important key points related to light interception in soybean and maize intercropping system. The effect of the interaction between light intensity and the rhizosphere of soybean under its intercropping system with maize was revised. Furthermore, the effect of the height of soybean plants and spacing between maize plants on light interception and consequently on productivity was also deliberated. The effect of different planting arrangements of soybean and maize under its intercropping system on light use efficiency and relative photosynthetic potential was also reviewed.

Keywords

Light quality · Light interference · Photosynthetically active radiation · Phenotypic simple correlation · Photosynthetic efficiency

T. I. Abdel-Wahab (✉) · S. I. Abdel-Wahab
Crop Intensification Research Department, Field Crops Research Institute, Agricultural Research Center (ARC), Giza, Egypt
e-mail: tamer.abdel-wahab@arc.sci.eg

E. I. Abdel-Wahab
Food Legumes Research Department, Field Crops Research Institute, Agricultural Research Center (ARC), Giza, Egypt

A. E.-H. Zohry, S. Ouda (eds.), *Resilient Agroecosystems*, Sustainability Sciences in Asia and Africa, https://doi.org/10.1007/978-981-96-9630-7_4

Introduction

Nowadays, minimizing environmental impact is essential to safeguard our planet for future generations. Adopting sustainable practices and minimizing waste can help in conserving natural resources and address the impacts of environmental stressors (Lodh 2024). By opting for renewable energy sources and advocating for eco-friendly alternatives, progress towards a more sustainable future can be attained. It is crucial for individuals, businesses, and governments to prioritize environmental conservation to secure a healthy food for future generations. Assessing the ecological impact on soil health and biodiversity is essential for gauging the overall well-being of an ecosystem (Rinot et al. 2019). In such analysis, researchers can make informed choices regarding conservation efforts and sustainable land management practices to protect the environment. Such assessments can also pinpoint areas at risk of degradation or biodiversity loss, enabling targeted interventions to avert further harm (Wang et al. 2024).

Sunlight is the energy source for photosynthesis in the plants. The sunlight delivers about 1360 joule (J) of power per square meter (m^2) per second (s) to Earth (Hansen et al. 2005). Solar radiation and light interception have a significant influence on the growth and development of plants (Abdel-Wahab and Abdel-Rahman 2016). Summer crops could not grow below the trophic sun radiation threshold of 8.4 MJ $m^{-2}d^{-1}$ as reported by the FAO (1990). Sandri et al. (2003) suggest that this limit may not be universally applicable. The maximum solar radiation threshold that plants can tolerate is not well-documented in the literature. According to Kozai et al. (2022), solar radiation has a wavelength ranging from approximately 300 to 2500 nm, with a peak wavelength around 500 nm (end of blue-green spectrum). Visible light is just a small portion of the entire electromagnetic spectrum. For humans, visible light ranges from 380 to 740 nm, while photosynthetically active radiation covers 400–700 nm (Mõttus et al. 2012). Violet light has a wavelength of around 400 nm and a frequency of $7.5*10^{14}$ Hz, while red light has a wavelength of about 700 nm and a frequency of 4.3*1014 Hz (Goldman and Kolmos 2023). Solar irradiance is the amount of power per unit area received from the sun (Goswami 2015).

Lux is a unit that measures light intensity when projected onto a surface (Morcillo 2024). Light meters indicate that the number of lux produced by full sunshine is around 100,000 lux (~10,000 foot candles) (Mead 2008). It is comparable to being outside on a clear day in brightness. Weather, season, and geographic location are some of the variables that might affect solar irradiance levels (Bhargawa and Singh 2019). Comprehending the quantity of solar energy accessible for diverse uses, including solar power production and farming, requires a thorough grasp of this measurement (Goswami 2015; NIST 2019).

Solar radiation is converted to photosynthetically active radiation using a fixed ratio of about 0.45–0.46 (Akitsu et al. 2022). Light below 400 nm and above 700 nm can stimulate photosynthetic activity, resulting in a photon flux yield (McCree 1972) that considers photosynthetic activity between 360 and 760 nm (Barnes et al. 1993). The photosynthetic photon flux density is the ratio of the number of photons in the photosynthetically active radiation waveband crossing a small surface element per unit time divided by the area of the element (Sager and Giger 1980).

The dispersal of leaves and photosynthetic capacity in the canopy directly impact the interception of light and the efficiency of light use in crops, which in turn affect dry matter accumulation and yield formation (Gao et al. 2010). Particularly, Ouda and Zohry (2024) found a significant positive correlation between seasonal solar radiation and maize yield. However, not all light that strikes the leaf surface is absorbed, as some is reflected and some is transmitted through the leaf lamina (Buschmann 2007; Pedrós et al. 2008; Falcioni et al. 2017). Furthermore, plants use light signals to adjust nutrient absorption based on changing conditions (Zhou et al. 2019). Light intensity and quality can affect nutrient uptake in plants (Yang et al. 2024). Research has shown a relation between light interception and nutrient absorption, emphasizing the importance of light in plant growth and development (Wu et al. 2025).

Intercropping, which involves planting different crops together in the same field, has many environmental impacts (Willey 1979). Studying the ecological impact of intercropping soybean and maize on biodiversity and soil health could provide valuable insights into sustainable farming practices. According to Zohry (2005), intercropping soybean and maize can help in reducing weed competition with the two crops and increase overall productivity of both of them. By investigating the interactions between these two crops, namely, soybean and maize under its intercropping system, researchers can discover ways to optimize crop yield while minimizing the negative environmental effects (Abdel-Wahab and Abdel-Wahab 2020a). The yields of both soybean and maize under their intercropping system are affected by light interception as a result of their planting density. This can affect the efficiency of capturing light energy, which is essential for optimizing photosynthesis and, in turn, crop productivity (Wang et al. 2023). Consequently, optimizing light harvesting techniques is crucial for maximizing soybean output while minimizing weed interference in its intercropping system with maize.

In this chapter, the interaction between light and intercropped soybeans with maize was thoroughly reviewed to enhance our understanding of their absorption capabilities of light and nutrients, as well as their biodiversity benefits.

Light Harvesting Efficiency in Soybean and Maize Intercropping System

The process of converting sunlight into energy through photosynthesis is referred to as light harvesting (Schlau-Cohen 2015). This energy is vital for plants to thrive and for their growth and development (Wu et al. 2025). Intercropping influences soil nutrient contents and microbial communities in the rhizosphere, with light intensity playing a key role in these alterations (Liu et al. 2024). These modifications in soil conditions can have diverse effects on plant growth and productivity. Intercropping soybeans with maize offers several benefits compared to growing soybeans alone, as illustrated in Fig. 4.1. When maize and soybeans are grown together, the soybean

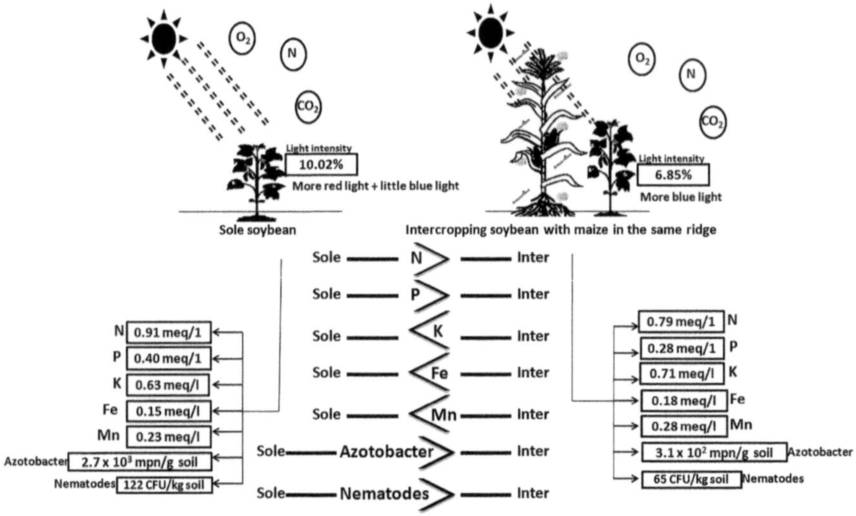

Fig. 4.1 Light intensity in the middle of soybean canopy and availability of chemical and biological soil contents by soybean plants under intercropping and sole planting

plants experience different light conditions compared to when soybeans are grown alone (Wu et al. 2016). Intercropping maize and soybeans leads to increased exposure to blue light and lower light intensity in the middle of the canopy (Yang et al. 2013). These differences in light exposure can affect the growth and development of both crops, potentially influencing soil nutrient absorption, nutrient uptake, and overall plant health and yield (Yang et al. 2024).

Light Intensity and Quality Interaction with Soil Nutrient Content in Soybean and Maize Intercropping System

The reduction in light intensity under intercropping with maize enhances nutrient availability in the soil, resulting in improved growth and yield of soybean plants (Metwally et al. 2019). It has been reported that there is a mutual exchange of nutrients between soybean and maize plants in intercropping systems that can enhance plant growth and soil fertility. Metwally et al. (2019) showed that, in intercropping soybean with maize, soybean plants release N and P into the soil, which are then absorbed by maize plants, while maize plants release K into the soil, which is absorbed by soybean plants. These findings indicate that the soybean and maize intercropping system is more effective in increasing soil K content compared to sole soybeans. Likewise, intercropping soybean with maize generally reduced the levels of N and P in the soybean rhizosphere, likely due to reduced light availability (Abdel-Wahab et al. 2024). However, the opposite effect was observed for soil K, iron (Fe), and manganese (Mn) (Metwally et al. 2019; Abdel-Wahab et al. 2024). Therefore, intercropping soybeans with maize can improve soil health and biodiversity, resulting in a net N benefit to the agroecosystem. It has been reported that higher values of soil N and P contents in rhizosphere of sole soybeans were found than under intercropped soybeans (Metwally et al. 2019).

Additionally, according to Abdel-Wahab et al. (2019a), soybeans could leverage the increased atmospheric carbon dioxide (CO_2) levels by trading excess C with N. Moreover, El-Shamy et al. (2015) reported that intercropping maize and soybeans on the same ridge, with reduced mineral N fertilizer application, can enhance maize root penetration into deeper soil layers, facilitating N uptake from the rhizosphere of both crops. They also stated that lower mineral N levels in maize can stimulate the growth of N-fixing bacteria near maize roots, leading to improved nodulation of adjacent soybean roots. Plant growth-promoting microorganisms, especially rhizobia, can significantly enhance the growth of maize (El-Shamy et al. 2015).

According to Ning et al. (2022), monochromatic red light significantly increased the levels of dissolved organic matter and hydrophilic fractions in the rhizosphere of plants, while monochromatic blue light had the opposite effect. Dissolved organic matter composition is known to influence its elemental composition, with O_2, H, N, P, and S being the main components associated with C (Bolan et al. 2010). This could explain why the rhizosphere of sole soybeans has higher soil N and P content compared to intercropped soybeans. The differences in dissolved organic matter composition between red and blue light treatments may also impact nutrient availability for plants in the rhizosphere (Ning et al. 2022).

Blue light is essential and sufficient for suppressing primary root growth under P deficiency (Zheng et al. 2019). Accordingly, blue light which is prevailing in soybean intercropping system with maize can prevent soybean main roots from growing when maize and soybeans are intercropped, particularly on the same ridge. This is thought to be an active cellular response that enhances the plant's capacity to absorb P from the soil by encouraging the growth of lateral roots (Gao et al. 2021). Wang et al. (2020) suggest that intercropped soybeans can enhance P absorption through this biological mechanism. Therefore, optimizing the use of blue light in intercropping systems has the potential to increase P uptake efficiency in soybean plants. Understanding the role of blue light in plant responses to P deficiency can contribute to the development of more sustainable agricultural practices (Pech et al. 2024).

Under intercropping system, maize plants are exposed to red and blue light and showed increase in soil K content in the rhizosphere, a benefit not observed in the rhizosphere of sole maize (Metwally et al. 2019). When soybeans and maize are grown together, particularly on the same ridge, the blue light emitted by the maize plants enhances the soybean plants' ability to absorb more K from the soil (Metwally et al. 2019). This is because the blue light stimulates the activity of K transporters in soybean roots, leading to increased K absorption. Thus, soybean intercropping system with maize could enhance K availability in the soil, potentially boosting plant growth and productivity.

Furthermore, higher soil Fe content in rhizosphere of intercropped soybeans than sole soybeans is explained by the intercropped plants exhibiting reduced Fe absorption under blue light exposure due to the formation of biomolecular condensates containing protein complexes induced by Fe deficiency (Trofimov et al. 2024). As a result, intercropped soybeans had lower Fe uptake than sole soybeans. Consequently, the rhizosphere of intercropped soybeans showed lower soil Fe content compared to sole soybeans (Abdel-Wahab et al. 2024). The presence of maize in intercropping systems appeared to impact the expression of key genes and enzymes involved in Fe absorption and utilization in soybeans, leading to reduced gene expression of Fe transporter and ferritin, as well as decreased ferrous concentrations, and catalase activity (Liu et al. 2025). Additionally, according to Kopittke and Menzies (2004), in a soil profile with sufficient N supply, a deficiency of Mn in the plants' rhizosphere worsened root-induced acidification. They added that rhizosphere acidification was more pronounced in inoculated plants compared to non-inoculated plants when the soil had higher N concentrations. The ability of inoculated plants to fix atmospheric N instead of relying on soil N stores, like non-inoculated plants, likely contributes to the difference in their capacity to lower the pH of the rhizosphere (Di Benedetto et al. 2017). Due to the higher soil N levels in the rhizosphere of sole soybeans, lower soil Mn content is exhibited.

Light Intensity and Quality Interactions with Soil Microbiome Under Soybean Intercropping System with Maize

According to Abdel-Wahab et al. (2024), intercropping soybeans with maize decreased the populations of nematodes and azotobacter compared to growing soybeans alone. It has been reported that the reduction in root exudation and alteration in root shape in low light conditions likely led to a reduction in soil nematode population as light intensity decreased (Wu et al. 2024). This shows that nematode behavior and reproduction are affected by light intensity, indicating that biological engineering, which involves changing light conditions through plant density and distribution of both species in the unit area, could be a viable approach in managing nematodes in agricultural systems (Abdel-Wahab et al. 2024). Moreover, intercropping soybeans with maize decreased nematode populations in the rhizosphere of soybean plants, possibly through the blue light photoreceptor cytochrome, which enhances resistance to nematodes (Abdel-Wahab et al. 2024). This reduction in nematodes could result in higher soybean yields and overall plant health improvement. Red light led to more galls and nematodes compared to blue light, and all light treatments resulted in significantly more galls and nematodes than plants kept in constant darkness (Wu et al. 2024).

However, intercropped soybeans show reduced populations of soil azotobacter (3.1×10^2 mpn/g soil) and nematodes (65 CFU/kg soil), while having higher content of soil K (0.71 meq/l), Fe (0.18 meq/l), and Mn (0.28 meq/l) in the rhizosphere compared to sole soybeans. These changes are attributed to the decreased light intensity within the plants (Abdel-Wahab et al. 2024). Furthermore, Szymanek et al. (2020) indicated that certain efficient photosynthetic bacteria that coexist with azotobacter, capable of binding atmospheric N, are likely enhanced by red light, possibly due to cereal's specific N requirements. These photosynthetic bacteria may provide additional nutrients to azotobacter or form a more favorable environment for its growth (Chen et al. 2003). The successful growth of maize under red light conditions may rely on this symbiotic relationship (Boomsma and Vyn 2008).

The Relationship Between Light Intensity, Soil Nutrient and Soil Microbiome Under Soybean Intercropping System with Maize

The relationship between soil nutrient content and light intensity in the middle of soybean canopy, as well as availability of chemical and biological soil contents by soybean plants under intercropping and sole planting after 80 days from soybean sowing, was studied by Abdel-Wahab et al. (2024). They stated that, under intercropping, light intensity had a negative correlation with soil N content, P content, and nematode count. Moreover, light intensity had a positive correlation with soil K content, Fe content, Mn content, and soil azotobacter count, whereas under sole soybean, a light intensity had a negative correlation with soil K content, Fe content, and Mn content. Furthermore, light intensity had a positive correlation with soil N content, P content, azotobacter count, and nematode count (Table 4.1).

These results suggest that intercropping soybean with maize may alter the relationship between light intensity and soil nutrients compared to sole soybean. Overall, these findings emphasize the potential benefits of intercropping in optimizing soil microbial communities in intensification systems. Similar findings were obtained by Sakuraba and Yanagisawa (2018), which showed a correlation between

Table 4.1 Phenotypic simple correlation coefficients between light intensity in the middle of soybean canopy and chemical and biological soil contents in the rhizosphere of soybean plants under intercropping and sole planting (average of the two seasons)

Traits	Cropping system	Correlation coefficient (r)
Light intensity × soil N content	Intercrop	−0.706*
	Sole	0.898**
Light intensity × soil P content	Intercrop	−0.228
	Sole	0.755*
Light intensity × soil K content	Intercrop	0.719*
	Sole	−0.327
Light intensity × soil Fe content	Intercrop	0.793*
	Sole	−0.715*
Light intensity × soil Mn content	Intercrop	0.809*
	Sole	−0.188
Light intensity × soil azotobacter count	Intercrop	0.083
	Sole	0.856*
Light intensity × soil nematode count	Intercrop	−0.939**
	Sole	0.746*

Source: Abdel-Wahab et al. (2024), * and ** indicate a significance level of 5% and 1%, respectively.

light perception and nutrient absorption impacting their capacity to absorb and utilize nutrients effectively.

Light Interception and Productivity of Intercropped Soybeans with Maize

Productivity of intercropping soybeans with maize depends on the spatial arrangement of the two crops within the field (Abdel-Galil et al. 2014). By strategically planting soybeans between rows of maize, farmers can maximize sunlight exposure for both crops, leading to increased overall productivity. Selecting complementary varieties of soybeans and maize that have similar growth rates can further enhance light interception and productivity in an intercropping system (Abdel-Galil et al. 2014). The light interception and productivity of intercropping system of soybeans with maize are dependent on several factors, namely, the height of the soybean plants affecting their ability to capture sunlight and the spacing between the maize hills which allows for more efficient use of available light (El-Shamy et al. 2015). Additionally, the complementary root systems of soybeans and maize can help improve soil structure and nutrient uptake, further enhancing overall productivity in intercropping systems.

The Height of Soybean Plants

The minimum amount of light needed for soybean growth and development has not been extensively researched, even though shading in intercropping systems can significantly decrease soybean seed yield and quality (Metwally et al. 2012). To ensure sustainable soybean production in these systems, it is crucial to determine the optimum light intensity (Metwally et al. 2012). Abdel-Wahab et al. (2019b) found that reduced light levels lead to lower photosynthesis and production in soybeans, which require a minimum of 50% full sunshine for optimal growth. They added that the height of soybean plants and their ability to capture sunlight are key factors in determining their yields. A measurement of sunlight lux can be used to calculate the photosynthetic photon flux density in micromoles per square meter per second (Thimijan and Heins 1983). The amount of light available for photosynthesis in soybean and maize plants can be estimated using photosynthetically active radiation and photosynthetic photon flux density. Photosynthetically active radiation measures the range of light wavelengths that plants can utilize for photosynthesis, while photosynthetic photon flux density quantifies the number of photons within this range that hit a specific area over a defined period (McCree 1972).

A study by Abdel-Wahab and Abdel-Rahman (2016) revealed that plant height and seed yield per m^2 of soybean were variable within different planting density of soybean and maize under their intercropping system (Fig. 4.2). They stated that both yield attributes were generally higher under intercropping systems compared to sole soybean planting. The highest value of plant height was found for 4 maize:2

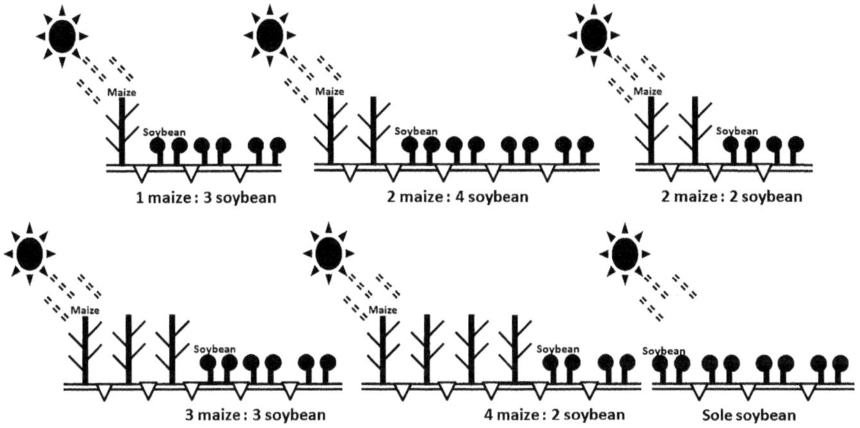

Fig. 4.2 A simple schematic of different intercropping systems of soybean and maize and sole planting of soybean

soybean intercropping system, whereas the highest value of soybean seed yield per m^2 was found for 1 maize:3 soybean intercropping system (Table 4.2). These results suggest that intercropping can enhance crop yields by improving light interception and resource efficiency and ultimately promoting agricultural sustainability.

Table 4.2 also showed that the sole planting of soybeans increased light intensity in the middle of the canopy (221 lux) at 80 days after soybean sowing, increased photosynthetically active radiation (94.95 $\mu mol{\cdot}m^{-2}{\cdot}s^{-1}$), increased photosynthetic photon flux density (204.42 $\mu mol{\cdot}m^{-2}{\cdot}s^{-1}$), and increased seed yield per m^2 (332 g), but decreased plant height (90.92 cm) values. Intercropping 1 maize:3 soybean and 2 maize:4 soybean ranked second in terms of light intensity in the middle of the canopy, photosynthetically active radiation, photosynthetic photon flux density, and seed yield per m^2.

Increasing light intensity often results in a faster rate of photosynthesis (Poorter et al. 2019) because light is essential for providing the energy needed to convert CO_2 and water (H_2O) into glucose through the photosynthetic process (Yang et al. 2010). However, the 4 maize:2 soybean intercropping system exhibited lower values of light intensity in the middle of the canopy (178 lux), photosynthetically active radiation (75.60 $\mu mol{\cdot}m^{-2}{\cdot}s^{-1}$), photosynthetic photon flux density (164.65 $\mu mol{\cdot}m^{-2}{\cdot}s^{-1}$), and soybean seed yield per m^2 (102 g), but higher value of plant height (103.77 cm). This indicates that increasing maize plant density per unit area in its intercropping system with soybean negatively impacted soybean yield and resource availability. Therefore, optimizing maize plant density in its intercropping systems with soybean is crucial to maximize resource utilization and soybean yield without compromising individual crop productivity. Balancing maize plant densities effectively can mitigate the negative effects on soybean yield while maximizing the benefits of intercropping.

Table 4.2 Effect of cropping system on different parameters of the soybean plants (average of the two seasons)

Cropping systems	Plant height (cm)	Light intensity in the middle of the plant (lux)	Photosynthetically active radiation ($\mu mol \cdot m^{-2} \cdot s^{-1}$)	Photosynthetic photon flux density ($\mu mol \cdot m^{-2} \cdot s^{-1}$)	Seed yield (g/m^2)
1 maize:3 soybean	94.14	204	87.30	188.70	247
2 maize:4 soybean	95.40	201	85.95	185.92	212
2 maize:2 soybean	98.12	191	81.45	176.67	158
3 maize:3 soybean	98.52	190	81.00	175.75	147
4 maize:2 soybean	103.77	178	75.60	164.65	102
Sole soybean	90.92	221	94.95	204.42	332
L.S.D. 0.05	4.69	14.38	6.82	13.03	92.16

Source: Abdel-Wahab and Abdel-Rahman (2016)

Optimize Spacing Between Maize Hills to Capture Sunlight

The spacing between maize hills under its intercropping system with soybean plays a crucial role in light interception and the productivity of both crops. Optimal spacing allows for better sunlight penetration to the soybean plants, enhancing their growth and yields (Abdel-Galil et al. 2014). Moreover, appropriate spacing helps minimize competition for resources between the two crops, resulting in improved overall productivity of the intercropping system (El-Shamy et al. 2015).

Abdel-Galil et al. (2014) studied the effect of three spacings between maize hills (one maize plant/hill spaced at 30 cm between hills, two maize plant/hill spaced at 60 cm between hills, and three maize plant/hill spaced at 90 cm between hills) in intercropping systems with soybeans on light intensity and soybean plant height and yield per plant (Fig. 4.3).

Their results indicated that soybean plant height was lower when the spacing between maize hills was 90 cm, whereas higher values of light intensity in the middle of soybean canopy at 80 days after soybean sowing, photosynthetically active radiation, photosynthetic photon flux density, and soybean seed yield per m^2 were found (Table 4.3). These high values helped to boost the crop's overall productivity.

The above results indicate that increasing the spacing between maize hills can enhance soybean growth and yield by improving light penetration and distribution.

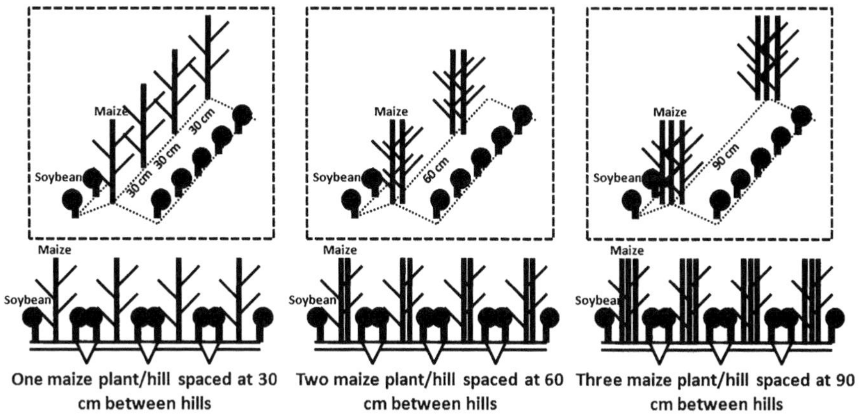

Fig. 4.3 A simple schematic of the three spacings between maize hills in an intercropping system with soybeans

Table 4.3 Effect of the three spacings between maize hills on soybean plant height, light intensity in the middle of soybean plant, photosynthetically active radiation, photosynthetic photon flux density, and soybean seed yield/m^2 (average of the two seasons)

Spacing between maize hills	Soybean plant height (cm)	Light intensity in the middle of soybean plant (lux)	Photosynthetically active radiation (μmol·m^{-2}·s^{-1})	Photosynthetic photon flux density (μmol·m^{-2}·s^{-1})	Soybean seed yield/m^2 (g)
30 cm	104.10	94	42.30	86.95	156
60 cm	101.10	109	49.14	100.82	169
90 cm	97.90	124	55.80	114.70	166
L.S.D. 0.05	5.89	0.55	5.17	12.88	7.62

Prepared by the authors from Abdel-Galil et al. (2014)

These findings underscore the significant impact of maize hill spacing on soybean growth and yield. Therefore, optimizing the spacing between maize hills is crucial to maximize light exposure and enhance soybean productivity. Both crop species benefited from intercropping in terms of light interception, as the different growth and development stages extended the period of light interception (Yu et al. 2015).

Light Interference and Use Efficiency in Soybean-Maize Intercropping with Different Planting Arrangement

Light interference and utilization efficiency in soybean and maize intercropping system are influenced by planting arrangement and crop ratio (Abdel-Wahab et al. 2019a). Shorter distance between maize and soybean plants can intensify light interference, encouraging sunlight access to lower leaves and potentially reducing crop yields due to resource competition (Abdel-Wahab et al. 2019b). Moreover, a higher soybean proportion in the intercropping system can enhance light interference and utilization efficiency, benefiting both crops (Abdel-Wahab et al. 2019a).

Light Interference

Light interference is the phenomenon that happens when two or more light waves overlap, either reinforcing or canceling each other (Vaidyanathan 2024). This phenomenon happens when different plant heights in an intercropping system improve light distribution within the canopy, leading to optimized sunlight capture and utilization. It occurs when light waves from intercropped plants intersect, resulting in a more efficient use of resources in the ecosystem (Vaidyanathan 2024). Light interference in soybean-maize intercropping can lead to improved photosynthetic efficiency, which results in higher yields for both crops (Abdel-Wahab et al. 2019a).

Figure 4.4 and Table 4.4 present simple schematic of soybean intercropping system with maize implemented by Abdel-Wahab et al. (2019b) in a field experiment. They find out that plant height and grain yield were higher in mixed and sole maize, while leaf area per plant, leaf area index, relative photosynthetic potential, leaf ear chlorophyll content, light intensity around the maize plant's leaf area at 80 days from sowing, and grain yield per plant were lower.

Furthermore, in the intercropping systems of 2 maize:2 soybean and 2 maize:4 soybean, plant height was lower than its value in mixed cropping and sole soybean cropping. Higher values of leaf area per plant, leaf area index, relative photosynthetic potential, leaf area chlorophyll content, light intensity around the maize plant's leaf area after 80 days from sowing, and grain yield per plant were found (Table 4.4). It seems that light interference increased in maize when transitioning from direct to diffuse light, potentially affecting photosynthetic efficiency and overall plant growth as suggested by Abdel-Wahab et al. (2019b). The distribution of chlorophyll and light absorption inside the plant's leaves may have changed, causing this occurrence (Metwally et al. 2021). Interference from soybean plants

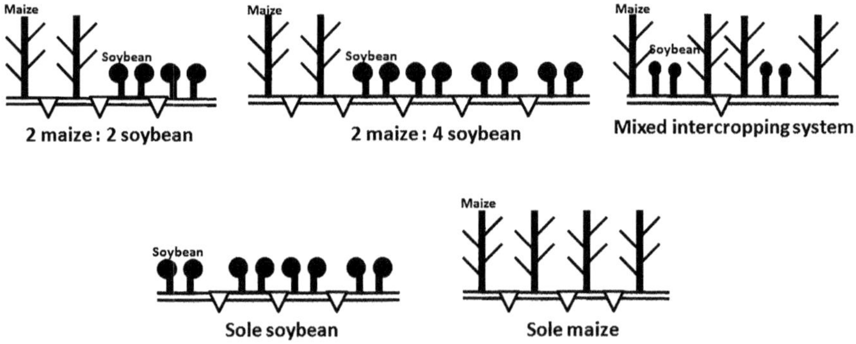

Fig. 4.4 A simple schematic of intercropping systems between soybean and maize and sole plantings of both crops

Table 4.4 Effect of cropping systems on plant height and light interference of maize plants (average of the two seasons)

Intercropping systems	Plant height (cm)	Leaf area/plant (cm^2)	Leaf area index	Relative photosynthetic potential (g/leaf area index unit)	Leaf ear chlorophyll content (SPAD value)
2 maize:2 soybean	254.20	7772	4.44	58.86	52.87
2 maize:4 soybean	236.80	8035	4.59	77.75	55.23
Mixed	278.72	7366	4.20	37.53	47.66
Sole maize	280.06	7402	4.22	40.72	48.13
L.S.D. 0.05	18.42	289.71	0.16	19.75	2.37
Intercropping systems	**Light intensity around the maize plant's leaf area (lux)**			**Grain yield/plant (g)**	**Grain yield (ton/ha)**
	Upper layer	**Middle layer**	**Lower layer**		
2 maize:2 soybean	1246.12	530.61	267.93	261.38	7.20
2 maize:4 soybean	1335.65	582.43	280.24	356.89	5.84
Mixed	1305.44	555.26	278.58	157.64	8.66
Sole maize	1066.45	372.63	241.60	171.86	9.16
L.S.D. 0.05	91.68	59.36	7.54	99.03	1.34

Source: Abdel-Wahab et al. (2019b)

enhanced light distribution within the maize canopy, leading to increased light penetration into the lower canopy (Abdel-Wahab et al. 2019b). The improvement in sunlight captures and utilization by the maize plants compensated for their lower relative photosynthetic potential in the mixed system (Abdel-Wahab et al. 2019a).

These findings suggest that intercropping soybean with maize could positively impact certain growth parameters and ultimately increase grain yield per plant. Through a deeper comprehension of the effects of light interference on maize development, farmers may enhance crop output by making better decisions (Abdel-Wahab et al. 2019b).

Additionally, Table 4.5 illustrates that the mixed intercropping system and sole soybean exhibited lower light intensity values around the soybean plant's leaf area at 80 days from soybean sowing (288.50, 92.69, and 70.60 lux at upper, middle, and lower layers, respectively) compared to the other systems. Sole soybean had a shorter plant (91.74 cm) but a higher number of branches/plant (3.77), leaf chlorophyll content (23.17 SPAD), leaf area index (6.39), and seed yield (3.69 ton). On the other hand, the mixed intercropping system had a taller soybean plant (107.13 cm) but a lower number of branches per plant (2.85), leaf chlorophyll content (19.04 SPAD), leaf area index (5.88), relative photosynthetic potential (2.93 g/leaf area index unit), seed yield per plant (17.25 g), and seed yield per ha (1.16 ton) (Abdel-Wahab et al. 2019b).

Table 4.5 Effect of cropping systems on plant height and light interference of soybean plants (average of the two seasons)

Intercropping systems	Plant height (cm)	Number of branches/ plant	Light intensity around the soybean plant's leaf area (lux)		
			Upper layer	Middle layer	Lower layer
2 maize:2 soybean	98.83	3.38	317.75	145.20	75.10
2 maize:4 soybean	94.31	3.42	332.39	154.84	83.80
Mixed	107.13	2.85	288.50	92.69	70.60
Sole soybean	91.74	3.77	292.23	93.06	71.81
L.S.D. 0.05	6.63	0.48	12.37	32.79	4.84
Intercropping systems	**Leaf chlorophyll content (SPAD value)**	**Leaf area index**	**Relative photosynthetic potential (g/leaf area index unit)**	**Seed yield/ plant (g)**	**Seed yield (ton/ha)**
2 maize:2 soybean	19.72	5.65	4.12	23.32	1.36
2 maize:4 soybean	21.82	6.23	4.03	25.14	1.63
Mixed	19.04	5.88	2.93	17.25	1.16
Sole soybean	23.17	6.39	3.77	24.11	3.69
L.S.D. 0.05	2.01	0.31	0.59	2.54	1.18

Source: Abdel-Wahab et al. (2019b)

These results suggest that while sole soybean may have certain advantages in terms of individual plant characteristics, the mixed intercropping system may provide a more balanced overall yield. Notably, there were no significant differences in seed yield between soybean and maize in all intercropping systems. Intercropping systems 2 maize:2 soybean and 2 maize:4 soybean showed higher light intensity values around the soybean plant's leaf area, relative photosynthetic potential, and seed yield per plant. Accordingly, this shows that the 2 maize:2 soybean and 2 maize:4 soybean intercropping systems may be more efficient in utilizing light energy and resources for seed production compared to the other systems.

Light interference from mixed intercropping system in soybeans, caused by increased maize plant densities, can reduce light availability, leading to decreased photosynthesis and, ultimately, lower soybean yield per plant. Conversely, light interference can positively impact light utilization efficiency in maize plants, especially in the mixed intercropping system with maize plants (Abdel-Wahab et al. 2019b). It seems that light interference increases in a soybean when transitioning from direct to diffuse light. Diffused light that penetrates through the upper layers of leaves is crucial for maximizing photosynthesis and overall plant development in mixed systems, as it contributes to light interference at the base of the soybean canopy (Abdel-Wahab et al. 2019b).

Diffuse light can penetrate deeper into the canopy, reaching more leaves and enhancing photosynthesis and overall plant growth (Li et al. 2014). The light use efficiency of a crop rises when it transitions from direct to diffuse light (Hammer and Wright 1994). However, the number of soybean plants per unit area varies depending on the intercropping system used, and this can have a positive interaction with light interference. Therefore, it is crucial to carefully consider the spacing and density of maize plants when intercropping with soybeans to manage light interference and enhance crop productivity. Further research is needed to fully understand the effects of light interference on plant growth and productivity.

Light Use Efficiency

A study by Abdel-Wahab and Abdel-Rahman (2016) on soybean intercropping system with maize under different plant spacing (Table 4.6) revealed that sole soybean plants showed higher values for all parameters included in the table compared to intercropped soybeans. Sole soybean exhibited higher light intensity per m^2 compared to intercropping with maize, with increases of 7.69%, 9.05%, 13.57%, 14.03%, and 19.46% over 1:3, 2:4, 2:2, 3:3, and 4:2 maize/soybean ratios, respectively. This indicates that planting sole soybean could lead to improved light utilization efficiency.

Sole soybean experienced higher solar irradiance levels compared to different soybean and maize planting ratios, with increases of 30.86%, 39.09%, 56.79%, 57.00%, and 73.46% in 1:3, 2:4, 2:2, 3:3, and 4:2 maize/soybean ratios, respectively. This indicates that soybean may have a greater ability to harness solar energy efficiently than maize across different planting combinations. Sole soybean had

Table 4.6 Effect of cropping system on different parameters of the soybean plants (average of the two seasons)

Cropping systems	Light intensity/m^2 (Lumen)	Solar irradiance ($J/s/m^2$)	The amount of energy ($J/s/m^2$)	Total dry matter (g/m^2)	Light use efficiency ($g \cdot MJ^{-1}$)	Seed yield (ton/ha)
1 maize:3 soybean	10,200	3.36	0.89	1055.57	0.76	2.40
2 maize:4 soybean	10,050	2.96	0.78	925.37	0.66	2.04
2 maize:2 soybean	9550	2.10	0.55	725.23	0.52	1.49
3 maize:3 soybean	9500	2.09	0.55	692.29	0.49	1.42
4 maize:2 soybean	8900	1.29	0.34	480.41	0.34	0.96
Sole soybean	11,050	4.86	1.29	1561.66	1.12	3.17
L.S.D. 0.05	1100	1.93	0.52	167.32	0.47	1.22

Source: Abdel-Wahab and Abdel-Rahman (2016)

higher energy content compared to different soybean/maize ratios, with increases of 31.01%, 39.53%, 57.36%, 57.36%, and 73.64% for 1:3, 2:4, 2:2, 3:3, and 4:2 maize/soybean ratios, respectively. This indicates that soybean is a more energy-dense crop than maize in various combinations. Furthermore, sole soybean exhibited a higher total dry matter increase of 32.41%, 40.74%, 53.56%, 55.67%, and 69.24% compared to the intercropping ratios of 1:3, 2:4, 2:2, 3:3, and 4:2 maize/soybean ratios, respectively. These findings indicate that increasing the soybean proportion in the intercropping system can greatly boost total dry matter production (Table 4.6).

In terms of light use efficiency, sole soybean exhibited higher light use efficiency compared to intercropping with maize. Sole soybean showed an increase by 32.14%, 41.07%, 53.57%, 56.25%, and 69.64% in light use efficiency compared to 1:3, 2:4, 2:2, 3:3, and 4:2 maize/soybean ratios, respectively. These results indicate that sole soybean may be more effective in utilizing light resources than intercropping with maize. Sole soybean achieved a higher seed yield per hectare compared to different soybean/maize intercropping ratios, with increases of 24.29%, 35.65%, 53.00%, 55.21%, and 69.72% over 1:3, 2:4, 2:2, 3:3, and 4:2 maize/soybean ratios, respectively. These results indicate that planting sole soybean could be a more effective utilization of land and resources than various soybean/maize intercropping combinations (Table 4.6).

Among the intercropping systems tested, 1:3 and 2:4 maize/soybean ratios ranked second in the performance, with no significant differences observed between them and sole soybean. These findings suggest that soybeans exhibit consistently superior performance under sole soybean, intercropping systems 1:3 and 2:4 maize/soybean ratios, compared to other intercropping systems in terms of the measured parameters (Table 4.6). The results also show that maize and soybean plants compete for sunlight, affecting light consumption efficiency in soybean and maize intercropping system at different maize plant densities. The light use efficiency in soybean and maize intercropping system is influenced by the differences in canopy structure and photosynthetic capacity between the two crops.

This can result in higher overall productivity and resource use efficiency in the intercropping system. Maize, with its taller and more erect canopy, is able to intercept and utilize light more effectively compared to soybeans. Studies in this area aim to maximize the ability of intercropped soybean plants to adapt to low-light conditions, such as reduced light penetration to the lower canopy (Liu et al. 2018), faster senescence of the lower leaves (Borrás et al. 2003), and decreased light use efficiency (Jia et al. 2018), which are all consequences of too high maize density. A metric suggested by Higashide (2022) quantifies how effectively plants convert light energy into biomass, in which total dry matter production is equals light use efficiency multiplied by light intercepted by the plants. A plant that covers one m^2 of ground gets 3.76 J of energy/s from photosynthesis (Highsmith 2017). Higher light use efficiency signifies that plants can generate more dry matter per unit of light intensity (Liang et al. 2024). Therefore, it is essential to consider plant density when aiming to optimize light usage efficiency in soybean and maize intercropping systems.

Relative Photosynthetic Potentials

Understanding the relative photosynthetic potentials of different cropping systems is crucial for comprehending their impact on plant productivity and ecosystem functioning (Smith et al. 2018). Researchers can gain valuable insights into the adaptability of plant species to changing cropping systems by quantifying and analyzing these potentials. When intercropping soybeans with maize, complementary and competitive photosynthetic potentials can influence photosynthetic efficiency, ultimately affecting the plant's growth and survival (Abdel-Wahab and Abdel-Wahab 2020b). Maize or soybean plants require light, CO_2, water, and nutrients to produce dry matter through photosynthesis. During this process, plants utilize sunlight to convert inorganic elements into organic chemicals (Yang et al. 2010).

Burgess and Pletschke (2011) demonstrate that the conversion of atmospheric 6 CO_2 and 6 H_2O into carbohydrates ($C_6H_{12}O_6$) and 6 O_2 is the key chemical pathway in photosynthesis. The rate of atmospheric CO_2 conversion to carbohydrates is crucial for photosynthesis when light is not a limiting factor. The enzyme ribulose-1,5-bisphosphate carboxylase/oxygenase (Rubisco) facilitates this process, known as C-fixation. Rubisco, one of the most abundant proteins worldwide, plays a vital role in the global carbon cycle (Bathellier et al. 2020).

Complementary Photosynthetic Potentials

Soybeans and maize grown together in an intercropping system on the same ridge can benefit from complementary photosynthetic potentials. Justes et al. (2021) suggest that productivity gains in intercropping occur when interspecific competition between the intercrops is lower than the intraspecific competition in sole crops, leading to complementarity, or when interspecific facilitation exceeds intraspecific facilitation. Abdel-Wahab and Abdel-Wahab (2020b) emphasize the significance of selecting the appropriate soybean plant density and variety to maximize the complementary interactions between maize and soybeans in their intercropping system. Specifically, certain soybean varieties are adept in capturing light and converting it into energy, while maize plants excel in fixing CO_2. Also, soybeans and maize can complement each other's photosynthetic capabilities by optimizing soybean plant density to improve light interception and resource utilization (Blessing et al. 2022). By carboxylation of atmospheric CO_2 with fixed N, optimizing soybean plant density and variety can enhance the complementary interactions between maize and soybeans, leading to improved photosynthesis efficiency (Abdel-Wahab and Abdel-Wahab 2020b). This symbiotic relationship enhances overall production and resource efficiency for both crops (Raza et al. 2022), which is considered as a strategic approach leading to increased yields for both crops, optimizing overall productivity and resource utilization.

To maximize the benefits of cereal and legume crops in intercropping systems, it is crucial to manage and optimize their spatial and temporal complementarity (Mugi-Ngenga et al. 2023). This helps minimize competition for resources between the two component crops (Vandermeer 1989). In an intercropping system, the synergy between maize and soybean helps maintain a stable and productive ecosystem.

Intercropping legumes with cereal species in the same ridge can enhance the photosynthetic efficiency of legumes and reduce the need for mineral N fertilizer in cereals (Abdel-Wahab et al. 2019b). Cereals lack photorespiratory CO_2 loss, making them more efficient in utilizing CO_2 for photosynthesis compared to legumes (Michelet et al. 2013). Adenosine triphosphate (ATP) and nicotinamide adenine nicotinamide phosphate (NADPH) produced during the light reactions of photosynthesis are crucial for converting CO_2 into organic compounds (Tong et al. 2025). Cereals concentrate CO_2 in leaf locations where the Rubisco enzyme incorporates CO_2 in the Calvin-Benson cycle (Michelet et al. 2013). Higher atmospheric CO_2 levels have minimal impact on photosynthesis in cereals due to their CO_2 concentration mechanisms (Tong et al. 2025).

Competitive Photosynthetic Potentials

When plants are not given enough growth resources, such as sunlight, water, and nutrients, competitive photosynthetic potentials may develop (Evans 2013). Intercropping soybeans with maize can enhance photosynthetic potential when intraspecific facilitation surpasses interspecific facilitation (Abdel-Wahab and Abdel-Wahab 2020b). In such situations, maize plants may outcompete soybean plants for resources and enhance their photosynthetic efficiency. This can lead to an overall increase in the fitness and growth rates of the plant population (Slattery and Ort 2020). The most significant disadvantages of intercropping systems are random intercropping, choosing incompatible soybean and maize varieties for above- and below-ground conditions, selecting planting dates that are not suitable for growing soybeans and maize together, and potential competition for resources like water, nutrients, and sunlight between the two species. These factors can result in reduced yields for both crops and increased management complexity.

However, it is crucial to manage resource competition effectively to avoid negative impacts on overall yields. In particular, different species of plants can have different photosynthetic potentials in intercropping systems because they have evolved to efficiently capture and use light, allowing them to use other essential growth resources (Metwally et al. 2021). According to Abdel-Wahab and Abdel-Wahab (2020a), decreasing the competitive pressure between maize and soybean for available growth resources requires taking into account the dynamics of interspecific interactions. It is important to comprehend these dynamics to enhance crop yield and sustainability. Therefore, the production of maize and soybeans can both benefit farmers using more efficient and sustainable agricultural practices that take these dynamics into consideration.

Conclusions and Future Perspectives

There are potential benefits of intercropping in enhancing soil nutrient dynamics and microbial communities in agricultural systems. Enhancing light availability in intercropping soybeans with maize can help achieve a balance between increasing crop output and reducing plant competition for resources. Future research on light interference in intercropping might involve examining various light wavelengths to maximize plant development and how light interference affects intercropping systems' ability to control pests and diseases.

References

Abdel-Galil AM, Abdel-Wahab TI, Abdel-Wahab SI (2014) Productivity of four soybean varieties as affected by intercropping and corn planting geometry. Soybean Res 12(1):36–58

Abdel-Wahab TI, Abdel-Rahman RA (2016) Response of some soybean cultivars to low light intensity under different intercropping patterns with maize. Int J Appl Agric Sci 2(2):21–31

Abdel-Wahab TI, Abdel-Wahab EI (2020a) Interspecies interactions dynamics between soybean and corn as affected by different soybean varieties. Int J Adv Sci Technol 29(11s):2864–2876

Abdel-Wahab SI, Abdel-Wahab EI (2020b) Competitive and facilitative effects of intercropping some soybean varieties with corn under different soybean plant densities. Plant Arch 20(Suppl 2):1631–1639

Abdel-Wahab TI, Abdel-Wahab SI, Abdel-Wahab EI (2019a) Benefits of intercropping legumes with cereals. Int J Conf Proc 1(2):ICP.000510.2019

Abdel-Wahab TI, Abdel-Wahab EI, Taha AM, Adel MM, Hussein HM (2019b) Varietal response of soybean to applied irrigation water and insects incidence under different intercropping systems with maize. Res Crops 20(Issue Suppl):S1–S25

Abdel-Wahab TI, Abdel-Wahab SI, Abdel-Wahab EI (2024) Chapter 15: biological engineering and its relationship to nematode resistance. In: Chaudhary KK, Meghvansi MK, Siddiqui S (eds) Sustainable management of nematodes in agriculture, vol. 2: role of microbes—assisted strategies, sustainability in plant and crop protection. Springer, Cham, pp 383–408

Akitsu TK, Nasahara KN, Ijima O, Hirose Y, Ide R, Takagi K, Kume A (2022) The variability and seasonality in the ratio of photosynthetically active radiation to solar radiation: a simple empirical model of the ratio. Int J Appl Earth Obs Geoinf 108:102724

Barnes C, Tibbitts T, Sager J, Deitzer G, Bubenheim D, Koerner G, Bugbee B (1993) Accuracy of quantum sensors measuring yield photon flux and photosynthetic photon flux. HortSci 28(12):1197–1200

Bathellier C, Yu J, Farquhar GD, Coote ML, Lorimer GH, Tcherkez G (2020) Ribulose 1,5-bisphosphate carboxylase/oxygenase activates O2 by electron transfer. Proc Natl Acad Sci USA 117(39):24234

Bhargawa A, Singh A (2019) Solar irradiance, climatic indicators and climate change—an empirical analysis. Adv Space Res 64(1):271–277

Blessing DJ, Gu Y, Cao M, Cui Y, Wang X, Asante-Badu B (2022) Overview of the advantages and limitations of maize-soybean intercropping in sustainable agriculture and future prospects: a review. Chilean J Agric Res 82(1):177–188

Bolan NS, Adriano DC, Kunhikrishnan A, James T, McDowell R, Senesi N (2010) Dissolved organic matter: biogeochemistry, dynamics, and environmental significance in soils. Adv Agron 110:1–75

Boomsma CR, Vyn TJ (2008) Maize drought tolerance: potential improvements through arbuscular mycorrhizal symbiosis? Field Crop Res 108(1):14–31

Borrás L, Maddonni GA, Otegui ME (2003) Leaf senescence in maize hybrids: plant population: row spacing and kernel set effects. Field Crop Res 82(1):13–26

Burgess JE, Pletschke BI (2011) fundamentals of biochemistry, cell biology and biophysics. Carbon fixation, vol. 1. In: Encyclopedia of life support systems. https://www.eolss.net/sample-chapters/c03/E6-51-01-02.pdf

Buschmann C (2007) Variability and application of the chlorophyll fluorescence emission ratio red/far-red of leaves. Photosynth Res 92:261–271

Chen Y, Wang X, Han J (2003) Effects of He-Ne laser pretreatment of Isatis indigotica seeds on yield and quality of Isatis indigotica leave. Chin Tradit Herb Drug 34:1054–1056

Di Benedetto NA, Corbo MR, Campaniello D, Cataldi MP, Bevilacqua A, Sinigaglia M, Flagella Z (2017) The role of plant growth promoting bacteria in improving nitrogen use efficiency for sustainable crop production: a focus on wheat. AIMS Microbiol 3(3):413

El-Shamy MA, Abdel-Wahab TI, Abdel-Wahab SI, Ragheb SB (2015) Advantages of intercropping soybean with maize under two maize plant distributions and three mineral nitrogen fertilizer rates. Adv Biosci Bioeng 3(4):30–48

Evans JR (2013) Improving photosynthesis. Plant Physiol 162(4):1780

Falcioni R, Moriwaki T, Bonato CM, de Souza LA, Nanni MR, Antunes WC (2017) Distinct growth light and gibberellin regimes alter leaf anatomy and reveal their influence on leaf optical properties. Environ Exp Bot 140:86–95

FAO (1990) Plant production and protection paper, 90. Protected cultivation in the Mediterranean climate. FAO, Rome, p 313

Gao Y, Duan A, Qiu X, Sun J, Zhang J, Liu H, Wang H (2010) Distribution and use efficiency of photosynthetically active radiation in strip intercropping of maize and soybean. Agron J 102(4):1149–1157

Gao Y, Bu L, Han M, Wang Y, Li Z, Liu H, Chao D (2021) Long-distance blue light signaling regulates phosphate deficiency-induced primary root growth inhibition. Mol Plant 14(9):1539–1553

Goldman MS, Kolmos E (2023) Blue spectral quality contributes to yield and ascorbic acid content in 'Micro Tom' tomato under sole-source lighting. BioRxiv, the Preprint Server For Biology, Cold Spring Harbor Laboratory. https://doi.org/10.1101/2023.11.27.568869

Goswami Y (2015) Principles of solar engineering. CRC Press

Hammer GL, Wright GC (1994) A theoretical analysis of nitrogen and radiation effects on radiation use efficiency in peanut. Aust J Agric Res 45(3):575–589

Hansen J, Nazarenko L, Ruedy R, Sato M, Willis J, Del Genio A, Koch D, Lacis A, Lo K, Menon S, Novakov T, Perlwitz J, Russell G, Schmidt GA, Tausnev N (2005) Earth's energy imbalance: confirmation and implications. Science 308:1431–1435

Higashide T (2022) Review of dry matter production and growth modelling to improve the yield of greenhouse tomatoes. Hortic J 91(3):247–266

Highsmith CM (2017) An October 2017 aerial view of a mixture of evergreens (Maine is known as the Pine Tree State) and deciduous trees at the height of their fall colors in the forests near Ogunquit, Maine. Library of Congress Prints and Photographs Division, Washington, DC. https://www.loc.gov/item/2017883331/

Jia Q, Sun L, Mou H, Ali S, Liu D, Zhang Y, Zhang P, Ren X, Jia Z (2018) Effects of planting patterns and sowing densities on grain-filling, radiation use efficiency and yield of maize (*Zea mays* L.) in semi-arid regions. Agric Water Manag 201:287–298

Justes E, Bedoussac L, Dordas C, Frak E, Louarn G, Boudsocq S, Journet E, Lithourgidis A, Pankou C, Zhang C, Carlsson G, Jensen ES, Watson C, Li L (2021) The 4C approach as a way to understand species interactions determining intercropping productivity. Front Agric Sci Eng 8(3):387–399

Kopittke PM, Menzies NW (2004) Effect of Mn deficiency and legume inoculation on rhizosphere pH in highly alkaline soils. Plant Soil 262:13–21

Kozai T, Niu G, Masabni J (2022) Terms related to PFALs. In: Plant factory basics, applications and advances. Academic, pp 11–23

Li T, Heuvelink E, Dueck TA, Janse J, Gort G, Marcelis LFM (2014) Enhancement of crop photosynthesis by diffuse light: quantifying the contributing factors. Ann Bot 114:145–156

Liang J, Ji F, He D (2024) Higher light intensity combined with early topping improves the yield and quality of pea shoots in LED plant factory. Horticulture 10(6):650
Liu X, Rahman T, Song C, Yang F, Su B, Cui L, Bu W, Yang W (2018) Relationships among light distribution, radiation use efficiency and land equivalent ratio in maize-soybean strip intercropping. Field Crop Res 224:91–101
Liu C, Wang X, Li X, Yang Z, Dang K, Gong X, Feng B (2024) Effects of intercropping on rhizosphere microbial community structure and nutrient limitation in proso millet/mung bean intercropping system. Eur J Soil Biol 122:103646
Liu W, Wang G, Wen S, Zhao Y, Ding Y, Yao B, Wang Z, Shu D, Wei G, Chen J, Shangguan Z (2025) Microbiome-mediated mechanisms regulating adaptability to iron deficiency in the intercropping system of soybean and maize. Agronomy 15(2):286
Lodh D (2024) Sustainable waste management practices and their profound impact on environmental conservation. In: Futuristic trends in agriculture engineering & food sciences, IIP series, vol 3, Part 1, pp 7–18
McCree KJ (1972) The action spectrum, absorbance and quantum yield of photosynthesis in crop plants. Agric Meteorol 9(3–4):191–216
Mead MN (2008) Benefits of sunlight: a bright spot for human health. Environ Health Perspect 116(4):A160
Metwally AA, Shafik MM, El-Habbak KE, Abdel-Wahab SI (2012) Yield and soybean characters under some intercropping patterns with corn. Soybean Res 10:24–42
Metwally AA, Abdel-Wahab TI, Abdel-Wahab SI (2019) Increasing land and water use efficiencies by intercropping summer legumes with corn in Egypt. Agric Biol Res 35(2):6–10
Metwally AA, Safina SA, Abdel-Wahab EI, Abdel-Wahab SI, Abdel-Wahab TI (2021) Screening thirty soybean genotypes under solid and intercropping plantings in Egypt. J Crop Sci Biotechnol 24:203–220
Michelet L, Zaffagnini M, Morisse S, Sparla F, Pérez-Pérez ME, Francia F, Danon A, Marchand CH, Fermani S, Trost P, Lemaire SD (2013) Redox regulation of the Calvin–Benson cycle: something old, something new. Front Plant Sci 4:470
Morcillo R (2024) Lux and lumen: what are they and what are the differences? Faro Barcelona. Available at https://faro.es/en/blog/es-lux-y-lumen/#:~:text=The%20concepts%20of%20lux%20and%20lumen%20are%20closely%20related.,one%20lumen%20per%20square%20meter
Mõttus M, Sulev M, Baret F, Lopez-Lozano R, Reinart A (2012) Photosynthetically active radiation: measurement and modeling. In: Meyers RA (ed) Encyclopedia of sustainability science and technology. Springer, New York
Mugi-Ngenga E, Bastiaans L, Anten N, Zingore S, Baijukya F, Giller K (2023) The role of interspecific competition for water in maize-legume intercropping systems in northern Tanzania. Agric Syst 207:103619
Ning W, Yang Y, Chen W, Li R, Cao M, Luo J (2022) Effect of light combination on the characteristics of dissolved organic matter and chemical forms of Cd in the rhizosphere of Arabidopsis thaliana involved in phytoremediation. Ecotoxicol Environ Saf 231:113212. https://doi.org/10.1016/j.ecoenv.2022.113212
NIST (2019) Realization of related photometric units. https://www.nist.gov/pml/sensor-science/optical-radiation/realization-related-photometric-units
Ouda SA, Zohry AA (2024) The strength of the relationship between maize productivity and the seasonal weather elements. In: Integration of legumes crops with cereal crops under changing climate. Springer. ISBN: 978-3-031-68101-1
Pech R, Volná A, Špunda V, Nezval J (2024) Blue light as an important factor increasing plant tolerance to acute photooxidative stress. Environ Exp Bot 226:105923
Pedrós R, Moya I, Goulas Y, Jacquemoud S (2008) Chlorophyll fluorescence emission spectrum inside a leaf. Photochem Photobiol Sci 7:498–502
Poorter H, Niinemets Ü, Ntagkas N, Siebenkäs A, Mäenpää M, Matsubara S, Pons T (2019) A meta-analysis of plant responses to light intensity for 70 traits ranging from molecules to whole plant performance. New Phytol 223(3):1073–1105

Raza MA, Yasin HS, Gul H, Qin R, Khalid MH, Hussain S, Gitari H, Saeed A, Wang J, Sabagh AE, Manzoor A, Fatima A, Ahmad S, Yang F, Skalicky M, Yang W (2022) Maize/soybean strip intercropping produces higher crop yields and saves water under semi-arid conditions. Front Plant Sci 13:1006720

Rinot O, Levy GJ, Steinberger Y, Svoray T, Eshel G (2019) Soil health assessment: a critical review of current methodologies and a proposed new approach. Sci Total Environ 648:1484–1491

Sager JC, Giger W Jr (1980) Re-evaluation of published data on the relative photosynthetic efficiency of intermittent and continuous light. Agric Meteorol 22:289–302

Sakuraba Y, Yanagisawa S (2018) Light signalling-induced regulation of nutrient acquisition and utilisation in plants. Semin Cell Dev Biol 83:123–132

Sandri MA, Andriolo JL, Witter M, Dal Ross T (2003) Effect of shading on tomato plants grow under greenhouse. Hortic Bras 21:642–645

Schlau-Cohen GS (2015) Principles of light harvesting from single photosynthetic complexes. Interface Focus 5(3):20140088

Slattery RA, Ort DR (2020) Perspectives on improving light distribution and light use efficiency in crop canopies. Plant Physiol 185(1):34

Smith MR, Rao IM, Merchant A (2018) Source-sink relationships in crop plants and their influence on yield development and nutritional quality. Front Plant Sci 9:420465

Szymanek M, Zarajczyk J, Michałek S, Tanaś W (2020) The influence of red light (RL) and effective microorganism (EM) application on soil properties, yield, and quality in wheat cultivation. Agronomy 10(8):1201

Thimijan RW, Heins RD (1983) Photometric, radiometric, and quantum light units of measure: a review of procedures for interconversion. HortSci 18(6):818–822

Tong T, Chen X, Tang K, Ma W, Gao C, Song W, Wu J, Wang X, Liu Q, Liu L (2025) A new-to-nature photosynthesis system enhances utilization of one-carbon substrates in *Escherichia coli*. Nat Commun 16:145

Trofimov K, Mankotia S, Ngigi M, Baby D, Satbhai SB, Bauer P (2024) Shedding light on iron nutrition: exploring intersections of transcription factor cascades in light and iron deficiency signaling. J Exp Bot:erae324. https://doi.org/10.1093/jxb/erae324

Vaidyanathan V (2024) What is interference of light? ScienceABC Website. Available at https://www.scienceabc.com/pure-sciences/interference-of-light.html

Vandermeer JH (1989) The ecology of intercropping. Cambridge University Press, Cambridge

Wang L, Zhou T, Cheng B, Du Y, Qin S, Gao Y, Xu M, Lu J, Liu T, Li S, Liu W, Yang W (2020) Variable light condition improves root distribution shallowness and P uptake of soybean in maize/soybean relay strip intercropping system. Plan Theory 9(9):1204

Wang Q, Ning Z, Awan SA, Gao J, Chen J, Lei Y, Tan X, Wu X, Wu Y, Liu C, Yang F, Yang W (2023) Far-red light mediates light energy capture and distribution in soybeans (*Glycine max* L.) under the shade. Plant Physiol Biochem 204:108130

Wang Z, Wang T, Zhang X, Wang J, Yang Y, Sun Y, Guo X, Wu Q, Nepovimova E, Watson AE, Kuca K (2024) Biodiversity conservation in the context of climate change: facing challenges and management strategies. Sci Total Environ 937:173377

Willey RW (1979) Intercropping: its importance and research needs. Part-2. Agronomy and research approaches. Field Crop Abstr 32:73–81

Wu Y, Gong W, Yang F, Wang X, Yong T, Yang W (2016) Responses to shade and subsequent recovery of soya bean in maize-soya bean relay strip intercropping. Plant Prod Sci 19:206–214

Wu B, Jia X, Zhu W, Gao Y, Tan K, Duan Y, Chen L, Fan H, Wang Y, Liu X, Xuan Y, Zhu X (2024) Light signaling regulates root-knot nematode infection and development via HY5-SWEET signaling. BMC Plant Biol 24(1):664

Wu W, Chen L, Liang R, Huang S, Li X, Huang B, Luo H, Zhang M, Wang X, Zhu H (2025) The role of light in regulating plant growth, development and sugar metabolism: a review. Front Plant Sci 15:1507628

Yang J, Gao H, Liu P, Li G, Dong S (2010) Effects of planting density and row spacing on canopy apparent photosynthesis of high-yield summer corn. Acta Agron Sin 36:1226–1235. https://doi.org/10.3724/SP.J.1006.2010.01226

Yang F, Huang S, Gao R, Liu W, Yong T, Wang X, Wu X, Yang W (2013) Growth of soybean seedlings in relay strip intercropping systems in relation to light quantity and red: far-red ratio. Field Crop Res 155:245–253

Yang J, Sun J, Wang X, Zhang B (2024) Light intensity affects growth and nutrient value of hydroponic barley fodder. Agronomy 14(6):1099

Yu Y, Stomph TJ, Makowski D, van der Werf W (2015) Temporal niche differentiation increases the land equivalent ratio of annual intercrops: a meta-analysis. Field Crop Res 184:133–144

Zheng Z, Wang Z, Wang X, Liu D (2019) Blue light-triggered chemical reactions underlie phosphate deficiency-induced inhibition of root elongation of arabidopsis seedlings grown in petri dishes. Mol Plant 12(11):1515–1523

Zhou T, Wang L, Yang H, Gao Y, Liu W, Yang W (2019) Ameliorated light conditions increase the P uptake capability of soybean in a relay-strip intercropping system by altering root morphology and physiology in the areas with low solar radiation. Sci Total Environ 688:1069–1080

Zohry AA (2005) Effect of preceding winter crops and intercropping on yield, yield components and associated weeds in maize. Ann Agric Sci Moshtohor 43(1):139–148

Mitigation of Shading Stress on Soybean Varieties Through Intercropping with Maize

5

Sherif Ibrahim Abdel-Wahab, Tamer Ibrahim Abdel-Wahab, and Eman Ibrahim Abdel-Wahab

Abstract

The successful soybean intercropping system with maize requires choosing tolerant soybean varieties to shading stress under intercropping, as well as appropriate plant population densities and distribution within the intercropping system. High competition between the two intercrops for essential growth resources may exist when incompatible soybean variety with the wrong plant density was chosen. In this chapter, we assessed the contribution of land equivalent ratio (LER) in defining the compatibility and competition in the intercropping system of soybean and maize. Furthermore, the role of system production index (SPI) in distinguishing between complementary and complimentary in the intercropping system of soybean and maize was also tackled. Lastly, the ability of five studied soybean varieties to tolerate shading stress under its intercropping system with maize was thoroughly discussed through previous research conducted in Egypt.

Keywords

Land equivalent ratio · System production index · Compatibility · Competition · Complementary · Complimentary

S. I. Abdel-Wahab (✉) · T. I. Abdel-Wahab
Crop Intensification Research Department, Field Crops Research Institute, Agricultural Research Center (ARC), Giza, Egypt
e-mail: sherif.abdel-wahab@arc.sci.eg

E. I. Abdel-Wahab
Food Legumes Research Department, Field Crops Research Institute, Agricultural Research Center (ARC), Giza, Egypt

A. E.-H. Zohry, S. Ouda (eds.), *Resilient Agroecosystems*, Sustainability Sciences in Asia and Africa, https://doi.org/10.1007/978-981-96-9630-7_5

Introduction

The Egyptian government aims at increasing crop production and improving its cropping systems since the beginning of the twentieth century. To achieve that, the Field Crops Research Institute has been founded in 1973; since then, it has released many crop varieties with tolerance capability to environmental stresses. Additionally, the Egyptian government has the target of increasing the yield of the major strategic field crops, producing improved seeds, researching production constraints, providing improved agronomic practices, and conducting extension activities throughout Egypt.

An important approach to increase productivity is to grow the two crops together in intercropping systems. Intercropping approach is popular around the world, particularly in areas where small farmers make extensive use of a small amount of land (Sayed-Galal et al. 1979); thus, it is wise that Egypt implement it to boost its overall agricultural output (Metwally et al. 2019a; Abdel-Wahab et al. 2024). Row intercropping, strip intercropping, mixed intercropping, and relay intercropping are the four primary forms of intercropping (Mousavi and Eskandari 2011). A crop is said to be row intercropped when it is grown in the same field as another crop, provided that at least one of the crops is grown in separate rows (Mousavi and Eskandari 2011). For example, crop A may be grown in one ridge alternating with crop B, or crop A may be grown in two ridges alternating with another crop B, or crop A may be grown in two ridges alternating with four ridges of crop B, or three ridges alternating with another crop B (Sayed-Galal and Metwally 1989). Meanwhile, a practice known as strip intercropping involves growing two crops alongside one another in wide strips (Toker et al. 2024). Growing two or more crops simultaneously with distinct line arrangements is known as mixed intercropping or mixed cropping (Sayed-Galal and Metwally 1989), whereas the practice of planting a crop alongside an established first crop before it is ready to be harvested is called relay cropping, or relay intercropping (Kline et al. 2001). However, proper management is needed to ensure the best results.

Intercropping can lead to more efficient use of resources like light, water, and nutrients, but it can also lead to disadvantages compared to growing the crops separately (Metwally 1999). The spatial arrangement of the intercropping system can affect the interaction between the intercrops and their use of environmental resources (Thompson et al. 2025). The density and distribution of the intercrops also play a role in their competition for resources. These intercrops often differ in terms of height, canopy structure, rooting ability, nutritional needs, and complementary use of growth resources (Blessing et al. 2022).

Intercropping summer legumes with maize has two opposing but equally important effects on legume crops: first, it protects summer legumes from high temperatures, and second, it reduces the effectiveness of summer legumes' photosynthesis because of the shading effects (Metwally et al. 2019a). After adopting the soybean intercropping system with maize, researchers compared various intercropping strategies, selected tolerant soybean varieties for each partner crop, and examined the effects of plant population densities and distribution within the unit area

(Abdel-Galil et al. 2014a). Furthermore, choosing the incompatible soybean variety with the wrong plant density to be intercropped with maize may make the intercrops more competitive with one another for essential growth resources (Abdel-Wahab and Abdel-Wahab 2020a).

Four local soybean varieties (Giza 21, Giza 22, Giza 35, and Giza 111), in addition to an American variety (Crawford), were studied thoroughly under the Egyptian conditions for its ability to withstand the shading stress under its intercropping system with maize (Abdel-Galil et al. 2014a, b; Lamlom et al. 2015; Abdel-Wahab and Abdel-Rahman 2016; Metwally et al. 2018, 2019b). These varieties have different morphology leading to different response to shading stress under intercropping with maize. For example, Giza 22 is characterized by having narrow leaves, which may increase the amount of light that reaches the lower leaves on the plant's stem (Metwally et al. 2012), whereas Giza 111 is characterized by having large leaves benefit from low light intensity (Abdel-Wahab and Abdel-Rahman 2016).

In this chapter, we assessed the contribution of LER in defining the compatibility and competition in soybean and maize intercropping system. Furthermore, the role of SPI in distinguishing between complementary and complimentary in soybean and maize intercropping system was also tackled. Lastly, the ability of five studied soybean varieties in tolerating shading stress under its intercropping system with maize was thoroughly discussed through previous research conducted in Egypt.

The Progress of Soybean Cultivation in Egypt

Legumes are in the second place to cereals, which represents 27% of global primary crop production and 33% of protein needs (Graham and Vance 2003). Soybean is one of the most important summer legume crops cultivated in Egypt. Egypt's commercial soybean plantations produced an average of 126 kg of seed per hectare in 1970, covering roughly 1260 hectare. The cultivated area has raised quickly; it covered 65,000 hectare and produced 400 kg of seed per hectare on average in 1983. The cultivated area of soybeans was gradually decreased to reach 13,000 hectares in 2018 due to the competition with other essential summer crops producing an average of 3.57 tons of seed per hectare, whereas, in 2022, the cultivated area was expanded to over 63,000 hectares, and the productivity per hectare reached 4.04 tons (Bulletin of Statistical Cost Production and Net Return 2023). Other reasons that negatively affected soybean cultivated area are the poor economic return and high production costs when compared to other summer crops like maize.

The Importance of Land Equivalent Ratio in Evaluating Intercropping Goodness

It is well known that biological efficiency is determined by the intercrop's overall yield, as well as the area of sole crops needed to produce the same component yields (Metwally et al. 2009). Furthermore, it is well acknowledged that intercropping

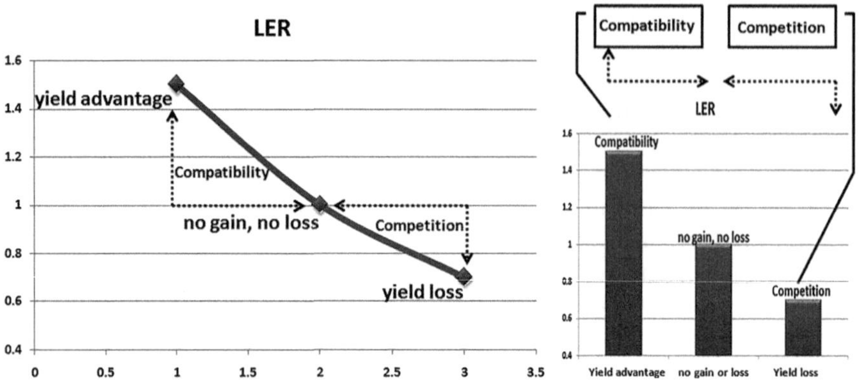

Fig. 5.1 Compatibility and competition are the implicit expression of LER

systems provide land use advantage over the sole plantings. Therefore, the value of land equivalent ratio (LER) has been the main focus of intercropping research (Abdel-Wahab and Abdel-Wahab 2020a). Vandermeer (1989) stated that the value of LER of an intercropping system is calculated using the yield of sole planting of both species and the intercropped yield. According to Mead and Willey (1980), LER is the area required for intercropping at the same management level as that required for sole planting in order to get an equivalent yield. The following formula is used to compute it:

$$\mathrm{LER} = \left(\mathrm{Yab} / \mathrm{Yaa}\right) + \left(\mathrm{Yba} / \mathrm{Ybb}\right) \tag{5.1}$$

where Yab is the intercrop yield of crop a, Yba is the intercrop yield of crop b, Yaa is the yield of sole crop a, and Ybb is the yield of sole crop b.

A greater value of LER than 1.00 for intercrops indicates that the intercropping system requires less land than sole plantings of both species (Vandermeer 1989). Furthermore, a yield advantage is attained if the value of LER is greater than 1.00. No gain or loss if LER is equal to 1.00, and yield loss is attained if LER is lower than 1.00 (Fetene 2003), as shown in Fig. 5.1.

Compatibility and Competition in Intercropping Systems

When the compatibility of the two crops for light and nutrients resources is higher than the effects of competition, intercropping usually yields these benefits (Ren et al. 2021). Niche differentiation is the process by which species that are in competition with one another use resources in a different way over time or space (Bazzaz and Catovsky 2001). It lowers interspecific competition, preserves species coexistence, and allows species to exploit resources complementarily (Loreau 2000). All the ways that different species affect each other's growth and resource consumption, from the negative effects of resource competition to the positive effects of resource facilitation, are collectively referred to as interspecies interaction (Abdel-Wahab and Abdel-Wahab 2020b). As a result, in intercropping systems, the yield advantage is frequently attained by coordinating interspecies interaction for above- and/or belowground competition (Li et al. 2006).

There is a consensus that intercropping makes better use of the resources in the environment, namely, light, water, and nutrients, than when the two crops are grown separately (Loreau 2000); thus, it yields higher when the crops are compatible (Maitra et al. 2021). Furthermore, interspecific competition is less intense than intraspecific competition, and component crops of an intercropping system have yield advantages when they compete only partially for the same plant growth resources (MacLaren et al. 2023). However, the planting pattern, which is an agronomic technique, may impact how the intercropping component crops interact with one another, how they use natural resources, and ultimately, how successful intercropping is compared to sole plantings (Toker et al. 2024).

Legume intercropping and cereal intercropping have higher efficiency in nutrient usage than sole plantings (Willey 1979). Planting cereals next to crops that fix N improved the ability of cereals to synthesize starch and sucrose while reducing gas emissions and N leaching (Abdel-Wahab et al. 2019b). Soybean intercropping system with maize is an example of that. The two crops compete with one another for resources and space (Vandermeer 1989). Maize plants benefited from the fixed N that soybeans made available to them during their vegetative and reproductive growth (El-Shamy et al. 2015).

Abdel-Wahab et al. (2019a) studied the effect of different planning densities of maize and soybean under their intercropping systems planted on two ridge widths in terms of their compatibility and competition. Two maize ridges were studied (70 cm wide) alternating with two soybean ridges (2:2 maize/soybean), two maize ridges alternating with four soybean ridges (2:4 maize/soybean), and growing soybean and maize in the same ridge, 140 cm wide (mixed system). One side of the ridge was used to grow sole maize, and two rows of soybeans were drilled on the ridges to grow sole soybean.

The findings of the study (Table 5.1) indicated that, when compared to the other treatments, mixed cropping of maize and soybean and 2:2 maize/soybean intercropping systems had the highest value of LER. A higher competitive pressure between maize and soybean on the available growth resources was found as the density of soybean plants increases in the intercropping system, which is most likely the cause of these results. In terms of the stability of soybean plant density, there was no discernible difference between 2:2 maize/soybean intercropping system and mixed maize and soybean systems for the value of LER. Alternatively, the intraspecific competition between the same species did not differ between the two intercropping systems (mixed cropping of maize and soybean and 2:2 maize/soybean intercropping systems), despite the higher plant density of maize per unit area. Thus, it might be feasible to grow 50% of the soybean plant density, which is comparable to its counterpart in 2:2 maize/soybean intercropping system, in order to enhance the number of maize plants per unit area up to 100% of the recommended maize plant density. Additionally, there was no discernible difference between sole planting of maize and mixed system for the value of LER (Table 5.1).

Table 5.1 Effect of different cropping systems on LER, economic yield, and 100-seed weight, averaged over two growing seasons

Ridge width (cm)	Cropping system Maize/soybean	LER	Economic yield (ton/ha)		100-seed weight (g)	
			Soybean seeds	Maize grains	Soybean seeds	Maize grains
0.7 m	2:2	1.13	1.14	7.16	15.87	38.05
	2:4	1.06	1.39	5.78	15.92	34.48
	Sole soybean	–	3.24	–	17.42	–
	Sole maize	1.00	–	9.16	–	39.23
1.4 m	Mixed	1.25	0.98	8.67	13.85	40.34
	Sole soybean	–	3.31	–	17.30	–
	Sole maize	1.00	–	8.98	–	39.36
L.S.D. 5%		0.14	2.07	0.62	2.37	2.47

Source: Abdel-Wahab et al. (2019a)

Besides, there were notable variations between sole planting of maize and 2:4 maize/soybean intercropping system for the values of LER. It is evident that in alternating ridges, the proportion of maize plants per unit area dropped from 100% in sole plantings of maize to 33% of maize planting density in 2:4 maize/soybean intercropping system (Table 5.1). This demonstrates that because one of the two cropping systems did not contain a leguminous component, decreasing the planting density of maize per unit area in alternating ridges resulted in the superiority of intraspecific competition over interspecific competition and the appearance of significance between the two cropping systems. Additionally, the presence of soybeans in alternating ridges may have also contributed to increased nutrient availability and reduced pest pressure for the maize plants. These factors could potentially explain the differences in growth and yield observed between the two cropping systems.

Besides, the significance between 2:2 and 2:4 maize/soybean intercropping systems disappeared as a result of the two components' varying plant densities when grown together in the alternating ridges, which caused confusion between intraspecific and interspecific competition (Abdel-Wahab et al. 2019a). Additionally, there was no discernible difference between 2:2 and 2:4 maize/soybean intercropping systems for the values of LER. Additionally, the behaviour of intraspecific competition (maize or soybeans) and interspecific competition (maize and soybeans) when they grow together in one row or alternating ridges.

These findings demonstrate that, despite changes in the planting densities of both species, intraspecific competition remained steady amongst the same species in 2:2 and 2:4 maize/soybean intercropping systems. Furthermore, these findings show that the most significant element in lowering intraspecific competition is reducing the number of maize plants per unit area in 2:2 maize/soybean intercropping system, with little effect on interspecific competition (Abdel-Wahab et al. 2019a).

Overall, these results suggest that adjusting plant densities within an intercropping system can impact intraspecific competition more than interspecific competition. This highlights the importance of carefully considering plant ratios to optimize crop yield and minimize competition. These findings are comparable to those of Metwally et al. (2021) and Abdel-Wahab and Abdel-Rahman (2016).

The findings of this study show that the intraspecific competition between maize and soybeans disappeared when they were planted together in maize ridges. It is evident that the number of maize plants per unit area remained constant under these intercropping systems. Thus, planting up to 50% more soybean plants per unit area in maize ridges did not increase the efficiency of either intraspecific competition or interspecific competition. This shows that the development and yield of the maize plants in this study were not substantially impacted by the presence of soybeans. The long-term impacts of growing maize and soybeans together in this way may require more investigation (Abdel-Wahab et al. 2019a).

It is noteworthy to emphasize that for the value of LER, there was a strong correlation between 2:4 maize/soybean intercropping system and mixed systems of maize and soybean. With varying planting densities of both components, it seems that the intraspecific competition between maize and soybeans in mixed intercropping system is entirely different from that of 2:4 maize/soybean intercropping system. These findings demonstrate that, in the two intercropping systems, interspecific competition may outcompete intraspecific competition by lowering the planting density of maize per unit area and increasing the planting density of soybeans per unit area (Abdel-Wahab et al. 2019c).

Furthermore, there were no appreciable variations between sole plantings of maize and 2:2 maize/soybean intercropping system. When maize was planted alone, the number of maize plants per unit area was 100%; however, when maize and soybeans were planted in alternating ridges, this number dropped to 50%. The presence of soybeans in the intercropping system can have a significant impact on the planting density of maize. These findings highlight the complex relationships between different plant species in agroecosystems and emphasize the importance of considering these interactions when managing crops. The absence of soybean plants in one of the cropping systems resulted in a noticeable difference between the two systems, with intraspecific competition becoming more dominant over interspecific competition as the planting density of maize per unit area decreased in alternating ridges (Abdel-Wahab et al. 2019a).

Complementary and Complimentary in Intercropping System

Complementary is defined as the interaction between different components in a system that enhances overall productivity, whereas complimentary refers to interactions that may not necessarily improve productivity (Brooker et al. 2015). This distinction is crucial for understanding how various elements within a production system can impact overall output. Complementary means satisfying or addressing the demands of both crops under the intercropping system (Wu et al. 2022); thus, an increase in their growth occurs, when compared to the single planting of each crop, and that results in a rather steady productivity (Prieto et al. 2015). When two crops are planted together instead of separately, one crop grows more efficiently than the other under intercropping systems, thus, a complementary crop is the one that uses basic growth resources negatively so that the other crop might profit from them

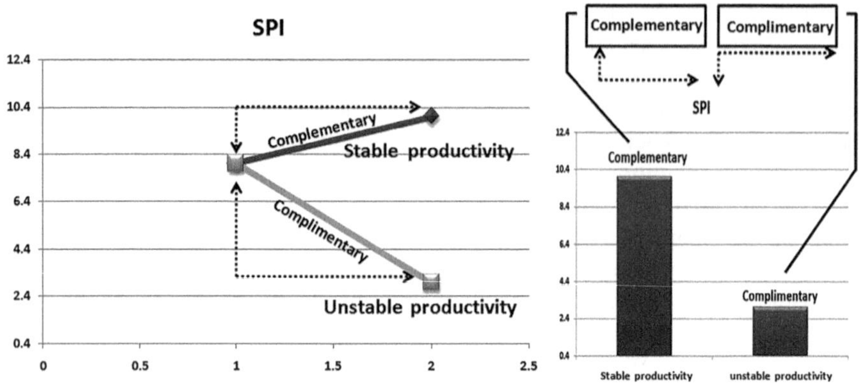

Fig. 5.2 Complementary and complimentary are the implicit expression of SPI

(Shanmugam et al. 2022). Figure 5.2 illustrates complementary and complimentary effects, which are the implicit expression of system production index (SPI).

SPI defines combinations that maximized growth resources and keep yield performance steady while standardizing the yield of the secondary crop in relation to the primary crop. SPI is calculated according to the formula provided by Odo (1991) as follows:

$$\text{SPI} = \left[\left(\text{SA}/\text{LB}\right) \times \text{Lb}\right] + \text{Sa} \tag{5.2}$$

where SA and LB are the yield of crop a and crop b in sole cropping, and Sa and Lb are the yield of crop a and crop b in intercropping.

In the earlier study of Abdel-Wahab et al. (2019a) and their results presented in Table 5.1, they found that transitioning from 2:4 maize/soybean intercropping system to 2:2 maize/soybean intercropping system to mixed intercropping system of maize and soybean resulted in a decrease in the yield of soybean and an increase the yield of maize, thus suggesting that the mixed system led to enhanced crop productivity through complementary effects. This indicates that intercropping could be advantageous for farmers aiming to increase yields and minimize environmental impact. Furthermore, they also stated that 2:4 maize/soybean intercropping system tended to increase complimentary effects. This increase in complimentary effects did not lead to improve overall crop yield.

Evaluation of Soybean Varieties Tolerant to Shading Stress

Shading stress in intercropping can harm crop growth and yield by blocking sunlight and limiting photosynthesis (Wang et al. 2021). This can lead to reduced productivity in the intercropped system, underscoring the need to effectively manage shading effects in intercropping systems (Moreira et al. 2024).

Previous research published in Egypt on soybean varieties planted in different intercropping systems with maize are presented in Table 5.2, namely, Crawford, Giza 21, Giza 22, Giza 35, and Giza 111. Two maize ridges (70 and 140 cm wide) alternating with two soybean ridges (2:2 maize/soybean), two maize ridges alternating with four soybean ridges (2:4 maize/soybean), and growing soybean and maize in the same ridge (mixed system of maize and soybean) are the three intercropping systems used in these studies. One side of the ridge was used to grow sole maize, and two rows of soybeans were drilled on the ridges to grow sole soybean.

The results in Table 5.2 showed that the seed yield per soybean plant was lower in intercropped Crawford or Giza 111 with maize in 2:2 maize/soybean intercropping system or in a mixed intercropping system than 2:4 maize/soybean intercropping system. These results verify that the shading from intercropped maize plants has a detrimental effect on Crawford or Giza 111 in 2:2 maize/soybean and mixed intercropping systems. It has been reported that the shade from maize plants reduces dry matter accumulation, leading to a decrease in seed production (Hussain et al. 2020). Intercropping Crawford or Giza 111 with maize at one-third of its recommended planting density (2:4 maize/soybean intercropping system) can reduce shading effects and improve the photosynthesis process in soybean. Accordingly, there is less intraspecific competition amongst Crawford or Giza 111 plants than what was found between Crawford and maize or Giza 111 and maize (Table 5.2).

The results also showed that intercropping Giza 21, Giza 22, or Giza 35 with maize in 2:2 maize/soybean intercropping system resulted in higher seed yields per plant compared to 2:4 maize/soybean and mixed intercropping systems. Accordingly, there is less intraspecific competition amongst Giza 21, Giza 22, or Giza 35 plants than what is found between Giza 21 and maize, between Giza 22 and maize, or between Giza 35 and maize in 2:4 maize/soybean and mixed intercropping systems (Table 5.2). The results also showed that intercropping Giza 21, Giza 22, or Giza 35 with maize in 2:4 and mixed intercropping systems had a negative impact on Giza 21, Giza 22, or Giza 35. Thus, the intraspecific competition between Giza 21, Giza 22, or Giza 35 plants is greater than the intraspecific competition between Giza 21 and maize, between Giza 22 and maize, or between Giza 35 and maize. It is therefore possible for Giza 21, Giza 22, or Giza 35 to coexist with maize plants in 2:2 maize/soybean intercropping system as opposed to if Giza 21, Giza 22, or Giza 35 were grown as a sole crop (Table 5.2).

With regard to Crawford, Table 5.2 showed that seed yield per hectare was negatively impacted by lowering the density of maize plants to one-third of the recommended rate for sole planting, according to the obtained results. This is probably because, in contrast to Egyptian varieties, Crawford was grown in overcast conditions in the United States, producing leaves with higher chlorophyll content (Abdel-Wahab et al. 2019a, 2020, 2024). Crawford was therefore grown in Egypt at higher temperatures, which accelerated photorespiration and decreased dry matter accumulation during its growth and development.

The results in Table 5.2 also revealed that intercropping Giza 21 with maize in 2:2 maize/soybean intercropping system resulted in higher seed yields per hectare compared to mixed and 2:4 maize/soybean intercropping systems. These studies

Table 5.2 Soybean and maize yields and their relative yields, LER, and SAPI of soybean varieties with maize under three intercropping systems

Soybean variety	Intercropping system	Soybean yield/plant (g)	Soybean yield (ton/ha)	Maize yield (ton/ha)	Relative yield of soybean	Relative yield of maize	LER	SPI	References
Crawford	2 maize: 2 soybean	17.36	1.80	7.07	0.50	0.96	1.46	10.75	Metwally et al. (2019b)
	2 maize: 4 soybean	18.12	1.05	5.77	0.38	0.63	1.01	9.25	Abdel-Wahab and Abdel-Wahab (2020a)
	Mixed intercropping	17.50	2.18	6.50	0.61	0.88	1.49	11.00	Metwally et al. (2019b)
Giza 21	2 maize: 2 soybean	14.83	1.71	6.51	0.51	0.90	1.41	10.19	Metwally et al. (2018)
	2 maize: 4 soybean	–	–	–	–	–	–	–	–
	Mixed intercropping	5.45	1.56	6.93	0.49	0.97	1.46	10.43	Abdel-Galil et al. (2014a)
Giza 22	2 maize: 2 soybean	15.60	1.74	6.56	0.48	0.91	1.39	10.02	Metwally et al. (2018)
		9.24	1.70	5.34	0.59	0.88	1.47	8.92	Lamlom et al. (2015)
		8.12	1.58	4.62	0.47	0.65	1.12	7.96	Abdel-Wahab and Abdel-Rahman (2016)
	Average	10.98	1.67	5.50	0.51	0.81	1.32	8.96	–
	2 maize: 4 soybean	8.35	2.19	3.26	0.66	0.46	1.12	7.93	Abdel-Wahab and Abdel-Rahman (2016)
		8.89	1.98	3.37	0.77	0.68	1.45	7.18	Abdel-Galil et al. (2014a)
	Average	8.62	2.08	3.31	0.71	0.57	1.28	7.43	–
	Mixed intercropping	6.57	1.81	7.07	0.52	0.99	1.51	10.78	Abdel-Galil et al. (2014a)

Giza 35	2 maize: 2 soybean	15.32	1.24	6.71	0.36	0.93	1.29	9.30	Metwally et al. (2018)
	2 maize: 4 soybean	6.62	1.59	3.24	0.70	0.66	1.36	6.67	Abdel-Galil et al. (2014b)
	Mixed intercropping	6.10	1.67	7.05	0.51	0.99	1.50	10.68	Abdel-Galil et al. (2014a)
Giza 111	2 maize: 2 soybean	7.39	1.35	4.59	0.44	0.64	1.08	7.74	Abdel-Wahab and Abdel-Rahman (2016)
	2 maize: 4 soybean	7.55	1.94	3.28	0.64	0.46	1.10	7.84	
		7.75	1.88	3.27	0.79	0.66	1.45	7.18	Abdel-Galil et al. (2014b)
	Average	7.65	1.91	3.71	0.62	0.58	1.27	7.67	–
	Mixed intercropping	5.22	1.50	6.85	0.48	0.96	1.44	10.27	Abdel-Wahab et al. (2019c)

also found that intercropping Giza 21 with maize in mixed and 2:4 systems had a negative impact on Giza 21, but this effect was reduced when intercropping Giza 21 with maize in 2:2 maize/soybean intercropping system.

The results in Table 5.2 revealed that intercropping Giza 22 or Giza 111 with maize in mixed and 2:2 maize/soybean intercropping systems produced a lower seed yield per hectare than 2:4 maize/soybean intercropping system. This may be explained by the narrow leaves of Giza 22 (Metwally et al. 2012), which increase the amount of light that reaches the lower leaves on the plant's stem. Giza 111, on the other hand, benefits from low light intensity because of its large leaves (Abdel-Wahab and Abdel-Rahman 2016), particularly the bottom leaves. The process of photosynthesis and the dry matter accumulation during growth and development were both accelerated by these biological variations. These traits enable these two soybean varieties to withstand intense shading in mixed system.

To calculate the values of LER and SPI, sole plantings of the two crops were utilized. Intercropping Crawford with maize in the mixed system resulted in the highest values of LER, namely, 1.49, and SPI of 11.00 compared to other treatments. This system showed potential for maximizing land use efficiency and productivity by effectively utilizing resources and space to increase yields for both crops. Intercropping Crawford with maize in 2:2 maize/soybean intercropping system ranked second in terms of the values of LER (1.46) and SPI (10.75). Conversely, intercropping Crawford with maize in 2:4 maize/soybean intercropping system resulted in lower values of LER and SPI, highlighting the importance of crop arrangement in intercropping systems.

Intercropping Giza 21 with maize in mixed intercropping system achieved higher values of LER (1.46) and SPI (10.43) compared to other treatments, showing its potential for maximizing land use efficiency and productivity. Intercropping Giza 21 with maize in 2:2 maize/soybean intercropping system ranked second in terms of the values of LER (1.41) and SPI (10.19). However, intercropping Giza 21 with maize in 2:4 maize/soybean intercropping system yielded lower values of LER and SPI, emphasizing the significance of crop arrangement in intercropping systems. Intercropping Giza 22 with maize in mixed system resulted in higher values of LER (1.51) and SPI (10.78) compared to other treatments, demonstrating its potential for maximizing land use efficiency and productivity. Intercropping Giza 22 with maize in 2:2 maize/soybean intercropping system ranked second in terms of the values of LER (1.32) and SPI (8.96). Conversely, intercropping Giza 22 with maize in 2:4 maize/soybean intercropping system led to lower values of LER and SPI, underscoring the importance of crop arrangement in intercropping systems. Intercropping Giza 35 with maize in a mixed system achieved higher values of LER (1.50) and SPI (10.68) compared to other treatments, highlighting its potential for maximizing land use efficiency and productivity. Intercropping Giza 35 with maize in 2:2 maize/soybean intercropping system ranked second in terms of values of LER (1.29) and SPI (9.30). However, intercropping Giza 35 with maize in 2:4 maize/soybean intercropping system resulted in lower values of LER and SPI, indicating the significance of crop arrangement in intercropping systems. Intercropping Giza 111 with maize in mixed system achieved higher values of LER (1.44) and SPI (10.27) compared to other treatments. Intercropping Giza 111 with maize in 2:2 maize/soybean intercropping system ranked second in terms of values of LER (1.08) and SPI (7.74). Conversely, intercropping Giza 111 with maize in a 2:4 maize/soybean intercropping system led to lower values of LER and SPI, emphasizing the importance of crop arrangement in intercropping systems.

Conclusions

Soybeans are crucial for sustainable food production systems, especially in cropping systems with limited external inputs. Complementary and complimentary in soybean and maize intercropping system are influenced by the choice of soybean variety. The tested soybean varieties were compatible with maize in a mixed system; however, compatibility declines when soybean and maize were intercropped in 2:2 maize/soybean intercropping system. Additionally, productivity tends to decline when soybean ridges surpass maize ridges because of the emergence of competition between the two species. Thus, it is suggested that while soybean and maize can be successfully grown together in a mixed system, careful consideration should be given to the spacing and arrangement of the crops to maximize productivity.

Additionally, the tested soybean varieties are complementary to maize in a mixed system; however, this complementary diminishes when soybean and maize are intercropped in 2:2 maize/soybean intercropping system, and productivity tends to be unstable when soybean ridges surpass maize ridges because of the emergence of complimentary traits between the two species. This indicates that further research is needed to optimize the intercropping system for soybean and maize to achieve consistent and high yields.

Thus, Giza 22 and Giza 35 can coexist with maize plants in mixed system, rather than when grown as a sole crop.

References

Abdel-Galil AM, Abdel-Wahab SI, Abdel-Wahab TI (2014a) Compatibility of some maize and soybean varieties for intercropping under sandy soil conditions. Soybean Res 12(special issue):22–44

Abdel-Galil AM, Abdel-Wahab TI, Abdel-Wahab SI (2014b) Productivity of four soybean varieties as affected by intercropping and corn planting geometry. Soybean Res 12(1):36–58

Abdel-Wahab TI, Abdel-Rahman RA (2016) Response of some soybean cultivars to low light intensity under different intercropping patterns with maize. Int J Appl Agric Sci 2(2):21–31

Abdel-Wahab SI, Abdel-Wahab EI (2020a) Competitive and facilitative effects of intercropping some soybean varieties with corn under different soybean plant densities. Plant Arch 20(2):1631–1639

Abdel-Wahab TI, Abdel-Wahab EI (2020b) Interspecies interactions dynamics between soybean and corn as affected by different soybean varieties. Int J Adv Sci Technol 29(11s):2864–2876

Abdel-Wahab TI, Abdel-Wahab EI, Taha AM, Adel MM, Hussein HM (2019a) Varietal response of soybean to applied irrigation water and insects incidence under different intercropping systems with maize. Res Crops 20(Issue Suppl):S1–S25

Abdel-Wahab TI, Abdel-Wahab SI, Abdel-Wahab EI (2019b) Benefits of intercropping legumes with cereals. Int J Conf Proc 1(2):ICP.000510.2019

Abdel-Wahab SI, Abdel-Wahab EI, Taha AM, Saied SM, Naroz MM (2019c) Evaluation of intercropped soybean cultivars with corn for water consumption and soybean mosaic virus infection under different soybean plant densities. Res Crops 20(Issue Suppl):S26–S46

Abdel-Wahab TI, Abdel-Wahab EI, Taha AM, Adel MM, Hussein HM (2020) Water consumptive use and insects incidence of three soybean cultivars under different intercropping systems with maize. Soybean Res 18(1):01–30

Abdel-Wahab TI, Abdel-Wahab SI, Abdel-Wahab EI (2024) Chapter 15: biological engineering and its relationship to nematode resistance. In: Chaudhary KK, Meghvansi MK, Siddiqui S (eds) Sustainable management of nematodes in agriculture, vol. 2: role of microbes—assisted strategies, sustainability in plant and crop protection. Springer, Cham, pp 383–408

Bazzaz FA, Catovsky S (2001) Resource partitioning. In: Encyclopedia of biodiversity, 2nd edn. Blackwell Publishing, pp 429–437

Blessing DJ, Gu Y, Cao M, Cui Y, Wang X, Asante-Badu B (2022) Overview of the advantages and limitations of maize-soybean intercropping in sustainable agriculture and future prospects: a review. Chilean J Agric Res 82(1):177–188

Brooker RW, Bennett AE, Cong F, Daniell TJ, George TS, Hallett PD, Hawes C, Iannetta PP, Jones HG, Karley AJ, Li L, McKenzie BM, Pakeman RJ, Paterson E, Schöb C, Shen J, Squire G, Watson CA, Zhang C, Zhang F, Zhang J, White PJ (2015) Improving intercropping: a synthesis of research in agronomy, plant physiology and ecology. New Phytol 206(1):107–117

Bulletin of Statistical Cost Production and Net Return (2023) Summer and Nili field crops and vegetables and fruit. Agriculture Statistics and Economic Sector, Ministry of Egyptian Agriculture and Land Reclamation, Part (2), August 2023

El-Shamy MA, Abdel-Wahab TI, Abdel-Wahab SI, Ragheb SB (2015) Advantages of intercropping soybean with maize under two maize plant distributions and three mineral nitrogen fertilizer rates. Adv Biosci Bioeng 3(4):30–48

Fetene M (2003) Intra-and inter-specific competition between seedlings of *Acacia etbaica* and a perennial grass (*Hyparrhenia hirta*). J Arid Environ 55:441–451

Graham PH, Vance CP (2003) Legume importance and constraints to greater use. Plant Physiol 131:872–877

Hussain S, Liu T, Iqbal N, Brestic M, Pang T, Mumtaz M et al (2020) Effects of lignin, cellulose, hemicellulose, sucrose and monosaccharide carbohydrates on soybean physical stem strength and yield in intercropping. Photochem Photobiol Sci 19:462–472. https://doi.org/10.1039/cspp00369j

Kline A, McCoy S, Vyn T, West T, Christmas E (2001) Management considerations for relay intercropping: I. wheat. Agronomy guide. Purdue University Cooperative Extension. Department of Agronomy, Purdue University

Lamlom MM, Abdel-Wahab SI, Abdel-Wahab TI, Gendy EK (2015) Residual effects of some preceded winter field crops on productivity of intercropped soybean with three maize cultivars. Am J Biosci 3(6):226–242

Li L, Sun J, Zhang F, Guo T, Bao X, Smith FA, Smith SE (2006) Root distribution and interactions between intercropped species. Oecologia 147:280–290

Loreau M (2000) Biodiversity and ecosystem functioning: recent theoretical advances. Oikos 91:3–17

MacLaren C, Waswa W, Aliyu KT, Claessens L, Mead A, Schöb C, Vanlauwe B, Storkey J (2023) Predicting intercrop competition, facilitation, and productivity from simple functional traits. Field Crop Res 297:108926

Maitra S, Hossain A, Brestic M, Skalicky M, Ondrisik P, Gitari H, Brahmachari K, Shankar T, Bhadra P, Palai JB, Jena J, Bhattacharya U, Duvvada SK, Lalichetti S, Sairam M (2021) Intercropping—a low input agricultural strategy for food and environmental security. Agronomy 11(2):343

Mead R, Willey RW (1980) The concept of a "land equivalent ratio" and advantages in yields from intercropping. Exp Agric 16:217–228

Metwally AA (1999) Intensive cropping system in the battle against food crises. In: Proceedings of the 1st conference in recent technologies, Faculty of Agriculture at Cairo University, 27–29 Nov, vol 11, pp 333–341
Metwally AA, Shafik MM, El-Habbak KE, Abdel-Wahab SI (2009) Yield and land equivalent ratio of intercropped soybean with maize under different intercropping patterns and high population densities. Egypt J Agron 31(2):199–222
Metwally AA, Shafik MM, El-Habbak KE, Abdel-Wahab SI (2012) Yield and soybean characters under some intercropping patterns with corn. Soybean Res 10:24–42
Metwally AA, Safina SA, Abdel-Wahab TI, Abdel-Wahab SI, Hefny YAA (2018) Productivity of soybean varieties under intercropping culture with corn in Egypt. Soybean Res 16(1&2):63–77
Metwally AA, Abdel-Wahab TI, Abdel-Wahab SI (2019a) Increasing land and water use efficiencies by intercropping summer legumes with corn in Egypt. Agric Biol Res 35(2):6–10
Metwally AA, Safina SA, Abdel-Wahab TI, Abdel-Wahab SI (2019b) Growing of twenty soybean genotypes in solid and intercropping systems with corn. Res Crops 20(Issue Suppl):S47–S57
Metwally AA, Safina SA, Abdel-Wahab EI, Abdel-Wahab SI, Abdel-Wahab TI (2021) Screening thirty soybean genotypes under solid and intercropping plantings in Egypt. J Crop Sci Biotechnol 24:203–220
Moreira B, Gonçalves A, Pinto L, Prieto MA, Carocho M, Caleja C, Barros L (2024) Intercropping systems: an opportunity for environment conservation within nut production. Agriculture 14(7):1149
Mousavi S, Eskandari H (2011) A general overview of intercropping and its advantages in sustainable agriculture. Appl Environ Biol Sci 1(11):482–486
Odo PE (1991) Evaluating short and tall sorghum varieties in mixtures with cowpea in Sudan savanna of Nigeria: LER, grain yield and system productivity index. Exp Agric 27:435–441
Prieto I, Violle C, Barre P, Durand JL, Ghesquiere M, Litrico I (2015) Complementary effects of species and genetic diversity on productivity and stability of sown grasslands. Nat Plants. https://doi.org/10.1038/NPLANTS.2015.33
Ren Y, Zhang L, Yan M, Zhang Y, Chen Y, Palta JA, Zhang S (2021) Effect of sowing proportion on above- and below-ground competition in maize–soybean intercrops. Sci Rep 11:15760
Sayed-Galal S, Metwally AA (1989) Intercropping corn with soybean as the only practical solution for increasing soybean production in Egypt. In: Proceedings of the fourth world soybean research conference, Buenos Aires, Argentina, 2004–2009
Sayed-Galal S., Hindi L, Abdalla MMF, Metwally AA (1979) Soybean and corn yields under different intercropping patterns. In: World soybean research conference II-raleigh, North Carolina, p 69
Shanmugam S, Hefner M, Pelck JS, Labouriau R, Kristensen HL (2022) Complementary resource use in intercropped faba bean and cabbage by increased root growth and nitrogen use in organic production. Soil Use Manag 38:729–740. https://doi.org/10.1111/sum.12765
Thompson JB, Döring TF, Bellingrath-Kimura SD, Grahmann K, Glemnitz M, Reckling M (2025) Spatial arrangement of intercropping impacts natural enemy abundance and aphid predation in an intensive farming system. Agric Ecosyst Environ 378:109324
Toker P, Canci H, Turhan I, Isci A, Scherzinger M, Kordrostami M, Yol E (2024) The advantages of intercropping to improve productivity in food and forage production—a review. Plant Prod Sci:1–15. https://doi.org/10.1080/1343943X.2024.2372878
Vandermeer J (1989) The ecology of intercropping. Cambridge University Press, New York
Wang Y, Huang R, Zhou Y (2021) Effects of shading stress during the reproductive stages on photosynthetic physiology and yield characteristics of peanut (*Arachis hypogaea* Linn.). J Integr Agric 20(5):1250–1265
Willey RW (1979) Intercropping—its importance and research needs. Part 2. Agronomy and research approaches. Field Crop Abst 32:73–85
Wu J, Bao X, Zhang J, Lu B, Zhang W, Callaway RM, Li L (2022) Temporal stability of productivity is associated with complementarity and competitive intensities in intercropping. Ecol Appl 33(1):e2731

6 Advanced Soybean Breeding Techniques to Alleviate Biotic and Abiotic Stresses Under Intercropping

Eman Ibrahim Abdel-Wahab, Sherif Ibrahim Abdel-Wahab, and Tamer Ibrahim Abdel-Wahab

Abstract

Soybean is one of the most important legume crops cultivated all over the world. Breeding of soybean plays a crucial role in developing new soybean varieties with improved traits, such as higher yield, as well as variety tolerance to pests, and shading stresses. This chapter discusses the key aspects of soybean breeding, including genetic breeding and molecular methods. Furthermore, the impact of intercropping soybean with maize in reducing the damage of soybean mosaic virus and other insects' infestation was reviewed. The effect of cropping systems on gene expression as a potential strategy for disease control in soybean fields was also tackled. Genetic analysis of some Egyptian varieties of soybean for its tolerance to soybean mosaic virus and other insects' infestation under intercropping was also provided.

Keywords

Soybean varieties · Soybean intercropping system with maize · Soybean mosaic virus · Insects · Pd1 and Pd2 genes · Pyramiding genes

---

E. I. Abdel-Wahab (✉)
Food Legumes Research Department, Field Crops Research Institute, Agricultural Research Center (ARC), Giza, Egypt
e-mail: eman.abdel-wahab@arc.sci.eg

S. I. Abdel-Wahab · T. I. Abdel-Wahab
Crop Intensification Research Department, Field Crops Research Institute, Agricultural Research Center (ARC), Giza, Egypt

A. E.-H. Zohry, S. Ouda (eds.), *Resilient Agroecosystems*, Sustainability Sciences in Asia and Africa, https://doi.org/10.1007/978-981-96-9630-7_6

Introduction

Environmental quality is crucial for the well-being of all living organisms (Vargas-Almendra et al. 2024). Investing in infrastructure, technology, and education can strengthen food systems, leading to improved nutrition, increased food security, and enhanced economic opportunities for communities (Abdel-Wahab et al. 2019a). This investment can help build resilience and sustainability within our food systems. Implementing sustainable farming practices can enhance yields and preserve soil health and biodiversity for future generations (Abdel-Wahab et al. 2020a). Controlling agricultural pests enhances the environment by bolstering food chains and reducing the reliance on harmful pesticides. This leads to the development of more sustainable agricultural practices and healthier ecosystems.

Soybean is one of the most important legume crops cultivated all over the world. It provides humans and animals with protein, oil, and phytochemical components (Alghamdi et al. 2018). The first domestication of soybean has been traced to the eastern half of North China in the eleventh century B.C. or perhaps a bit earlier (Gibson and Benson 2005). Cultivated on approximately 6% of the world's arable land, soybean is often referred to as the "golden miracle bean" due to its unique seed composition, making up around 70% of total protein meal and over 60% of global oilseed production (Hartman et al. 2011; Vieira and Chen 2021). After the oil is extracted, soymeal provides a cheap and abundant source of proteins with a balanced amino acid composition (Krishnan and Jez 2018). Among oilseed crops, soybean production ranks second only to oil palm, comprising over half of global production from 2012 to 2016 (El-Hamidi and Zaher 2018). In 2021, global soybean production reached 371.7 million metric tons (Mt), with Brazil (134.9 Mt), the United States of America (120.7 Mt), and Argentina (46.2 Mt) contributing to 81.2% of the total production (FAO 2023).

Soybeans suffer for several biotic stressors, namely, insects and diseases. The aphid (*Aphis gossypii*) is one of the insects that spreads several plant viruses, including soybean mosaic viruses (SMV), while also causing damage by feeding on phloem sap (Todd et al. 2022). Because it may reduce soybean yield by up to 58%, the soybean aphid is a pest of considerable economic importance (Wang et al. 1994). Beckendorf et al. (2008) found that the feeding habits of soybean aphids can greatly impact the economic value and seed quality of soybean plants. Their research showed that just one soybean aphid on a plant at the V5 growth stage (when the plant is 20–30 cm tall with five fully developed leaf nodes) can reproduce and result in over 4000 aphids per plant. According to Hill and Whitham (2014), SMV can lead to a decrease in soybean yield ranging from 8% to 35%. This is primarily due to the reduced pod set and seed weight.

Moreover, soybean production in the Mediterranean region faces significant yield losses due to direct damage from whitefly (*Bemisia tabaci*) pests, both adult and nymphs (Gulluoglu et al. 2010). Research by Murgianto and Hidaya (2017) indicates that whitefly infestations can lead to a reduction in soybean yield of up to 80%. Furthermore, according to Higley and Boethel (1994), soybean leaf miners (*Liriomyza trifolii*) have traditionally targeted soybeans, while Gouge et al. (1999)

argue that thrips (*Thrips* sp.) are not a major economic pest in soybeans. Gent et al. (2004) point out that thrips can be a significant pest in various soybean-producing regions due to the damage caused by both larvae and adult feeding, as well as the indirect harm caused by the transmission of viruses.

To enhance sustainability and productivity in soybean farming, researchers are exploring innovative pest control methods, including biological engineering (Abdel-Wahab et al. 2019b). Biological engineering can help manage pests by utilizing various planting geometry techniques and reducing the need for synthetic pesticides (Naroz et al. 2018). This leads to the development of more sustainable agricultural practices and healthier ecosystems. Implementing sustainable farming practices can enhance yields and preserve soil health and biodiversity for future generations (Naroz et al. 2019).

Intercropping provides solutions to the challenges of sole cropping, offering a viable ecological alternative that is fully compatible with modern agriculture (Horwith 1985). Intercropping increases diversity, which enhances biological pest control (Stephen 2009). By utilizing the power of genetics, soybean breeders can make great strides in producing varieties that are compatible for intercropping with maize (*Zea mays* L.) and for pest control management (Chu et al. 2021). Furthermore, significant progress has been made in agricultural pest control. Evaluating innovative techniques such as intercropping and soybean varieties for tolerance to common pests such as SMV, aphids, whiteflies, leaf miner flies, and thrips. Tolerance refers to a plant's ability to maintain its fitness despite insect damage by continuing to reproduce and grow (Rosenthal and Kotanen 1994).

In the soybean and maize intercropping system, soybean plants experience shading throughout their growth period with maize (Hussain et al. 2020). To mitigate the effect of the abiotic stress of shading, researchers compared various intercropping strategies, selected tolerant soybean varieties for each partner crop, and examined the effects of plant population densities and distribution within the unit area (Abdel-Galil et al. 2014a). Effective management of light availability under intercropping system of soybean with maize is an important practice to reduce the impact of shading on the overall productivity of intercropped soybean plants.

This chapter discusses key aspects of soybean breeding, including genetic breeding and molecular methods. Furthermore, the role of intercropping soybean with maize to reduce soybean mosaic virus (SMV) infection and insect damage was also presented. The intensity of shading and shading tolerance of soybean varieties under its intercropping system with maize were also deliberated. The impact of cropping system on gene expression was presented, as well as genetic insights in the tolerance of some of local soybean varieties to SMV infection.

Genetic Breeding of Soybean

Knowledge of genetic diversity and relationships among soybean genotypes commonly grown in Egypt may play a significant role in breeding programs to improve seed yield, as well as oil and protein content (Metwally et al. 2018, 2019, 2021,

2024). Soybean was cultivated in Egypt only 40 years ago and the yield of local soybean varieties was a resultant of interaction between physiological function and genetically makeup through this period (Abdel-Wahab et al. 2019c, 2024a, 2024b). Physiological function of soybean plant is mainly affected by environmental factors. Consequently, the yield of soybean plant could be increased through modification of its genetic makeup and improving adverse environmental conditions (Abdel-Wahab 2015; Abdel-Wahab et al. 2019c, 2024a, 2024b).

Advancements in genetic engineering have led to the rapid development of new soybean varieties with improved qualities such as higher yield, disease resistance (Abdel-Wahab et al. 2019b, 2020b), and insect tolerance (Naroz et al. 2018, 2019). Despite significant progresses in traditional breeding techniques, there remains a gap in our understanding of the genetic and molecular pathways that confer tolerance to biotic and abiotic stresses (Parmar et al. 2017). With the help of genomics, a deeper understanding of the genetic architecture and molecular mechanisms that govern complex traits exists (Sedivy et al. 2017). Genomics-assisted breeding has shown promising results by speeding up the development of new crop varieties with desired traits such as increased yield and pest tolerance (Ćeran et al. 2024). By implementing suitable cropping systems, this technology has the potential to enhance agricultural productivity and resilience to climate change, ultimately addressing global food security concerns (Ouda and Zohry 2022). This has enabled to establish marker-trait associations for economically important traits, thereby enhancing the efficiency of breeding selection (Bandillo et al. 2023).

Molecular Methods in Soybean Breeding

Advancements in breeding techniques have allowed for more precise selection of desirable traits in soybeans, resulting in increased yields and improved pest tolerance in soybean varieties (Abdel-Wahab et al. 2019c). Advances in sequencing technology, particularly long-read sequencing, have led to the production of improved assembly genomes (Burton et al. 2013; Yang and Huang 2018). The use of molecular methods has accelerated the breeding process, leading to the development of new soybean genotypes with superior traits (Vargas-Almendra et al. 2024). These genotypes offer higher yields, increased resilience to pests and diseases, and adaptability to diverse environmental conditions in cropping systems (Abdel-Wahab et al. 2019b, 2020b). Molecular breeding enables the selection of specific genes that contribute to desirable traits, resulting in more precise and efficient crop improvement (Abdul Aziz and Masmoudi 2024). This technology has revolutionized soybean breeding by allowing breeders to develop improved varieties more quickly than traditional methods (Vargas-Almendra et al. 2024). Additionally, molecular breeding reduces the time and resources needed for field trials, as genotypes can be screened at the molecular level before advancing to field testing (Sinha et al. 2023). This speeds up the breeding process and ensures that only the most promising genotypes are selected for further evaluation and commercialization (Krishnappa et al. 2021).

Intercropping Soybean with Maize to Reduce SMV Infection and Insect Infestation

Reduction the Infection of SMV

SMV is a significant threat to soybean production, and growing resistant varieties is a crucial strategy to reduce virus-related damage (Chu et al. 2021). SMV can negatively affect soybean growth and development under intercropping and sole plantings (Abdel-Wahab et al. 2019b). Intercropping soybean with maize has been shown to reduce SMV infection in soybean leaves more effectively than sole soybean, as reported by Abdel-Wahab et al. (2019b) and Taha and Abdel-Wahab (2024). Maize acts as a natural barrier against the virus and helps regulate sunlight exposure for soybean plants (Abdel-Wahab et al. 2020b). According to Abdel-Wahab et al. (2020b), intercropping resulted in lower SMV infection compared to sole planting, indicating that intercropping could be a promising approach to reduce SMV in soybean cultivation.

Furthermore, different soybean varieties exhibit varying responses to SMV infection. Choosing SMV-tolerant soybean varieties can further enhance yield in intercropping and sole plantings (Abdel-Wahab et al. 2019b). These differences in SMV tolerance may be attributed to genetic variations among the soybean varieties, resulting in variations in leaf pubescence length, pubescence density, and leaf nitrogen content (Abdel-Wahab et al. 2019b). Eid et al. (2023) found that Giza 21 showed tolerance to SMV infection, whereas soybean Giza 82 was the least tolerant variety. While soybean variety Giza 82 was susceptible to SMV infection, soybean varieties Giza 21 and Giza 111 showed lower susceptibility to SMV infection than soybean variety Giza 82 under intercropping and sole plantings. Particularly, Giza 21 and Giza 111 plants showed a significant degree of tolerance to SMV infection because of the dense pubescence on their leaves (Abdel-Wahab et al. 2019b, 2020b) under intercropping and sole plantings (Fig. 6.1). This discouraged aphids from landing and feeding on leaves because of these traits and consequently decreased the spread of SMV. Abdel-Wahab et al. (2019b) demonstrated that soybean varieties Giza 21 and Giza 111 exhibited tolerance to SMV infection. Using these varieties have reduced the reliance on chemical pesticides, leading to increased agricultural output and improved sustainable farming systems.

Additionally, the virus's spread within soybean plants may be restricted by the small core cavity of Giza 111, hindering the movement of photosynthetic organisms (Chiang and Norris 1983). Despite being infected, Giza 111's canopy structure maintains photosynthetic efficiency and provides energy for transport (Silva et al. 2024), ensuring a continuous supply of photosynthetic products. This distinct feature of Giza 111 may enhance its tolerance to viral infections compared to other plant varieties under intercropping and sole plantings.

The effective energy supply within the canopy structure supports the overall health and productivity of the plant. Meanwhile, the tolerance of Giza 21 to SMV may be attributed to the activation of Pd1 and Pd2 genes under intercropping, which are associated with pubescence density (Liu et al. 2020). The Giza 21 variety is

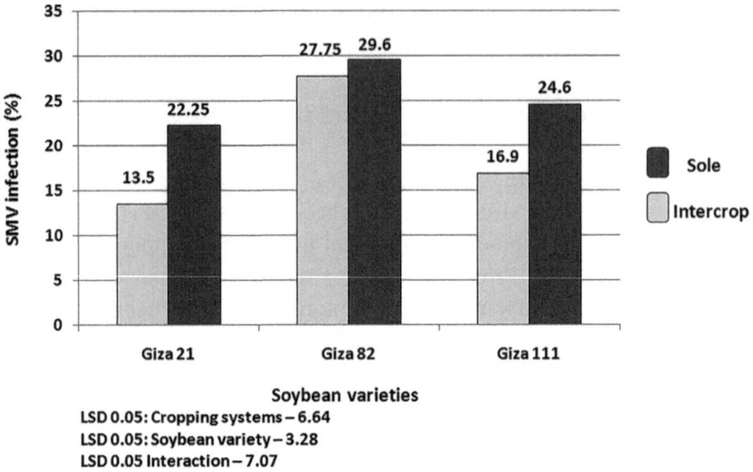

Fig. 6.1 Effect of cropping system, soybean variety, and their interactions on SMV infection. (Source: Abdel-Wahab et al. 2019b)

known for its taller plants, higher pubescence density, and less branching compared to Giza 111 and Giza 82; specifically Giza 21 carries the Pd1 and Pd2 genes (Abdel-Wahab et al. 2019b), which are associated with reduced seed yield and increased plant height (Specht et al. 1985; Abdel-Wahab et al. 2019b).

However, there were no noticeable impacts on viral infection susceptibility for Giza 82 under intercropping and sole planting (Eid et al. 2023). These studies underscore the role of genetic factors in soybean tolerance to SMV and suggest that breeding programs could leverage this knowledge to develop soybean varieties with enhanced SMV tolerance.

Reduction Insects' Infestation

With respect to insect infestation, intercropping legumes with cereals increased insect diversity and increased the number of pest predators, which had the advantage of being a natural pest control method (Hamdalla et al. 2014). Different crop arrangement patterns in intercropping such as row intercropping, strip intercropping, mixed planting, and relay intercropping have been effective for insect management (Kugbe et al. 2018). Abdel-Wahab et al. (2020a) discovered that mixed intercropping soybeans with maize effectively reduce aphids, whiteflies, leaf miner flies, and thrips when compared to sole soybean. Whiteflies may necessitate effective pest control measures in soybean fields, especially in warm climates (Abdel-Wahab et al. 2019a). Breeders should consider this factor when developing new soybean varieties (Almeida et al. 2021). By incorporating these innovative techniques with appropriate soybean varieties, farmers can enhance sustainable farming practices. Lourenção et al. (2004) demonstrated that using insect-tolerant soybean varieties can reduce pesticide usage and promote sustainable agriculture.

Different varieties of soybean were tested for its tolerance to insect infestation by Abdel-Wahab et al. (2019a). They found that Crawford soybean variety exhibited high susceptibility to infestation by aphids, whiteflies, leaf miners, and thrips, while Giza 22 and Giza 35 soybean varieties displayed lower susceptibility to these pests under intercropping and sole plantings (Fig. 6.2). Insect tolerance was higher in Giza 22 than in Giza 35. Despite Giza 35 having a high leaf pubescence density and Giza 22 having a medium leaf pubescence density, Giza 22 exhibited higher insect tolerance compared to Giza 35 (Abdel-Wahab et al. 2020a). This might be because Giza 22 had more genes from traditional breeding than Giza 35. The process of stacking several genes into a single genotype in order to combine desired traits through conventional breeding or recombinant DNA technology is known as "gene pyramiding" (Lombardo et al. 2016).

Furthermore, the multi-pest tolerance of Giza 22 can reduce the reliance on chemical insecticides, promoting sustainable agricultural practices (Abdel-Wahab et al. 2020a). This genetic trait makes Giza 22 a valuable choice for farmers seeking to minimize insect damage in their soybean crops. On the other hand, the high chlorophyll content of the Crawford variety made it susceptible to the insects being studied (Abdel-Wahab et al. 2020a). The study showed that intercropping Giza 22 and Giza 35 soybean varieties with maize effectively reduced insect infestation, while there was no significant difference in susceptibility of Crawford to insects between intercropping and sole planting.

Shading Intensity and Yield Attributes

Shading can impact soybean growth and development in intercropping. In the soybean-maize intercropping system, soybean plants experience shading throughout the concurrent growth period with maize (Hussain et al. 2020). Liu et al. (2008)

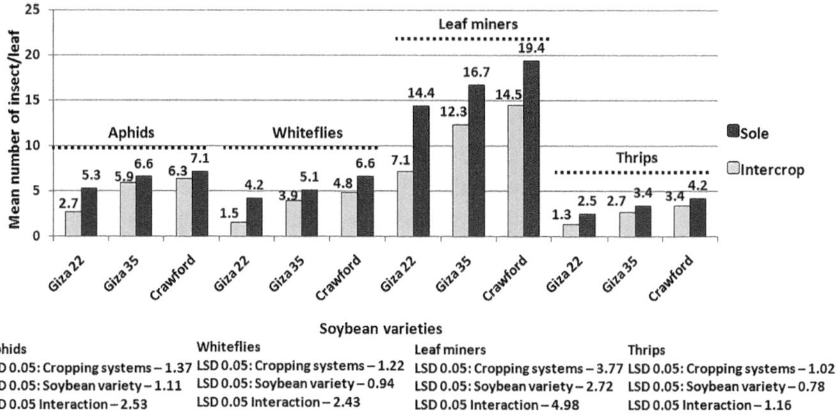

Fig. 6.2 Effect of cropping system, soybean variety, and their interactions on insect infestation. (Source: Abdel-Wahab et al. 2019a)

proposed that modifying soybean planting density is crucial for optimal crop growth, maximizing biomass and grain yield. High-density planting of soybeans can result in stem elongation due to the combined effects of photoperiods and shading, ultimately reducing yields (Qin et al. 2023). When soybeans are intercropped with maize at high densities, the prolonged photoperiods during growth cause soybean plants to grow taller (Hussain et al. 2020). Excessive stem elongation in high-density planting and intercropping can weaken the stem's mechanical strength and increase susceptibility to diseases (Yang et al. 2015). Effective management of light availability in mixed cropping systems is essential, as demonstrated by the significant impact of shading on the overall productivity of intercropped soybean plants.

Abdel-Wahab and Abdel-Rahman (2016) reported that shading intensity was higher in the system of intercropping soybean with maize compared to sole planting by 7.70% in the middle of the canopy and 4.12% in the bottom of the canopy. This outcome is likely due to increased competition for sunlight among intercropped plants compared to those planted individually. With respect to soybean varieties, Giza 111 was superior in shading intensity where it had 93.26% of full shading in the middle of the plant and 96.45% of full shading in the bottom of the plant. Meanwhile, Giza 82 was inferior in shading intensity where it had 92.45% of full shading in the middle of the plant and 95.50% of full shading in the bottom of the canopy. These results may be attributed to variations in the penetration of solar radiation among the studied varieties, which are influenced by their genetic makeup and canopy morphological characteristics (Metwally et al. 2021, 2024).

Abdel-Wahab et al. (2019b) reported that intercropping Giza 111 with maize had higher shading intensity at the middle of the canopy (96.78% of full shading) and the bottom of the plant (98.37% of full shading), as compared with the other treatments. Conversely, sole planting of Giza 82 had lower shading intensity in the middle of the canopy (89.01% of full shading) and the bottom of the canopy (93.55% of full shading), as compared with the other treatments (Table 6.1).

The results in Table 6.1 also show taller plant in intercropping compared to sole planting by 12.06%. The shading effects of intercropping formed unfavorable conditions for soybean plants in the early stages of growth and development. As a result, soybean plants grew taller due to increased levels of plant hormones (Metwally et al. 2021). With respect to soybean varieties, Giza 21 produced the tallest plants; meanwhile, Giza 82 plants were shorter. With respect to the interaction between cropping systems and soybean varieties, intercropping Giza 21 with maize had taller plants, as compared with the other treatments. Conversely, sole planting of Giza 82 had lower plant height, as compared with the other treatments. When soybeans are intercropped with maize, the prolonged photoperiods during growth cause soybean plants to grow taller (Metwally et al. 2021). Moreover, seed yields per plant and per hectare were decreased in intercropping than sole planting by 32.64% and 55.71%, respectively. It seems that early-stage light enrichment and shading can impact the availability of assimilates for developing reproductive structures, influence flowering, and affect the number of flowers and pods that abscise, ultimately impacting seed yields per plant and per hectare (Abdel-Wahab and Abdel-Rahman 2016).

Table 6.1 Effect of cropping system, local soybean varieties, and their interactions on shading intensity and seed yields per plant and per hectare

Cropping system	Soybean variety	Shading intensity in Middle of the canopy (%)	Bottom of the canopy (%)	Plant height (cm)	Seed yield/ plant (g)	Seed yield (ton/ha)
Intercrop	Giza 21	96.17	97.93	117.97	14.38	1.56
	Giza 82	95.89	97.46	100.96	12.37	1.23
	Giza 111	96.78	98.37	104.49	15.29	1.75
Mean		96.28	97.92	107.80	14.01	1.51
Sole	Giza 21	89.42	94.05	104.91	20.30	3.56
	Giza 82	89.01	93.55	87.62	19.11	2.86
	Giza 111	89.75	94.53	96.06	22.99	3.80
Mean		89.39	94.04	96.19	20.80	3.41
Average	Giza 21	92.79	95.99	111.44	17.34	2.56
	Giza 82	92.45	95.50	94.29	15.74	2.04
	Giza 111	93.26	96.45	100.27	19.14	2.77
L.S.D. 0.05 cropping system		1.52	1.41	8.33	4.27	0.88
L.S.D. 0.05 soybean variety		0.73	0.54	6.92	3.34	0.17
L.S.D. 0.05 interaction		1.87	1.68	10.11	5.63	1.13

Source: Abdel-Wahab et al. (2019b)

With respect to soybean varieties, Giza 111 had higher seed yields per plant and per hectare where it had 19.14 g and 2.77 ton/ha, respectively. Also, Giza 21 had higher seed yields per plant and per hectare where it had 17.34 g and 2.56 ton/ha, respectively. Meanwhile, Giza 82 plants had lower seed yields per plant and per hectare where it had 15.74 g and 2.04 ton/ha, respectively. This outcome may be attributed to the extended vegetative growth period of the late-maturing varieties. The late-maturing soybean varieties, Giza 111 and Giza 21, exhibited superior growth and yield compared to the early-maturing Giza 82 variety, as the late varieties were more efficient in utilizing environmental resources (Abdel-Wahab et al. 2019b). With respect to the interaction between cropping systems and soybean varieties, sole planting of Giza 111 had higher seed yield per plant (22.99 g), as compared with the other treatments. Conversely, intercropping Giza 82 with maize had lower seed yield per plant (12.37 g), as compared with the other treatments (Table 6.1). These results indicate that the genetic potential of the studied varieties interacts with environmental resources during the vegetative and reproductive stages, ultimately affecting seed yield (Metwally et al. 2019).

Regarding the interaction between cropping systems and soybean varieties, sole planting of Giza 111 had higher seed yield (3.80 ton/ha), as compared with the other treatments. Conversely, intercropping Giza 82 with maize had lower seed yield per hectare (1.23 ton/ha), as compared with the other treatments. The effects of shading on Giza 21 seem to have positively impacted the PH13H3 protein, causing increased stem elongation and reduced seed yield (Abdel-Wahab et al. 2019b). They also found that the soybean variety Giza 111 had higher seed yields per plant and per hectare in both mixed intercropping and sole plantings compared to other soybean varieties (Table 6.1). This could be due to Giza 111 having a higher potential to counterbalance both intraspecific and interspecific competition, leading to a more positive impact on seed productivity (Metwally et al. 2021).

In another study by Abdel-Wahab et al. (2019a), shading intensity was higher in intercropping system of soybean with maize compared to sole planting by 6.49% in the middle of the canopy and 3.57% in the bottom of the canopy. Intercropping led to increased competition for sunlight among surrounding plants, resulting in higher shading intensity in the middle (6.49%) and bottom (3.57%) of the canopy compared to sole planting (Table 6.2). With respect to soybean varieties, Giza 35 exhibited superior shading intensity with 92.11% of full shading in the middle of the canopy and 95.23% of full shading in the bottom of the canopy. Meanwhile, Crawford was inferior in shading intensity where it had 90.63% of full shading in the middle of the canopy and 94.60% of full shading in the bottom of the canopy (Table 6.2). Genetic variations among these varieties, particularly in their ability to produce larger, denser leaves, can play a crucial role in capturing more sunlight (Abdel-Wahab et al. 2020a).

With respect to the interaction between cropping systems and soybean varieties, intercropping Giza 35 with maize had higher shading intensity in the middle of the canopy (95.03% of full shading) and the bottom of the canopy (96.79% of full shading), as compared with the other treatments. Intercropping Giza 35 with maize can significantly increase shading intensity levels, which may affect the growth and development of plants in the field (Abdel-Wahab et al. 2020a). Conversely, sole

Table 6.2 Effect of cropping system, soybean varieties, and their interactions on shading intensity and seed yields per plant and per hectare

Soybean plant density	Soybean variety	Shading intensity in Middle of the canopy (%)	Shading intensity in Bottom of the canopy (%)	Plant height (cm)	Seed yield/ plant (g)	Seed yield (ton/ha)
Intercrop	Giza 22	94.31	96.47	110.61	17.25	1.16
	Giza 35	95.03	96.79	102.88	15.16	1.03
	Crawford	93.56	96.31	109.60	13.17	0.77
Mean		94.30	96.52	107.69	15.19	0.98
Sole	Giza 22	88.76	93.02	107.13	25.66	3.69
	Giza 35	89.19	93.68	97.20	22.81	3.27
	Crawford	87.71	92.89	101.67	21.77	2.77
Mean		88.55	93.19	102.00	23.41	3.24
Average	Giza 22	91.53	94.74	108.87	21.45	2.42
	Giza 35	92.11	95.23	100.04	18.98	2.15
	Crawford	90.63	94.60	105.63	17.47	1.77
L.S.D. 0.05 cropping system		1.16	0.91	5.87	6.65	1.07
L.S.D. 0.05 soybean variety		0.87	0.32	4.42	3.94	0.35
L.S.D. 0.05 interaction		1.35	1.16	7.11	8.23	1.29

Source: Abdel-Wahab et al. (2019a)

planting of Crawford had lower shading intensity in the middle of the canopy (87.71% of full shading) and the bottom of the canopy (92.89% of full shading), as compared with the other treatments (Table 6.2).

Furthermore, Table 6.2 showed that plant height was taller in intercropping than sole planting by 5.57%. The shading effects of sole planting formed favorable conditions for soybean plants in the early stages of growth and development. As a result, the stability levels of plant hormones prevented soybean plants from growing taller (Metwally et al. 2021).

With respect to soybean varieties, Giza 22 produced taller plants where it had 108.87 cm. Meanwhile, Giza 35 plants were shorter where it had 100.04 cm. When Giza 35 is grown in sole planting, the shorter photoperiods during growth reduced plant height (Metwally et al. 2021). Regarding the interaction between cropping systems and soybean varieties, intercropping Giza 22 with maize had higher value of plant height (110.61 cm), as compared with the other treatments. Conversely, sole planting of Giza 35 had lower plant height (97.20 cm), as compared with the other treatments (Table 6.2). An increase in internode length and number may have contributed to these results, as evidenced by more interspecific competition between the two species for available solar radiation (Abdel-Galil et al. 2014a, b).

Seed yields per plant and per hectare were decreased in intercropping than sole planting by 35.11% and 69.75%, respectively. It seems that shading can impact the availability of assimilates for developing reproductive structures, ultimately impacting seed yields per plant and per hectare (Abdel-Wahab and Abdel-Rahman 2016). With respect to soybean varieties, Giza 22 and Giza 35 had higher seed yields per plant where it had 21.45 and 18.98 g, respectively. Meanwhile, Crawford plants had lower seed yield per plant where it had 17.47 g (Table 6.2). The genetic potential of Giza 22 and Giza 35 shows a positive interaction with environmental resources compared to Crawford during the vegetative and reproductive stages, ultimately impacting seed yield (Abdel-Wahab et al. 2020a).

With respect to the interaction between cropping systems and soybean varieties, Table 6.2 shows that sole planting of Giza 22 had higher seed yields per plant (25.66 g) and per hectare (3.69 ton/ha), compared to the other treatments. Conversely, intercropping Crawford with maize had lower seed yield per plant (13.17 g) and per hectare (0.77 ton/ha), as compared with the other treatments. The presence of shading from maize plants had a significant impact on the growth and development of soybeans in intercropping systems, highlighting the importance of considering light availability for optimal crop management and yield potential.

Impact of Cropping Systems on Gene Expression and Soybean Growth

Investigating gene expression in cropping systems could offer potential strategies for disease control in soybean fields (Chen et al. 2020). To assess genotypic variance in protein production in response to pathogen infection (SMV), SDS-PAGE analysis of total soluble proteins is the way to attain it (Eid et al. 2023). Furthermore, the expression of genes conferring tolerance to the virus was previously analyzed using the SDS-protein PAGE technique (Eid et al. 2023). Abdel-Wahab et al. (2019b) conducted SDS-PAGE analysis of total soluble proteins between four soybean varieties planted under sole and intercropping conditions. They indicated that, under sole soybean planting, the differences observed in Fig. 6.3 were mainly cosmetic changes or band disappearances. Ten distinct bands ranging from 245 to 20 kDa were identified, with one monomorphic band (10.0%) and nine polymorphic bands (90.0%) under sole planting. The Giza 82 soybean variety exhibited a single band, while Giza 21 and Giza 111 displayed 7 and 6 protein bands, respectively. Unique bands at 245, 180, 135, and 100 kDa were observed in Giza 21 and at 20, 25, and 35 kDa was observed in Giza 111, with additional bands at 48, 63, and 75 kDa in these two varieties. A 63 kDa band was specific to Giza 82, indicating susceptibility to SMV infection under sole planting.

The absence of the protein bands at 245, 180, 135, and 100 kDa in Giza 82 and Giza 111 indicates that the proteins responsible for the virus resistance genes Pd1 and Pd2 have high molecular weights, similar to those found in Giza 21. This disparity in protein presence suggests that the tolerance mechanisms in Giza 111 may differ from those in Giza 21 under sole planting. The protein profile variations suggest a defense mechanism against SMV infection. The appearance of new bands in Giza 21 and Giza 111 may indicate enhanced gene expression in response to SMV under sole plantings. These findings highlight the importance of protein synthesis in plant defense mechanisms and adaptation to pathogen challenges (Fig. 6.3).

However, under intercropping system, Fig. 6.4 revealed that the total soluble proteins were separated electrophoretically using the SDS-PAGE technique to find out a genotypic variation in intercropping induced protein upon pathogen infection with SMV. The major variations are expressed as changes in appearance or disappearance of some bands. A total of 10 bands were detected with different molecular weights ranging from 245 to 20 kDa. These protein bands have distributed into 3 monomorphic bands (30.0%) and 7 polymorphic bands (70.0%) under mixed

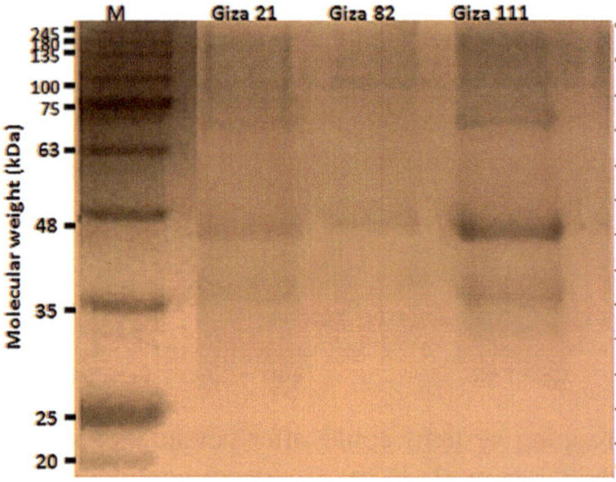

M.W. (kDa)	Giza 21	Giza 82	Giza 111
245	+	-	-
180	+	-	-
135	+	-	-
100	+	-	-
75	+	-	+
63	+	+	+
48	+	-	+
35	-	-	+
25	-	-	+
20	-	-	+
Total	7	1	6

(+) presence of band; (-) absence of band; M.W= Molecular weight.

Fig. 6.3 SDS-protein PAGE banding pattern analysis of soybean varieties infected with SMV under sole planting. (Source: Abdel-Wahab et al. 2019b)

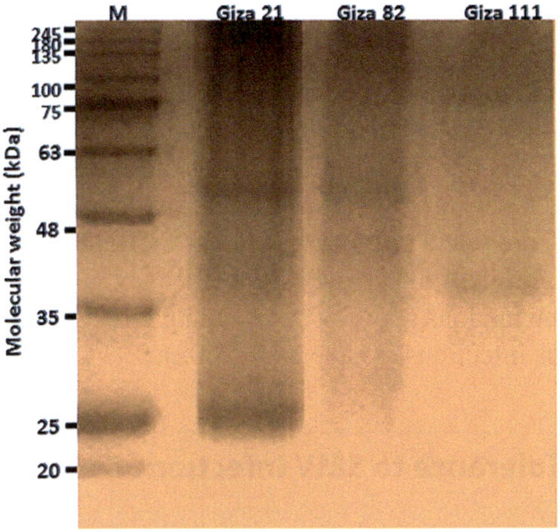

M.W. (kDa)	Giza 21	Giza 82	Giza 111
245	+	-	-
180	+	-	-
135	+	-	-
100	+	-	-
75	-	-	+
63	+	-	-
48	+	+	+
35	+	-	+
25	+	+	+
20	+	+	+
Total	9	3	5

(+) presence of band; (-) absence of band; M.W= Molecular weight.

Fig. 6.4 SDS-protein PAGE banding pattern analysis of soybean varieties infected with SMV under intercropping. (Source: Abdel-Wahab et al. 2019b)

intercropping. Giza 21 and Giza 111 soybean cultivars showed the highest number of protein bands (9 and 5 bands, respectively) and Giza 82 soybean variety had the lowest number of protein bands (3 bands).

Unique bands at 245, 180, 135, 100, and 63 kDa were observed in Giza 21 and 75 kDa was observed in Giza 111, with additional bands at 20, 25, 35, and 48 kDa in these two varieties. Bands at 20, 25, and 48 kDa were specific to Giza 82, indicating susceptibility to SMV infection under intercropping (Fig. 6.4).

Furthermore, the absence of the protein bands at 245, 180, 135, 100, and 63 kDa in Giza 82 and Giza 111 indicates that the proteins responsible for the virus resistance genes Pd1 and Pd2 have high molecular weights, similar to those found in Giza 21 (Fig. 6.4). This disparity in protein presence suggests that the tolerance mechanisms in Giza 111 may differ from those in Giza 21 under intercropping. The protein profile variations suggest a defense mechanism against SMV infection. The appearance of new bands in Giza 21 and Giza 111 may indicate enhanced gene expression in response to biotic stress. These findings highlight the importance of protein synthesis in plant defense mechanisms and adaptation to pathogen challenges (Heath 2000). The expression of genes that confer SMV tolerance increased as a result of intercropping (Abdel-Wahab et al. 2020a). In a study by Abdel-Wahab et al. (2019b), Giza 21 showed lower levels of SMV infection under intercropping and sole plantings. The soybean Giza 21 variety is likely to possess resistance genes for SMV.

It has been reported that soybean leaves suffered oxidative damage from increased light intensity, which aided the virus's transmission (Shang et al. 2023). When exposed to strong light, infected soybeans had a reduction in defense gene expression, which lessened oxidative damage (Shang et al. 2023).

Furthermore, poor soybean growth was caused by a disruption in the plant-pathogen interaction pathway at low light levels. Similar results align with the findings of Zhang et al. (2019), who discovered that intercropped soybeans had significantly lower rates of SMV infections, compared to sole crops. Soybeans adjust their resource allocation between growth and defense to cope with low light intensity in intercropping systems (Abdel-Wahab et al. 2020b). The Pd1 and Pd2 genes are essential for breeding programs aiming to enhance soybean production and environmental resilience (Pfeiffer et al. 2003). Previously, it was known that single dominant alleles in soybeans provided genetic resistance to RNA virus (SMV), while Pd1 and Pd2 genes controlled higher pubescence density (Pfeiffer et al. 2003). The complex molecular interactions between the invading virus and the host plant result in systemic plant virus infections (Maule et al. 2002).

Genetic Insights into Giza 21 Tolerance to SMV Infection in Intercropping

The activation of Pd1 and Pd2 genes, which are associated with pubescence density and stress tolerance in soybeans (Liu et al. 2020), may play a key role in the tolerance of Giza 21 to SMV under low light conditions. The high pubescence density in Giza 21 likely contributes to reduced SMV transmission, with Pd1 and Pd2 genes being activated in low light to enhance resistance. The phytochrome signaling pathway is believed to be responsible for the activation of Pd1 and Pd2 genes in low light, promoting plant growth and development (Qiu et al. 2023). Thus, this pathway can be essential for regulating plant responses to light quality and quantity, ultimately influencing SMV tolerance in Giza 21 in intercropping (Fig. 6.5).

Genetic and Physiological Insights into Giza 22 Tolerance to Insect Infestation in Intercropping

Tolerance entails combining multiple insect-resistance genes in a single crop variety to offer comprehensive protection against pests (Nagaraj et al. 2024). Pyramiding gene, which involves combining multiple main resistance genes, was found to

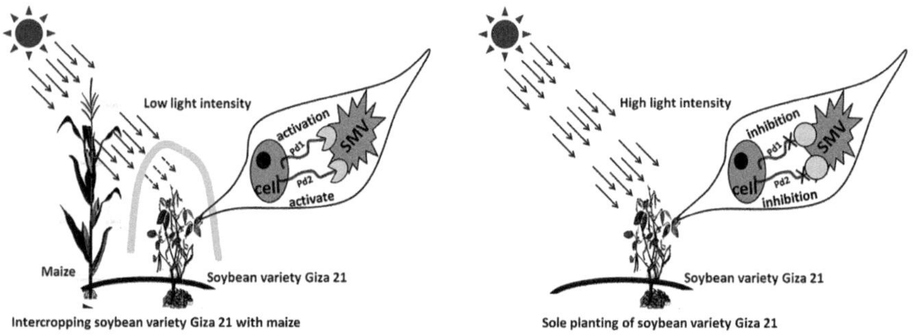

Fig. 6.5 The relationship of Pd1 and Pd2 genes of soybean variety Giza 21 against SMV infection under intercropping and sole plantings

enhance tolerance (Bansal et al. 2013). Pyramiding genes have been successful in managing various insect pests in soybeans (MacIntosh 2019). Pyramiding gene across biotic stresses is the perfect way to increase host-plant resistance (Ramalingam et al. 2020).

The soybean variety Giza 22, likely a pyramided variety, exhibited a significant decrease in aphids, whiteflies, thrips, and leaf miner attacks under intercropping, leading to higher seed yield compared to variety Giza 35 (Abdel-Wahab et al. 2020a). In response to increased shading, the pyramiding genes of Giza 22 that are involved in photosynthesis can be enhanced (Abdel-Wahab et al. 2019b). Jing et al. (2023) found that genes related to photosynthesis were upregulated with increasing shading levels. This indicates that soybean plants may have a mechanism to improve photosynthetic efficiency when exposed to low light conditions (Metwally et al. 2021).

The anatomical characteristics of leaves in soybean varieties indicate that Giza 22 exhibited higher osmotic pressure in the phloem compared to the other varieties when grown under intercropping conditions (Fig. 6.6). The plants' adaptation to low light levels for enhancing photosynthetic efficiency is likely the reason for this upregulation, which could have a positive effect on the accumulation of metabolites from photosynthesis (Abdel-Wahab and Abdel-Rahman 2016).

Consequently, it is expected that this response will help maintain the high osmotic pressure in the phloem or water balance, leading insects to feed on xylem sap (Spiller et al. 1990). In cases where the host plant is unsuitable, whiteflies may limit their phloem consumption and instead feed on xylem for extended periods (Milenovic et al. 2019). These findings reveal that the tolerance of Giza 22 to the Milanovic studied insects can be due to pyramiding genes. Whether in intercropping or sole planting, all insects consume more phloem sap than xylem sap, as shown by the anatomical characteristics of the leaves of the soybean varieties Crawford and Giza 35 (Fig. 6.6). This suggests that some soybean varieties could be more susceptible to insects that feed on phloem.

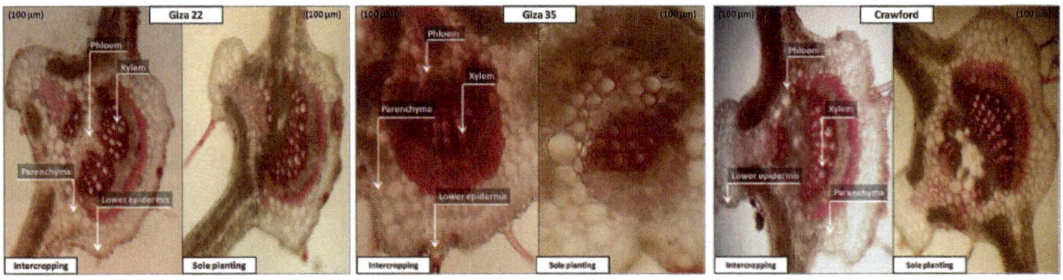

Fig. 6.6 Anatomical characteristics of leaves in soybean varieties under intercropping and sole plantings

Conclusions and Directions for Future Study

Breeding programs that aim to develop tolerance to pest and shading stresses will be successful with genetic diversity. Future research should focus on identifying specific genes that play a role in stress tolerance in plants and exploring different gene editing techniques to enhance these traits. Utilizing the Pd1 and Pd2 alleles with gene pyramids can decrease dependency on chemical pesticides, limit crop damage, and ultimately increase yields. This approach can help develop tolerance to pest damage and shading challenges through genetic modification.

References

Abdel-Galil AM, Abdel-Wahab TI, Abdel-Wahab SI (2014a) Productivity of four soybean varieties as affected by intercropping and corn planting geometry. Soybean Res 12(1):36–58

Abdel-Galil AM, Abdel-Wahab SI, Abdel-Wahab TI (2014b) Compatibility of some maize and soybean varieties for intercropping under sandy soil conditions. Soybean Res 12(special issue):22–44

Abdel-Wahab EI (2015) Genetic and molecular studies for characters related to cotton leaf worm resistance in soybean. PhD thesis, Genetics Department, Faculty of Agriculture, Ain Shams University, Egypt

Abdel-Wahab TI, Abdel-Rahman RA (2016) Response of some soybean cultivars to low light intensity under different intercropping patterns with maize. Int J Appl Agric Sci 2(2):21–31

Abdel-Wahab TI, Abdel-Wahab EI, Taha AM, Adel MM, Hussein HM (2019a) Varietal response of soybean to applied irrigation water and insects incidence under different intercropping systems with maize. Res Crops 20(Issue Suppl):S1–S25

Abdel-Wahab SI, Abdel-Wahab EI, Taha AM, Saied SM, Naroz MH (2019b) Evaluation of intercropped soybean cultivars with corn for water consumption and soybean mosaic virus infection under different soybean plant densities. Res Crops 20(Issue Suppl):S26–S46

Abdel-Wahab EI, Naroz MH, Abd El-Rahman SF (2019c) Potential of some soybean varieties for resistance to lima bean pod borer (*Etiella zinckenella*) under field conditions. Res Crops 20(2):389–398

Abdel-Wahab TI, Abdel-Wahab EI, Taha AM, Adel MM, Hussein HM (2020a) Water consumptive use and insects incidence of three soybean cultivars under different intercropping systems with maize. Soybean Res 18(1):01–30

Abdel-Wahab SI, Abdel-Wahab EI, Taha AM, Saied SM, Naroz MH (2020b) Water consumptive use and soybean mosaic virus infection in intercropped three soybean cultivars with maize under different soybean plant densities. Soybean Res 18(1):31–59

Abdel-Wahab TI, Abdel-Wahab SI, Abdel-Wahab EI (2024a) Chapter 15: biological engineering and its relationship to nematode resistance. In: Chaudhary KK, Meghvansi MK, Siddiqui S (eds) Sustainable management of nematodes in agriculture, vol. 2: role of microbes—assisted strategies, sustainability in plant and crop protection. Springer, Cham, pp 383–408

Abdel-Wahab EI, Mohamed MKA, Baheeg MA, Abdel-Rahman SF, Naroz MH (2024b) Response of some soybean genotypes to insect infestation under three mineral nitrogen fertilizer rates. Agric Sci Dig 44(1):122–138

Abdul Aziz M, Masmoudi K (2024) Molecular breakthroughs in modern plant breeding techniques. Hortic Plant J 11(1):15–41

Alghamdi S, Khan M, El- Harty E, Ammar M, Farooq M, Migdadi H (2018) Comparative phytochemical profiling of different soybean (*Glycine max* (L.) Merr) genotypes using GC–MS. Saudi J Biol Sci 25(1):15–21

Almeida MF, Tavares CS, Araújo EO, Picanço MC, Oliveira EE, Pereira EJG (2021) Plant resistance in some modern soybean varieties may favor population growth and modify the stylet penetration of *Bemisia tabaci* (Hemiptera: Aleyrodidae). J Econ Entomol 114(2):970–978

Bandillo NB, Jarquin D, Posadas LG, Lorenz AJ, Graef GL (2023) Genomic selection performs as effectively as phenotypic selection for increasing seed yield in soybean. Plant Genome 16(1):e20285

Bansal R, Jun TH, Mian MAR, Michel AP (2013) Developing host-plant resistance for hemipteran soybean pests: lessons from soybean aphid and stink bugs, soybean-pest resistance. Intech Open. https://doi.org/10.5772/54597

Beckendorf EA, Catangui MA, Riedell WE (2008) Soybean aphid feeding injury and soybean yield, yield components, and seed composition. Agron J 100:237–246

Burton JN, Adey A, Patwardhan RP, Qiu R, Kitzman JO, Shendure J (2013) Chromosome-scale scaffolding of de novo genome assemblies based on chromatin interactions. Nat Biotechnol 31:1119–1125

Ćeran M, Miladinović D, Đorđević V, Trkulja D, Radanović A, Glogovac S (2024) Genomics-assisted speed breeding for crop improvement: present and future. Front Sustain Food Syst 8:1383302

Chen F, Zhou W, Yin H, Luo X, Chen W, Liu X, Wang X, Meng Y, Feng L, Qin Y, Zhang C, Yang F, Yong T, Wang X, Liu J, Du J, Liu W, Yang W, Shu K (2020) Shading of the mother plant during seed development promotes subsequent seed germination in soybean. J Exp Bot 71(6):2072–2084

Chiang HS, Norris DM (1983) Physiological and anatomical stem parameters of soybean resistance to agromyzid beanflies. Entomol Exp Appl 33:203–212

Chu J, Li W, Piao D, Kong Y, Du H, Lin F, Li X, Zhang C (2021) Mining of a major QTL and novel genes conferring resistance to SC3 and SC7 strains in soybean. Plant Breed 140(5):851–859

Eid MA, Momeh GN, El-Shanshoury AR, Allam NG, Gaafar RM (2023) Comprehensive analysis of soybean cultivars' response to SMV infection: genotypic association, molecular characterization, and defense gene expressions. J Genet Eng Biotechnol 21(1):1–11

El-Hamidi M, Zaher FA (2018) Production of vegetable oils in the world and in Egypt: an overview. Bull Natl Res Cent 42:19

FAO (2023) Food and Agriculture Organization of the United Nations FAOSTAT. https://www.fao.org/faostat/en/#data/QCL

Gent DH, Schwartz HR, Khosla R (2004) Distribution and incidence of IYSV in Colorado and its relation to onion plant population and yield. Plant Dis 88:446–452

Gibson L, Benson G (2005) Origin, history, and uses of soybean (*Glycine max*). Iowa State University, Department of Agronomy, USA

Gouge DH, Way MO, Knutson A, Cronhom G, Patrick C (1999) Managing soybean insects. Texas A&M AgriLife Extension Service B-1501, p 36

Gulluoglu K, Arioglu H, Kurt C (2010) Field evaluation of soybean cultivars for resistance to whitefly (*Bemisia tabaci* Genn.) infestations. Afr J Agric Res 5(7):555–560

Hamdalla WA, Shalably EM, Dawood RA, Zohry AA (2014) Effect of cowpea with maize intercropping on yield and its components. Int J Biol Vet Agric Food Eng 8(11):1170–1176

Hartman GL, West ED, Herman TK (2011) Crops that feed the world 2. Soybean—worldwide production, use, and constraints caused by pathogens and pests. Food Secur 3:5–17

Heath MC (2000) Nonhost resistance and nonspecific plant defenses. Curr Opin Plant Biol 3:315–319

Higley LG, Boethel DJ (eds) (1994) Handbook of soybean insect pests. Entomol Soc Amer, Lanham, pp 15–116

Hill JH, Whitham SA (2014) Control of virus diseases in soybeans. Adv Virus Res 90:355–390

Horwith B (1985) A role for intercropping in modern agriculture. Bioscience 35(5):286–291

Hussain S, Liu T, Iqbal N, Brestic M, Pang T, Mumtaz M et al (2020) Effects of lignin, cellulose, hemicellulose, sucrose and monosaccharide carbohydrates on soybean physical stem strength and yield in intercropping. Photochem Photobiol Sci 19:462–472. https://doi.org/10.1039/cspp00369j

Jing X, Chen P, Jin X, Lei J, Wang L, Chai S, Yang X (2023) Physiological, photosynthetic, and transcriptomics insights into the influence of shading on leafy sweet potato. Genes (Basel) 14(12):2112

Krishnan HB, Jez JM (2018) Review: the promise and limits for enhancing sulfur-containing amino acid content of soybean seed. Plant Sci 272:14–21

Krishnappa G, Savadi S, Tyagi BS, Singh SK, Mamrutha HM, Kumar S, Mishra CN, Khan H, Gangadhara K, Uday G, Singh G, Singh GP (2021) Integrated genomic selection for rapid improvement of crops. Genomics 113(3):1070–1086

Kugbe XJ, Yaro RN, Soyel JK, Kofi ES, Ghaney P (2018) Role of intercropping in modern agriculture and sustainability: a review. Br J Sci 16(2):67–75

Liu XB, Jin J, Wang GH, Herbert SJ (2008) Soybean yield physiology and development of high-yielding practices in Northeast China. Field Crop Res 105:157–171

Liu S, Fan L, Liu Z, Yang X, Zhang Z, Duan Z, Liang Q, Imran M, Zhang M, Tian Z (2020) A Pd1–Ps–P1 feedback loop controls pubescence density in soybean. Mol Plant 13(12):1768–1783

Lombardo L, Coppola G, Zelasco S (2016) New technologies for insect-resistant and herbicide-tolerant plants. Trends Biotechnol 34(1):49–57

Lourenção AL, Braga N, de Miranda M, Reco P, Valle G, Pereira J (2004) Insect damage and powdery mildew severity in soybean cultivars and lines. Sci Agric (Piracicaba, Braz) 61(6):584–592

MacIntosh GC (2019) Gene pyramids and the balancing act of keeping pests at bay. J Exp Bot 70(18):4591–4593

Maule A, Leh V, Lederer C (2002) The dialogue between viruses and hosts in compatible interactions. Curr Opin Plant Biol 5:279–284

Metwally AA, Safina SA, Abdel-Wahab TI, Abdel-Wahab SI, Hefny YAA (2018) Productivity of soybean varieties under intercropping culture with corn in Egypt. Soybean Res 16(1&2):63–77

Metwally AA, Safina SA, Abdel-Wahab TI, Abdel-Wahab SI (2019) Growing of twenty soybean genotypes in solid and intercropping systems with corn. Res Crops 20(Issue Suppl):S47–S57

Metwally AA, Safina SA, Abdel-Wahab EI, Abdel-Wahab SI, Abdel-Wahab TI (2021) Screening thirty soybean genotypes under solid and intercropping plantings in Egypt. J Crop Sci Biotechnol 24:203–220

Metwally AA, Saleh NA, Abdel-Wahab EI, Hefny YAA (2024) Intercropping some soybean genotypes with corn for producing soybean seeds, quality and land equivalent ratios. Asian J Biol Sci 17(3):331–350. https://ajbs.scione.com/cms/abstract.php?id=865

Milenovic M, Wosula EN, Rapisarda C, Legg JP (2019) Impact of host plant species and whitefly species on feeding behavior of *Bemisia tabaci*. Front Plant Sci 10:1

Murgianto F, Hidaya P (2017) Whitefly infestation and economic comparison of two different pest control methods on soybean production. J Agron Sci 5(2):110–115

Nagaraj S, Rajasekaran R, Palaniappan J, Rangasamy S, Narayanasamy C, Narayanan B (2024) Emerging technological developments to address pest resistance in Bt cotton. J Cotton Res 7:30

Naroz MH, Abdel-Wahab EI, Saied SM, Taha AM, Abdel-Wahab SI (2018) Response of soybean insect communities to different plant densities of some soybean cultivars under two cropping systems. Soybean Res 16(1&2):78–107

Naroz MH, Abd El-Rahman SF, Abdel-Wahab EI (2019) The relationship between agronomic characters of certain soybean varieties and infestation resistance of *Etiella zinckenella* (Lepidoptera: Pyralidae) under natural conditions. J Entomol Zool Stud 7(2):69–77

Ouda S, Zohry A (2022) Water-smart practices to manage water scarcity. In: Climate-smart agriculture: reducing food insecurity. Springer. 978-3-030-93110-0

Parmar N, Singh KH, Sharma D, Singh L, Kumar P, Nanjundan J, Khan YJ, Chauhan DK, Thakur AK (2017) Genetic engineering strategies for biotic and abiotic stress tolerance and quality enhancement in horticultural crops: a comprehensive review. 3 Biotech 7(4):239

Pfeiffer TW, Peyyala R, Ren Q, Ghabrial SA (2003) Increased soybean pubescence density: yield and soybean mosaic virus resistance effects. Crop Sci 43:2071–2076

Qin C, Li Y, Li D et al (2023) *PH13* improves soybean shade traits and enhances yield for high-density planting at high latitudes. Nat Commun 14:6813. https://doi.org/10.1038/s41467-023-42608-5

Qiu X, Sun G, Liu F, Hu W (2023) Functions of plant phytochrome signaling pathways in adaptation to diverse stresses. Int J Mol Sci 24(17):13201

Ramalingam J, Raveendra C, Savitha P, Vidya V, Chaithra TL, Velprabakaran S, Saraswathi R, Ramanathan A, Arumugam Pillai MP, Arumugachamy S, Vanniarajan C (2020) Gene pyramiding for achieving enhanced resistance to bacterial blight, blast, and sheath blight diseases in rice. Front Plant Sci 11:591457

Rosenthal JP, Kotanen PM (1994) Terrestrial plant tolerance to herbivory. Trends Ecol Evol 9:145–148

Sedivy EJ, Wu F, Hanzawa Y (2017) Soybean domestication: the origin, genetic architecture and molecular bases. New Phytol 214(2):539–553

Shang J, Zhao L, Yang X, Qi X, Yu J, Du J, Li K, He C, Wang W, Yang W (2023) Soybean balanced the growth and defense in response to SMV infection under different light intensities. Front Plant Sci 14:1150870

Silva LB, Almeida LC, Silva MCE, Brito RR, Miranda RD, Rodrigues RH, Santos CM, Ribeiro GD, Morais JV, Silva AF, Mayard H, Oliveira TR, Costa Lima VM, Silva Fontes LD, Rocha NS, Jesus Lacerda JD, Pavan BE (2024) Change in the physiological aspects of soybean caused by infestation by *Bemisia tabaci* MEAM1. Agronomy 14(3):481

Sinha D, Maurya AK, Abdi G, Majeed M, Agarwal R, Mukherjee R, Ganguly S, Aziz R, Bhatia M, Majgaonkar A, Seal S, Das M, Banerjee S, Chowdhury S, Adeyemi SB, Chen T (2023) Integrated genomic selection for accelerating breeding programs of climate-smart cereals. Gene 14(7):1484

Specht JE, Williams JH, Pearson DR (1985) Near-isogenic analyses of soybean pubescence genes. Crop Sci 25:92–96

Spiller NJ, Koenders L, Tjallingii WF (1990) Xylem ingestion by aphids—a strategy for maintaining water balance. Entomol Exp Appl 55:101–104

Stephen M (2009) Does intercropping have a role in modern agriculture? J Soil Water Conserv 64(2):55A–57A

Taha AM, Abdel-Wahab EI (2024) Water used by soybean and maize intercropping patterns as affected by different soybean cultivars and plant population. J Soli Crop 34(1):29–39

Todd JC, Stewart LR, Redinbaugh MG, Wilson JR (2022) Soybean aphid (Hemiptera: Aphididae) feeding behavior is largely unchanged by soybean mosaic virus but significantly altered by the beetle-transmitted bean pod mottle virus. J Econ Entomol 115(4):1059–1068

Vargas-Almendra A, Ruiz-Medrano R, Núñez-Muñoz LA, Ramírez-Pool JA, Calderón-Pérez B, Xoconostle-Cázares B (2024) Advances in soybean genetic improvement. Plan Theory 13:3073

Vieira CC, Chen P (2021) The numbers game of soybean breeding in the United States. Crop Breed Appl Biotechnol 21:387521–387531

Wang XB, Fang YH, Lin SZ, Zhang LR, Wang HD (1994) A study on the damage and economic threshold of the soybean aphid at the seedling stage. Plant Prot 20:12–13

Yang J, Huang X (2018) A new high-quality genome sequence in soybean. Sci China Life Sci 61:1604–1605

Yang F, Wang X, Liao D, Lu F, Gao R, Liu W, Yang W (2015) Yield response to different planting geometries in maize-soybean relay strip intercropping systems. Agron J 107:296–304. https://doi.org/10.2134/agronj14.0263

Zhang L, Shang J, Wang W, Du J, Li K, Wu X, Yu L, Liu C, Khaskheli MI, Yang W (2019) Comparison of transcriptome differences in soybean response to soybean mosaic virus under normal light and in the shade. Viruses 11(9):793

Part III

Untraditional Approaches to Maximize Agricultural Production

7 Opportunities and Challenges of Greenhouses and Soilless Agricultural Systems in Combating Food Insecurity

M. A. Mahmoud and M. A. M. Moursy

Abstract

Water shortages pose a significant challenge to global food security, particularly in arid and semiarid regions. Greenhouses and soilless agricultural systems represent a promising approach to addressing food production challenges and promoting sustainable agriculture. They provide opportunities to enhance resource efficiency, increase crop yields, and improve food accessibility. Greenhouses offer precise control over environmental factors including extreme weather events, facilitating year-round cultivation and reducing greenhouse gas emissions associated with conventional farming methods. Soilless systems like hydroponics and aeroponics conserve water and land resources while maximizing crop productivity. Despite their advantages, these systems face barriers such as high initial investment costs, energy use, technological and management complexity, and market accessibility issues. Overcoming these challenges is crucial for their widespread adoption and long-term sustainability. The objective of this work is to highlight the key concerns and emphasize the importance of greenhouses andsoilless systems in contributing to food security. Furthermore, the revision of previous scientific efforts to innovate new technologies for overcoming obstacles to their widespread adoption and to develop resilient and sustainable agroecosystems for future generations was also provided.

M. A. Mahmoud (✉)
Water Requirements and Field Irrigation Research Department, Soils, Water and Environment, Research Institute, Agricultural Research Center,
Giza, Egypt

M. A. M. Moursy
Water Management Research Institute, National Water Research Center,
El-Qanater El-Khairiya, Egypt

A. E.-H. Zohry, S. Ouda (eds.), *Resilient Agroecosystems*, Sustainability Sciences in Asia and Africa, https://doi.org/10.1007/978-981-96-9630-7_7

Keywords

Controlled agriculture · Hydroponics · Aeroponics · Aquaponics · Food security · Water productivity

Introduction

The agricultural industry is essential to both global food security and economic stability. Food security is a key global issue that encompasses the availability, stability, utilization, and accessibility of food sources. A growing population increases the need for food, yet declining natural factors like water scarcity and climate change make agriculture less productive. The local population's food needs are frequently not met by traditional farming practices, which increases hunger and malnutrition (FAO 2017). Economic inequality, environmental deterioration, and geopolitical instability are interconnected challenges that make food security a significant issue, even with breakthroughs in agriculture and food delivery (Podkolzina et al. 2023). To solve these issues and guarantee a steady supply of food, modern agricultural techniques are crucial.

There is growing recognition that climate change and its syndrome, namely, climate variability, poses a serious threat to agricultural systems and exacerbates food poverty globally, especially in areas already at risk (Mahmoud 2019). Rising temperatures, altered precipitation patterns, and an increase in the incidence of extreme weather events, which disrupt traditional farming systems, all pose threats to food supply (Toromade et al. 2024). Climate variability and water scarcity pose significant concerns to global food security, particularly in arid and semiarid regions. These concerns jeopardize crop yield and food supply, necessitating creative solutions to maintain and boost agricultural output.

Greenhouses and soilless agriculture systems are important alternatives to traditional farming practices. These systems provide significant benefits in minimizing the negative consequences of climate change and water scarcity by boosting agricultural productivities, improving food accessibility, and increasing resource efficiency (Dalai et al. 2020). Greenhouses have become essential components of contemporary agriculture, providing controlled settings that boost crop output and extend growing seasons (McCartney and Lefsrud 2018). By regulating temperature, humidity, and light, greenhouse systems enable year-round cultivation and an increase crop yields (Ghoulem et al. 2019). They address many challenges faced in conventional agriculture, such as weather variability, resource scarcity, and pest management. While greenhouses offer numerous benefits for crop production and resources management, they also have disadvantages that must be carefully considered. These drawbacks can impact the overall efficiency and sustainability of greenhouse operations, such as high initial investment costs (Maraveas 2019; Hopwood et al. 2024), high energy consumption (Jin et al. 2023; Hopwood et al. 2024), technical expertise and maintenance requirements (Vox et al. 2010; Ghani et al. 2019), environmental impacts (Ghani et al. 2019; Sayadi-Gmada et al. 2019), limited crop diversity (McCartney and Lefsrud 2018; Shamshiri et al. 2018), and economic viability and market access (Gatahi 2020).

Furthermore, soilless agriculture, namely, hydroponics and aeroponics, can significantly increase crop productivity and resource efficiency, making it ideal for areas with limited water resources and arable land. Hydroponics and aeroponics handle crucial challenges like soil degradation and water scarcity. By utilizing nutrient-rich solutions or other media, these systems provide a controlled environment for plant growth, emerging as viable alternatives to conventional soil-based farming (Hussain et al. 2014; AlShrouf 2017; Fussy and Papenbrock 2022; Mir et al. 2022; Sousa et al. 2024).

In this chapter, we presented the opportunities and challenges of greenhouses and soilless agricultural systems in enhancing crop productivity to cope with water shortages and climate variability, ultimately contributing to increase food security.

Food Security Problems

The capacity for everyone to have access physically, socially, and economically to enough safe, nourishing food to satisfy their dietary needs and preferences for an active and healthy life is known as food security (FAO 2020). In 2020, about one-third of the world's population (2.37 billion) lacked access to enough food. Both developed and developing countries are currently facing urgent and interconnected challenges that undermine efforts to secure food for the growing population and restore the environment and natural resources (Mrabet 2023). There are numerous issues affecting food security, including economic inequality, environmental consequences, and policy concerns (World Bank 2024). Research was implemented to highlight key concerns and investigate viable solutions for increasing global food security. Economic inequities present a substantial impediment to food security. In many locations, poverty hinders access to food because people and families cannot buy enough nutritious food (Sen 1999). Recent studies suggest that income disparity worsens food insecurity, with poorer people disproportionately affected by swings in food costs and economic downturns (FAO et al. 2021). For instance, the COVID-19 epidemic has exposed and amplified these differences, resulting in increasing food insecurity in several low-income nations (Daniels and Morton 2023). Environmental problems such as climate variability and resource depletion are also serious risks to food security. Because of changing precipitation patterns, an increase in the frequency of extreme weather events, and rising temperatures, climate change has an impact on agricultural productivity (Toromade et al. 2024). These changes have the potential to reduce crop yields, disrupt food supply chains, and enhance food system vulnerability (IPCC 2021). Furthermore, resource depletion, such as soil degradation and water scarcity, undermines food production and security (Mrabet 2023).

Food production and delivery networks are disrupted by geopolitical instability and conflicts, which makes food poverty worse. Conflicts can destroy agricultural infrastructure, displace populations, and disrupt trade routes (FAO 2020). Areas

Table 7.1 The global food security situation, the proportion of undernourished individuals, and their numbers, 2005–2023

	2005		2010		2015		2020[a]		2023[a]	
	%	Millions	%	Millions	%	Millions	%	Millions	%	Millions
Africa	19.9	184.4	15.9	167.4	16.0	192.1	18.8	256.5	20.4	298.4
Asia	13.9	552.6	9.3	391.4	7.5	336.3	7.8	361.7	8.1	384.5
Oceania	6.9	2.3	7.3	2.7	6.9	2.8	6.7	2.9	7.3	3.3
Latin America and Caribbean	8.9	49.8	6.1	36.0	5.2	32.5	6.5	47.2	6.2	57.8
Northern America and Europe	<2.5	n.r.	<2.5	n.r.	<2.5	n.r.	<2.5	n.r.	<2.5	n.r.
World	12.2	798.3	8.7	604.8	7.7	570.2	8.5	669.3	9.1	733.4

Source: FAO (2024)
Notes: *n.r.* not reported, as the prevalence is less than 2.5%
[a]Values are based on the point estimates.

affected by continuous hostilities frequently face significant food shortages and humanitarian disasters (Podkolzina et al. 2023). For example, the conflict in Syria has resulted in considerable agricultural losses and a sharp increase in regional food insecurity (Ibrahim et al. 2024). To address the complex interplay of economic, environmental, and social elements, effective food security requires strong policies and institutions (Mrabet 2023). Improving food security necessitates stronger collaboration among governments, local communities, and international organizations to develop context-specific solutions (Carrasco Azzini et al. 2022). Increasing agricultural productivity through sustainable practices, strengthening food distribution networks, and tackling economic inequities are all critical initiatives. Investments in agricultural research, infrastructure, and education can boost food production and resilience (Mrabet 2023). Furthermore, incorporating climate adaptation methods into agricultural policies can reduce the influence of environmental changes on food security (Toromade et al. 2024). Table 7.1 illustrates the prevalence of food insecurity globally.

Challenges of Climate Change on Agriculture

Crop productivity, food security, and agricultural practices are all impacted by climate change, which presents serious problems for global agriculture. Rising global temperatures have both direct and indirect impacts on crop growth and productivity (Mahmoud 2019). Higher temperatures can accelerate crop development, potentially resulting in lower yields if crops mature before their ideal growth; otherwise, heat stress can damage plant physiology, limit photosynthesis, and affect pollination (Kim and Lee 2023). According to research, certain crops are more sensitive to temperature rises, which may exacerbate yield variability and food poverty (Janni et al. 2024). Changes in precipitation patterns present another key challenge. More frequent droughts and floods can disrupt planting and harvest schedules, reduce soil moisture, and negatively impact agricultural growth (Kim and Lee 2023). Regions dependent on rainfed agriculture are especially vulnerable to these changes, resulting in lower agricultural production and an increased risk of crop failure (Chen et al. 2024). The rising incidence of severe weather conditions, like hurricanes, storms, and heat waves, poses new hazards to agriculture. These catastrophes can directly destroy crops and infrastructure, leading to significant economic losses (Grigorieva et al. 2023). Furthermore, extreme weather can exacerbate soil erosion, nutrient loss, and pest outbreaks, reducing agricultural productivity (Yang et al. 2024).

To alleviate the effects of climate change, several adaptation measures are being investigated, including breeding for climate-resilient crop types, upgrading water management methods, and implementing conservation agriculture practices (Mahmoud 2019). Precision agriculture and the application of climate-smart technologies can also help optimize resources utilization and reduce farm system vulnerability (Mrabet 2023). Mitigating climate change impacts entails lowering greenhouse gases emissions and increasing carbon sequestration in agriculture. Agroforestry, cover cropping, and reduced tillage practices can help store soil carbon and reduce emissions (Tanveer et al. 2019). Policy interventions and support for sustainable agriculture practices are critical for increasing the adoption of these measures (Toromade et al. 2024).

Greenhouses Agricultural Systems

Greenhouse agriculture systems offer innovative methods to increase crop productivity by creating controlled environments that improve growing conditions and protect plants from external weather (Dalai et al. 2020). Compared to open-field agriculture, they provide consistent crop quality and year-round output through the precise management of temperature, humidity, light, and CO_2 (Ahmed et al. 2024). Greenhouse technology has advanced significantly over the years, evolving from simple structures that extend growing seasons to sophisticated systems equipped with climate control technologies (Ghoulem et al. 2019). Innovations have consistently increased efficiency and production under greenhouses systems. Recent improvements in greenhouse technology include the use of automation and data analytics to increase efficiency and lower labor costs (Shamshiri et al. 2018). Smart sensors, climate control systems, and artificial intelligence are enhancing the management of greenhouses (Kavga et al. 2021). Furthermore, research into renewable energy sources and energy-efficient designs is addressing the problem of excessive energy usage (McCartney and Lefsrud 2018; Iddio et al. 2020).

Types of Greenhouses

According to Dalai et al. (2020), greenhouse systems play an important role in modern agriculture, offering regulated settings that promote plant growth and output. They provide various technologies for optimizing plant development by controlling environmental variables such as light, temperature, and humidity. Greenhouses are grouped into numerous categories, including design, application, and technology improvements. Each greenhouse system has unique qualities that cater to certain agricultural needs and environmental conditions (Ghoulem et al. 2019). Figure 7.1 illustrates various types of greenhouses.

Traditional Glass Greenhouses

Traditional glasshouses, also known as glass greenhouses, use transparent materials, such as glass or polycarbonate to permit sunlight into the growth chamber (Maraveas 2019). Glass greenhouses have excellent light penetration, making them ideal for areas with limited sunlight (Fadel et al. 2016). These structures are typically used for high-value crops and ornamental plants. Furthermore, they are appropriate for producing vegetables, flowers, and fruits. The construction consists of glass panels or polycarbonate sheets supported by metal frames (Maraveas 2019). This type of greenhouse allows for increased interior light intensity, higher air infiltration rates, lower interior humidity, and excellent disease prevention quality, which supports robust plant growth and photosynthesis (Dalai et al. 2020). It is also effective at moderating internal temperatures and protecting crops from external weather conditions (Ghoulem et al. 2019). However, they have high energy requirements for heating and cooling, they are vulnerable to extreme weather events, and they have greater initial setup and maintenance expenses compared to alternative systems (Hemming et al. 2018; Maraveas 2019).

Hoop Greenhouses

Hoop houses, also known as high tunnels, are simple structures with a frame of hoops covered by a layer of polyethylene film. Their structure consists of arched metal or plastic hoops covered with a single or double layer of polyethylene film. They are used to extend the growing season and protect crops from environmental stresses, making them ideal for vegetables, herbs, and flowers in temperate climates (Kaiser and Ernst 2012). Hoop greenhouses have lower initial investment and operational costs, as well as more cost-effective and versatile compared to other types of greenhouses. They are suitable for various climates, protect crops from frost, and extend the growing season. However, polyethylene film requires replacement and may be less durable than glass, with reduced light quality compared to glasshouses (Fitzgerald and Hutton 2012; Kaiser and Ernst 2012; Ernst et al. 2020).

High-Tech Controlled Environment Greenhouses

High-tech controlled environment greenhouses use modern technologies to precisely regulate environmental conditions and maximize crop growth. These systems frequently incorporate automation, sensors, and climate control technology (Burchi

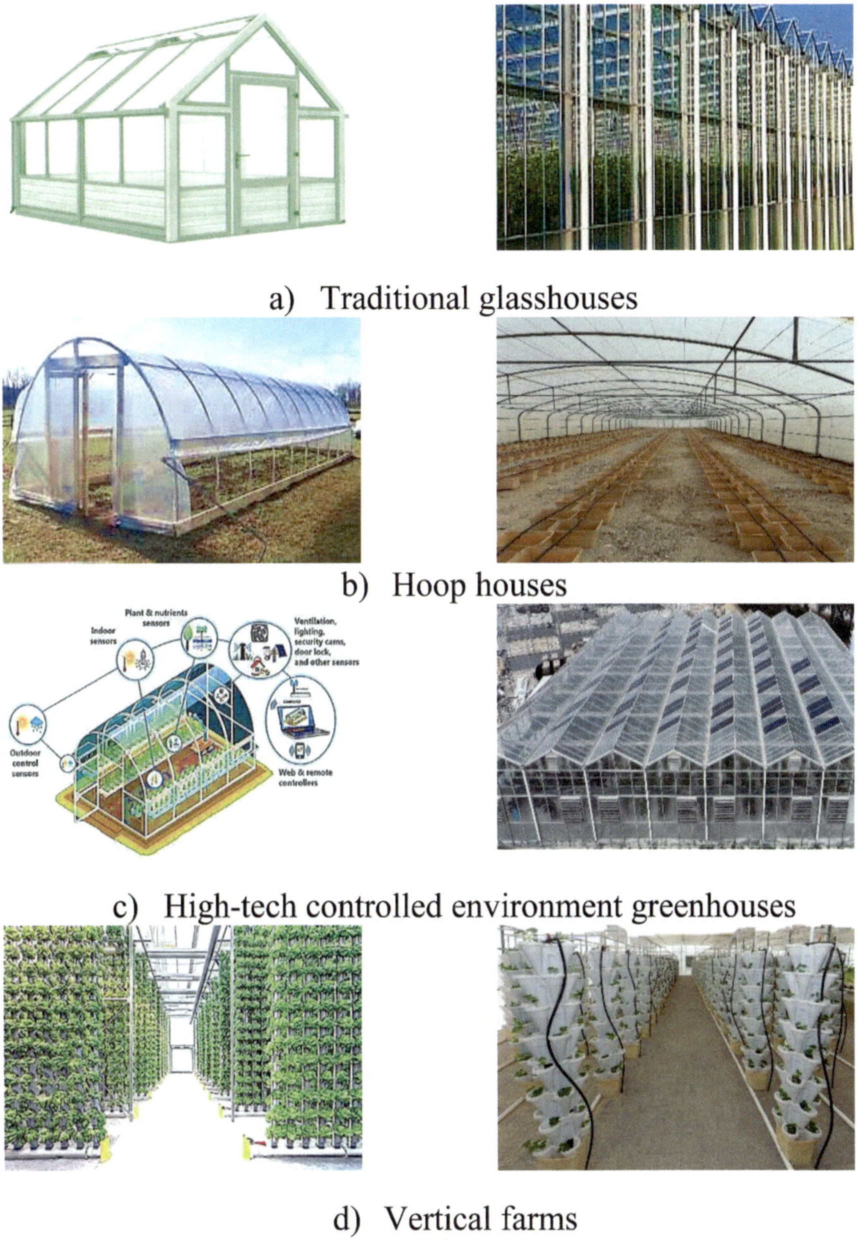

Fig. 7.1 Types of greenhouse systems

et al. 2018). High-tech greenhouses use computerized climate management, irrigation systems, and artificial lighting to maximize crop growth. The structure is often a fully contained environment with automated climate control systems, making it ideal for high-value crops and research applications that require precise control (Ahmed et al. 2024). High-tech greenhouses enable precise control of temperature, humidity, CO_2 levels, and light, as well as efficient water and nutrient utilization. While they require a large initial investment and operational expenditures, they also necessitate professional operators, regular maintenance, and substantial energy usage (Hopwood et al. 2024).

Vertical Greenhouses

Vertical greenhouses are a novel concept involving stacking layers of plants vertically to maximize space usage and increase productivity per square meter (van Delden et al. 2021). Vertical farming can improve resources efficiency and reduce transportation costs; the structure consists of multilayered or stacked systems that often incorporate hydroponics or aeroponics. It is ideal for urban agriculture and high-density crop production because it maximizes space use while reducing the need for long-distance transportation (Despommier 2013). However, there are large initial expenditures, extensive infrastructure needs, and significant energy requirements for lighting and climate management (Jin et al. 2023; Vatistas et al. 2022).

Advantages of Greenhouse Cultivation

Greenhouse systems have become a vital part of contemporary agriculture, promoting sustainable agriculture by conserving water and reducing the demand for chemical inputs. Furthermore, greenhouses can enhance local food security by enabling production near metropolitan areas and offer various benefits in addressing climate change, crop productivity, sustainability, and food security as follows:

- *Extended growing seasons*
- One of the major advantages of greenhouse systems is their capacity to extend agricultural growing seasons by maintaining appropriate temperature and humidity levels, allowing for year-round cultivation and increased food production and availability. In addition, they overcome the constraints imposed by seasonal variations. Greenhouses allow crops to be grown outside their typical growth seasons, improving overall productivity and availability. They also protect plants from severe weather events such as frost, heavy rainfall, and extreme temperatures, lowering the risk of crop failure (Ernst et al. 2020; Hopwood et al. 2024).
- *Increased crop yields*
- Greenhouses help increase agricultural yields by creating an optimal growing environment that promotes plant health and productivity. The regulated settings of greenhouses enable better management of elements such as light, temperature, and nutrients. Greenhouses can control light, temperature, and humidity to fit the

individual needs of different crops, resulting in increased growth rates and yields. Enclosed settings minimize insect and disease exposure, lowering the requirement for chemical interventions while enhancing overall output (Ghoulem et al. 2019; Dalai et al. 2020; Ahmed et al. 2024).

- *Efficient resources use*
- Greenhouses improve the use efficiency of resources by utilizing water and nutrients more effectively than traditional soil-based agriculture. The regulated environment enables the precise application of inputs, reducing waste and encouraging sustainability. Greenhouses can combine modern irrigation methods, such as drip irrigation and hydroponics, to reduce water usage and prevent runoff. Nutrient solutions in hydroponic systems can be carefully monitored and modified, resulting in more efficient fertilizer use and a lower environmental impact (Marcelis et al. 2019; Hopwood et al. 2024).
- *Enhanced control over growing conditions*
- Greenhouse systems offer enhanced ability to manage environmental conditions, allowing for the optimization of factors that influence plant growth. This control can result in optimal growing circumstances, lessening the influence of adverse weather, boosting crop quality, and shortening production time. Automated systems may adjust temperature, humidity, and CO_2 levels to ensure that plants receive the optimum growth conditions. Consistent and optimal growing conditions contribute to higher-quality produce with superior taste, appearance, and nutritional value (Vox et al. 2010; Badji et al. 2022).
- *Reduced environmental impact*

- Greenhouses contribute to more sustainable agricultural practices by reducing the need for pesticides and fertilizers, as well as the carbon footprint correlated with food production and transportation. They also allow for the use of organic growing methods, which reduces pest and disease incidence. Greenhouses can be positioned near urban areas, minimizing the carbon footprint involved with moving food over great distances, making them more efficient in resource usage and potentially lower in emissions than conventional farming (Niggli et al. 2009; Rebolledo-Leiva et al. 2017).

Challenges of Greenhouse Agriculture

Greenhouse systems are renowned for generating controlled environments that enhance agricultural productivity. However, they have drawbacks that can jeopardize their feasibility and sustainability, including high initial costs, significant energy consumption, maintenance requirements, and potential environmental consequences. Energy consumption is a serious concern, particularly in traditional and large-scale greenhouses, where heating and cooling costs can be substantial. Understanding these limitations is essential for stakeholders to make informed decisions about greenhouse adoption and maintenance.

- *High initial investment costs*
- One of the most significant disadvantages of greenhouse systems is the substantial capital required for construction and initial setup. The investment needed for constructing and equipping a greenhouse can be considerable, especially for high-tech controlled environment systems. The cost of building a greenhouse, including materials and labor, can be prohibitive, particularly for advanced systems with automated controls and climate regulation. High-tech greenhouses require expensive equipment such as sensors, climate control systems, and lighting, which can further increase the initial investment (Maraveas 2019; Hopwood et al. 2024).
- *Significant energy consumption*
- Greenhouses often require considerable energy to maintain optimal growing conditions, particularly in regions with extreme temperatures. High energy requirements for climate control (cooling or heating) and lighting systems can lead to high operational costs and environmental concerns (Jin et al. 2023; Hopwood et al. 2024). This energy demand can further increase the initial investment (Maraveas 2019; Hopwood et al. 2024).
- *Technical expertise and maintenance requirements*
- Greenhouse systems require ongoing maintenance to ensure effective operation and longevity. Specialized knowledge is necessary for their operation and maintenance, which can involve significant time and financial resources (Vox et al. 2010; Ghani et al. 2019). Regular cleaning of glass panels or polyethylene film is necessary to maintain light transmission, and repairs are needed to address wear and tear (Badgery-Parker 2015). High-tech greenhouses with complex climate control systems require skilled personnel to operate and maintain the equipment (Vox et al. 2010; Hopwood et al. 2024).
- *Environmental impact*
- While greenhouses can reduce certain environmental impacts, they can also contribute to others. The significant energy requirements for heating, cooling, and lighting can add to a greenhouse's overall carbon footprint, especially if nonrenewable energy sources are used. The use of synthetic materials and nutrient solutions in greenhouses can lead to waste management challenges and potential environmental contamination if not handled properly (Ghani et al. 2019; Sayadi-Gmada et al. 2019).
- *Limited crop diversity*
- Certain greenhouse systems may not be suitable for all types of crops. The specific conditions required by some plants may not be easily accommodated within standard greenhouse designs. Some crops require specific growing conditions that are difficult to achieve within a greenhouse environment, limiting the range of crops that can be effectively cultivated (Vox et al. 2010). Greenhouses designed for certain types of crops may lack flexibility, making it challenging to adapt to different or changing agricultural needs (McCartney and Lefsrud 2018; Shamshiri et al. 2018).
- *Economic viability and market access*

- Greenhouse production faces limited market access for small-scale farmers due to high production costs. Economic viability is further challenged by fluctuations in market prices and competition from outdoor farming, which often has lower production costs (Gatahi 2020).

Soilless Agricultural Systems

Soilless agriculture systems, such as hydroponics, aeroponics, and aquaponics, are innovative ways to modern farming that have grown in popularity as alternatives to traditional soil-based farming. These methods solve various issues connected with traditional soil-based agriculture, such as soil deterioration, limited arable area, and water scarcity. They can improve plant development conditions and address a variety of agricultural challenges by reducing the need for soil (Hussain et al. 2014; Fussy and Papenbrock 2022).

Advantages of Soilless Agriculture Systems

Soilless agriculture offers various advantages over traditional soil-based farming systems. These systems are gaining popularity due to their efficiency, adaptability, and sustainability. This discussion explores the key benefits of these innovative farming methods, the different types of soilless systems, their unique characteristics, and their impact on modern agriculture.

- *Resource use efficiency*

- One of the most significant benefits of soilless agriculture systems is the efficient use of water, fertilizers, and land. Efficient water uses and recycling lower overall demand. Hydroponic systems, which provide nutrients directly to plant roots, can save up to 90% more water than typical soil-based farming. Aeroponics, which includes spraying a fertilizer mist directly onto plant roots, improves water and nutrient efficiency (Hussain et al. 2014; AlShrouf 2017; Fussy and Papenbrock 2022; Mir et al. 2022; Sousa et al. 2024). This efficient resource management is especially advantageous in locations with restricted water resources since it improves water use efficiency, as shown in Table 7.2, promoting more sustainable farming practices.
- *Space optimization*
- Soilless agriculture systems are highly adaptable to various spatial layouts, making them excellent for urban and indoor farming. Vertical farming and rooftop gardens, which use soilless techniques, are useful methods for maximizing space in highly crowded regions (Fussy and Papenbrock 2022). Vertical hydroponic and aeroponic systems are ideal for urban environments with limited space. Growing plants in stacked layers or confined spaces increases food production

Table 7.2 Water use efficiency of some crops in soilless systems

Country	Soilless systems	Cultivated crops	WUE, kg/m^3	References
India	Hydroponic (drip)	Cucumber (Cucumis sativus)	34.5–51.4	Singh et al. (2019)
Iran	Hydroponic (NFT)	Tomato (Solanum lycopersicum)	33.7	Fayezizadeh et al. (2021)
Palestine	Hydroponic (NFT)	Lettuce (Lactuca sativa)	121	Shtaya and Qubbaj (2022)
Iran	Aeroponic	Cucumber (Cucumis sativus)	94.4	Jamshidi et al. (2020)
Egypt	Aeroponic	Lettuce (Lactuca sativa)	8	Bedair (2017)
India	Aeroponic	Lettuce (Lactuca sativa)	66.7	Nigadi et al. (2024)
Egypt	Hydroponic (DWC)	Lettuce (Lactuca sativa)	64.40	El-Saeidy et al. (2022)
	Hydroponic (NFT)		66.86	

without requiring additional land, which is specifically beneficial in urban environments where space is scarce (AlShrouf 2017; Sousa et al. 2024).

- *Reduction in pest and disease issues*
- The absence of soil in soilless agriculture systems reduces the risk of diseases and pests associated with soil (Mir et al. 2022). This decreased insect and disease pressure lessens the need of herbicides and pesticides, which can harm the environment and human health (Gebreegziher 2023). Soilless methods thus contribute to cleaner and more sustainable agricultural operations by reducing the use of potentially hazardous pesticides (Fussy and Papenbrock 2022).
- *Enhanced control and productivity*
- Soilless systems provide precise control over growing parameters such as nutrient levels, pH, light exposure, and water availability, resulting in optimal plant growth (Gorbe and Calatayud 2010). This promotes ideal plant growth conditions, leading to increased growth rates and yields (Fussy and Papenbrock 2022). Hydroponic methods enable year-round production and faster growth cycles, ensuring a steady supply of fresh products and increased overall output (Rajaseger et al. 2023).
- *Environmental benefits*
- Soilless agricultural technologies offer significant environmental benefits. These techniques help prevent soil erosion and degradation caused by traditional farming methods. Furthermore, controlling the application of water and fertilizers reduces runoff and pollution in the environment. Soilless systems contribute to larger environmental conservation goals by promoting sustainable agricultural practices (Fussy and Papenbrock 2022).

Challenges of Soilless Agriculture Systems

While soilless agriculture systems have tremendous benefits, they also confront a number of challenges that may restrict their effectiveness and widespread adoption. These barriers include expensive start-up costs, technical complexities, and potential environmental issues.

- *High initial costs*

- One of the most major barriers to soilless agriculture systems is the expensive initial setup cost, which includes equipment and infrastructure (Sharma et al. 2018; Mir et al. 2022). Hydroponic and aeroponic systems necessitate specific infrastructure like as fertilizer supply systems, pumps, and lighting, which can be expensive (Ampim et al. 2022; Wang et al. 2022). This significant financial investment may be a barrier to entry for small-scale or unskilled growers, limiting their usage of soilless systems.
- *Technical complexity*
- Soilless agriculture techniques often use complex technologies and require extensive maintenance to ensure optimal growing conditions. For example, hydroponic systems demand precise control over fertilizer concentrations, pH levels, and environmental conditions (Bugbee 2003). Maintaining ideal conditions in soilless systems generally requires substantial technical knowledge and ongoing maintenance (Mir et al. 2022). Additionally, continuous monitoring and maintenance are necessary to prevent system failures and preserve plant health (Hussain et al. 2014). This technical complexity can make system management difficult, especially for those lacking specialized knowledge or experience. Technical faults or mismanagement can result in significant crop losses and reduced output (Fussy and Papenbrock 2022).
- *Nutrient management*
- Plants rely completely on the nutrient solutions delivered; hence, good nutrient management is critical in soilless agriculture. Imbalances or deficiencies in the nutrient solution might result in poor plant development and production loss. Furthermore, the necessity for constant nutrient monitoring and adjustment complicates and increases the work necessary to manage soilless systems. Maintaining consistent nutrition solution quality is difficult and time-consuming (Sambo et al. 2019).
- *Environmental impact*
- While soilless agriculture technologies are typically thought to be more environmentally friendly than traditional soil farming methods, they do not come without environmental repercussions. For example, the production of synthetic nutritional solutions and the energy necessary for artificial lighting in indoor systems can both contribute to environmental degradation if not controlled appropriately. Furthermore, inappropriate disposal of spent growth media and system components may have negative environmental consequences (Fussy and Papenbrock 2022).

- *Dependence on technology*
- Soilless agriculture systems are primarily reliant on technology, leaving them susceptible to system failures and power outages. For example, a nutrient delivery system failure or a power outage that affects artificial lighting can swiftly result in plant health decline and crop loss (Mir et al. 2022). Farmers must have backup systems and contingency plans in place to reduce the chance of system failure.

Types of Soilless Agriculture Systems

Hydroponics

Hydroponics is an appropriate method to plant crops without soil by providing nutrients elements directly to the roots via a water-based solution or mist (Kannan et al. 2022). This technique allows for precise control of pH levels, nutrient delivery, and water quality, ensuing in earlier plant growth and higher yields than traditional soil-based methods (Khatri et al. 2024). It has won significant interest due to its potential to revolutionize agriculture. This approach conserves water and reduces soil requirements (Bradley and Marulanda 2000; Pineda-Pineda et al. 2009), enabling fine control over fertilizer delivery and water utilization (Grewal et al. 2011). This system is divided into five types: nutrient film technology (NFT), deep water culture (DWC), Ebb and flow (flood and drain), drip systems, and wick system. Several types of hydroponic exist as followed:

- *Nutrient film technique (NFT)*
- In NFT, a thin film of nutritional solution flows constantly over the roots of plants contained in a slanted trough. This continuous flow delivers vital nutrients and oxygen (Burrage 1992). The technology makes optimal use of nutrients and water by reducing waste and optimizing nutrient uptake, encouraging rapid plant growth and development (Geilfus 2019; Savvas and Gruda 2018). However, it is not ideal for large plants or those with extensive root systems, and there is a risk of system failure (Geilfus 2019; Savvas and Gruda 2018). Because of pauses in nutrient supply, plants can be quickly damaged (Graves 1983).
- *Deep water culture (DWC)*
- DWC suspends plant roots in a nutrient-rich oxygenized water solution. An air pump delivers oxygen to the roots, preventing them from suffocating and fostering healthy growth (Nursyahid et al., 2021). It has a simple design that is relatively easy to set up and maintain, as well as high oxygen availability, which supports robust root development and plant growth (Thakur et al. 2023); however, it is temperature sensitive and has the potential for root diseases, which could be caused by poor water quality or insufficient aeration (Geilfus 2019; Savvas and Gruda 2018; Mohammed 2018).
- *Ebb and flow (flood and drain)*

- The Ebb and flow system, also called flood and drain, alternates between flooding plant roots with a nutrient solution and draining it back into a reservoir. This cycle provides the plants with essential nutrients and oxygen (Geilfus 2019; Savvas and Gruda 2018; Reshma and Joseph 2016). This technique has the advantage of being versatile; it may be applied to a wide range of plant species and growth phases, ensuring optimum nutrient distribution. However, it has significant downsides, such as pump reliance and the possibility of root saturation. Overflooding, if not controlled effectively, can result in root rot.
- *Drip system*
- The drip system delivers nutrition solution directly to the base of each plant via a network of tubes and emitters. This strategy enables for perfect control over the amount of nutrients that each plant receives (Geilfus 2019; Savvas and Gruda 2018; Sharma et al. 2018). The advantage of this system is customizable nutrient delivery, which allows for tailored nutrient supply to individual plants and reduced water use through efficient application that minimizes water waste (Geilfus 2019; Savvas and Gruda 2018; Sharma et al. 2018). However, it has constraints such as emitter clogging, which may affect nutrient delivery and higher maintenance requirements, and it requires regular monitoring and maintenance of the irrigation system.
- *Wick system*
- The wick system employs a wick to transport nutritional solution from a tank to plant roots. This passive hydroponic approach uses capillary action to distribute nutrients (Savvas and Gruda 2018; Sinha et al. 2020). It requires little maintenance due to its simplistic design with few moving components and is less expensive than more sophisticated systems. However, its usefulness is restricted; it is best suited for smaller plants and less demanding crops, and nutrient delivery is slower than in active systems (Savvas and Gruda 2018; Sinha et al. 2020). Figure 7.2 depicts several types of hydroponic systems.

Aeroponics

Aeroponics, an innovative soilless cultivation method combining "aero" (air) and "ponos" (labor), is a cutting-edge agricultural technique that involves growing plants in a soil-free environment by suspending their roots in air and delivering nutrients through a fine mist (Osvald et al. 2001; Kumari and Kumar 2019), as shown in Fig. 7.3.

This design maximizes the contact between the roots and oxygen, resulting in faster development and higher yields than standard and hydroponic systems (Mir et al. 2022). Aeroponics is increasingly being utilized in commercial agriculture to grow high-value crops including lettuce, herbs, and strawberries. It improves root oxygenation, promotes faster plant development, and increases fertilizer and water efficiency, making it an appealing option for growers looking to maximize production (Ampim et al. 2022). Minimal water is used compared to other soilless systems (Fussy and Papenbrock 2022). While aeroponics have some challenges as system sensitivity, it requires precise control of misting and nutrient delivery, and any

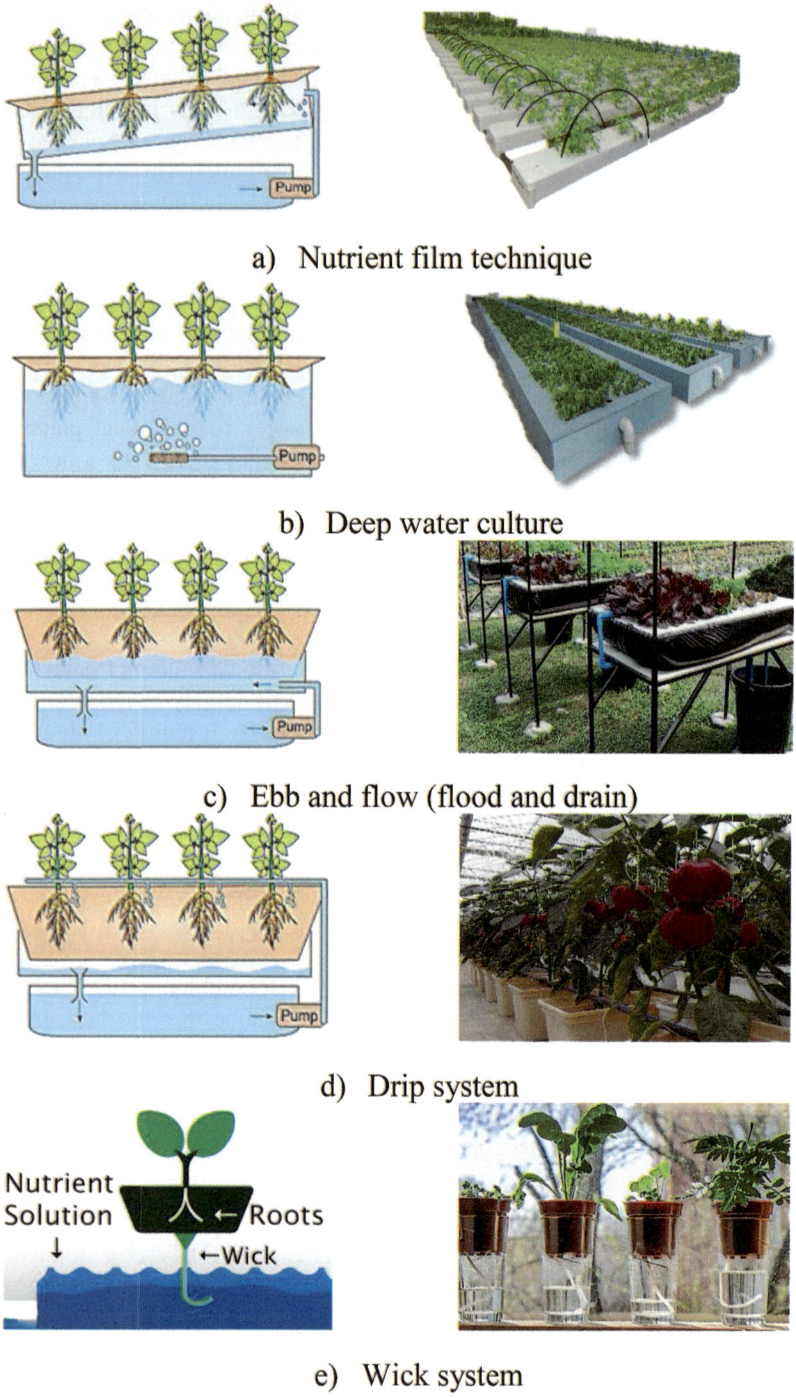

Fig. 7.2 Types of hydroponic systems

Fig. 7.3 Aeroponic system

failure can lead to rapid plant stress (Lakhiar et al. 2018). High maintenance and frequent monitoring and maintenance are needed to prevent clogging and ensure system functionality (Kumari and Kumar 2019).

Types of Aeroponic Systems

- *High-pressure aeroponics (HPA)*

- High-pressure aeroponics use high-pressure pumps to produce a fine mist of nutrient solution, providing effective nutrient and oxygen delivery. High-pressure aeroponics is the process of producing a thin mist of nutrient solution with high-pressure pumps. This method produces a very fine spray, which ensures efficient nutrient and oxygen delivery to the plant roots; the fine mist provides uniform nutrient distribution and excellent oxygenation, which aid in optimal nutrient and oxygen delivery (Kumari and Kumar 2019); and high-pressure systems frequently result in quicker plant growth and higher yields (Sreekumar 2020). However, high-pressure pumps and nozzles are expensive and require careful maintenance, and fine nozzles can become clogged, reducing misting system performance (Sreekumar 2020; Faisal et al. 2024).
- *Low-pressure aeroponics (LPA)*
- Low-pressure aeroponics employs lower pressure to produce a less fine mist, making it a simpler and more inexpensive system than HPA. Lower-pressure systems are less difficult, are more economical to implement, and have fewer clogging issues than high-pressure systems (Kumari and Kumar 2019; Sreekumar 2020). However, it has less efficient nutrient delivery since the mist created is not as fine, which may impair nutrient absorption and growth rates, resulting in lower growth rates and yields than high-pressure systems (Sreekumar 2020).
- *Fogponics*

- Fogponics is a type of aeroponics that use ultrasonic foggers to produce a fine fog of nutrient solutions. This approach is like HPA but uses ultrasonic technology to generate the fog; it delivers great fertilizer and oxygen delivery to the roots (Sreekumar 2020). Ultrasonic foggers consume less energy than high-pressure pumps (Yeang 2008; Imran et al. 2018). However, the downside of this

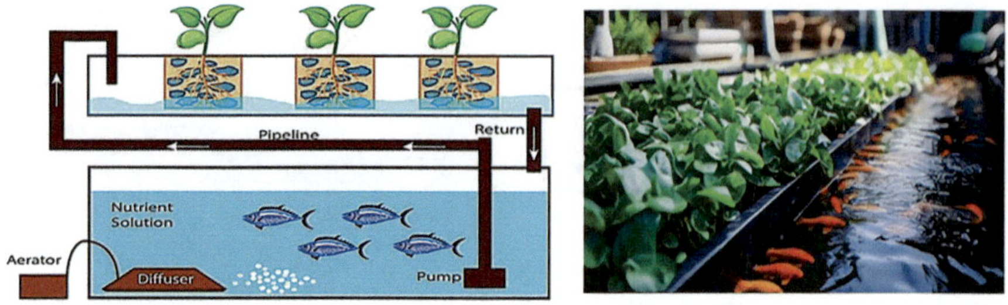

Fig. 7.4 Aquaponics system

technology is that it requires regular power use for mist delivery, and if this is disrupted, the plants may suffer greatly (Yield 2013).

Other Soilless Systems

Aquaponics

Aquaponics combines hydroponics with aquaculture, where fish and plants are grown together in a mutually beneficial system. The fish waste provides nutrients for the plants, and the plants help filter the water for the fish. This symbiotic system creates a sustainable environment where fish and plants support each other (Fussy and Papenbrock 2022). Aquaponics has many advantages, including efficient recycling of nutrients and water, reduced waste, integrated waste recycling, reduced need for synthetic fertilizers, and the simultaneous cultivation of fish and plants (Timmons and Ebeling 2012; Palm et al. 2018). However, aquaponic systems face several challenges. Managing the balance between the needs of fish and plants can be difficult due to the complexity of system management and the potential for imbalanced nutrient levels. Further research is needed on the role of plant growth-promoting microbes in nutrient uptake (Goddek et al., 2015; Yep; Zheng, 2019). Additionally, high initial investment costs are a barrier, as setting up and managing these systems requires significant upfront capital (Goddek et al. 2015) (Fig. 7.4).

Applications of Soilless Systems

Commercial agriculture: Soilless systems are rapidly being employed in commercial agriculture to cultivate a diverse range of crops, from leafy greens to fruiting vegetables. Hydroponics is becoming increasingly used in commercial agriculture for growing high-value crops like lettuce, herbs, and tomatoes. Their great water and nutrient efficiency make them ideal for high-density and urban farming. Its efficiency and high yields make it an appealing option for producers looking to boost productivity (Sharma et al. 2018).

Urban and vertical farming: Soilless systems are perfect for urban and vertical farming projects, where space is restricted and maximum production is required.

These technologies allow for the development of fresh produce in urban areas, minimizing transportation and promoting local food security (Benke and Tomkins 2017).

Research and education: Soilless systems are utilized in research and teaching to investigate plant growth, fertilizer management, and system optimization. Educational institutions also use hydroponics to teach students about modern agricultural methods; this research helps to build more efficient and sustainable farming practices (Fields et al. 2023).

Future Prospects

The future of sustainable agriculture could depend on the continuous development and use of greenhouse and soilless technologies. Overcoming the obstacles associated with these technologies, such as high costs, energy needs, and market barriers, will necessitate ongoing innovation, investment, and collaboration across multiple industries. Integrating renewable energy sources, improving automation technologies, and finding cost-effective solutions will be critical to make these systems more accessible and sustainable. Furthermore, the following topics are vital for realizing this future vision:

- *Technological Advancements and Research Priorities*
- Future developments in soilless systems are expected to focus on enhanced automation, improved nutrient management technologies, and overall performance and sustainability. Innovations such as smart sensors and data analytics will play a crucial role in optimizing system performance (Sambo et al. 2019; Mir et al. 2022). To support these advancements, future research should prioritize increasing the efficiency and sustainability of greenhouse and soilless systems. Collaborative efforts among academia, industry, and government are essential to drive this innovation forward (Fuentes-Peñailillo et al. 2024). By combining technological advancements with targeted research initiatives, the potential for significant improvements in soilless agriculture can be realized.
- *Scalability, Global Impact, and Policy Support*
- To fully realize the worldwide impact of soilless agricultural technologies, large-scale implementation techniques are required. Fuentes-Peñailillo et al. (2024) suggest developing cost-effective methods to improve market access. Furthermore, favorable policies and investments in R&D are critical to the advancement of these technologies (Sambo et al. 2019; Sousa et al. 2024). Government incentives and corporate support can help overcome adoption barriers and encourage wider use of novel soilless systems (Ghanayem et al. 2022). These aspects work together to increase the global impact of soilless agriculture.
- *Sustainability and Long-Term Resilience in Soilless Systems*
- Increasing the sustainability and scalability of soilless systems is essential for their broader adoption. Efforts to reduce energy consumption, integrate renew-

able resources, and improve system designs are key to making these systems viable for global food production and security (Fuentes-Peñailillo et al. 2024). Additionally, ensuring long-term sustainability and resilience requires adapting to future climate scenarios and committing to continuous. Together, these approaches will enhance the viability of soilless agriculture in the face of changing environmental conditions.

Conclusions

Both climate change and water scarcity are serious challenges to global food security, especially in arid and semiarid regions. Greenhouses and soilless agriculture systems are considered viable alternatives for enhancing resource efficiency, crop yields, and food accessibility. Despite initial high costs for establishment and energy demands, overcoming these barriers is critical for wider acceptance and long-term sustainability. Continued innovation, investment, and collaboration are essential for creating resilient agroecosystems that will support future generations.

Food security is a complicated and multifaceted issue that requires coordinated actions from all sectors. Economic imbalances, environmental obstacles, and geopolitical instability all lead to long-term food insecurity. Addressing these concerns requires comprehensive policy, international collaboration, and targeted actions aimed at improving access to food and guaranteeing sustainable food systems. It is critical to execute integrated policies that address the underlying causes of food insecurity, with continual study and collaboration among stakeholders required to develop successful solutions.

Climate change presents multiple problems to global agriculture, necessitating a holistic strategy. While adaptation and mitigation techniques provide potential solutions, their efficacy varies by region and is dependent on successful large-scale implementation. We can improve agricultural systems' resilience and ensure food availability in a changing climate by investing in research, technology, and policy.

Greenhouse systems offer various benefits that increase agricultural output and sustainability. They lengthen growing seasons, enhance crop yields, and optimize resource utilization, all of which contribute to more efficient farming techniques. However, they confront issues like as expensive costs, large energy use, and the need for maintenance. Understanding these disadvantages is critical for determining the viability of greenhouse systems and developing mitigating solutions.

Soilless agriculture techniques, such as hydroponics and aeroponics, provide novel solutions to many of the problems associated with traditional soil-based farming. They enable better control over growing conditions and resource efficiency, making them appealing for long-term food production. Nonetheless, these systems have drawbacks such as high starting prices and technological complexity. Recent technology improvements are assisting in mitigating these concerns, opening the door for greater use.

Finally, soilless agricultural technologies and novel farming practices provide realistic solutions to today's agricultural difficulties. Their particular advantages and limitations make them appropriate for a wide range of applications. Continued technological breakthroughs, sustainability, and supportive legislation will be key to realizing these systems' full potential, ultimately addressing global food security and environmental sustainability. Moving forward, researchers, policymakers, and practitioners will need to work together to shape resilient and sustainable agricultural landscapes.

References

Ahmed N, Zhang B, Deng L, Bozdar B, Li J, Chachar S, Chachar Z, Jahan I, Talpur A, Gishkori MS, Hayat F (2024) Advancing horizons in vegetable cultivation: a journey from age old practices to high-tech greenhouse cultivation—a review. Front Plant Sci 15:1357153

AlShrouf A (2017) Hydroponics, aeroponic and aquaponic as compared with conventional farming. Am Sci Res J Eng Technol Sci 27(1):247–255

Ampim PA, Obeng E, Olvera-Gonzalez E (2022) Indoor vegetable production: an alternative approach to increasing cultivation. Plan Theory 11(21):2843

Badgery-Parker J (2015) Keep it CLEAN: reducing costs and losses in the management of pests and diseases in the greenhouse. NSW Agriculture

Badji A, Benseddik A, Bensaha H, Boukhelifa A, Hasrane I (2022) Design, technology, and management of greenhouse: a review. J Clean Prod 373:133753

Bedair OM (2017) Vertical-farming irrigation system appropriate for lettuce and celery crops. Misr J Agric Eng 34:1701–1718

Benke K, Tomkins B (2017) Future food production systems: vertical farming and controlled-environment agriculture. Sustain Sci Pract Policy 13(1):13–26

Bradley P, Marulanda C (2000) Simplified hydroponics to reduce global hunger. In: World congress on soilless culture: agriculture in the coming millennium, vol 554, pp 289–296

Bugbee B (2003) Nutrient management in recirculating hydroponic culture. In: South Pacific soilless culture conference-SPSCC, vol 648, pp 99–112

Burchi G, Chessa S, Gambineri F, Kocian A, Massa D, Milazzo P, Rimediotti L, Ruggeri A (2018) Information technology-controlled greenhouse: a system architecture. In: 2018 IoT vertical and topical summit on agriculture-Tuscany (IOT Tuscany). IEEE, pp 1–6

Burrage SW (1992) Nutrient film technique in protected cultivation. In: Symposium on soil and soilless media under protected cultivation in mild winter climates, vol 323, pp 23–38

Carrasco Azzini G, Conti V, Holleman C, Smulders M (2022) Best practices in addressing the major drivers of food security and nutrition to transform food systems, Background paper for the state of food security and nutrition in the world 2021. FAO agricultural development economics technical study, no 23. FAO, Rome

Chen X, Wang L, Cao Q, Sun J, Niu Z, Yang L, Jiang W (2024) Response of global agricultural productivity anomalies to drought stress in irrigated and rainfed agriculture. Sci China Earth Sci 67(11):3579–3593

Dalai S, Tripathy B, Mohanta S, Sahu B, Palai JB (2020) Greenhouses: types and structural components. In: Protected cultivation and smart agriculture. New Delhi Publishers, New Delhi

Daniels GE, Morton MH (2023) COVID-19 recession: young adult food insecurity, racial disparities, and correlates. J Adolesc Health 72(2):237–245

Despommier D (2013) Farming up the city: the rise of urban vertical farms. Trends Biotechnol 31(7):388–389

El-Saeidy EA, Elgeziry AA, Eissa AHA (2022) Study the effect of some engineering factors on the performance of an integrated aquaponic unit. J Soil Sci Agric Eng Mansoura Univ 13(8):253–263

Ernst M, Woods T, Butler A, Wolff B, Jacobsen K (2020) High tunnel production and marketing survey: data summary. University of Kentucky Cooperative Extension Service CCD-SP-17. https://www.uky.edu/ccd/sites; www.uky.edu.ccd/files/HTsurvey.pdf
Fadel MA, Hammad BAB, AlHosany F, Iwaimer O (2016) Greenhouses covering materials: a comparative study. Agric Eng Int CIGR J 18(1):48–57
Faisal A, Rajashekar V, Biswal P, Mukherjee A, Bhowmick GD (2024) Advancing agriculture with aeroponics: a critical review of methods, benefits, and limitations. In: Hydroponics: the future of sustainable farming. Springer, pp 311–322
FAO (2017) The future of food and agriculture—trends and challenges. Annual report, 296. Food and Agriculture Organization (FAO). https://doi.org/10.4060/cb4474en
FAO (2020) The state of food security and nutrition in the world 2020. Food and Agriculture Organization of the United Nations. Retrieved from FAO https://doi.org/10.4060/cb4474en
FAO (2024) FAOSTST: suite of food security indicators. https://www.fao.org/faostat/en/data/FS. Accessed on 24 July 2024
FAO, IFAD, UNICEF, WFP, WHO (2021) The state of food security and nutrition in the world 2021. Food and Agriculture Organization of the United Nations. Retrieved from FAO. https://openknowledge.fao.org/handle/20.500.14283/cb4474en
Fayezizadeh MR, Ansari NAZ, Albaji M, Khaleghi E (2021) Effects of hydroponic systems on yield, water productivity and stomatal gas exchange of greenhouse tomato cultivars. Agric Water Manag 258:107171
Fields JS, Owen JS, Lamm A, Altland J, Jackson B, Oki L et al (2023) Surveying North American specialty crop growers' current use of soilless substrates and future research and education needs. Agriculture 13(9):1727
Fitzgerald CB, Hutton M (2012) Production practices and challenges with high tunnel systems in Maine. J NACAA 5(2):1–8
Fuentes-Peñailillo F, Gutter K, Vega R, Silva GC (2024) New generation sustainable technologies for soilless vegetable production. Horticulturae 10(1):49
Fussy A, Papenbrock J (2022) An overview of soil and soilless cultivation techniques—chances, challenges and the neglected question of sustainability. Plan Theory 11(9):1153
Gatahi DM (2020) Challenges and opportunities in tomato production chain and sustainable standards. Int J Hortic Sci Technol 7(3):235–262
Gebreegziher WG (2023) Soilless culture technology to transform vegetable farming, reduce land pressure and degradation in drylands. Cogent Food Agric 9(2):2265106
Geilfus CM (2019) Controlled environment horticulture. In: Improving quality of vegetables and medicinal plants. Springer, Cham, pp 1–233
Ghanayem AA, Almohamed S, Al Assaf A, Majdalawi M (2022) Socioeconomic analysis of soilless farming system—an comparative evidence from Jordan. Middle East Int J Food Agric Econ 10(3):205–223
Ghani S, Bakochristou F, ElBialy EMAA, Gamaledin SMA, Rashwan MM, Abdelhalim AM, Ismail SM (2019) Design challenges of agricultural greenhouses in hot and arid environments—a review. Eng Agric Environ Food 12(1):48–70
Ghoulem M, El Moueddeb K, Nehdi E, Boukhanouf R, Calautit JK (2019) Greenhouse design and cooling technologies for sustainable food cultivation in hot climates: review of current practice and future status. Biosyst Eng 183:121–150
Goddek S, Delaide B, Mankasingh U, Ragnarsdottir KV, Jijakli H, Thorarinsdottir R (2015) Challenges of sustainable and commercial aquaponics. Sustain For 7(4):4199–4224
Gorbe E, Calatayud Á (2010) Optimization of nutrition in soilless systems: a review. Adv Bot Res 53:193–245
Graves CJ (1983) The nutrient film technique. Hortic Rev 5:1–44
Grewal HS, Maheshwari B, Parks SE (2011) Water and nutrient use efficiency of a low-cost hydroponic greenhouse for a cucumber crop: an Australian case study. Agric Water Manag 98(5):841–846
Grigorieva E, Livenets A, Stelmakh E (2023) Adaptation of agriculture to climate change: a scoping review. Climate 11(10):202

Hemming S, Mohammadkhani V, Kempkes F, Peters A, Lieffrig V, Harada K, Spronsen M, Werff T, Louter C (2018) Feasibility study thin glasses for greenhouse roof designs (No. WPR-732). Wageningen Plant Research

Hopwood W, Lopez-Reyes Z, Bantan A, Vietti C, Al-Shahrani D, Al-Harbi A, Qaryouti M, Davies P, Tester M, Wing R, Waller R (2024) Benchmarking techno-economic performance of greenhouses with different technology levels in a hot humid climate. Biosyst Eng 244:177–199

Hussain A, Iqbal K, Aziem S, Mahato P, Negi AK (2014) A review on the science of growing crops without soil (soilless culture)—a novel alternative for growing crops. Int J Agric Crop Sci 7(11):833

Ibrahim K, Bavorova M, Zhllima E (2024) Food security and livelihoods in protracted crisis: the experience of rural residents in Syria's war zones. Food Secur 16(3):659–673

Iddio E, Wang L, Thomas Y, McMorrow G, Denzer A (2020) Energy efficient operation and modeling for greenhouses: a literature review. Renew Sust Energ Rev 117:109480

Imran AL, Jianmin G, Tabinda NS, Farman AC, Noman AB (2018) Modern plant cultivation technologies in agriculture under controlled environment: a review on aeroponics. J Plant Interact 13(1):338–352

Climate Change 2021: The Physical Science Basis. Contribution of Working Group I to the Sixth Assessment Report of the Intergovernmental Panel on Climate Change[Masson-Delmotte, V., P. Zhai, A. Pirani, S.L. Connors, C. Péan, S. Berger, N. Caud, Y. Chen, L. Goldfarb, M.I. Gomis, M. Huang, K. Leitzell, E. Lonnoy, J.B.R. Matthews, T.K. Maycock, T. Waterfield, O. Yelekçi, R. Yu, and B. Zhou (eds.)]. Cambridge University Press, Cambridge, United Kingdom and New York, NY, USA, In press, doi:10.1017/9781009157896.

Islam M, Yasmin R (2017) Impact of aquaculture and contemporary environmental issues in Bangladesh. Int J Fish Aquat Stud 5:100–107

Jamshidi AR, Moghaddam AG, Ghoraba FM (2020) Simultaneous optimization of water usage efficiency and yield of cucumber planted in a columnar aeroponic system. Int J Hortic Sci Technol 7:365–375

Janni M, Maestri E, Gullì M, Marmiroli M, Marmiroli N (2024) Plant responses to climate change, how global warming may impact on food security: a critical review. Front Plant Sci 14:1297569

Jin W, Formiga Lopez D, Heuvelink E, Marcelis LF (2023) Light use efficiency of lettuce cultivation in vertical farms compared with greenhouse and field. Food Energy Secur 12(1):e391

Kaiser C, Ernst M (2012) High tunnel overview. Agric Nat Resour 2012:1–7

Kannan M, Elavarasan G, Balamurugan A, Dhanusiya B, Freedon D (2022) Hydroponic farming—a state of art for the future agriculture. Mater Today Proc 68:2163–2166

Kavga A, Thomopoulos V, Barouchas P, Stefanakis N, Liopa-Tsakalidi A (2021) Research on innovative training on smart greenhouse technologies for economic and environmental sustainability. Sustainability 13(19):10536

Khatri L, Kunwar A, Bist DR (2024) Hydroponics: advantages and challenges in soilless farming. Big Data Agric 6:81–88

Kim KH, Lee BM (2023) Effects of climate change and drought tolerance on maize growth. Plan Theory 12(20):3548

Kumari R, Kumar R (2019) Aeroponics: a review on modern agriculture technology. Indian Farmer 6(4):286–292

Lakhiar IA, Gao J, Syed TN, Chandio FA, Buttar NA (2018) Modern plant cultivation technologies in agriculture under controlled environment: a review on aeroponics. J Plant Interact 13(1):338–352

Nursyahid A, Setyawan TA, Sa'diyah K, Wardihani ED, Helmy H, & Hasan A (2021) Analysis of Deep Water Culture (DWC) hydroponic nutrient solution level control systems. In IOP conference series: Materials science and engineering 1108(1):012032 IOP Publishing

Mahmoud MA (2019) Impact of climate change on the agricultural sector in Egypt. In: Conventional water resources and agriculture in Egypt, pp 213–227. https://doi.org/10.1007/698_2017_48

Maraveas C (2019) Environmental sustainability of greenhouse covering materials. Sustain For 11(21):6129

Marcelis LF, Costa JM, Heuvelink E (2019) Achieving sustainable greenhouse production: present status, recent advances and future developments. In: Achieving sustainable greenhouse cultivation. Burleigh Dodds Science Publishing, London, pp 1–14

McCartney L, Lefsrud M (2018) Protected agriculture in extreme environments: a review of controlled environment agriculture in tropical, arid, polar, and urban locations. Appl Eng Agric 34(2):455–473

Milstein A (2005) Polyculture in aquaculture. Anim Breed Abstr 73(12):15N–41N

Mir YH, Mir S, Ganie MA, Shah AM, Majeed U, Chesti MH et al (2022) Soilless farming: an innovative sustainable approach in agriculture. Pharma Innov J 11(6):2663–2675

Mohammed S (2018) Tomorrow's agriculture: "NFT hydroponics"-grow within your budget. Springer, Berlin/Heidelberg, p 50

Mrabet R (2023) Sustainable agriculture for food and nutritional security. In: Sustainable agriculture and the environment. Academic, pp 25–90

Nigadi RR, Ramalingannanavar N, Kumari N, Bhagwat SM, Rawat G, Kadam GG, Khadatare MY, Keskar PK, Jagtap DN (2024) Assessment of water use efficiency and fertilizer use efficiency of solar-powered aeroponic system. Int J Res Agron 7(6):220–223

Niggli U, Fließbach A, Hepperly P, Scialabba N (2009) Low greenhouse gas agriculture: mitigation and adaptation potential of sustainable farming systems. Ökologie Landbau 141:32–33

Osvald J, Petrovic N, Demsar J (2001) Sugar and organic acid content of tomato fruits (*Lycopersicon lycopersicum* Mill.) grown on aeroponics at different plant density. Acta Aliment 30(1):53–61

Palm HW, Knaus U, Appelbaum S, Goddek S, Strauch SM, Vermeulen T et al (2018) Towards commercial aquaponics: a review of systems, designs, scales and nomenclature. Aquac Int 26:813–842

Pineda-Pineda J, Ramírez-Arias A, Sánchez del Castillo F, Castillo-González AM, Valdez-Aguilar LA, Vargas-Canales JM (2009) Extraction and nutrient efficiency during the vegetative growth of tomato under hydroponics conditions. In: International symposium on high technology for greenhouse systems: GreenSys, vol 893, pp 997–1005

Podkolzina I, Tenishchev A, Gornostaeva Z, Tekeeva H, Tandelova O (2023) Ecological and food security in the conditions of the geopolitical situation in the world global digital transformation trends in real sectors of the economy. In SHS web of conferences, EDP sciences, p 02041

Rajaseger G, Chan KL, Tan KY, Ramasamy S, Khin MC, Amaladoss A, Haribhai PK (2023) Hydroponics: current trends in sustainable crop production. Bioinformation 19(9):925

Rebolledo-Leiva R, Angulo-Meza L, Iriarte A, González-Araya MC (2017) Joint carbon footprint assessment and data envelopment analysis for the reduction of greenhouse gas emissions in agriculture production. Sci Total Environ 593:36–46

Reshma T, Joseph S (2016) Hydroponic cultivation of tomatoes—an attempt for Kerala conditions. J Trop Agric 54(2):164–164

Sambo P, Nicoletto C, Giro A, Pii Y, Valentinuzzi F, Mimmo T, Lugli P, Orzes G, Mazzetto F, Astolfi S, Terzano R, Cesco S (2019) Hydroponic solutions for soilless production systems: issues and opportunities in a smart agriculture perspective. Front Plant Sci 10:923

Savvas D, Gruda N (2018) Application of soilless culture technologies in the modern greenhouse industry—a review. Eur J Hortic Sci 83(5):280–293

Sayadi-Gmada S, Rodríguez-Pleguezuelo CR, Rojas-Serrano F, Parra-López C, Parra-Gómez S, García-García M, Rosana García-Collado R, Lorbach-Kelle M, Manrique-Gordillo T (2019) Inorganic waste management in greenhouse agriculture in Almeria (SE Spain): towards a circular system in intensive horticultural production. Sustain For 11(14):3782

Sen A (1999) Development as freedom. Oxford University Press, Oxford

Shamshiri R, Kalantari F, Ting KC, Thorp KR, Hameed IA, Weltzien C, Ahmad D, Shad ZM (2018) Advances in greenhouse automation and controlled environment agriculture: a transition to plant factories and urban agriculture. Int J Agric Biol Eng 11(1):1–22

Sharma N, Acharya S, Kumar K, Singh N, Chaurasia OP (2018) Hydroponics as an advanced technique for vegetable production: an overview. J Soil Water Conserv 17(4):364–371

Shtaya MJY, Qubbaj T (2022) Effect of different soilless agriculture methods on irrigation water saving and growth of lettuce (*Lactuca sativa*). Res Crops 23:156–162

Singh M, Singh G, Singh J (2019) Nutrient and water use efficiency of cucumbers grown in soilless media under a naturally ventilated greenhouse. J Agric Sci Technol 21:193–207
Sinha A, Singh D, Swapnil D (2020) Hydroponic—a new method of growing crops without soil. Sci Agric Allied Sect Mon Newsl 2(3):AL202052
Sousa R, Bragança L, Da Silva MV, Oliveira RS (2024) Challenges and solutions for sustainable food systems: the potential of home hydroponics. Sustain For 16:817
Sreekumar V (2020) Droplet size analysis of high pressure aeroponic system. Massey University
Tanveer SK, Lu X, Hussain I, Sohail M (2019) Soil carbon sequestration through agronomic management practices. In: CO_2 sequestration. IntechOpen
Thakur P, Wadhwa H, Kaushal S (2023) Nutrient dynamics for hydroponic production system. Int J Plant Soil Sci 35(21):982–993
Timmons M, Ebeling JM (2012) Recirculating aquaculture systems. In: Tidwell J (ed) Aquaculture production systems. Wiley-Blackwell, Ames, pp 245–277
Toromade AS, Soyombo DA, Kupa E, Ijomah TI (2024) Reviewing the impact of climate change on global food security: challenges and solutions. Int J Appl Res Soc Sci 6(7):1403–1416
Van Delden SH, SharathKumar M, Butturini M, Graamans LJA, Heuvelink E, Kacira M, Kaiser E, Klamer RS, Klerkx L, Kootstra G, Loeber A, Schouten RE, Stanghellini C, van Ieperen W, Verdonk JC, Vialet-Chabrand S, Woltering EJ, van de Zedde R, Zhang Y, Marcelis LFM (2021) Current status and future challenges in implementing and upscaling vertical farming systems. Nat Food 2:944–956
Vatistas C, Avgoustaki DD, Bartzanas T (2022) A systematic literature review on controlled-environment agriculture: how vertical farms and greenhouses can influence the sustainability and footprint of urban microclimate with local food production. Atmosphere 13(8):1258
Vox G, Teitel M, Pardossi A, Minuto A, Tinivella F, Schettini E (2010) Sustainable greenhouse systems. In: Sustainable agriculture: technology, planning and management. Nova Science Publishers, p 14
Wang S, Adekunle A, Raghavan V (2022) Exploring the integration of bioelectrochemical systems and hydroponics: possibilities, challenges, and innovations. J Clean Prod 366:132855
World bank (2024) Food security update. December 13, 2024. www.worldbank.org
Yang Y, Tilman D, Jin Z, Smith P, Barrett CB, Zhu YG, Burney J, D'Odorico P, Fantke P, Fargione J, Finlay JC, Rulli MC, Sloat L, Groenigen KJ, West PC, Ziska L, Michalak AM, Team CA, David B, Lobell DB (2024) Climate change exacerbates the environmental impacts of agriculture. Science 385(6713):eadn3747
Yeang K (2008) Ecoskyscrapers and ecomimesis: new tall building typologies. In: Proceedings of the 8th CTBUH world congress on tall & green: typology for a sustainable urban future, Dubai, UAE, 3–5 March 2008, pp 84–94
Yep B, Zheng Y (2019) Aquaponic trends and challenges—a review. J Clean Prod 228:1586–1599
Yield M (2013) Fogponics. https://www.maximumyield.com/definition/160/fogponics

Using Fishponds Wastewater to Enhance Agricultural Crop Productivity and Sustainability

8

M. A. M. Moursy and M. A. Mahmoud

Abstract

Utilizing fishpond wastewater in agriculture sustainably enhances crop production by recycling nutrients and reducing waste. The nutrient-rich wastewater of fishponds acts as a fertilizer, improving soil fertility and encouraging plant growth while promoting water conservation, especially in drought-prone regions. It also boosts farm production and profitability by reducing input costs and increasing crop yields. However, careful management is crucial to address challenges like nutrient imbalance and regulatory compliance. Overall, integrating fishpond wastewater into agriculture fosters a symbiotic relationship between aquaculture and agriculture, enhancing system resilience and advancing food security, environmental preservation, and rural development. The objective of this chapter is to provide a comprehensive assessment of current practices, technologies, and research in aquaculture systems, with a particular focus on enhancing production efficiency, promoting environmental sustainability, and ensuring economic viability. By integrating insights from recent studies, this review aims to serve as a valuable resource for researchers, policymakers, and stakeholders, offering guidance to drive the advancement of sustainable aquaculture practices.

M. A. M. Moursy (✉)
Water Management Research Institute, National Water Research Center, El-Qanater El-Khairiya, Egypt
e-mail: mohamed_anter@nwrc.gov.eg

M. A. Mahmoud
Water Requirements and Field Irrigation Research Department, Soils, Water and Environment Research Institute, Agricultural Research Center, Giza, Egypt

A. E.-H. Zohry, S. Ouda (eds.), *Resilient Agroecosystems*, Sustainability Sciences in Asia and Africa, https://doi.org/10.1007/978-981-96-9630-7_8

Keywords

Aquaculture · Environmental impacts · Socioeconomic impacts · Economic feasibility · Water productivity

Introduction

As the global population grows, the demand for food and water is increasing. Water is essential for life and crucial in agricultural production. Agriculture is the largest global water user, accounting for 70% of total withdrawals, and around 30% of global energy is used in food production and its supply chain (UN 2014). Water, food, and energy are interconnected, and their sustainable management is essential for preserving biodiversity and protecting the environment (Hejnowicz et al. 2022). Growing urban and industrial water demands are straining irrigation supplies and threatening food security. With a global shift toward sustainable development, enhancing resource efficiency, particularly in agriculture, is increasingly prioritized (Boyd et al. 2008). With the rising global population, especially in developing countries, advancing technology to increase food production is crucial.

Aquaculture, the fastest-growing agricultural sector, supplies fish and meets global demand for valuable animal protein (FAO 2014). It now provides 46% of global fish production, up from 25.7% in 2000, and is expected to meet future fish demand. Global fish consumption per capita grew from 9.9 kg in 1960 to 18.9 kg in 2010, with early 2012 estimates suggesting it could reach 19.2 kg. By 2030, global fish production is projected to reach 204 million tons, with aquaculture production expected to rise by 32% to 109 million tons (FAO 2014, 2020). Africa produces only 1.74 million tons, less than 2% of global aquaculture output. Egypt leads with 1.1 million tons, followed by Nigeria and Uganda with 313,000 and 111,000 tons, respectively. Nile tilapia makes up 43.6% of Africa's aquaculture production, with common carp and African catfish also significant. Aquaculture supplies nutrient-dense water for crop irrigation, enhancing food security and sustainable water resource use (Rana et al. 2005). Aquaculture experienced an average annual growth rate of 5.3% between 2001 and 2018 and is the most rapidly expanding sector in food production. The top 10 food fish aquaculture-producing countries in the world are shown in Fig. 8.1.

Integrated farming systems are highly effective for boosting productivity among small-scale farmers in resource-limited rural areas and should be promoted (Yuan et al. 2019). By integrating livestock, aquatic species, fruit-bearing trees, and vegetables, a farmer can achieve consistent and efficient farm production with optimized resource use and ecological preservation (Lightfoot and Gonzalves 2001). The technology supports integrating fish, livestock, and crops either in a single system or separately, utilizing waste to produce affordable food, generate jobs, and increase household earnings (Singh et al. 2021). To address water scarcity and food insecurity, we must enhance water use and optimize food production (Mancosu et al. 2015).

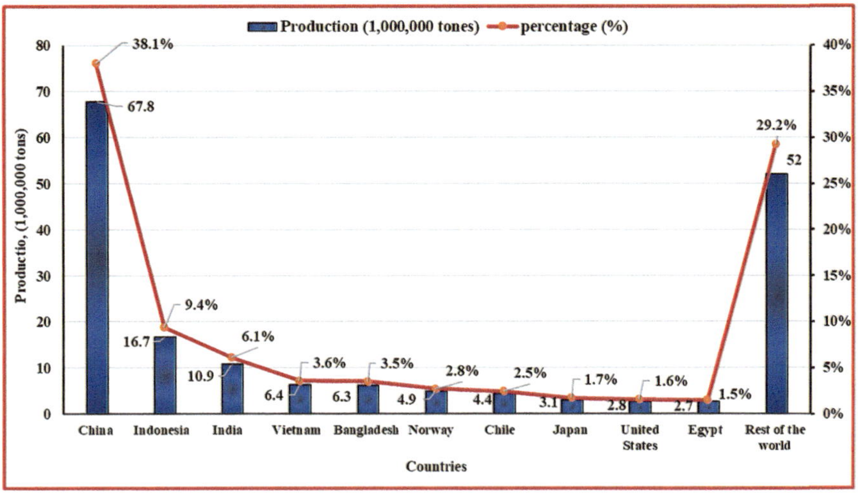

Fig. 8.1 The top 10 food fish aquaculture-producing countries in the world. (Source: FAO 2022)

Fish excrete nutrients that crops absorb through the breakdown of organic waste, with most nutrients for crop growth originating from fish feed (Abdelraouf and Hoballah 2014; Moursy et al. 2022). Aquaculture water reuse creates new irrigation resources, allowing plants to use water more efficiently and reducing the need for fertilizers, while providing beneficial nutrients like nitrogen and phosphorus for plant growth (Mustapha et al. 2014; Yildiz et al. 2017). Fishpond wastewater boosts crop productivity and soil quality by supplying organic matter and irrigation water, reducing fertilizer needs. It enhances soil nutrient exchange and supports faster plant growth with nutrients from fish excretion and microbial breakdown (Elnwishy et al. 2006).

The objective of this chapter is to provide a comprehensive assessment of current practices, technologies, and research in aquaculture systems, with a particular focus on enhancing production efficiency, promoting environmental sustainability, and ensuring economic viability. By integrating insights from recent studies, this review aims to serve as a valuable resource for researchers, policymakers, and stakeholders, offering guidance to drive the advancement of sustainable aquaculture practices.

Fisheries and Aquaculture

Background

Inland aquaculture is one of the most rapidly expanding sectors within animal production worldwide (FAO 2012). Global fishery and aquaculture production rose from 19 million tons in 1950 to 178.9 million tons in 2018, growing at a rate of 3.3% annually. After a 0.84% decline in 2019, production slightly increased by

0.23% in 2020, reaching 177.8 million tons (FAO 2022). In 2020, Asia, the Americas, Europe, Africa, and Oceania produced 70%, 12%, 10%, 7%, and 1% of global production, respectively (FAO 2022). Egypt, Nigeria, and Uganda are the Africa's top three major producers, accounting for approximately 1.92% of global aquaculture production (Halwart 2020). Despite challenges such as limited markets, inputs, and extension services, Africa's aquaculture production nearly doubled from 1,424,400 tons in 2000 to 2,354,300 tons in 2020. Northern Africa contributed only 0.38% to global aquaculture production in 2020 (FAO 2022).

Fish Culture Systems

Freshwater aquaculture is distinct for its integration with agricultural systems, utilizing water recycling, waste as fertilizer, and agricultural byproducts as fish food (Zajdband 2011). It primarily involves cultivating short-food chain fish like tilapia. Fish culture systems can be classified into extensive, semi-intensive, and intensive categories, reflecting varying levels of intensity according to some parameters as shown in Table 8.1.

Table 8.1 Fish culture systems

Items	Fish type	Extensive	Semi-intensive	Intensive level
Stocking densities (No/ha)	*Tilapia*	5000	10,000	20,000–25,000
	Shrimps	5000–10,000	50,000–1,00,000	300,000–500,000
Feeds	–	No	Supplemental	Formulated
Water replacement (%)	–	No	20/week	25/30 day
Production (ton/ha)	*Tilapia*	1–3	5–7	12–15
	Shrimps	Less than 0.5	1.5	8–10
Need for labor	–	Low	More	More
Initial and operating cost	–	Low	More	More
Risk	–	Less	High (disease outbreak, management issues, and aerator failures during power outages lead to oxygen depletion)	
Financial return	–	Low	High	High

Source: Egna and Boyd (1997)

Crops Suitable for Aquaculture

Agricultural activity is shaped by land and water characteristics, as well as crop supply and demand. For aquaculture farming, crop selection depends on factors like water quality, soil type, and salinity. According to Elewa and Nasr-Allah (2023), certain crops, such as mangoes and vegetables, are vulnerable to water salinity, whereas others, like olives and pomegranates, can tolerate higher salinity levels.

Table 8.2 illustrates the link between total salt concentration in water and its suitability for irrigation.

The integration of fishponds, crops, and animals is known as aquiculture (Luu 2021). Using fishpond water to irrigate vegetables like onions, tomatoes, and cabbage lowers the farmer's need to purchase chemical fertilizers (Luu 2021). Instead of having to buy expensive, specially formulated diets, small-scale farmers can use vegetable wastes as fish food (Maulu et al. 2019). In Vietnam, animal manure and pond muck are used annually to fertilize crops and fruit trees. This method, applied in both highlands and lowlands, integrates fish and vegetable farming to recycle waste and utilizes existing supplies and machinery for small-scale operations (Anschell and Salamanca 2021). With the help of an integrated system, a farmer can cultivate vegetables in the pond dykes and fish in the pond (Luu 2021). Using fishponds to grow crops like alfalfa and wheat can enhance land productivity and economic returns. Successful trials demonstrate its viability. Fruits such as bananas, lemons, and papayas can be grown on pond dams, while vegetables like beans and squash are cultivated on pond slopes with bamboo structures (Dey et al. 2010). Several studies have been conducted on the utilization of fish farming effluent in crop production for growing diverse species, including tomato, lettuce, radish, petunia and melon (Rodrigues et al. 2010; Baumgartner et al. 2007; Abdul-Rahman et al. 2011; Danaher et al. 2013; Medeiros et al. 2010). Some aquatic weeds, like Azolla, duckweed, water hyacinth, and water spinach are used as fodder for fish and livestock (FAO 2000; Poot et al. 2010).

Table 8.2 Salt concentration and its irrigation suitability

Salt concentration (ppm)	Suitability for irrigation
<160	Appropriate for irrigating all crops across all land types
160–500	Suitable for irrigating crops sensitive to low salinity
500–1000	Ideal for low-salinity sensitive crops with proper drainage to wash away salt residue from the roots
1000–1500	Suitable for irrigating well-drained sandy land by increasing water amount gradually and selecting crops tolerant to salinity
>1500	Drip-spill systems handle up to 7000 ppm salinity when diluted with fresh water.

Source: Elewa and Nasr-Allah (2023)

Fishponds Wastewater

Fishponds are artificial bodies of water used for fish farming and recreational fishing. They've been utilized since the Middle Ages and are a common method of aquaculture. Sizes vary from 100 to 100,000 m^2, with depths typically ranging from 1.2 to 1.5 m (Ngo et al. 2017).

Utilizing fishpond wastewater for agriculture is a sustainable practice that enhances crop production and protects the environment. Benefits include improved irrigation, integration into aquaponics systems, and enhanced soil fertility (Hanjra et al. 2012). Effective management is crucial for maximizing benefits and ensuring sustainability.

Fishponds can be categorized according to several criteria, as shown in Fig. 8.2. Tank construction material is a key classification criterion, with options including concrete, plastic, and earthen ponds (Olagunju 2021). Farmers should choose based on corrosion resistance, ease of cleaning, and durability to improve yields and production efficiency.

Sustainable aquaculture development requires maximizing benefits while minimizing environmental impacts (Anamarija and Carl 2003). Water quality of fishpond effluent is vital for maximizing fish production in ponds. However, high nutrient concentrations from fish aquaculture can degrade water resources when discharged (Yusoff et al. 2024). Temperature, dissolved oxygen (DO), pH, total dissolved solids (TDS), electrical conductivity (EC), phosphorus, nitrate, and nitrogen are the main parameters affecting the quality of the water (Misman et al. 2023), as shown in Table 8.3. Fishpond aquaculture has the ability to affect water quality. According to Pulatsu (2004), aquaculture significantly impacts dissolved oxygen (DO), nitrate nitrogen (NO_3-N), and total phosphorus by increasing salt and

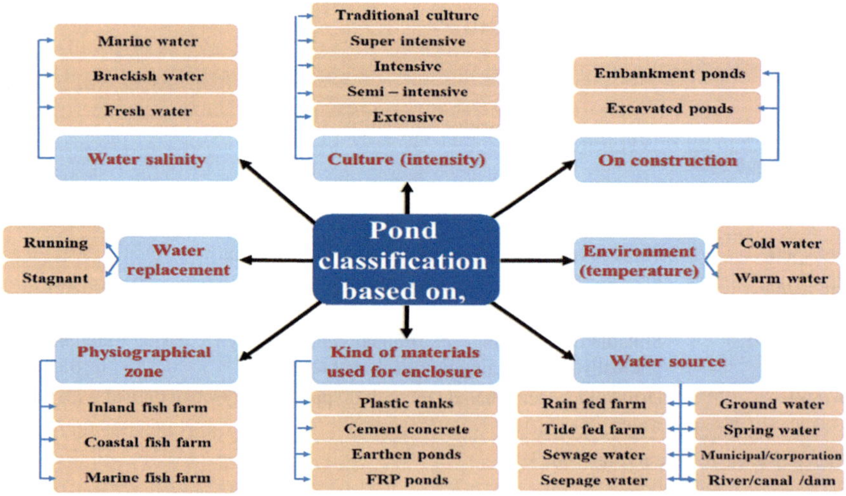

Fig. 8.2 Fishpond classification

Table 8.3 Fishpond effluent water quality

Parameter	Impact/value/range	References
Temperature	Ideal for tropical fish: 28–32 °C. Aquaculture range: 15–35 °C	Bhatnagar and Pooja (2013) and Agbaire (2015)
DO	Optimum >5 mg/L. Tolerable for catfish and tilapia: 3 mg/L and higher	Makori et al. (2017) and Keremah (2014)
pH	<5 mg/L results in decreased feed intake, higher FCR, sluggish development, stress, and illness.	Bhatnagar and Pooja (2013)
	Fatal at <1 mg/L. Recommended: 4–6 mg/L	WHO (2009)
	Ideal range: 6.5–9.0. Ideal for fishponds: 7–8	Adedeji and Adeniyi (2013) and Bhatnagar and Pooja (2013)
	Tolerable range: 6.5–8.5. Extreme values lead to death, pH 9–11 slows development, and <4 and 11 is fatal.	WHO (2009) and WRC (2010)
TDS	EMCA (2006) norm: 1200 mg/L, WHO (2009) max: 1000 mg/L, freshwater: 50–1500 mg/L	EMCA (2006), Nwinyimagu et al. (2016), FEPA (1991), Niels et al. (2016), and WHO (2009)
	Excess kills fish and restricts development; sign of pollution	Weber and Duffy (2007)
EC	Range: 30–5000 μS/cm. Preferred: 100–2000 μS/cm	Bhatnagar and Pooja (2013), FEPA (1991) and WHO (2009)
Nitrate	Freshwater max: 10 mg/L. Contamination >4 mg/L. Aquaculture standard: 50 mg/L	US-EPA (2017), WHO (2009), and ANZECC (2000)
Phosphorus	Permitted for aquaculture: 0.03–3 mg/L. Ideal for fish development: 0.005–0.2 mg/L	WHO (2009) and Boyd (2003)

suspended matter in the water. However, variations in pH and ammonia-nitrogen (NH_3-N) levels are minimal.

Using Fishponds Wastewater in Agriculture

Integrated agriculture-aquaculture enhances food security and sustainability by leveraging synergies between agriculture and livestock farming, recycling farming waste as a resource for other activities (Walia and Kaur 2013). In these systems, water is reused on site to support agriculture, avoiding direct wastage as seen in other fish systems (Corner et al. 2020). The Food and Agriculture Organization (FAO) emphasized their significance, reporting on integrated irrigated agriculture, wherein water productivity is enhanced by cultivating fish in irrigation canals and using the water for crop cultivation (FAO 2020). Mustapha et al. (2014) proposed dual use of irrigation water storage basins, producing fish first and then utilizing the effluent-rich water as fertilizers for irrigation. Similarly, Van der Heijden et al. (2012) suggested fish farming in water storage reservoirs, maintaining the water's usability for irrigation. The use of fish farming ponds involves not only the use of fish wastewater to irrigate crops but also the use of pond sediments to fertilize crops. Pond sediments can serve as a source of important nutrients for plant growth (Dróżdż et al. 2020), which could be reduced cost and needs of chemical fertilizers by 25–50%, or even up to 75% under cucumber and water spinach plants (Da et al. 2021). This approach improves environmental sustainability and reduces ecosystem impacts. Integrated aquaculture systems, combining tilapia and vegetables, boost fish yield by 60% and increase net income by 175% due to reduced costs and higher yields. This integration enhances water use efficiency and reduces reliance on synthetic fertilizers (Mulokozi 2021). Desalinated water is used to irrigate high-value crops and support aquaculture (Martinez-Alvarez et al. 2020; Tanner and Drabsch 2021). Numerous worldwide studies report the benefits of using sediments and wastewater from fishponds on crop water productivity as shown in Tables 8.4 and 8.5.

Potential Effects of Fishpond Wastewater on Soils, Crops Growth, and Productivity

In 2006, the FAO proposed "integrated irrigated agriculture," combining fish farming in irrigation canals utilizing their slightly saline drainage water for irrigation, and suggested growing fish in water reservoirs for dual irrigation use (Van der Heijden et al. 2012). This approach is used in the Czech Republic, where large fishponds are part of the environment (Adamek et al. 2012). Fishponds act as irrigation water reservoirs, attract wildlife, and allow for recreational activities. Reservoirs that supply water for irrigation and hydropower can also be used to grow fish (Finlayson et al. 2013). Table 8.6 illustrated the production of fish and crops at different countries in the world.

Utilizing fish wastewater for irrigation positively influences crop characteristics by supplying essential nutrients. Naddafi et al. (2005) and Moursy (2018) observed increased growth rates and improved quality in various crops, such as sugar beet. Da et al. (2021) noted improvements in soil properties with pond effluent application, including improved infiltration, soil stability, water retention, and nutrient retention. Osaigbovo et al. (2010) found significant increases in plant height, stem girth, and dry weight with higher fishpond effluent concentrations in maize crop. Water from fish farming increases production by 75–175% compared to tap water when used for cultivating cabbage and amaranth, respectively (Mulokozi et al. 2022). Mariscal-Lagarda et al. (2012) found that tomato yield was 30% higher with irrigation using shrimp effluent compared to groundwater. Similarly, Zohry et al. (2020) observed that fish farm effluent increased faba bean yield by 11% and chickpea yield by 29% compared to Nile water. Integrated aqua-agriculture systems boost crop yields and efficiency (Dey et al. 2010; Mariscal-Lagarda et al. 2012; Moursy et al. 2019, 2022).

Table 8.4 International experience for using fishponds wastewater and sediment on crop water productivity

	Methodology	Crop productivity (ton/ha)		References
Vietnam	T1, control using 100% C.F.	18.96		Da et al. (2021)
	T2, 75% of C.F., and 25% of organic fertilizers (pond sediment)	25.63		
	T3, 50% of C.F., and 50% of pond sediment	28.65		
	T4, 25% of C.F., and 75% of pond sediment	25.47		
	T5, 0% of C.F., and 100% of pond sediment	12.97		
Vietnam		Malabar spinach	Amaranthus cruentus	Thanh et al. (2023)
	T1, 100% inorganic fertilizer (I.O.F.)	17.7 ± 0.5	9.9 ± 0.1	
	T2, 75% I.O.F. + 25% organic fertilizer (O.F)	17.7 ± 0.6	10 ± 0.1	
	T3, 50% I.O.F. + 50% O.F.	17.8 ± 0.7	10 ± 0.1	
	T4, 25% I.O.F + 75% O.F.	17.7 ± 0.5	9.9 ± 0.1	
	T5, 100% O.F.	17.5 ± 0.3	9.7 ± 0.6	
Poland	Fresh fishpond sediments were mixed with compost in three variations: M1 with cardboard waste (1:0.09), M2 with woodchips-derived biochar (1:0.02), and M3 with both cardboard waste and biochar (1:0.09:0.02). All mixtures were then combined with soil.	Wet weight (gm)	Dry weight (gm)	Dróżdż et al. (2020)
	T1, soil	13.94 ± 1.09	2.98 ± 0.71	
	T2, soil + compost (1:0.03)	17.68 ± 1.68	3.14 ± 0.38	
	T3, soil + M1 (1:0.03)	13.99 ± 2.93	2.21 ± 0.56	
	T4, soil + M2 (1:0.03)	17.27 ± 5.18	3.09 ± 0.91	
	T5, soil + M3 (1:0.03)	11.64 ± 3.31	1.74 ± 0.80	
Bangladesh	Control, no external fertilizer	28		Haquea et al. (2016)
	T1, 25% urea and 75% fishpond sediment (FPS)	53		
	T2, 50% urea and 50% FPS	44		
	T3, 75% urea and 25% FPS	49		
	T4, 100% urea and 0% FPS	41		
	T5, 0% urea and 100% FPS	58		

Environmental Impacts of Fishpond Wastewater

Aquaculture is an interdisciplinary field that encompasses environmental engineering, soil mechanics, chemistry, and biology (Edwards 2015; Wu and Song 2021). Food production impacts the environment by altering soil, water, and landscapes,

Table 8.5 International experience for using fishponds wastewater on crop water productivity

	Methodology	Crop productivity (ton/ha)					Crop water productivity (kg/m^3)					Water productivity (\$/m^3)					References
Tanzania	A study on various fish stocking densities and fertilizer applications:	Cabbage			Amaranth		Cabbage		Amaranth			Cabbage		Amaranth			Mulokozi (2021)
	Low fish stocking (LFS), no fertilizer	26.1			12.3		3.33		1.55			1.3		0.61			
	LFS, partially fertilized	34.5			18		4.36		2.33			1.71		0.91			
	Medium fish stocking (MFS), no fertilizer	40.1			25.5		5.07		3.23			1.99		1.27			
	MFS, partially fertilized	45.02			28.4		5.69		3.59			2.23		1.41			
	High fish stocking (HFS), no fertilizer	44.1			32.8		5.56		4.14			2.18		1.62			
	HFS, partially fertilized	51.2			38.1		6.44		4.76			2.04		1.14			
	Tap water with full fertilization	41.3			30.3		5.21		2.91			2.53		1.87			
	Tap water with no fertilizer	14.9			4.5		1.92		0.56			0.75		0.22			
	Evaluate the impact of varying water resources and nitrogen levels:		Well		Aquaculture			Well		Aquaculture			Well		Aquaculture		
	Two water sources (well and aquaculture) with nitrogen levels (20, 60, 100%) under surface (SD) and subsurface (SSD) drip irrigation	N%	SD	SSD	SD	SSD	N%	SD	SSD	SD	SSD	N%	SD	SSD	SD	SSD	Soliman et al. (2020)
		20	0.82	0.74	20	0.82	20	0.17	0.16	0.18	0.17	20	2.13	1.94	4.56	4.48	
		60	0.98	0.88	60	0.98	60	0.2	0.19	0.21	0.2	60	2.52	2.29	5.01	4.91	
		100	1.1	0.99	100	1.1	100	0.22	0.21	0.22	0.21	100	2.78	2.53	4.94	4.82	
	Two water resources (ground water and aquaculture) under drip irrigation systems (SD and SSD)	Well		Aquaculture			Well		Aquaculture			–		–			Soliman et al. (2020)
		SD	17.6	19.2			SD	6.6	6.9								
		SSD	14.5	17.03			SSD	5.7	6.5								

Ethiopia	Compared between different fertilizer resources:	Cabbage	Onion	Tomato	–	–	–	–	Lemma (2017)
	T11, river water (not part of the integrated system) + 3 types of horticultural crops	9885	16,567	17,150					
	T12, T11 + recommended inorganic fertilizer +3 horticultural crop varieties	21,042	35,531	45,832					
	T21, pond water (water in the integration system) + 3 horticultural crop varieties	22,183	37,160	46,914					
	T22, T21 + recommended inorganic fertilizer +3 horticultural crop varieties	23,634	38,140	47,661					
Lebanon	Compared between 2 water resources:	Maize		Radis	Maize	Radis	Maize	Radis	Abdul-Rahman et al. (2011)
	Aquaculture effluent and no fertilizer	70.7		0.373	1.96	8.24	7.07	10.3	
	Aquaculture effluent and inorganic fertilizer	69.6		0.3733	1.91	8.24	6.69	10.3	
	Well water and no fertilization	42.1		0.3093	1.16	6.83	4.21	8.53	
	Well water with inorganic fertilizer	61.0.5		0.3303	1.69	7.29	6.15	9.11	
USA	Compared between 2 water resources:	Cotton		Barley	–	–	–	–	Stevenson et al. (2010)
	Ground water with no chemical fertilizer (C.F.)	1.890 ± 0.18		2.184 ± 0.34					
	Ground water with C.F.	1.808 ± 0.08		3.688 ± 0.03					
	Fishpond effluent with no C.F.	1.989 ± 0.04		2.779 ± 0.39					
	Fishpond effluent with C.F.	2.033 ± 0.03		4.242 ± 0.2					

Table 8.6 Productivity of fish and crops in some countries in integrated agriculture-aquaculture systems

Country	Crop	Fish used	Productivity (ton/ha)		References
			Fish	Crops	
South Africa	Tomato Spinach Lettuce	Common carp, bighead carp, silver carp, and grass carp	19.5	32.18	Prinsloo and Schoonbee (1987)
Bangladesh	Cucumber	Prawn, carp, and mola	0.6, 1.3, and 0.2	1.5	Ahmed et al. (2007)
Vietnam	No data	Common carp, silver barb, kissing gourami, tilapia, and catfish	0.48	1.62	Phong et al. (2008)
Malawi	Maize	Tilapia	2.02	1.98	Tran et al. (2013)
Ethiopia	Onion (Allium cepa)	O. niloticus, C. carpio, and C. gariepinus	8	10.8	Tugie et al. (2017)
Madagascar	Orchards	–	0.7–1.3	–	Mortillaro et al. (2017)
Nigeria	Cucumber	–	–	5.86	Ikechukwu et al. (2023)
Tanzania	Chinese cabbage (Brassica spp)	Tilapia (78%), catfish (16%), and a polyculture of tilapia and catfish (6%)	2.46	28	Mulokozi (2021)

causing habitat loss, and reducing wildlife diversity. Integrated agriculture-aquaculture systems, being semi-intensive, use lower inputs compared to intensive systems, which reduces reliance on feed and fertilizers, lower organism density, and minimized pollution and disease risks. The rapid expansion of aquaculture can have both positive and negative environmental effects. Its impact depends on factors such as the species farmed, management practices, production location, and feed quality and management (Wang et al. 2005; Podemski and Blanchfield 2006).

- *Advantages of Aquaculture*
  - *Improved Resource Use*
  - Aquaculture systems use lower inputs compared to intensive systems, reducing the reliance on feed and fertilizers, which helps minimize pollution and

lowers disease risks. This efficient use of resources contributes to sustainability (Wang et al. 2005; Podemski and Blanchfield 2006).

- *Soil Improvement*
- Fish wastewater contains essential nutrients such as phosphorus, potassium, calcium, and magnesium, which can enhance soil fertility. Fish effluent, when used properly, can be an excellent fertilizer for nearby farmlands (Musa et al. 2020). Ojobor and Tobih (2015) and Ndubuisi (2019) report that fishpond effluent positively affects soil chemical properties due to its high nutrient content.
- *Pollution Reduction*
- Aquaculture can help reduce pollution by recycling nutrients and organic waste. It also enhances food production while minimizing the negative environmental impacts associated with traditional farming and intensive aquaculture systems (Ahmed and Garnett 2011).
- *Sustainability and Low Impact*
- The integration of agriculture and aquaculture in semi-intensive systems leads to more sustainable farming practices that reduce the impact on the environment by enhancing nutrient recycling and minimizing the need for external inputs (Edwards 2015). Using pond sediment in agriculture mitigates aquaculture's environmental impact, as it enriches soil with organic matter and phosphorus, enhancing fertility and sustainability (Rahman et al. 2004).

• *Disadvantages of Aquaculture*

- *Pollution from Waste*
- Intensive aquaculture systems produce significant amounts of waste, including medications, chemicals, solid waste, and uneaten food. If not properly managed, solid waste can deplete oxygen and release toxic ammonia, negatively affecting water quality and nearby ecosystems (Liu et al. 2002).
- *Chemical and Pathogen Contamination*
- Wastewater from aquaculture systems contains various pollutants, including chemicals, antimicrobial agents, and bacteria. If untreated, this wastewater can contaminate receiving water bodies, posing a threat to aquatic life and local ecosystems (Rico et al. 2012; Jegatheesan et al. 2011).
- *Soil Degradation*
- The use of fish effluent as fertilizer can sometimes lead to soil degradation, including a decrease in organic carbon, nitrogen, and cation exchange capacity over time. Studies show that the long-term use of fish wastewater may negatively affect soil properties, especially when not balanced with other agricultural inputs (Osaigbovo et al. 2010; Musa et al. 2020).
- *Spread of Diseases*
- Aquaculture systems can contribute to the spread of diseases and pathogens such as bacteria, viruses, and parasites, which may escape into surrounding environments and affect wild aquatic species (Jegatheesan et al. 2011).

Socioeconomic Scopes Fishpond Wastewater

Aquaculture impacts communities beyond food security and environmental effects. Currently, 18.7 million people are employed as fish farmers, with numbers tripling to quadrupling when including postharvest and secondary jobs (FAO 2016). There has been increased training and workforce participation for women in aquaculture, a more planned and labor-intensive method compared to traditional capture fishing. One of the main sources of income for many households and a means of support for many farmers is aquaculture (Bhari and Visvanathan 2018). Integrated fishponds, which produce fish, crops, and protein, boost farm production by 11% and net income per hectare by 134% compared to nonintegrated fishponds (Shoko et al. 2019). Rice-fish farming is more cost-effective and efficient than alternate methods, with production comparable to rice monoculture (Dwiyana and Mendoza 2008). It helps meet food demand but struggles with low acceptance, high costs, and environmental risks (Ahmed et al. 2011). Many people in desert and remote rural areas are not accustomed to eating fish. This technique can provide them with nutritious, protein-rich fish, improving their nutrition and health, particularly for children, while reducing deficiencies and malnutrition. It also creates job opportunities in the sector (Corner et al. 2020).

Water Productivity for Fishpond Wastewater

Water productivity (WP) measures efficiency by comparing water used to output of water use (FAO 2022). Improving WP is crucial for sustainable water management, especially in water-stressed areas, through technologies, effective irrigation, recycling, and conservation. Water use efficiency and WP are key metrics to reduce agricultural water consumption (Hussain et al. 2007). Physical water productivity is the ratio of crop (fish, rice, and other aquatic foods) output to water used (kg product/m^3 water), while economic water productivity measures the revenue per unit of water consumed ($/$m^3$ water) (Mohanty et al. 2014).

Pond aquaculture, one of the oldest fish farming methods, maintains conditions for fish and is nearly as efficient as crop production in producing edible dry matter and protein (Pillay 1990). In dry regions, it is more effective than agriculture only if it is highly intensive or integrated with other agricultural activities to optimize water and nutrient use (Brummett 2007). Table 8.7 showed the water productivity for fish and crops at different countries.

Economic Feasibility of Fishpond Wastewater

Integrated aquaculture can boost food production and self-sufficiency, enhancing economic empowerment for rural families. It provides year-round income through consistent fish and plant production (Nishan et al. 2015). Integrated aquaculture often yields higher income than traditional farming due to the profitability of both

Table 8.7 Water productivity for fish and crop at different countries using fishpond wastewater

			Aquaculture WP (kg/m^3)		Agriculture WP (kg/m^3)	
Country	Fish used	Crop	Fish	Crop	Crop	References
Cameroon	Nile tilapia	Wheat	0.32	0.59	–	Brummett (2007)
Lebanon	Nile tilapia	Maize	0.22	1.91	–	Abdul-Rahman et al. (2011)
		Vegetable	0.22	8.24	–	
Egypt	Nile tilapia	Cabbage	8.4	12.38–20.73	6.64–1274	Elsbaay and Darwesh (2022)
Tanzania	Nile tilapia	Cabbage	–	1.92 ± 0.1	0.56 ± 0.04	Mulokozi et al. (2022)
		Amaranth	–	3.33 ± 0.09	1.55 ± 0.14	

plant and fish products. Farmers leverage these higher profits to achieve food security, as the integrated system is more profitable (Oladimeji and Isah 2019). This technique can also lower water acquisition costs and enhance soil fertility (Mafwila et al. 2019).

Irrigation with fish farm wastewater improves the cost-benefit ratio of both aquaculture and agriculture. Revenue from fish sales can cover water costs, and in some areas, small-scale fishing can be a key revenue source (Schuhbauer and Sumaila 2016; Watanabe and Tahara 2016). Azari and Udoh (2023) found that fish farming was more profitable than crop farming, with a net income of Rs. 252,426 per acre compared to Rs. 58,612 for crops. Table 8.8 showed the fishpond integrated farming systems have varying profitability and economic returns in different countries.

Optimal Approaches for Utilizing Fishponds Wastewater

Methods of Water Purification

Efficient water treatment techniques such as sedimentation, filtration, aeration, and biological treatments enhance water quality by removing hazards and contaminants from fishpond wastewater before irrigation. Investing in proper infrastructure and modern technologies is key to ensuring water quality and crop safety (Mukherjee and Chakraborty 2017).

Guidelines for Crop Selection and Rotation

Selecting suitable crops for fishpond wastewater irrigation is vital for maximizing productivity and preventing soil depletion. Systematic crop rotation helps maintain soil fertility and avoid nutrient imbalances (Riya et al. 2023).

Strategies for Surveillance and Management

Ongoing monitoring of soil and water quality, productivity, and environmental impacts is crucial for effective fishpond wastewater irrigation. Developing adaptive

Table 8.8 Economic analysis of fishpond farming at different countries in the world

Country items	Nigeria (₦)		Kenya (KSh)		Bangladesh (৳)	Malawi (MK)	Egypt (£E)	
	Fish farming	Vegetable production	Fish farming	Vegetable production	Fish farming	Fish farming	Fish farming	Rice production
Fishpond area, m^2	625	–	4000	–	10,000	299.5	400m^3	–
Crop	–	Pumpkin	–	Maize	–	–	Rice	–
Total variable cost, $	246.62	8.87	–	–	14,879	29	–	–
Total fixed cost, $	14.73	1.78	–	–	755	–	–	–
Total cost, $	261.35	10.65	297.90	182.60	15,634	–	35,937	1026
Total revenue, $	306.96	28.09	412.54	344.83	17,396	98	40,924	12,273
Gross margin, $	60.34	19.22	–	–	2517	69	4987	11,247
Net farm income, $	45.61	17.44	114.64	324.46 = (162.23*2)	1763	58	–	–
Profit margin (%)	14.86	62.1	27.8	94.1	10.13	–	–	–
Gross ratio	19.66	37.9	–	–	14.47	–	–	–
Return on investment	1.17	2.64	1.38	1.89	–	–	–	–
Net margin to cost ratio	0.18	1.63	–	–	–	–	–	–
BCR(B/C)	–		–	–	1.11	–	–	11.96
References	Oladimeji and Isah (2019)		Wambua (2015)		Mukta et al. (2019)	Munthali et al. (2022)	Goda et al. (2024)	

management strategies based on this data enhances resource efficiency, protects crops and soil, and mitigates risks (Iglesias and Garrote 2015).

Monitoring and Evaluation of Integrated Aquaculture Systems

A performance assessment is crucial for influencing policy decisions, yet holistic evaluations of aquaculture systems are scarce (Rossignoli et al. 2021). Monitoring and evaluating integrated aquaculture systems are crucial for ensuring their sustainability, productivity, and environmental compatibility.

Monitoring

A monitoring framework organizes data into inputs, activities, and outputs to meet policymakers' and project managers' needs. It utilizes specific indicators, such as the adoption of new technologies by fish farmers, to streamline monitoring in a cost-effective manner (Zeller et al. 2022). Key data for integrated aquaculture systems include water quality (DO, NH_3, nitrite, NO_3, pH, EC), fish health (growth rates, mortality rates, disease prevalence), crop productivity, and waste management (efficiency of treatment and nutrient recycling) (Yusoff et al. 2024). Economic performance is evaluated through production costs, revenues, and profitability, while social aspects assess the system's impact on local communities, including employment and food security (Tunde et al. 2015).

Evaluation

The performance evaluation of the integrated aquaculture system involves analyzing data to assess effectiveness against predefined goals and standards. An Environmental Impact Assessment (EIA) evaluates potential impacts like water pollution and biodiversity loss (Glasson and Therivel 2013). A comprehensive sustainability assessment focuses on ecological, economic, and social criteria, while a risk assessment identifies potential risks such as disease outbreaks and market fluctuations (Lind et al. 2015). Engaging stakeholders, including farmers and policymakers, offers valuable feedback on the system's performance. Recommendations are then developed to enhance efficiency, sustainability, and overall performance based on the assessment finding.

Tools and Methods

To optimize integrated aquaculture systems for sustainability, profitability, and minimal environmental impact, a multifaceted approach is essential (Kamleshbhai et al. 2023). This involves using various data collection tools, including sensors, water sampling, surveys, and interviews, to gather comprehensive information (Yusoff et al. 2024). Geographic information systems and remote sensing technologies aid in spatial mapping and monitoring, while modeling and simulation tools predict system behavior and optimize resource allocation (Kapetsky and Aguilar-Manjarrez 2007). Capacity-building programs provide training for farmers and stakeholders to improve skills in monitoring and sustainable management (Mohanty et al. 2024). Together, these strategies establish an effective framework for the monitoring and evaluation of integrated aquaculture systems, ensuring their long-term viability and contribution to sustainable development.

Worldwide Case Studies on Integrated Aquaculture

Successful Implementation

Successful implementation of integrated aquaculture systems has been observed in several countries, highlighting innovative approaches that enhance sustainability and productivity. In Chile, the salmon industry adopted recirculation systems in the 1990s, allowing for cost-effective smolt production and enabling four batches annually instead of one (Bregnballe 2015). Denmark excels in eco-friendly trout farming by using groundwater in closed systems, maintaining consistent water temperatures, and reducing costs (Bregnballe 2015). Algeria's "Pescado de la Duna" produces 300 tons of catfish annually alongside crops like tomatoes and cucumbers using a flow-through system (Corner et al. 2020). Oman has implemented recirculating aquaponic systems for growing salad vegetables and herbs, repurposing wastewater for irrigation and reducing fertilizer use by about 50% (Corner et al. 2020). Vietnam utilizes integrated aquaculture in artificial basins to enhance crop production and soil health, significantly contributing to its economy (Nhan et al. 2007). Finally, Thailand employs treated water basins for shrimp production, integrating fishpond wastewater into agriculture to boost crop yields and minimize environmental impact (Szuster 2003). These examples demonstrate the potential of integrated aquaculture systems to promote economic growth and sustainability.

Unsuccessful Studies and Lessons Learned

Unsuccessful studies in various countries have provided important lessons for improving the integration of fishpond wastewater in agriculture. In India, challenges arose from poor water quality management, underscoring the necessity of regular water testing and treatment before irrigation (Patel and Dudhagara 2023). Similarly, Indonesia encountered issues such as nutrient imbalances and soil compaction, highlighting the critical need for effective nutrient management and soil health assessments (Leimona et al. 2015). In Kenya, the utilization of fishpond wastewater was impeded by water scarcity and inadequate treatment infrastructure, emphasizing the importance of sustainable water management and investment in proper treatment technologies (Mulwa et al. 2021). Ghana also faced challenges with water contamination and poor nutrient management when using fishpond wastewater, which reinforced the need for rigorous water quality monitoring and balanced nutrient management practices (Asiedu et al. 2016). Collectively, these cases illustrate that while integrating fishpond wastewater in agriculture has potential benefits, success requires addressing critical issues related to water quality, nutrient management, and infrastructure development.

Challenges of Using Fishponds Wastewater on Crop Production

Integrated aquaculture-agriculture is a promising strategy for sustainable food production by combining aquaculture with traditional agriculture. However, it faces several challenges that need to be addressed for successful implementation:

- Integrated aquaculture-agriculture systems need to diversify resources and activities to produce high-quality, affordable products. However, there is a risk that farmers might choose specialized, less sustainable systems for higher income from premium products.
- Inconsistent nutrient composition in fishpond wastewater can harm plant health and growth.
- Contaminants in wastewater pose potential risks of introducing pathogens to crops.
- Improper handling of integrated aquaculture-agriculture systems can result in environmental pollution, leading to detrimental effects on ecosystems and water bodies.
- Adhering to water quality and environmental regulations is challenging for integrated aquaculture-agriculture practitioners, requiring investment in infrastructure and adherence to best practices.
- Crops in integrated aquaculture systems can be vulnerable to waterborne diseases from wastewater, impacting yield and quality.

Conclusions

Using fishpond wastewater in agriculture enhances crop productivity by recycling nutrients, which reduces waste and environmental impact. The wastewater provides essential nutrients like nitrogen, phosphorus, and potassium, improving soil fertility and plant growth. It also promotes water conservation and reduces the need for freshwater irrigation, benefiting drought-prone areas. This approach supports sustainable agriculture and the circular economy by closing nutrient loops and minimizing environmental nutrient losses.

Recycling wastewater for agricultural irrigation helps retain nutrients, reducing environmental degradation and pollution compared to discharging it into water bodies. This practice can enhance farm profitability by lowering input costs and increasing crop yields, as it eliminates the need for chemical fertilizers and freshwater irrigation. To ensure successful implementation, it is essential to address potential issues such as nutrient imbalance, pathogen risks, and regulatory compliance. Effective monitoring and nutrient management, along with adherence to environmental standards, are key to sustainable outcomes. Proper filtration, waste management, and pollution control measures are necessary to minimize disease spread and environmental impacts.

Using fishpond wastewater to boost crop productivity benefits society, the environment, and the economy. This approach enhances food security, environmental preservation, and rural development through integrated management. It fosters a symbiotic relationship between aquaculture and agriculture, improving system resilience, stability, and sustainability by maximizing resource use and spreading risks.

References

Abdelraouf R, Hoballah ES (2014) Impact of irrigation systems, fertigation rates and using drainage water of fish farms in irrigation of potato under arid regions conditions. Int J Sci Res Agric Sci 1(5):67–79

Abdul-Rahman S, Saoud IP, Owaied MK, Holail H, Farajalla N, Haidar M, Ghanawi J (2011) Improving water use efficiency in semi-arid regions through integrated aquaculture/agriculture. J Appl Aquac 23(3):212–230. https://doi.org/10.1080/10454438.2011.600629

Adamek Z, Linhart O, Kratochvil M, Flajshans M, Randak T, Policar T, Kozak P (2012) 2012: a prosperous and modern European sector based on a thousand-year history of pond culture. World Aquacult 1012(68):20–27

Adedeji AA, Adeninji IF (2013) Effects of pond fertilization on the physico—chemical water quality of selected earthen fish ponds in Osun state, Nigeria. Zoologist 11:13–20

Agbaire PO, Akporido SO, Emoyan OO (2015) Determination of some physico-chemical parameters of water from artificial concrete fish ponds in Abraka and its environs, Delta state, Nigeria. Int J Plant Anim Environ Sci 5(3):70–76

Ahmed N, Garnett ST (2011) Integrated rice-fish farming in Bangladesh: meeting the challenges of food security. Food Secur 3(1):81–92

Ahmed N, Wahab MA, Thilsted SH (2007) Integrated aquaculture-agriculture systems in Bangladesh: potential for sustainable livelihoods and nutritional security of the rural poor. Aquacult Asia 12:14–22

Ahmed N, Zander KK, Garnett ST (2011) Socioeconomic aspects of rice-fish farming in Bangladesh: opportunities, challenges and production efficiency. Aust J Agric Resour Econ 55:199–219

Anamarija F, Carl H (2003) Sustainable aquaculture: developing the promise of aquaculture. Aquac Int 11:517–530

Anschell N, Salamanca A (2021) Integrated agriculture-aquaculture systems for climate change adaptation, mitigation and new livelihood opportunities, ASEAN climate-smart land use insight brief 1. Deutsche Gesellschaft für International Zusammenarbeit (GIZ), Jakarta

ANZECC (2000) The Australian and New Zealand guidelines for fresh and marine water quality, National water quality management strategy 4(1). Australian and New Zealand Environment and Conservation Council/Agriculture and Resource Management Council of Australia and New Zealand, Canberra, Chapter 1–7

Asiedu B, Asase A, Iddrisu S, Ayisi LC (2016) Management, challenges and developments of aquaculture in Ghana, West Africa. Nova Science, New York, pp 1–28. ISBN: 978-1-63485-563-1

Azari AA, Udoh FE (2023) Economic viability of fish farming compared to crop cultivation. Int Contemp J Sci Technol 2(2). https://doi.org/10.5281/zenodo.7859463
Baumgartner D, Sampaio SC, Silva TD, Teo CRPA, Vilas Boas MA (2007) Reúso de águas-residuárias da piscicultura e da suinoculturanairrigação da cultura da alface. Eng Agric 27(1):152–163
Bhari B, Visvanathan C (2018) Sustainable aquaculture: socio-economic and environmental assessment. Springer
Bhatnagar A, Pooja D (2013) Water quality guidelines for the management of pond fish culture. Int J Environ Sci 3(6):1980–1993
Boyd EC (2003) Guidelines for aquaculture effluent management at the farm level. Elsevier Aquacult 226:101–112
Boyd EC, Lim C, Queiroz J, Salie K, De-Wet L, Mc-Nevin A (2008) Best management practices for responsible aquaculture. In: USAID: aquaculture collaborative research support program. Oregon State University, Corvallis, p 47
Bregnballe J (2015) A guide to recirculation aquaculture. An introduction to the new environmentally friendly and highly productive closed fish farming systems. The Food and Agriculture Organization of the United Nations (FAO) and EUROFISH International Organisation
Brummett RE (2007) Comparative analysis of the environmental costs of fish farming and crop production in arid areas. In: Bartley DM, Brugère C, Soto D, Gerber P, Harvey B (eds) Comparative assessment of the environmental costs of aquaculture and other food production sectors: methods for meaningful comparisons, FAO/WFT expert workshop, 24–28 April 2006, Vancouver, Canada. FAO fisheries proceedings no. 10. FAO, Rome, pp 221–228
Corner R, Fersoy H, Crespi V (2020) Integrated agri-aquaculture in desert and arid lands: learning from case studies from Algeria, Egypt and Oman, Fisheries and aquaculture circular no. 1195. FAO, Cairo. https://doi.org/10.4060/ca8610en
Da CT, Vu TH, Duy DT, Ty NM, Thanh DT, Nguyen-Le MT, Berg H, Nguyen QH, Bui XT (2021) Recycled pangasius pond sediments as organic fertilizer for vegetables cultivation: strategies for sustainable food production. Clean Techn Environ Policy 9:369–380. https://doi.org/10.1007/s10098-021-02109-9
Danaher JJ, Pickens JM, Sibley JL, Chappell JA, Hanson TR, Boyd CE (2013) Petunia growth response to container substrate amended with dewatered aqua-culture effluent. Hortic Technol 23(1):57–63
Dey MM, Paraguas FJ, Kambewa P, Pemsl DE (2010) The impact of integrated aquaculture-agriculture on small-scale farms in Southern Malawi. Agric Econ 41:67–79. https://doi.org/10.1111/j.1574-0862.2009.00426.x
Dróżdż D, Malińska K, Kacprzak M, Mrowiec M, Szczypiór A, Postawa P, Stachowiak T (2020) Potential of fish pond sediments composts as organic fertilizers. Waste Biomass Valorization 11:5151–5163. https://doi.org/10.1007/s12649-020-01074-6
Dwiyana E, Mendoza TC (2008) Determinants of productivity and profitability of rice-fish farming systems. Asia Life Sci 17(1):21–42
Edwards P (2015) Aquaculture environment interactions: past, present and likely future trends. Aquaculture 447:2–14
Egna HS, Boyd CE (1997) Dynamics of pond aquaculture. CRC Press, Boca Raton. 437 pp. [A complete review of pond ecology and management of trophic food webs]
Elewa A, Nasr-Allah A (2023) Integrated aquaculture-agriculture: fish culture and plant crops module for arid areas, Guidelines: 2023-11. WorldFish, Penang
Elnwishy N, Salh M, Zalat S (2006) Combating desertification through fish farming. The future of drylands. In: Proceedings of the international scientific conference on desertification and drylands research, pp 19–21
Elsbaay AM, Darwesh MR (2022) Effect of fish effluent on cabbage yield under organic mulching conditions. Agric Eng Int CIGR J 24(3):1–11
EMCA (2006) Environmental management and co-ordination (water quality) regulations, Legal notice no. 121. National Environmental Management Authority. Ministry of Environment and Natural Resources, Kenya

FAO (2000) Small ponds make a big difference: integrating fish with crop and livestock farming. Food and Agriculture Organization of the United Nations (UN), Rome
FAO (2012) The state of world fisheries and aquaculture. Food and Agriculture Organization of the United Nations, Rome. https://www.fao.org/3/i2727e/i2727e.pdf
FAO (2014) The state of world fisheries and aquaculture. FAO, Rome
FAO (2016) The state of world fisheries and aquaculture 2016. Contributing to food security and nutrition for all. FAO, Rome, 200 pp. http://www.fao.org/3/a-i5555e.pdf
FAO (2020) The state of world fisheries and aquaculture 2020. Food and Agriculture Organization of the United Nations
FAO (2022) The state of world fisheries and aquaculture 2022. Towards blue transformation. FAO, Rome. https://doi.org/10.4060/cc0461en
FEPA (1991) Guidelines and standard for environmental pollution control in Nigeria, vol 27. Federal Environmental Protection Agency, p 20
Finlayson M, Bunting SW, Beveridge M, Tharme RE, Nguyen-Khoa S (2013) Wetlands. In: Boelee E (ed) Managing water and agroecosystems for food security. CAB International, Wallingford, pp 82–103
Glasson J, Therivel R (2013) Introduction to environmental impact assessment. Routledge, p 13
Goda AMA, Aboseif AM, Mohammedy EY, Taha MKS, Mansour AIA, Ramadan EA, Aboushabana NM, Zaher MM, Otazua NI, Ashour M (2024) Earthen pond-based floating beds for rice-fish co-culture as a novel concept for climate adaptation, water efficiency improvement, nitrogen and phosphorus management. Aquaculture 579:740215. https://doi.org/10.1016/j.aquaculture.2023.740215
Halwart M (2020) Fish farming high on the global food system agenda in 2020. FAO Aquacult Newslett 61:II–III
Hanjra MA, Blackwell J, Carr G, Zhang F, Jackson TM (2012) Wastewater irrigation and environmental health: implications for water governance and public policy. Int J Hyg Environ Health 215(3):255–269
Haquea MM, Beltonb B, Alamd MM, Ahmeda AG, Alama MR (2016) Reuse of fish pond sediments as fertilizer for fodder grass production in Bangladesh: potential for sustainable intensification and improved nutrition. Agric Ecosyst Environ 216:226–236. https://doi.org/10.1016/j.agee.2015.10.004
Hejnowicz AP, Thorn JP, Giraudo ME, Sallach JB, Hartley SE, Grugel J, Pueppke SG, Emberson L (2022) Appraising the water-energy-food nexus from a sustainable development perspective: a maturing paradigm? Earth's Future 10(12):e2021EF002622
Hussain I, Turral H, Molden D, Ahmad MD (2007) Measuring and enhancing the value of agricultural water in irrigated basins. Irrig Sci 25:263–282
Iglesias A, Garrote L (2015) Adaptation strategies for agricultural water management under climate change in Europe. Agric Water Manag 1(155):113–124
Ikechukwu CC, Okolie H, Edeh IC (2023) The efficacy of fish pond water effluent on the growth and yield of two varieties of cucumber (*Cucumis sativus*). Int J Water Res 5(4):157. https://doi.org/10.36266/IJWR/157
Jegatheesan V, Shu L, Visvanathan C (2011) Aquaculture effluent: impacts and remedies for protecting the environment and human. Health:123–135. https://doi.org/10.1016/B978-0-444-52272-6.00340-8
Kamleshbhai BP, Iqbal G, Bambhaniya I (2023) Integrated multi-trophic aquaculture system (IMTA). In: Traditional & recent aquaculture practices. AkiNik Publications, pp 109–125
Kapetsky JM, Aguilar-Manjarrez J (2007) Geographic information systems, remote sensing and mapping for the development and management of marine aquaculture (No. 458). Food & Agriculture Organisation
Keremah RI, Davies OA, Abezi ID (2014) Physico-chemical analysis of fish pond water in freshwater areas of Bayalsa state Nigeria. Green J Biol Sci 4(2):033–038
Leimona B, Amaruzaman S, Arifin B, Yasmin F, Hasan F, Agusta H, Sprang P, Jaffee S, Frias J (2015) Indonesia's 'green agriculture' strategies and policies: closing the gap between aspirations and application, Occasional paper 23. World Agroforestry Centre, Nairobi

Lemma AH (2017) Evaluation of integrated poultry-fish-horticulture production in Arsi zone, Ethiopia. Int J Fish Aquat Stud 5:562–565

Lightfoot C, Gonzalves M (2001) Introduction. In: Integrated agriculture-aquaculture: a primer, FAO fisheries technical paper 407. FAO. https://www.fao.org/3/y1187e/y1187e03.htm#c

Lind CE, Dana GV, Perera RP, Phillips MJ (2015) Risk analysis in aquaculture: a step-by-step introduction with worked examples. WorldFish

Liu CF, Qi ZR, He J, Zhang JX (2002) Environmental friendship aquaculture zero discharge integrated recirculating aquaculture systems. J Dalian Ocean Univ 17:220–226

Luu LT (2021) The VAC system in northern Vietnam. In: Integrated agriculture-aquaculture: a primer, FAO fisheries technical paper 407. FAO. https://www.fao.org/3/y1187e/y1187e10.htm#l

Mafwila KP, Mutiaka BK, Dochain D, Rollin X, Mafwila J, Bindelle J (2019) Smallholders' practices of integrated agriculture aquaculture system in peri-urban and rural areas in Sub Saharan Africa. Tropicultura 37:4. https://popups.uliege.be:443/2295-8010/index.php?id=1396

Makori AJ, Abuom PO, Kapiyo R, Anyona DN, Dida GO (2017) Effects of water physico-chemical parameters on tilapia (Oreochromis niloticus) growth in Teso North sub-County, Busia County. Fish Aquat Sci 20:30

Mancosu N, Snyder RL, Kyriakakis G, Spano D (2015) Water scarcity and future challenges for food production. WaterSA 7:975–992

Mariscal-Lagarda MM, Páez-Osuna F, Esquer-Méndez JL, Guerrero-Monroy I, Vivar AR, Félix-Gastelum R (2012) Integrated culture of white shrimp (*Litopenaeus vannamei*) and tomato (*Lycopersicon esculentum* Mill) with low salinity groundwater: management and production. Aquaculture 366:76–84

Martinez-Alvarez V, Bar-Tal A, Diaz Peña FJ, Maestre-Valero JF (2020) Desalination of seawater for agricultural irrigation. WaterSA 12(6):1712

Maulu S, Munganga BP, Hasimuna OJ, Haambiya LH, Seemani B (2019) A review of the science and technology developments in Zambia's aquaculture industry. J Aquacult Res Dev 10:1–6

Medeiros DC, Marques LF, Dantas MRS, Moreira JN, Azevedo CMSB (2010) Produção de mudas de melo-eiro com efluente de pisciculturaemdiferentestipos de substratos e bandejas. Rev Bras Agroecol 5(2):65–71

Misman NA, Sharif MF, Chowdhury AJ, Azizan NH (2023) Water pollution and the assessment of water quality parameters: a review. Desalin Water Treat 294:79–88

Mohanty RK, Kumar A, Mishra A, Panda DK, Patil DU (2014) Water budgeting and management: enhancing aquacultural water productivity, Bulletin no. 63. Directorate of Water management, Indian Council of Agricultural Research, Chandrasekharpur, Bhubaneswar, India, 62p

Mohanty B, Sinha A, Phand S, Das S, Basade Y (2024) Capacity building in fisheries and aquaculture. NIPA, New Delhi

Mortillaro JM, Rakotoarinoro M, Andria-Mananjara DE, Rakotomalala T, Raliniaina M, Mikolasek O, Martel P, Oswald M, Dabbadie L (2017) The Tanjona from Fenoarivo: integrated fish farming systems in Madagascar, World Aquaculture Society meetings, Le Cap, Afrique du Sud, 26–30 June. World Aquaculture Society, Cape Town, South-Africa

Moursy MAM (2018) Influence of agri-aquaculture and seed treatment techniques on sugar beet crop. J Soil Sci Agric Eng Mansoura Univ 9:63–68. https://doi.org/10.21608/jssae.2018.35531

Moursy MAM, Rady M, Ali AM, Elfetyany M, Elbably W (2019) Potentials of the use of brackish ground water in integrated aqua-agriculture systems. J Egypt Acad Soc Environ Dev D Environ Stud 20:117–130. https://doi.org/10.21608/jades.2019.68501

Moursy MAM, Negm A, Ghanem HG, Wasfy KI (2022) Enhancing water use efficiency and yield of pomegranate crop by using fish drainage water with bio-fertilizer under drip irrigation system. Agriculture 12:1376. https://doi.org/10.3390/agriculture12091376

Mukherjee S, Chakraborty D (2017) Demand for infrastructure investment for water services: key features and assessment methods. In: Charting the water regulatory, 27. Edward Elgar Publishing, pp 257–296

Mukta MA, Khan MA, Uddin Mian MR, Juice RA (2019) Is tilapia farming financially profitable and efficient? Policy options for sustainable farming. J Bangladesh Agric Univ 17(1):92–98

Mulokozi D (2021) Integrated agriculture and aquaculture systems (IAA) for enhanced food production and income diversification in Tanzania. PhD thesis, Department of Physical Geography, Stockholm University, Stockholm, Sweden

Mulokozi D, Berg H, Tamatamah RA, Lundh T, Onyango PO (2022) Integration of Nile tilapia (*Oreochromis niloticus*) and vegetables (*Amaranthus hybridus* and *Brassica rapa* pekinensis) for improved production, water use efficiency and nutrient recycling. Aqua Fish Fish 2:493–506. https://doi.org/10.1002/aff2.76

Mulwa F, Li Z, Fangninou FF (2021) Water scarcity in Kenya: current status, challenges and future solutions. Open Access Lib J 8(1):1–15

Munthali MG, Chilora L, Nyirenda Z, Salonga D, Wineman A, Muyanga M (2022) Challenges and opportunities for small-scale aquaculture development in Malawi. Report, MwAPATA Institute. www.mwapata.mw

Musa JJ, Dada POO, Adewumi JK, Akpoebidimiyen OE, Musa ET, Otache MY, Yusuf S (2020) Fish pond effluent effect on physicochemical properties of soils in Southern Guinea Savanna, Nigeria. Open Access Lib J 7:e5990. https://doi.org/10.4236/oalib.1105990

Mustapha ABA, Driss B, Mohamed B (2014) The main species of freshwater fish aquaculture interest in Morocco, current status and prospects. Int J Fish Aquat Stud 2:216–218

Naddafi F, Jaafarzadeh S, Mokktari M, Zakizadeh B, Sakiran MR (2005) Effect of wastewater stabilization pond effluent on agricultural crops. Int J Environ Sci Technol 4:273–277

Ndubuisi NL (2019) Response of fish pond effluent on soil chemical properties and growth of cucumber (*Cucumis sativus*) in Igbariam South Eastern, Nigeria. Int J Curr Microbiol App Sci 8(2):2799–2807

Ngo HH, Guo W, Tram VTP, Nghiem LD, Hai FI (2017) Aerobic treatment of effluents from the aquaculture industry. In: Current developments in biotechnology and bioengineering. Biological treatment of industrial effluents. Elsevier, pp 35–77

Nhan DK, Van Be N, Hieu Trung N (2007) Chapter 4: water use and competition in the Mekong Delta, Vietnam. In: Challenges to sustainable development in the Mekong Delta: regional and national policy issues and research needs. Stockholm Environment Institute, pp 146–188

Niels DT, Seid TM, Peter LM, Pieter B (2016) Water quality assessment of streams and wetlands in South West Ethiopia. WaterSA 8(4):123

Nishan MH, Mahmud MAI, Chowdhury MMI, Rahman AFMA (2015) An overview on sustainable aqua-farming integration in the mid coastal region of Bangladesh. Asian J Poult Sci 9:50–56

Nwinyimagu AJ, Nwonumara GN, Ani C, Ukaje IO (2016) Evaluation and management of the physico-chemical variables of Asu River, South East Nigeria. J Sustain Dev 9(2):54–66

Ojobor SA, Tobih FO (2015) Effects of fish Pond effluent and inorganic fertilizer on Amaranthus yield and soil chemical properties in Asaba, Delts State Nigeria. J Agric Environ Sci 4(1):237–244

Oladimeji YU, Isah SA (2019) Economic and potential benefits of using fish pond water for vegetable production among fish farmers in Kwara State, Nigeria. Nigerian J Fish Aquacult 7(2):11–19

Olagunju D (2021) 6 types of ponds for fish farming. https://medium.com/@Agricfy.com/6-types-of-ponds-for-fish-farming-882a1042afbc

Osaigbovo AU, Orhue ER, Nwaoguala NC (2010) Effect of fish pond effluent on some soil chemical properties and vegetative growth of maize (*Zea mays* L.). J Sustain Agric Eviron 12(2):123–131

Patel SJ, Dudhagara CR (2023) Water management practices and users conflicts of the village pond aquaculture. Pharma Innov J 12(7):16–20

Phong LT, Udo HMJ, van Mensvoort MEF, Tri LQ, Bosma RH, Nhan DK, van der Zijpp AJ (2008) Integrated agriculture aquaculture systems in the Mekong Delta, Vietnam: an analysis of recent trends. Asian J Agric Dev 4:51–66

Pillay T (1990) Aquaculture, principles and practices. Fishing News Books, A Division of Blackwell Publishing Ltd, Oxford

Podemski CL, Blanchfield PJ (2006) Overview of the environmental impacts of Canadian freshwater aquaculture. In: A scientific review of the potential environmental effects of aquaculture in aquatic ecosystems, Canadian technical report of fisheries and aquatic sciences, 138, vol V. Fisheries and Oceans Canada

Poot-López GR, Hernandez JM, Gasca-Leyva E (2010) Input management in integrated agriculture-aquaculture systems in Yucatan: tree spinach leaves as a dietary supplement in tilapia culture. Agric Syst 103:98–104

Prinsloo JF, Schoonbee H (1987) Investigations into the feasibility of a duck-fish-vegetable integrated agriculture-aquaculture system for developing areas in South Africa. Water SA 13:109–118

Pulatsu S, Rad F, Koksal G, Aydin F, Caian A, Benli K (2004) The impact of Rainbow trout farm effluents on water quality. J Fish Aquat Sci 4:9–15

Rahman MM, Yakupitiyage A, Ranamukhaarachchi SL (2004) Agricultural use of fish pond sediment for environmental amelioration. Thammasat Int J Sci Technol 9(4):1–10

Rana K, Anyile J, Salie K, Mahika C, Heck S, Young J (2005) The role of aqua farming in feeding African cities. Urban Agric 14:36–38

Rico A, Satapornvanit K, Haque MM, Min J, Nguyen PT, Telfer TC, Van Den Brink PJ (2012) Use of chemicals and biological products in Asian aquaculture and their potential environmental risks: a critical review. Rev Aquac 4:75–93

Riya P, Behera SR, Bhatt H, Tewari V (2023) Chapter 1: crop rotation strategies for sustainable vegetable farming. In: Vegetable chronicles: vegetable science compendium. Stella International Publication

Rodrigues DS, Leonardo AFG, Nomura ES, Tachibana L, Garcia VA, Correa CF (2010) Produção de mudas de tomateiroemsistemasflutuantes com adubosquímicos e águaresiduária de viveiros de piscicultura. Rev Bras Ciênc Agrár 5(1):32–35. https://doi.org/10.5039/agraria.v5i1a5

Rossignoli C, Phillips M, Beveridge M, Marwaha N (2021) Measuring impact of innovations on aquaculture system performance, CGIAR research program on fish agri-food systems. Program brief: FISH-2021-25. WorldFish, Penang

Schuhbauer A, Sumaila UR (2016) Economic viability and small-scale fisheries—a review. Ecol Econ 124:69–75

Shoko AP, Limbu SM, Lamtane HA, Kishe-Machumu MA, Sekadende B, Ulotu EE, Mgaya YD (2019) The role of fish-poultry integration on fish growth performance, yields and economic benefits among smallholder farmers in sub-Saharan Africa, Tanzania. Afr J Aquat Sci 44:15–24

Singh S, Prabjeet S, Verma N, Kumar D (2021) Integrated fish farming—rationale and scope. College of Fisheries, G. B. Pant University of Agriculture and Technology, Pantnagar, India

Soliman AIE, Morad MM, Wasfy KI, Moursy MAM (2020) Utilization of aquaculture drainage for enhancing onion crop yield under surface and subsurface drip irrigation systems. Agric Water Manag 239:106244. https://doi.org/10.1016/j.agwat.2020.106244

Stevenson KT, Fitzsimmons KM, Clay PA, Alessa L, Kliskey A (2010) Integration of aquaculture and arid lands agriculture for water reuse and reduced fertilizer dependency. Exp Agric 46(2):173–190. https://doi.org/10.1017/S0014479709990871

Szuster BW (2003) Shrimp farming in Thailand's Chao Phraya River delta: boom, bust and echo, Research report. IWMI, Comprehensive Assessment of Water Management in Agriculture, 57 p

Tanner JE, Drabsch S (2021) Literature review of potential impacts of desalination discharges in Boston Bay, with particular reference to aquaculture, SARDI publication no. F2021/000299–1. SARDI research report series no. 1105. South Australian Research and Development Institute (Aquatic Sciences), Adelaide

Thanh DT, Ty NM, Hien NV, Berg H, Nguyen TKO, Vu PT, Minh VQ, Da CT (2023) Effects of organic fertilizers produced from fish pond sediment on growth performances and yield of Malabar and Amaranthus vegetables. Front Sustain Food Syst 7:1045592. https://doi.org/10.3389/fsufs.2023.1045592

Tran N, Crissman C, Chijere A, Chee HM, Jiau TS, Valdivia RO (2013) Ex-ante assessment of integrated aquaculture agriculture adoption and impact in Southern Malawi, Working paper AAS-2013-03 CGIAR. Research Program Focused on Integrated Aquaculture and Agriculture Systems, Penang

Tugie D, Abebe A, Endebu M (2017) Potential of integrated fish-poultry-vegetable farming system in mitigating nutritional insecurity at small scale farmer's level in East Wollega, Oromia, Ethiopia. Int J Fish Aquat Stud 5(4):377–382

Tunde AB, Kuton MP, Oladipo AA, Olasunkanmi LH (2015) Economic analyze of costs and return of fish farming in Saki-East Local Government Area of Oyo State, Nigeria. J Aquac Res Dev 6(2):5

UN (2014) World population prospects: THaquhe 2012 revision. Population Division of the Department of Economic and Social Affairs of the United Nations Secretariat, New York

US-EPA (2017) Water quality standards: Regulations and resources. United States Environmental Protection Agency

Van der Heijden PGM, Alla AN, Kenawy D (2012) Water use at integrated aquaculture–agriculture farms. Global Aquacult Adv July/August:28–31

Walia SS, Kaur N (2013) Integrated farming system-an ecofriendly approach for sustainable agricultural environment—a review. Greener J Agron For Hortic 1(1):001–011

Wambua MM (2015) A cost–benefit analysis of the fish farming enterprise productivity program project in Kenya. The case of implementation of the aquaculture development component in Meru County. United Nations University Fisheries Training Programme, Iceland [final project]

Wang YB, Xu ZR, Guo BL (2005) The danger and renovation of the deteriorating pond sediment. Feed Ind 26(4):47–49

Watanabe K, Tahara K (2016) Life cycle inventory analysis for a small-scale trawl fishery in Sendai Bay, Japan. Sustainability 8:399

Weber-Scannell PK, Duffy LK (2007) Effects of Total Dissolved Solids on aquatic organisms. A review of literature and recommendation for salmonid spp. Am J Environ Sci 3(1):1–6

WHO (2009) World Health Organisation. Guidelines for drinking water quality. WHO, Geneva

WRC (2010) Manual for rural freshwater aquaculture, Water Research Commission report no. TT463/P/10. Department of Ichthyology and Fisheries Science, Rhodes University

Wu Y, Song K (2021) Source, treatment, and disposal of aquaculture solid waste: a review. J Environ Eng 147:03120012

Yildiz HY, Domínguez D, Robaina L, Pirhonen J, Mente E, Parisi G (2017) Fish welfare in aquaponic systems: its relation to water quality with an emphasis on feed and faeces—a review. WaterSA 9:13

Yuan D, Edwards P, Halwart M (2019) Current practices and future prospects for integrated agriculture and aquaculture in Asia—a regional review. In: Report of the special session on advancing integrated agriculture-aquaculture through agroecology, Montpellier, France, 25 August 2018, FAO fisheries and aquaculture report no. 1286. Food and Agriculture Organization of the United Nations, Rome, pp 69–92

Yusoff FM, Umi WAD, Ramli NM, Harun R (2024) Water quality management in aquaculture. Cambridge Prisms: Water 2(e8):1–22. https://doi.org/10.1017/wat.2024.6

Zajdband AD (2011) Integrated agri-aquaculture systems. In: Genetics, biofuels and local farming systems. Springer, pp 87–127

Zeller M, Khor LY, Tran N, Shikuku K, Rajaratnam S, Chan CY, Aung YM, Campos N (2022) A technical guideline on integrated aquaculture performance assessment, Guidelines: 2022-15. WorldFish, Penang

Zohry A, Hefny Y, Ouda S (2020) Interplanting four legume crops under orange trees using different irrigation water and fertilizer sources in sandy soil. In: Proceedings of the 16th international conference on crop science, Cairo, Egypt, pp 341–346

Part IV

Climate Change Impacts and Mitigation

Adaptation Techniques to Irrigation Water Shortages Due to Climate Change

9

Mohamed Hefzy, Ghada Abd-Elmonsef Mahmoud,
Nehad Mohamed Ahmed Soltan,
and Ahmed Ali Abdelrhman

Abstract

Climate change influences water availability across various locations by altering the spatiotemporal patterns of precipitation and evaporation, hence affecting water supplies. Alterations in precipitation and evaporation characteristics resulting from climate change may cause variations in water availability, impacting both human populations and ecosystems. Research indicates that changes in the average and seasonal patterns of precipitation and evaporation can profoundly affect water supply systems, hydrology, agricultural productivity, and ecosystems. In this chapter, we discussed potential solutions to mitigate the consequences of climate change on water availability encompass: enhancing water storage and management, improving drought monitoring and early warning systems, adopting climate-smart agriculture and water reuse practices, safeguarding natural buffers, promoting rainwater harvesting, and increasing irrigation water productivity. Implementing these solutions necessitates a synthesis of technological advances, policy initiatives, and community-based strategies to improve water security and foster resilience against the consequences of climate change on water supply.

M. Hefzy (✉)
Water Requirement and Field Irrigation Research Department, Soils, Water and Environment Research Institute, Agricultural Research Center, Giza, Egypt
e-mail: mohamed.hefzy@arc.sci.eg

G. A.-E. Mahmoud
Botany and Microbiology Department, Faculty of Science, Assiut University, Assiut, Egypt

N. M. A. Soltan
Geology Department, Faculty of Science, Al-Azhar University, Assiut, Egypt

A. Ali Abdelrhman
Department of Soils and Water, Faculty of Agriculture, Al-Azhar University, Assiut, Egypt

A. E.-H. Zohry, S. Ouda (eds.), *Resilient Agroecosystems*, Sustainability Sciences in Asia and Africa, https://doi.org/10.1007/978-981-96-9630-7_9

Keywords

Greenhouse gases emission · High temperature · Precipitation fluctuation · Water resources · Water management

Introduction

The contemporary world is confronting multiple issues, one of which is climate change (Botzen et al. 2021). Climate change has emerged as one of the most pressing global concerns of our days as it fundamentally affects the environment for all living things on Earth, and its variations have profound implications for every aspect of human society (Aini et al. 2023). Climate change denotes modifications in atmospheric conditions, temperature, precipitation, and many climatic factors. Global warming is a result of climate change, linked to the rise in the average temperature near the Earth's surface. This phenomenon leads to increasingly severe and potentially irreversible changes to the Earth's biological and geological systems (Buytaert et al. 2011; Santos et al. 2015).

Over the past 60 years, climate has experienced distinct changes in the amount, variability, timing, type, and intensity of precipitation, as well as in temperature rates, whether yearly, monthly, seasonal, or daily (Seneviratne et al. 2021). This exacerbates the intensity of climatic drought, subsequently impacting global water resources, resulting in both beneficial and detrimental effects, presenting significant difficulties to human welfare, economic progress, and environmental integrity (Seneviratne et al. 2021). Climate change can induce numerous environmental issues impacting human health, including ozone layer depletion, the proliferation of infectious diseases, strain on natural resources, and global freshwater scarcity in terms of availability, quality, and management practices (Kinney 2018). Researchers and the public have long acknowledged the implications of substantial alterations in the long-term average climate and its variability (Ma et al. 2020). The various repercussions continually highlight the vulnerability of the hydrological cycle and water resources, directly affecting ground and surface water systems through recharge and evaporation, while indirectly altering water demand (Kareem et al. 2021), resulting in modifications to water quality, quantity, and availability.

The aim of this chapter is to discuss the potential solutions of mitigation and adaptation practices to irrigation water deficiency under climate change encompass: Technologies to enhance water use efficiency, precision irrigation, and precision farming implement deficit irrigation, enhance soil management to improve soil health, and utilize water reuse strategies by water recycling.

Climate Changes Causes

The Earth's climate has experienced several alterations throughout millions of years and these severe changes occur for two essential reasons:

The Natural Reasons

Occurring over an extended period, natural reasons include climate change, with volcanic activity being a significant factor that emits substantial carbon dioxide quantities into the atmosphere, alongside the emissions of solar energy (Hegerl et al. 2019).

Human Activity

While climate undergoes natural changes over extended periods, investigations and research indicate a swift alteration in climate attributable to human activities (Druckman and McGrath 2019; Zscheischler et al. 2018).

The International Panel on Climate Change (IPCC) attributes the majority of heat recorded in the past 50 years to human activity (Nesmith et al. 2021) that modifies the composition of the global atmosphere and is supplementary to the natural climate variability reported across analogous timeframes primarily induced by the increasing levels of greenhouse gases, which causes the Earth's temperature to increase (Besley and Peters 2020). This evident transformation has coincided with the economic and technological advancements that have intensified from the mid-twentieth century (Khairullina et al. 2019).

Moreover, industrial activities that underpin the new nature of life have significantly elevated the concentrations of greenhouse gases, including methane, carbon dioxide, and nitrogen oxide, in the atmosphere (Al-Ghussain 2019). According to Prevedello et al. (2019), the atmospheric concentration of greenhouse gases has risen from a pre-industrial level of 280 ppm to the current 430 ppm, with an annual increase of around 2.5 ppm. Besides the emissions of industrial gases, forest harvesting releases carbon deposited in the soil into the atmosphere, hence exacerbating climate change. The amount of CO_2 sequestered in the soil is around 2000 billion tons (Lejeune et al. 2018). However, following Letcher (2020), the growth in greenhouse gas emissions is also attributed to the combustion of fossil fuels, which fulfills 80% of global energy requirements. Fossil fuels are derived from fossilized elements, including coal, natural gas, and petroleum. These elements are subsequently removed and combusted to generate heat utilized throughout several sectors (IPCC 2022). The global carbon cycle encompasses billions of tons of carbon in the form of CO_2 (Fig. 9.1).

CO_2 is sequestered by oceans and living biomass and is released into the atmosphere annually by natural processes. In a state of equilibrium, the transfer of carbon among these several reservoirs is balanced. CO_2 emissions have been recognized as the principal driver of climate change (Chiu et al. 2009; Gao et al. 2021). This has led to increased global average temperatures and erratic rainfall patterns (Anjum et al. 2012; Jin et al. 2021).

Solar radiation traverses the atmosphere and elevates the temperature of the Earth's surface. A portion of solar radiation is reflected by the Earth and its atmosphere. Atmospheric greenhouse gases, including, capture heat and thereby elevate the Earth's surface temperature. This phenomenon is referred to as the greenhouse effect (WMO 2024) (Fig. 9.2).

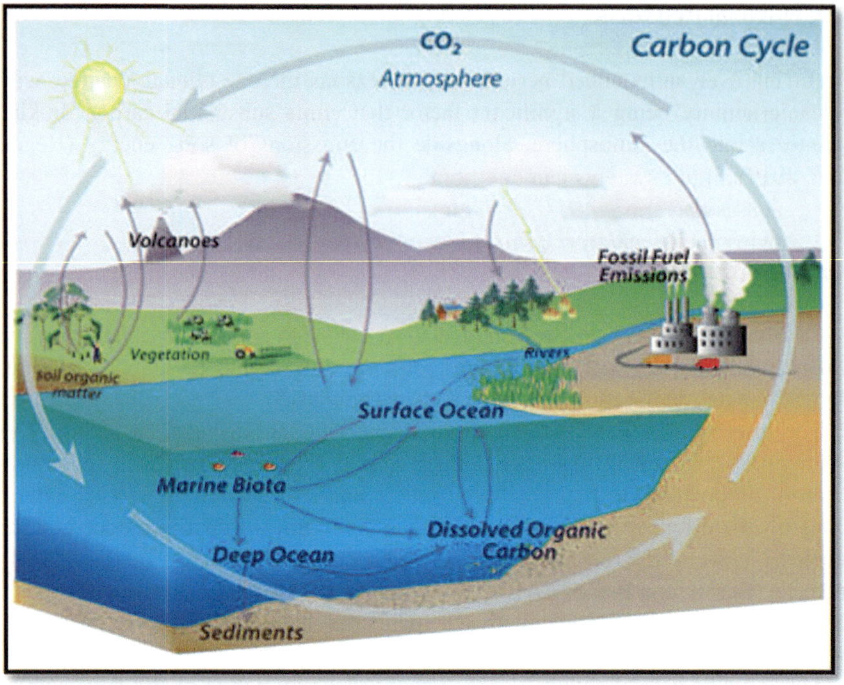

Fig. 9.1 The global carbon cycle. (Source: NOAA 2019)

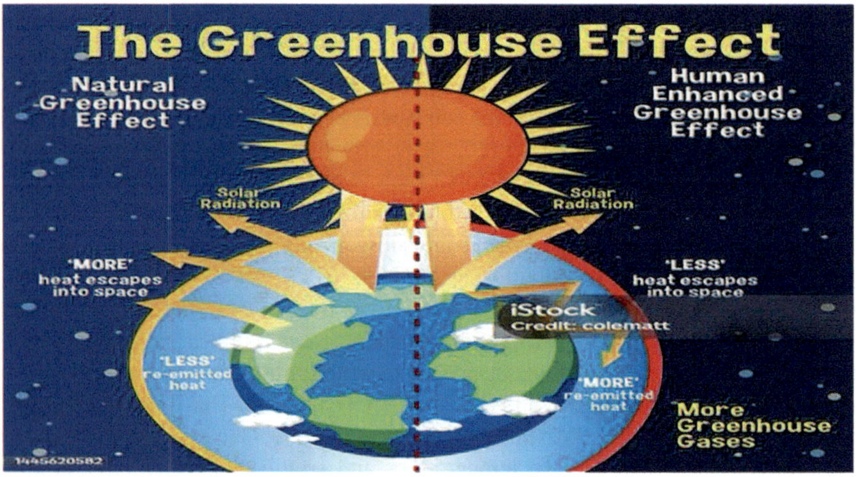

Fig. 9.2 The greenhouse effect. (Source: Colematt 2022)

The emission of greenhouse gases into the atmosphere results in the entrapment of heat that would typically be released into the space, leading to an increase in Earth's temperature and erratic rainfall patterns (Anjum et al. 2012). These factors exacerbate the impact of greenhouse gases in the atmosphere, resulting in detrimental consequences on the Earth's climate due to global warming (Bolan et al. 2024).

Climate Change-Induced Factors Affecting Water Resources

The study of IPCC (2023) predicts that fluctuations in temperature and precipitation would directly influence the hydrological cycle, altering the distribution and availability of water. Hydrological extremes, including droughts and floods, are classified as disasters (Wahidah et al. 2023). Insufficient rainfall and elevated surface temperatures cause water scarcity, leading to extensive and lengthy droughts, whereas excessive precipitation can contribute to flooding (Chaturvedi et al. 2021).

Temperature Changes

Temperature is a critical factor in determining hydrological cycles, profoundly impacting water availability. Climate change is resulting in elevated global temperatures, hence elucidating the intricate relationship between temperature variations and water supplies (Gentine et al. 2019). Temperature variations affect numerous natural processes, hence impacting the availability and quality of our water resources (Brooks et al. 2015).

The IPCC (2023) report indicates that climate change would lead to increased temperatures and reduced precipitation, with the annual average temperature projected to increase by 2–2.5 °C by 2100. The anticipated increase in temperatures contributes to heightened evaporation, affecting surface water and soil moisture (Meerhoff et al. 2022). A reduction in water supplies is expected to result in heightened demand due to rising temperatures, particularly affecting ecosystems, agriculture, and energy production, the primary users of water, along with overall water availability for municipal, industrial, and other applications (Ingrao et al. 2023). A significant effect of increasing temperatures is the enhancement of evaporation rates from surface water bodies (Kiparsky et al. 2012). The heightened evaporation results in an augmented demand for water in ecosystems, complicating the maintenance of water levels. Water levels in lakes, reservoirs, and rivers may decline, resulting in reduced overall water availability. This effect jeopardizes the delicate balance of aquatic ecosystems and the communities reliant on these water supplies (Zohary and Ostrovsky 2011). Mountainous regions, which generally rely on accumulated snow for water retention, encounter specific obstacles confronting certain challenges (Caretta et al. 2022). Elevated temperatures result in accelerated snowmelt, disrupting the usual timing of water flow. The modification in the release pattern influences downstream water availability, affecting the quantity of water accessible downstream, with repercussions for ecosystems and communities

(Caretta et al. 2022). Furthermore, elevated temperatures influence the pace of groundwater replenishment, resulting in further consequences for water supply (Caretta et al. 2022). Temperature variations can intensify saltwater intrusion problems in coastal aquifers, reducing both the quality and quantity of freshwater (Abd-Elaty et al. 2019). Comprehending the intricate connections between temperature fluctuations and water accessibility is crucial for formulating adaptive methods to tackle the emerging challenges of climate change. A thorough and systematic approach is crucial for addressing the intricacies of the hydrological system to provide effective water management amid climate change (Abd-Elaty et al. 2022).

Precipitation Patterns

Climate change is modifying precipitation patterns, significantly impacting global water supply. A crucial component of several economies, agriculture is susceptible to fluctuations in precipitation patterns. Changes in rainfall distribution affect crop development, productivity, and water requirements, hence jeopardizing food security (Mbow et al. 2019). The irregular distribution of precipitation can aggravate water scarcity in agricultural areas, amplifying current challenges and necessitating the implementation of adaptive farming techniques to sustain food production (Sinore and Wang 2024). Variations in precipitation patterns, including alterations in time, intensity, and spatial distribution, are increasingly evident and will impact water supply and quality. Elevated flood danger in regions experiencing intensified precipitation can lead to greater runoff (Sophocleous 2004). This event underscores the imperative of adopting adaptable methods in vulnerable places to mitigate and alleviate the potential effects of heightened rainfall (Sophocleous 2004). Conversely, changes in precipitation patterns affect groundwater replenishment and river flow, hence intensifying water shortages in specific locations and prolonging drought conditions in others. Reduced precipitation may lower surface water levels, thereby impacting agricultural operations and freshwater availability (Sinore and Wang 2024).

Climate Change Impacts on Water Resources

Climate change can significantly affect the hydrological cycle, as well as the quality and amount of water supplies. This subsequently influences demand and sources of water supply: rivers, lakes, springs, and groundwater (Shawul and Chakma 2020). Communities reliant on dependable water sources for survival and economic activities are threatened by these circumstances (Scanlon et al. 2007). Furthermore, urban water distribution systems are susceptible to the effects of changing precipitation patterns. The increasing water demands of growing populations are becoming difficult to provide due to climate change affecting conventional rainfall patterns (McGrane 2016). This requires the creation of resilient and flexible water distribution systems in metropolitan regions to provide a reliable and safe water supply for growing populations. The analysis of global climate change and precipitation patterns provides essential insights into regional vulnerability.

The consequences of climate change on water resources are evident in the irregular fluctuations in precipitation patterns, the temporal and spatial distribution of air temperatures globally, and a concomitant rise in water temperatures (Mandal et al. 2021). The repercussions include intensified storms, heavy rainfall, flooding, and droughts, alongside diminishing snow cover due to elevated temperatures, leading to the rapid melting of the polar snowpack and the consequent rise in sea levels (Das and Goyal 2021).

Climate variability and change directly impact ground and surface water supplies through recharge and evaporation, and indirectly through modifications in water consumption (Kareem et al. 2021). Increased evaporation rates, a higher proportion of precipitation falling as rain rather than snow, earlier and shorter runoff seasons, elevated water temperatures, and reduced water quality in both inland and coastal areas lead to both positive and negative consequences (Klove et al. 2014). To tackle these difficulties, it is crucial to develop targeted policies that include flexible measures, infrastructure improvements, and sustainable water conservation techniques to alleviate the effects of climate change on water supplies (Ssekyanzi et al. 2024). Furthermore, strategies for adaptation must be developed to ensure the resilience of these regions in the face of ongoing water constraints. Furthermore, there are negative physical and economic consequences explained as followed:

- *Projected water scarcity in several regions:* By 2050, billions may experience significant water shortages as a result of climate-induced alterations in supply and demand dynamics. This scarcity jeopardizes food security, economic stability, and public health. Climate change is resulting in more frequent and severe droughts in numerous regions globally (Williams et al. 2015), as precipitation levels remain constant or decline, leading to a reduction in net water supply. This may result in water scarcity for agriculture, industry, and residential consumption. Decreased precipitation may lead to an increased incidence of wildfires, and regions affected by wildfires are more susceptible to soil erosion (Fernando et al. 2019).
- *Snowpack diminution:* Numerous towns depend on snowpack as an essential freshwater resource. Increased temperatures are diminishing snow accumulation and prompting earlier snowmelt, thereby altering the timing and volume of water accessible for consumption and agriculture. This leads to spatial alterations in storm trajectories (Das and Goyal 2021) and engenders numerous environmental hazards, including sea-level rise, which may compromise water quality and availability in coastal regions. Rising sea levels may also directly impact groundwater quality through saltwater intrusion. Significant alterations to the freshwater hydrology of coastal zones, resulting from saltwater intrusion, would jeopardize the freshwater resources of numerous coastal locations (Fecht 2019).
- *Degraded water quality:* Variations in the timing, intensity, and length of precipitation can adversely impact water quality. Flooding, caused by heightened precipitation and severe rainstorms, transports sediments, nutrients, and pathogens

into aquatic systems. This may lead to detrimental algal blooms that pollute potable water supplies and adversely affect aquatic organisms. Flooding can overwhelm storm, combined sewer, and wastewater systems, causing untreated pollutants to enter waterways directly (Zhao and Boll 2022). In areas experiencing heightened rainfall frequency and intensity, runoff from agricultural land may generate increased pollution and sedimentation, contaminating rivers and streams with nutrients and pesticides. Water quality may also be affected by rising sea levels, which contaminates coastal aquifers with seawater. As water temperatures rise, the chemistry and biology of aquatic resources will be influenced by heightened primary production, accelerated organic matter decomposition, and increased nutrient cycling rates in lakes and streams, leading to diminished dissolved oxygen levels (Arnbjerg-Nielsen 2012). Lakes and wetlands linked to return flows from irrigated agriculture are particularly concerning; water pollution issues will escalate, adversely affecting numerous aquatic habitats and promoting the proliferation of harmful microorganisms, thereby further jeopardizing water quality and safety for human consumption. Elevated temperatures can enhance evaporation rates, resulting in diminished groundwater replenishment. This can exhaust groundwater supplies and complicate the fulfillment of water demand (OECD 2013).

- *Increased intensity of tropical storms:* Might compromise water infrastructure due to heightened flooding, which may overload systems and permit pollutants to infiltrate streams, contaminating water supplies. Coastal erosion frequently results from storms. The aging water infrastructure in numerous locations is not prepared to manage the exacerbated effects of climate change. Severe weather events can impair treatment facilities and distribution networks, resulting in service interruptions and heightened expenses for repairs and enhancements (OECD 2013).
- *Economic challenges:* As water scarcity increases, the expenses associated with obtaining clean water may escalate, disproportionately impacting low-income populations. Approximately 40 million American households may have challenges regarding water affordability as a result of climate effects. Water scarcity will result in increased water prices, either in monthly bills or one-time connection fees for new residences and enterprises. An adequately substantial price increase may influence the magnitude and distribution of urban development in the United States. Expensive water supply initiatives, including desalination facilities, pipelines, and reservoirs, will increasingly gain economic viability (OECD 2013).

However, there are some positive impacts reported previously by several authors, which might be detrimental in certain instances as followed:

- Climate change alters precipitation patterns, leading to increased rainfall in certain areas and a rise in the frequency and severity of extreme weather events, including intense rainfall. This can improve water accessibility for agriculture, drinking water, and more uses (Bolan et al. 2024).

- Increased snowmelt in mountainous regions due to elevated temperatures is inducing early and accelerated melting of snow and glaciers, potentially augmenting water flow in rivers and streams during the spring and summer seasons. Increasing sea levels may indirectly impact water availability in coastal regions by elevating water tables in groundwater aquifers, perhaps leading to enhanced surface runoff at the detriment of aquifer recharge (Bolan et al. 2024).

Mitigation and Adaption Strategies

Climate change presents substantial difficulties to water resources, necessitating the adoption of mitigation and adaptation techniques to guarantee sustainability and resilience. These measures can be categorized into distinct techniques to mitigate the effects of climate change on water availability, quality, and management (Bartlett and Dedekorkut-Howes 2023).

Mitigation Strategies

Reducing Greenhouse Gas Emission

- Transitioning to energy-efficient technologies in water treatment facilities can significantly diminish emissions associated with water management practices.
- Promoting sustainable agricultural practices that utilize reduced water consumption contributes to overall mitigation efforts by decreasing the carbon footprint of food production (Bartlett and Dedekorkut-Howes 2023).

Restoring the Ecosystem

- Preserving and restoring wetlands and forests enhances their efficacy as carbon sinks, contributing to climate change mitigation and improving water quality and availability (Schuster et al. 2024).

Education and Awareness Programs

- Informing the public about the need of sustainable water usage can foster community engagement in mitigation and adaptation efforts, leading to enhanced local practices (Lebu et al. 2024).

Adaption Strategies

Infrastructure Investments

- *Reservoirs and Water Storage:* The construction of new reservoirs and the enhancement of existing reservoirs help alleviate variations in annual rainfall. Managed aquifer recharge is progressively employed as a method for subterranean storage of surplus water, hence improving drought resilience (Bartlett and Dedekorkut-Howes 2023).

- *Irrigation Systems:* Enhancing irrigation systems, such as drip irrigation, can augment agricultural water efficiency in the face of increasing water scarcity due to climate change (Sidhu et al. 2021; Mahmoud et al. 2024).

Enhanced Water Management Techniques

- *Integrated Water Resources Management:* Emphasizes the coordination of water, land, and associated resources management to improve economic and social welfare while ensuring ecosystem sustainability (Ciampittiello et al. 2024).
- *Water Conservation Strategies:* Implementing rainwater collection and promoting reduced water consumption can significantly alleviate pressure on water resources (Richards et al. 2021).

Policy and Regulation Adjustments

- *Water Allocation Policies:* Updating water allocation policies to account for shifting precipitation patterns ensures that water distribution matches present circumstances.
- *Stakeholder Engagement:* By including local communities in decision-making processes, adaptation strategies are more effective as they incorporate local knowledge and needs (OECD 2015).

Technological Innovations

- *Desalination and Wastewater Reuse:* These technologies provide alternative water sources, particularly in arid regions with scarce freshwater resources (Williams et al. 2023).
- *Monitoring Systems:* Enhanced surveillance of water quality and quantity might facilitate improved resource management under variable climatic circumstances by more efficiently tracking these parameters (Bartlett and Dedekorkut-Howes 2023).

Practices to Reduce Irrigation Water Deficiency Under Climate Change

Technologies to Enhance Water Use Efficiency

Water use efficiency (WUE) is defined differently across several disciplines. WUE was initially defined by crop physiologists as the ratio of crop yield and carbon assimilation to transpiration (Viets 1962). Ultimately, it was redefined as the biomass quantity or yield output generated by the crop per unit of water utilized (Hatfield et al. 2001). Crop researchers define WUE as the ratio of grain production achieved to the volume of water accessible to the crop, which includes both soil moisture and water from precipitation and irrigation. Irrigation researchers define WUE as the ratio of total irrigation water transpired to the water extracted from the source (Israelsen an Hansen 1962). Ultimately, WUE refers to the quantity of grain or yield generated by a crop in relation to a defined supply of water. WUE has garnered significant attention due to the impacts of climate change and human use on the finite availability and uneven regional distribution of water resources. Enhancing crop water utilization is a very efficient strategy to ensure sustainable agricultural development by increasing water usage efficiency, particularly in arid and semiarid regions (Gu et al. 2024). Given the escalating strain on water resources and the impacts of climate change on rain-fed and irrigated agriculture in arid and semiarid countries, which consume around 70% of available water, the enhancement of WUE remains a critical issue for agriculture (Alharbi et al. 2024).

WUE in irrigated crops can be improved through modified management strategies, encompassing crop selection and other activities including fertilization, tillage, irrigation scheduling, and the enhancement of water usage and transportation systems (Gu et al. 2024). The physical and chemical features of the soil significantly influence the calculation of WUE. To optimize water use efficiency in agricultural systems, the integration of mulching, organic amendments, and irrigation methods appears to be highly advantageous (Alharbi et al. 2024).

Improving WUE necessitates careful planning and the selection of appropriate irrigation systems. Contemporary irrigation techniques such as drip irrigation and soil moisture sensing can lead to significant water conservation and enhanced agricultural yield (Levidow et al. 2017). Investigations indicate that contemporary irrigation techniques, such as drip or sprinkler irrigation, can enhance WUE by 70–90% relative to flood or furrow irrigation. Nonetheless, advancements in conventional irrigation techniques, such as field settlement and surge flow, can enhance water use efficiency by reducing water loss (Eisenhauer et al. 2021). WUE can be significantly enhanced by 20–40% with efficient scheduling that considers crop water needs, growth phases, and climatic conditions, in addition to employing sensors for soil moisture monitoring (Levidow et al. 2017).

Improving Agricultural Practices

Environmental change profoundly affects agriculture, and agricultural practices can be adapted to suit the evolving environment. Altering agricultural methods, such as irrigation timing, fertilizer use, tillage, and selecting appropriate crops, may aid in addressing water deficiency induced by climate change (Levidow et al. 2017; Thakur et al. 2022).

Tillage is the mechanical agitation of soil to prepare it for planting, aerate it, manage weeds, disintegrate soil crust, and integrate fertilizer and crop residues into the soil (Lamandé et al. 2023). Tillage influences several management practices, including irrigation and residue management, and alters the soil pore network, specifically macro- and mesopores. Tillage, in contrast to a residue-covered or undisturbed surface, improves water retention by soil infiltration while increasing water loss through evaporation. Numerous prior research indicated that tillage methods exerted varied impacts on soil properties, particularly soil porosity and bulk density, contingent upon soil texture (Strudley et al. 2008). Ali et al. (2017) stated that the implementation of conservation tillage technologies is essential for enhancing and optimizing WUE and minimizing water losses. It is an economical method that necessitates minimal knowledge. Moreover, conservation techniques may improve soil structure, promote soil aggregate formation, and augment the soil's water retention capacity. Liebhard et al. (2022) examined long-term soybean tillage studies, focusing on the partitioning of evapotranspiration by a stable isotope methodology in silt-loam soil under water-limited conditions. It was disclosed that, in contrast to conventional tillage, no-tillage (direct planting) and limited tillage (utilizing cultivators and quadrennial subsoiling) yield enhanced crop water productivity. On the contrary, inefficient tillage procedures may result in soil erosion caused by wind and water, and excessive tilling degrades the soil structure, facilitating air infiltration that leads to oxidation and a depletion of organic matter, hence diminishing crop yield and reducing WUE (Strudley et al. 2008). Deep tillage is frequently beneficial, however at times unsuccessful, as demonstrated by practice, yielding minor impacts on expansive clay soils. This strategy is useful contingent upon the meteorological conditions, especially in clay soil. During arid seasons, evaporation rates escalate, therefore diminishing water storage; conversely during humid seasons, water storage in deeper strata is augmented (Strudley et al. 2008). Table 9.1 displays various farming techniques and their effects on water use efficiency.

Use of Organic Materials

The use of organic material is crucial for soil carbon sequestration, aiding in the alleviation of climate change impacts. Furthermore, it is a more cost-effective method to enhance water retention, especially in regions with limited water supplies. Multiple research studies have demonstrated that improved soil management practices, including the application of organic material and mulching of crop residue, can effectively retain soil moisture and reduce water demands due to climate change (Zhang et al. 2017, 2022; Hatfield and Dold 2019; Liao et al. 2021). Organic matter inputs as organic amendments enhance soil structure by improving aggregation, aeration, and root penetration (Franco et al. 2020; Mahmoud et al. 2025). This enhances water infiltration and retention, augmenting the soil's capacity to retain moisture; it also generates additional pore spaces and a stable pore network, facilitating more effective water access for plants (Singh et al. 2020). Ghosh et al. (2015) found that the application of compost to sandy soil enhanced its moisture retention compared to untreated soils, leading to a 30% decrease in irrigation water requirements. The application of compost improved soil moisture retention and increased WUE in wheat, leading to a 25% reduction in irrigation needs, as demonstrated by Evanylo et al. (2008), Abd El-Mageed et al. (2018), and Meena et al. (2020). Chen et al. (2023) indicated that the application of organic fertilizers and mulching enhances WUE by 30%, underscoring their importance in sustainable agriculture practices.

The stability of soil organic carbon is mostly influenced by feedstock and its decomposition rate (Song et al. 2019). Sukkariyah et al. (2007) found that aeration and heat resulted in a greater decomposition of organic matter in sandy soils

Table 9.1 Various farming techniques and their effects on water use efficiency

Crop	Location	Practice	Soil	Climate	Observation	WUE kg.m^{-3}	Best pra.	References
Wheat (Triticum aestivum L.)	Iraq	Tillage	Silty clay loam	Subtropical	Irrigation methods, tillage, seed rate	0.78–1.10	Sprinkler irrigation, minimum tillage, seed rate (240 kg•ha^{-1})	Akol et al. (2021)
Cotton, ground nuts, wheat and sorghum	Sudan	Tillage	Dark gray clay	Semiarid	Conservation tillage methods, crops	0.12–0.85	Tillage, wheat, and sorghum	Ali et al. (2017)
Wheat and maize	China	Tillage	Silt loam	Semiarid	Tillage and residue	20.83–24.22	Conventional tillage with residue retention	He et al. (2022)
Soybean	Austria	Tillage	Silt loam	Dry subhumid			Conventional tillage	Liebhard et al. (2022)
Maize	China	Tillage	Silt loam	Arid	Tillage and green manures	2.2–3.3 WPc	Tillage and no tillage with green manures	Zhang et al. (2023)
Wheat-maize system	China	Irrigation schedule	Silty loam	Monsoon climate	Tillage and irrigation schedule	1.81–2.62	Preplanting + jointing	Wang et al. (2021)
Maize	China	Irrigation	Silty clay loam	Semi-humid	Irrigation management	5.54–11.13	Deficit drip irrigation	Yan et al. (2020)
Maize	China	Irrigation	Loam	Semi-humid monsoon	Irrigation management	1.9–2.4	Irrigation methods	Cao et al. (2022)
–	China	Biochar	Sandy loam	Continental climate	Application rate	2–9.43%	Biochar level of 6 kg·m^{-2}	Wang et al. (2022)
Soybean	China	Biochar	Sandy loam	Monsoon climate	Pot experiment	15.6%	Biochar level of 6 10 g kg^{-1}	Zhang et al. (2020)

compared to clayey soils. The application methods of organic material promote elevated microbial biomass, resulting in increased carbon input. Börjesson et al. (2016) and Six et al. (2000) assert that soil organic matter (SOM) exerts three primary effects: It stabilizes soil structure, sequesters organic carbon that indirectly sustains microorganisms, and directly provides nutrients that enhance crop vitality and improve water utilization efficiency. Furthermore, the microbially mediated breakdown of organic material produced the adhesive organic molecules essential for the stability of particulate organic matter (POM) (Mahmoud et al. 2023). Due to microbial activity, organic molecules with a low carbon to nitrogen ratio decompose more rapidly than those with a high carbon to nitrogen ratio. Blanco-Canqui and Lal (2009) demonstrated that soybean residues disintegrate more rapidly than maize residues due to a lower C:N ratio, influencing the SOC material. They determined that an increase in maize residue inputs was accountable for a rise in carbon levels. Long-term application of organic modifications is advantageous for agricultural production and sustainability objectives.

Biochar Application

Biochar application is another significant soil management technique that enhances water use efficiency in irrigated agriculture. This amendment is promising due to its multiple benefits, including the promotion of stable soil aggregate formation and heightened microbial activity, which foster deeper root systems and improved drought tolerance (Semida et al. 2019). With additional research, the advantages of biochar will be elucidated, offering solutions for climate change adaption and food security. The type of feedstock utilized and the production method employed in biochar manufacturing significantly influence its impact on WUE (Semida et al. 2019) (Fig. 9.3).

Biochar possesses a substantial surface area and an extensive pore network, enabling it to retain nutrients and moisture (Zhang et al. 2021). This enhances soil fertility and diminishes leaching, hence improving WUE. The hydrophobic properties of biochar significantly reduce infiltration capacity and hydraulic conductivity, leading to increased surface runoff from surface application, which restricts plant water availability in deeper soil layers (Fischer et al. 2019). The incorporation of biochar into deeper soil strata modifies the matric potential and the soil water retention curve (SWRC), enhancing water utilization efficiency. Cayuela et al. (2024) indicated that biochar significantly enhances water retention in dry and semiarid locations, rendering it a crucial instrument for augmenting agricultural yield in these areas.

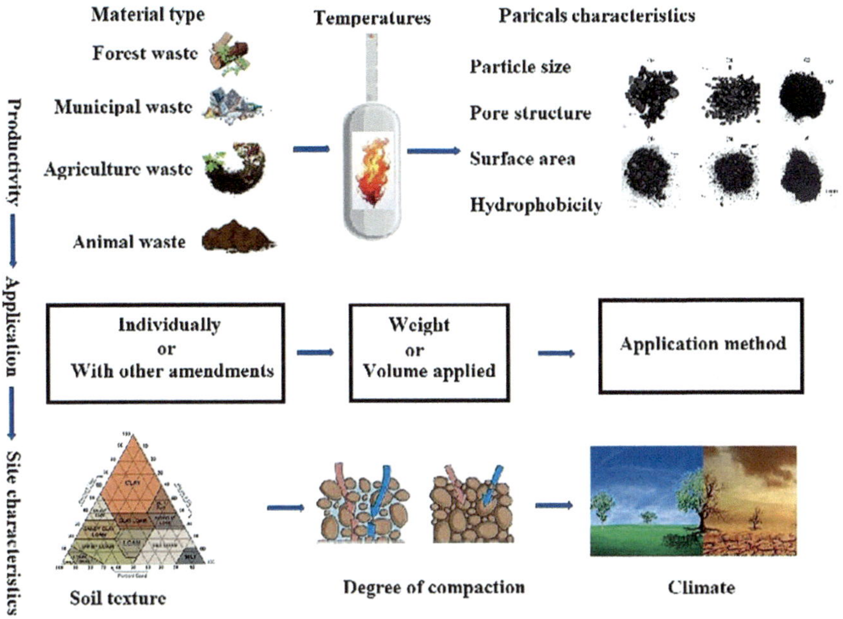

Fig. 9.3 Factors influence how well biochar is effective

Precision Irrigation and Precision Farming Mitigate the Effects of Climate Change

Precision Irrigation

Precision irrigation methods are widely acknowledged as essential instruments in alleviating the effects of climate change on agriculture. These strategies tackle difficulties related to fluctuating weather patterns, water scarcity, and heightened agricultural demands by maximizing water utilization and improving crop resilience (Li et al. 2024). Accurate water distribution is both the essential method and objective of precision irrigation, rendering the technology a critical component of precision agriculture (Li et al. 2024). Precision and smart irrigation water-saving systems (PISs/SISs) are essential for several reasons: They enable effective management and exploitation of water resources, improve crop yield, and reduce environmental consequences. Efficient water management strategies can significantly decrease water use in agricultural fields and enhance crop yields (Cotera, et al. 2024). Previous research suggests that enhancements in irrigation efficiency can fulfill 50% of the anticipated rise in water demand (Johansson, et al. 2002). Furthermore, efficient water management is crucial for the sustained profitability of global food production and processing. It is essential for equitable water allocation, enhancing resilience to climate change and other disturbances, especially in alleviating water scarcity. Furthermore, it is considered an essential component for achieving sustainable development as its progress corresponds with multiple Sustainable Development Goals set by the United Nations, including clean water and sanitation (SDG 6), climate action (SDG 13), and sustainable cities and communities (SDG 11) (Katila et al. 2019).

Precision irrigation water conservation systems are a sustainable approach to water management in the context of escalating water scarcity and variable climatic conditions (Xing and Wang 2024). PISs may integrate all challenges and improvements to deliver innovative and enhanced solutions and services to all stakeholders (Rumanti et al. 2023). Therefore, traditional irrigation water application and management techniques necessitate improvement to meet the increasing demands (Ahmed et al. 2023). Agriculturalists must adopt the latest advancements in irrigation water application and management to satisfy this growing demand. The primary aim of implementing precision irrigation water-saving technology in agriculture is to improve water conservation by utilizing advanced instruments such as soil moisture sensors, irrigation controllers, and real-time climate data collection and analysis tools. This method offers numerous advantages in water conservation for intelligent irrigation (Lakhiar et al. 2024):

1. SISs can deliver continuous and instantaneous monitoring of several field characteristics, including climate, soil, water, and vegetation. They enable farmers to receive updates on field conditions and make real-time decisions on irrigation systems, fertilizer management, and pest control applications.
2. SISs can eliminate the need for human data collection of various variables. They can reduce the effort required for field data collecting, mitigate data collection errors, and improve the overall operational efficiency of the system.
3. SISs can diminish multiple input costs by reducing water use, hence significantly lowering water utility expenses.
4. SISs can assist in preventing both over- and under-watering, ensuring that plants receive the requisite water and nutrients when needed.
5. SISs represent an eco-friendly technology. Their application with existing irrigation systems helps mitigate water losses resulting from over-irrigation (runoff and seepage). These losses are the principal conduits for applied fertilizers and pesticides into adjacent water bodies (Lakhiar et al. 2024).

Numerous research has examined the potential benefits of PISs/SISs; for instance, Gabuya et al. (2024) reported that SISs conserved water by 19% relative to manual watering methods while also enhancing growth rates and increasing the leaf count of coffee seedlings. Zeng et al. (2023) implemented SISs for rice agriculture and demonstrated that the application of modern technology methods for the effective management of water resources has attracted increased attention from researchers. They suggested SISs amalgamate monitoring, control, and cloud-based systems to optimize efficiency and alleviate the grower's workload. It may also implement automated irrigation systems to optimize water consumption and improve WUE. Lakshmi et al. (2023) observed a 46% decrease in water consumption and enhanced agricultural yield with an SIS compared to a conventional irrigation system. Laphatphakkhanut et al. (2021) utilized an IoT-based SISs, resulting in a 40.29% reduction in water usage for alternating wet and dry irrigation and a 29.22% reduction for basin irrigation compared to traditional approaches. Barkunan et al. (2019) reported a 41.5% decrease in water use for paddy agriculture using conventional floods and a 13% decrease with an automated drip irrigation method. Mason et al. (2019) employed CROPWAT 8.0 software to model several agricultural scenarios, illustrating how SISs could enhance and diminish water usage while maintaining tomato plant productivity. The simulation results demonstrated that the SISs reduce water consumption by an average of 59% while maintaining tomato plant output. González-Briones et al. (2019) reported a significant decrease in water use, reducing usage by 15.06% compared to traditional vehicle irrigation methods to improve irrigation efficiency in potato farming.

As climate change generates ever-erratic weather patterns, SISs will be essential for improving drought resilience. They will aid farmers in adapting to variable conditions by optimizing water usage during both drought and humid periods (Ahmed et al. 2023). Despite the initial expenses of an intelligent irrigation system seeming substantial, the enduring advantages make it a wise financial decision. Enhancing water efficiency and minimizing waste can significantly reduce water and electricity costs over time. The future of smart irrigation systems that conserve water is optimistic, offering significant possibilities for scientific progress, environmental benefits, and commercial advantages. As these systems progress, they will be crucial in addressing global water challenges and promoting sustainable water use. It is imperative to inform farmers and homeowners about the benefits and operation of smart irrigation systems to facilitate their widespread adoption and deployment in the industry (Lakhiar et al. 2024).

Benefits of Precision Irrigation

Efficient Water Usage

Precision irrigation emphasizes the direct delivery of water to crops in exact quantities, guaranteeing that plants have moisture precisely when required. This technique markedly lowers water wastage, which is essential in areas experiencing drought or inconsistent precipitation patterns. Research demonstrates that precision irrigation can conserve significant water resources, especially in high-value crops such as fruits and vegetables, with potential annual savings reaching €30 per hectare (Alam 2015; Zhao and Boll 2022).

Enhanced Crop Resilience

Precision irrigation employs technologies like remote sensing and soil moisture monitoring to enable farmers to proactively address environmental constraints. These technologies provide real-time modifications to irrigation schedules according to prevailing meteorological conditions and soil moisture levels, thereby safeguarding crops from yield reductions caused by droughts or heavy rains (Alam 2015; Hassaballah 2023; Hamed et al. 2025).

Reduced Environmental Impact

Implementing precision irrigation conserves water and reduces the utilization of fertilizers and pesticides. This reduction results in diminished greenhouse gases emissions and reduced nutrient discharge into rivers, hence enhancing environmental health (Ghazouani et al. 2014; Zhao and Boll 2022). Variable rate irrigation systems can augment these advantages by delivering inputs with greater precision according to particular field circumstances (Ghazouani, et al. 2014).

Improved Economic Viability

Farmers implementing precision irrigation may achieve enhanced yields and diminished operational expenses. By reducing waste and maximizing resource utilization, farmers may improve output while also decreasing their carbon footprints. Precision agriculture approaches have demonstrated a significant reduction in nitrogen oxide emissions by the judicious application of fertilizers (Ghazouani et al. 2014; Hassaballah 2023).

Adaptation to Climate Change

Precision irrigation functions as a responsive approach to climate change by enhancing water management in diverse climatic scenarios. Methods like deficit irrigation and efficient scheduling enable farmers to optimize crop yield while preserving water resources (El-Nashar and Elyamany 2023). The use of intelligent technologies allows farmers to precisely predict water requirements, thus improving resistance to climate-related disturbances (El-Nashar and Elyamany 2023; Hassaballah 2023).

Precision Farming

Precision farming (PF) denotes the forthcoming significant shift in the agricultural sector, commonly termed "Agriculture 4.0." This movement is a continuation of the "Green Revolution" or "Third Agricultural Revolution" that took place in the 1950s and 1960s (FAO 2017). The "Green Revolution" encompassed the enhancement of agricultural systems through high-yielding crops and livestock, facilitated by advanced genetic knowledge, alongside the expanded utilization and accessibility of large-scale mechanization and irrigation. This breakthrough enabled food production for a global population that had more than doubled while only expanding the cultivated land area by 30% (FAO 2017). Agriculture 4.0 emphasizes precision farming at the individual asset level, utilizing current technologies to enhance efficiency while reducing labor intensity. This focused strategy will enable continuous monitoring of individual animals and plants, enhancing productivity and minimizing inputs across the entire agricultural system. PF fundamentally enables the execution of appropriate actions in the correct context and at the optimal moment (Cutress 2020).

Generally, there are three primary components of PF technology applications: (1) Hardware and sensors, which are responsible for data collection or manipulation. (2) Data analysis and decision support systems encompass software, cloud platforms, and machine learning techniques that transform data into valuable insights. (3) Emphasis on commodities and comprehensive farm management (the approach in which decision support tools and models are commercialized and applied throughout entire farms rather than being restricted to specific fields) (Cutress 2020).

In addition to the economic advantages of PF, there is growing discourse over their essential contribution to a sustainable agricultural future. Numerous research indicate that PF facilitates comparable or enhanced productivity/yields while decreasing overall inputs (such as fuel, feed, minerals, or water) (FAO 2017). A decrease in greenhouse gases emissions, whether directly or indirectly, should be taking place. Although there is some clear evidence of emission reductions, few direct greenhouse gases mitigation studies have been conducted for PF technology too far (Cutress 2020). The efficiency of sprinkler and drip irrigation systems can be enhanced by incorporating current features or integrating precise technology to transform them into intelligent systems (Bhavsar et al. 2023). Furthermore, the integration of these systems with surface irrigation techniques might enhance their efficacy and facilitate equitable water distribution among farmers, such as through the use of water cards (Bhavsar et al. 2023). Intelligent methodologies can enable these systems to execute their functions autonomously. The sophisticated equipment continuously evaluates the water level in a field by predicting and comparing the existing moisture content in soil and vegetation with optimal water requirements. If the plant needs water, the system will automatically send commands to the sprinklers or drip irrigation to start functioning until the crop's water needs are met (Sinwar et al. 2020).

Moreover, the digital transformation across practically all sectors has emerged in recent years as the primary objective of achieving optimal output with minimal input. This digital shift has favorably impacted conventional agriculture and improved efficiency (Mondejar et al. 2021). The findings revealed that we have now addressed the shift from conventional agriculture to digital agriculture, commonly referred to as Agriculture 4.0. The apex of modern technologies related to digital transformation is to facilitate the acquisition of in situ measurements remotely and in real time without supplementary human interaction (Lakhiar et al. 2024). This modern technological flagship shifts the grower's attention to utilizing precision agricultural tools to tackle issues and conceptualizing agriculture within the context of cyber-physical systems to improve understanding and modeling of physical processes (Mondejar et al. 2021). The main aim of incorporating precision technology into traditional agriculture are to improve decision-making and to include renewable energy sources into smart agricultural practices. Achieving these objectives necessitates the incorporation of various modern technological tools, such as wireless sensors, the Internet of Things, global positioning systems, remote sensing, and an array of automation, monitoring, controlling, and management information systems, into the conventional agricultural sector (Ragazou et al. 2022).

Improving Irrigation Efficiency to Mitigate Climate Change Impacts

With climate change impacting water supply and agricultural output, enhancing irrigation efficiency is crucial for sustainable farming methods. Several solutions can be employed to improve irrigation systems and alleviate the effects of climate change:

Transition to Modern Irrigation Systems

Transitioning from conventional gravity-fed irrigation to contemporary pressured systems, such as drip and spray irrigation, can substantially decrease water consumption. These technologies enhance conveyance efficiency and reduce losses from evaporation and runoff, facilitating more accurate water distribution directly to plant roots (Climate-ADAPT 2024).

Implement Deficit Irrigation

This method entails utilizing less water than the complete crop-water demand throughout specific growth phases, hence optimizing water usage while preserving output. Studies indicate that this approach can enhance water productivity, rendering it a feasible solution in water-deficient areas. Nonetheless, it necessitates a profound comprehension of agricultural reactions to water scarcity (El-Nashar and Elyamany 2023).

Optimize Irrigation Scheduling Using Precision Irrigation

Employing modern technology, like meteorological forecasting, hydrological surveillance, and intelligent sensor networks, can improve irrigation scheduling. This guarantees the application of water at optimal times and quantities, minimizing waste and enhancing crop yields (El-Nashar and Elyamany 2023).

Adopt Intermittent Irrigation

Methods such as alternate watering and drying are very efficacious for some crops, including rice. This technique conserves water, decreases labor expenses, and may enhance yields (Climate-ADAPT 2024).

Enhance Soil Management to Improve Soil Health

Techniques like mulching, conservation tillage, and cover cropping enhance soil moisture retention and diminish evaporation rates. Healthy soils enhance agricultural resilience during droughts, hence improving irrigation efficiency (El-Nashar and Elyamany 2023).

Utilize Water Reuse Strategies by Water Recycling

Establishing systems for the reutilization of agricultural runoff or treated wastewater can furnish supplementary resources for irrigation. This conserves freshwater resources and mitigates water scarcity challenges intensified by climate change (Michel 2020).

Foster Community Engagement and Capacity Building

Equipping farmers with training in effective irrigation techniques and financial support for system upgrades can promote the adoption of innovative technology. Community engagement is essential for the effective implementation of sustainable irrigation systems (Michel 2020; Cotera et al. 2024).

Conclusions

Ultimately, climate change presents a substantial challenge to global water resources. Mitigating the effects necessitates collaborative efforts from governments, companies, and communities to formulate effective adaptation strategies and minimize further environmental damage. The pressing need for action is underscored by the connection between climate change and water resources to ensure sustainable access to this vital resource. Mitigating irrigation water shortages caused by climate change necessitates a comprehensive strategy that includes technology advancements, strategic agricultural practices, robust policy frameworks, and community education. Implementing these adaptation measures can enhance the resilience of agricultural systems to the difficulties presented by a changing climate.

Modifying irrigation techniques and implementing sophisticated technology are essential for alleviating the effects of climate change on agriculture. By improving water utilization efficiency and adopting new practices, farmers may more effectively address the difficulties presented by a changing climate while supporting food security. Policymakers must prioritize investments in these domains to facilitate sustainable agricultural development amid persistent environmental changes.

Precision irrigation methods provide a comprehensive strategy to mitigate the negative impacts of climate change on agriculture. These technologies are vital for contemporary agricultural practices focused on responding to an increasingly erratic climate by improving water efficiency, developing crop resilience, minimizing environmental consequences, and fostering economic sustainability. Enhancing irrigation efficiency is essential for adjusting agricultural operations to the challenges presented by climate change. Farmers can save water resources and sustain or augment agricultural yield by embracing contemporary technologies, optimizing management techniques, and improving soil health. These techniques enhance food security while fostering sustainable agriculture practices amid a changing climate.

References

Abd El-Mageed TA, El-Samnoudi IM, Ibrahim AEAM, Abd El Tawwab AR (2018) Compost and mulching modulates morphological, physiological responses and water use efficiency in sorghum (bicolor *L. Moench*) under low moisture regime. Agric Water Manag 208:431–439. https://doi.org/10.1016/j.agwat.2018.06.042

Abd-Elaty I, Abd-Elhamid HF, Negm AM (2019) Investigation of saltwater intrusion in coastal aquifers. In: Groundwater in the Nile Delta. Springer, pp 329–353

Abd-Elaty I, Kushwaha N, Grismer ME, Elbeltagi A, Kuriqi A (2022) Cost-effective management measures for coastal aquifers affected by saltwater intrusion and climate change. Sci Total Environ 836:155656

Ahmed Z, Gui D, Murtaza G, Yunfei L, Ali S (2023) An overview of smart irrigation management for improving water productivity under climate change in drylands. Agronomy 13:2113. https://doi.org/10.3390/agronomy13082113

Aini D, Farhaini A, Putra BK (2023) Community participation in improving health in remote areas: a literature review. Int J Educ Inform Technol Other 6(2):27–43. https://doi.org/10.5281/zenodo.7798056

Akol AM, Nassif N, Jaddoa KA et al (2021) Effect of irrigation system, tillage system, and seeding rates on wheat (*Triticum aestivum* L.) growth, grain yield and its water consumption and efficiency. Engineering 13:574–594. https://doi.org/10.4236/eng.2021.1311042

Al-Ghussain L (2019) Global warming: review on driving forces and mitigation. Environ Prog Sustain Energy 38(1):13–21. https://doi.org/10.1002/ep.13041

Alam K (2015) Farmers' adaptation to water scarcity in drought-prone environments: a case study of Rajshahi District, Bangladesh. Agric Water Manag 148:196–206

Alharbi S, Felemban A, Abdelrahim A, Al-Dakhil M (2024) Agricultural and technology-based strategies to improve water-use efficiency in arid and semiarid areas. WaterSA 16:1842. https://doi.org/10.3390/w16131842

Ali AB, Elshaikh NA, Hong L et al (2017) Conservation tillage as an approach to enhance crops water use efficiency. Acta Agric Scand Sect B Soil Plant Sci 67:252–262. https://doi.org/10.1080/09064710.2016.1255349

Anjum SA, Farooq M, Xie XY, Liu XJ, Ijaz MF (2012) Antioxidant defense system and proline accumulation enables hot pepper to perform better under drought. Sci Hortic 140:66–73

Arnbjerg-Nielsen K (2012) Quantification of climate change effects on extreme precipitation used for high resolution hydrologic design. Urban Water J 9(2):57–65.2

Barkunan SR, Bhanumathi V, Sethuram J (2019) Smart sensor for automatic drip irrigation system for paddy cultivation. Comput Electr Eng 73:180–193. https://doi.org/10.1016/j.compeleceng.2018.11.013

Bartlett JA, Dedekorkut-Howes A (2023) Adaptation strategies for climate change impacts on water quality: a systematic review of the literature. J Water Clim Change 14(3):651–675. https://doi.org/10.2166/wcc.2022.279

Besley T, Peters MA (2020) Life and death in the Anthropocene: educating for survival amid climate and ecosystem changes and potential civilization collapse. Educ Philos Theory 52(13):1347–1357

Bhavsar D, Limbasia B, Mori Y, Aglodiya MI, Shah M (2023) A comprehensive and systematic study in smart drip and sprinkler irrigation systems. Smart Agric Technol 5:100303. https://doi.org/10.1016/j.atech.2023.100303
Blanco-Canqui H, Lal R (2009) Crop residue management and soil carbon dynamics. Soil carbon sequestration. Greenhouse Effect 57:291–309. https://doi.org/10.2136/sssaspecpub57.2ed.c17
Bolan S, Padhye LP, Jasemizad T, Govarthanan M et al (2024) Impacts of climate change on the fate of contaminants through extreme weather events. Sci Total Environ 909:168388. https://doi.org/10.1016/j.scitotenv.2023.168388
Börjesson G, Menichetti L, Thornton B et al (2016) Seasonal dynamics of the soil microbial community: assimilation of old and young carbon sources in a long-term field experiment as revealed by natural 13C abundance. Eur J Soil Sci 67:79–89
Botzen W, Duijndam S, van Beukering P (2021) Lessons for climate policy from behavioral biases towards COVID-19 and climate change risks. World Dev 137:105214
Brooks PD, Chorover J, Fan Y, Godsey SE, Maxwell RM, McNamara JP, Tague C (2015) Hydrological partitioning in the critical zone: recent advances and opportunities for developing transferable understanding of water cycle dynamics. Water Resour Res 51(9):6973–6987
Buytaert W, Cuesta-Camacho F, Tobón C (2011) Potential impacts of climate change on the environmental services of humid tropical alpine regions. Glob Ecol Biogeogr 20(1):19–33
Cao Y, Cai H, Sun S et al (2022) Effects of drip irrigation methods on yield and water productivity of maize in Northwest China. Agric Water Manag 259:107227. https://doi.org/10.1016/j.agwat.2021.107227
Caretta MA, Mukherji A, Arfanuzzaman M, Betts RA, Gelfan A, Hirabayashi Y, Lissner TK, Liu J, Gunn EL, Morgan R, Mwanga S, Supratid S (2022) Water. In: Portner H-O, Roberts DC, Tignor M, Poloczanska ES, Mintenbeck K, Alegria A, Craig M, Langsdorf S, Loschke S, Moller V, Okem A, Rama B (eds) Climate change 2022: impacts, adaptation and vulnerability. Contribution of Working Group II to the Sixth Assessment Report of the Intergovernmental Panel on Climate Change. Cambridge University Press, Cambridge/New York, pp 551–712. https://doi.org/10.1017/9781009325844.006
Cayuela ML, Spott O, Pascual MB et al (2024) Key biochar properties linked to denitrification products in a calcareous soil. Biochar 6:90
Chaturvedi A, Pandey B, Yadav AK, Saroj S (2021) An overview of the potential impacts of global climate change on water resources. In: Water conservation in the era of global climate change. Elsevier, pp 99–120. https://www.researchgate.net/publication/349658342_An_overview_of_the_potential
Chen R, Chang H, Wang Z, Lin H (2023) Determining organic-inorganic fertilizer application threshold to maximize the yield and quality of drip-irrigated grapes in an extremely arid area of Xinjiang, China. Agric Water Manag 276:108070. https://doi.org/10.1016/j.agwat.2022.108070
Chiu SY, Kao CY, Tsai MT, Ong SC, Chen CH, Lin CS (2009) Lipid accumulation and CO2 utilization of Nannochloropsis oculata in response to CO2 aeration. Bioresour Technol 100:833–838
Ciampittiello M, Marchetto A, Boggero A (2024) Water resources management under climate change: a review. Sustain For 16(9):3590. https://doi.org/10.3390/su16093590
Climate-ADAPT (2024) Improvement of irrigation efficiency. https://climate-adapt.eea.europa.eu/en/metadata/adaptation-options/improvement-of-irrigation-efficiency
Colematt 2022. Stock illustrations| Climate change. https://www.istockphoto.com/vector/diagram-showing-the-greenhouse-effect-gm1445620582-483966759
Cotera RV, Egerer S, Nam C, Lierhammer L, Moors L, Costa MM (2024) Resilient agriculture: water management for climate change adaptation in Lower Saxony. J Water Clim Change 15(3):1034. https://doi.org/10.2166/wcc.2024.455
Cutress D (2020) Can precision farming help mitigate climate change? Farming Connect:1–8
Das J, Goyal MK (2021) Current trends and projections of water resources under climate change in Ganga River basin. In: Chauhan MS, Ojha CSP (eds) The Ganga River basin: a hydrometeorological approach. Springer, Cham, pp 233–256
Druckman JN, McGrath MC (2019) The evidence for motivated reasoning in climate change preference formation. Nat Clim Chang 9(2):111–119

Eisenhauer DE, Martin DL, Heeren DM, Hoffman GJ (2021) Irrigation systems management. American Society of Agricultural and Biological Engineers (ASABE)

El-Nashar W, Elyamany A (2023) Adapting irrigation strategies to mitigate climate change impacts: a value engineering approach. Water Resour Manag 37:2369–2386. https://doi.org/10.1007/s11269-022-03353-4

Evanylo G, Sherony C, Spargo J et al (2008) Soil and water environmental effects of fertilizer-, manure-, and compost-based fertility practices in an organic vegetable cropping system. Agric Ecosyst Environ 127:50–58. https://doi.org/10.1016/j.agee.2008.02.014

FAO (2017) The future of food and agriculture—trends and challenges. FAO, Rome

Fecht S (2019) How climate change impacts our water. https://news.climate.columbia.edu/2019/09/23/climate-change-impacts-water/

Fernando DN et al (2019) A process-based statistical seasonal prediction of May–July rainfall anomalies over Texas and the Southern Great Plains of the United States. Clim Serv 16:100133

Fischer BMC, Manzoni S, Morillas L et al (2019) Improving agricultural water use efficiency with biochar—a synthesis of biochar effects on water storage and fluxes across scales. Sci Total Environ 657:853–862. https://doi.org/10.1016/j.scitotenv.2018.11.312

Franco ALC, Cherubin MR, Cerri CEP et al (2020) Linking soil engineers, structural stability, and organic matter allocation to unravel soil carbon responses to land-use change. Soil Biol Biochem 150:107998. https://doi.org/10.1016/j.soilbio.2020.107998

Gabuya AQ, Mangubat FN, Patindol VH, Paglinawan JM, Catubis KML (2024) Improved growth of coffee seedlings (*Coffea canephora*) under SMART irrigation system. J Saudi Soc Agric Sci 23:103–111. https://doi.org/10.1016/j.jssas.2023.09.007

Gao M, Yang H, Xiao Q, Goh M (2021) A novel fractional grey Riccati model for carbon emission prediction. J Clean Prod 282:124471

Gentine P, Green JK, Guérin M, Humphrey V, Seneviratne SI, Zhang Y, Zhou S (2019) Coupling between the terrestrial carbon and water cycles—a review. Environ Res Lett 14(8):083003

Ghazouani W, Molle F, Swelam A, Rap E, Abdo A (2014) Understanding farmers' adaptation to water scarcity: a case study from the western Nile Delta, Egypt, IWMI research report 160. International Water Management Institute (IWMI), 31p, Colombo. https://doi.org/10.5337/2015.200

Ghosh S, Ow LF, Wilson B (2015) Influence of biochar and compost on soil properties and tree growth in a tropical urban environment. Int J Environ Sci Technol 12:1303–1310

González-Briones A, Mezquita Y, Castellanos-Garzón JA, Prieto J, Corchado JM (2019) Intelligent multi-agent system for water reduction in automotive irrigation processes. Procedia Comput Sci 151:971–976. https://doi.org/10.1016/j.procs.2019.04.136

Gu Z, Qib Z, Ma L, Guid D, Xue J, Fangf Q, Yuana S, Fengg G (2024) Development of an irrigation scheduling software based on model predicted crop water stress. Comput Electron Agric 143(2017):208–221. https://doi.org/10.1016/j.compag.2017.10.023

Hamed HA, Mahmoud GA-E, Abeed AHA (2025) Unraveling growth and metabolic dynamics in drought-stressed spinach plants: Exploring the contribution of biological gibberellin. Scientia Horticulturae, 340:113924. https://doi.org/10.1016/j.scienta.2024.113924

Hassaballah K (2023) Top 5 reasons why irrigation might not be your go to adaptation solution. https://www.icpac.net/news/top-5-reasons-why-irrigation-might-not-be-your-go-to-adaptation-solution/

Hatfield JL, Dold C (2019) Water-use efficiency: advances and challenges in a changing climate. Front Plant Sci 10:1–14. https://doi.org/10.3389/fpls.2019.00103

Hatfield JL, Sauer TJ, Prueger JH (2001) Managing soils to achieve greater water use efficiency: a review. Agron J 93:271–280. https://doi.org/10.2134/agronj2001.932271x

He C, Wang YQ, Yu WB et al (2022) Comprehensive analysis of resource utilization efficiency under different tillage systems in North China Plain. J Clean Prod 347:131289. https://doi.org/10.1016/j.jclepro.2022.131289

Hegerl GC et al (2019) Causes of climate change over the historical record. Environ Res Lett 14(12):123006

Ingrao C, Strippoli R, Lagioia G, Huisingh D (2023) Water scarcity in agriculture: an overview of causes, impacts and approaches for reducing the risks. Heliyon 9(8):e18507. https://doi.org/10.1016/j.heliyon.2023.e18507

IPCC (2022) Climate change 2022: mitigation of climate change. Contribution of Working Group III to the Sixth Assessment Report of the Intergovernmental Panel on Climate Change (Shukla PR, Skea J, Slade R, Al Khourdajie A, van Diemen R, McCollum D, Pathak M, Some S, Vyas P, Fradera R, Belkacemi M, Hasija A, Lisboa G, Luz S, Malley J (eds)). Cambridge University Press, Cambridge/New York. https://doi.org/10.1017/9781009157926

IPCC (2023) Climate CHANGE 2023: SYNTHESIS REPORT. Contribution of Working Groups I, II and III to the Sixth Assessment Report of the Intergovernmental Panel on Climate Change (Core Writing Team, Lee H, Romero J (eds)). IPCC, Geneva, 184 pp. https://doi.org/10.59327/IPCC/AR6-9789291691647

Israelsen OW, Hansen VE (1962) Irrigation principles and practices, 3rd edn. Wiley, New York

Jin S, Hao Z, Zhang K, Yan Z, Chen J (2021) Advances and challenges for the electrochemical reduction of CO2 to CO: from fundamentals to industrialization. Angew Chem Int Ed 60:20627–20648

Johansson RC, Tsur Y, Roe TL, Doukkali R, Dinar A (2002) Pricing irrigation water: a review of theory and practice. Water Policy 4:173–199. https://doi.org/10.1016/S1366-7017(02)00026-0

Kareem SL et al (2021) Water quality assessment and phosphorus effect using water quality indices: Euphrates River-Iraq as a case study. Groundw Sustain Dev 14:100630

Katila P, Colfer CJP, De Jong W, Galloway G, Pacheco P, Winkel G (2019) Sustainable development goals. Cambridge University Press, Cambridge. https://doi.org/10.1017/9781108765015

Khairullina ER et al (2019) Global climate change: cyclical nature of natural and permanent nature of man-made processes. EurAsian J BioSci 13(2):2311–2316

Kinney PL (2018) Interactions of climate change, air pollution and human health. Curr Environ Health Rep 5(1):179–186

Kiparsky M, Milman A, Vicuna S (2012) Climate and water: knowledge of impacts to action on adaptation. Annu Rev Environ Resour 37:163–194

Klove B et al (2014) Climate change impacts on groundwater and dependent ecosystems. J Hydrol 518:250–266

Lakhiar IA, Yan H, Zhang C, Wang G, He B, Hao B, Han Y, Wang B, Bao R, Syed TN, Chauhdary JN, Rakibuzzaman M (2024) A review of precision irrigation water-saving technology under changing climate for enhancing water use efficiency, crop yield, and environmental footprints. Agriculture 14:1141. https://doi.org/10.3390/agriculture14071141

Lakshmi GDP, Asha PN, Sandhya G, Sharma SV, Shilpashree S, Subramanya SG (2023) An intelligent IOT sensor coupled precision irrigation model for agriculture. Meas Sens 25:100608. https://doi.org/10.1016/j.measen.2022.100608

Lamandé M, Munkholm LJ, Børresen T (2023) Soil tillage. In: Goss MJ, Oliver M (eds) Encyclopedia of soils in the environment, 2nd edn. Academic, pp 433–438. https://doi.org/10.1016/B978-0-12-822974-3.00189-0

Laphatphakkhanut R, Puttrawutichai S, Dechkrong P, Preuksakarn C, Wichaidist B, Vongphet J, Suksaroj C (2021) IoT-based smart crop-field monitoring of rice cultivation system for irrigation control and its effect on water footprint mitigation. Paddy Water Environ 19:699–707. https://doi.org/10.1007/s10333-021-00868-1

Lebu S, Lee A, Salzberg A, Bauza V (2024) Adaptive strategies to enhance water security and resilience in low- and middle-income countries: a critical review. Sci Total Environ 925:171520. https://doi.org/10.1016/j.scitotenv.2024.171520

Lejeune Q et al (2018) Historical deforestation locally increased the intensity of hot days in northern mid-latitudes. Nat Clim Chang 8(5):386–390

Letcher TM (2020) 1—introduction with a focus on atmospheric carbon dioxide and climate change. In: Letcher TM (ed) Future energy, 3rd edn. Elsevier, pp 3–17

Levidow L, Zaccaria D, Maia R, Vivas E, Todorovic M, Scardigno A (2017) Improving water-efficient irrigation: prospects and difficulties of innovative practices. Agric Water Manag 146:84–94. https://doi.org/10.1016/j.agwat.2014.07.012

Li M, Zhou S, Shen S, Wang J, Yang Y, Wu Y, Chen F, Lei Y (2024) Climate-smart irrigation strategy can mitigate agricultural water consumption while ensuring food security under a changing climate. Agric Water Manag 292:108663. https://doi.org/10.1016/j.agwat.2023.108663

Liao Y, Cao HX, Liu X et al (2021) By increasing infiltration and reducing evaporation, mulching can improve the soil water environment and apple yield of orchards in semiarid areas. Agric Water Manag 253:106936. https://doi.org/10.1016/j.agwat.2021.106936

Liebhard G, Klik A, Neugschwandtner RW, Nolz R (2022) Effects of tillage systems on soil water distribution, crop development, and evaporation and transpiration rates of soybean. Agric Water Manag 269:107719. https://doi.org/10.1016/j.agwat.2022.107719

Ma J et al (2020) Hydrological cycle changes under global warming and their effects on multiscale climate variability. Ann N Y Acad Sci 1472(1):21–48. https://doi.org/10.1111/nyas.14335

Mahmoud GA-E, Abeed AHA, Mostafa HHA, Abdel Monsef O (2023) Responses of Pea (Pisum sativum L.) to Single and Consortium Bio-Fertilizers in Clay and Newly Reclaimed Soils. Plants, 12:3931. https://doi.org/10.3390/plants12233931

Mahmoud GAE, Hefzy M & Zahran MMAA (2024) Synergistic Effects of Microbial Gibberellic Acid and Vitamins on Onion (Allium cepa L.) Yield, and Quality in Low-Fertility Soil. J Soil Sci Plant Nutr. 24:5342–5359. https://doi.org/10.1007/s42729-024-01911-w

Mahmoud, G.AE., Sabra, M.A., Mohamed, A.E. et al. (2025) Optimization of Gibberellic Acid Production from Fusarium incarnatum and Its Effect on Zea mays Growth Promotion and Antioxidant Activity. J Plant Growth Regul 44:3219–3235. https://doi.org/10.1007/s00344-024-11612-4

Mandal T et al (2021) Rainfall insight in Bangladesh and India: climate change and environmental perspective. In: Rukhsana et al (eds) Habitat, ecology and ekistics: case studies of human-environment interactions in India. Springer, Cham, pp 53–74

Mason B, Rufí-Salís M, Parada F, Gabarrell X, Gruden C (2019) Intelligent urban irrigation systems: saving water and maintaining crop yields. Agric Water Manag 226:105812. https://doi.org/10.1016/j.agwat.2019.105812

Mbow C, Rosenzweig C, Barioni LG, Benton TG, Herrero M, Krishnapilla M, Liwenga E, Pradhan P, Rivera-Ferre MG, Sapkota T, Tubiello FN, & Xu Y (2019) Chapter 5 "Food Security". In N. Benkeblia, A. Challino, A. Khan, & J. R. Porter (Eds.), Climate Change and Land: An IPCC special report on climate change, desertification, land degradation, sustainable land management, food security, and greenhouse gas fluxes in terrestrial ecosystems (pp. 439–550). Cambridge University Press. https://doi.org/10.1017/9781009157988.007

McGrane SJ (2016) Impacts of urbanisation on hydrological and water quality dynamics, and urban water management: a review. Hydrol Sci J 61(13):2295–2311

Meena RS, Lal R, Yadav GS (2020) Long-term impacts of topsoil depth and amendments on soil physical and hydrological properties of an Alfisol in Central Ohio, USA. Geoderma 363:114164. https://doi.org/10.1016/j.geoderma.2019.114164

Meerhoff M, Audet J, Davidson TA, De Meester L, Hilt S, Kosten S, Scheffer M (2022) Feedback between climate change and eutrophication: revisiting the allied attack concept and how to strike back. Inland Waters 12(2):187–204

Michel S (2020) Increased efficiency of irrigation water use to address climate change related water shortage. https://panorama.solutions/en/solution/increased-efficiency-irrigation-water-use-address-climate-change-related-water-shortage

Mondejar ME, Avtar R, Diaz HLB, Dubey RK et al (2021) Digitalization to achieve sustainable development goals: steps towards a Smart Green Planet. Sci Total Environ 794:148539. https://doi.org/10.1016/j.scitotenv.2021.148539

Nesmith AA et al (2021) Climate change, ecology, and justice. In: Nesmith AA et al (eds) The intersection of environmental justice, climate change, community, and the ecology of life. Springer

NOAA (2019) Climate literacy, carbon capture and storage, education. https://www.noaa.gov/education/resource-collections/climate/carbon-cycle

OECD (2013) Water and climate change adaptation: policies to navigate uncharted waters. OECD studies on water. OECD Publishing. https://doi.org/10.1787/9789264200449-en

OECD (2015) Stakeholder engagement for inclusive water governance. OECD studies on water. OECD Publishing, Paris. https://doi.org/10.1787/9789264231122-en

Prevedello JA et al (2019) Impacts of forestation and deforestation on local temperature across the globe. PLoS One 14(3):e0213368

Ragazou K, Garefalakis A, Zafeiriou E, Passas I (2022) Agriculture 5.0: a new strategic management mode for a cut cost and energy efficient agriculture sector. Energies 15(9):3113. https://doi.org/10.3390/en15093113

Richards S, Rao L, Connelly S, Raj A, Raveendran L, Shirin S, Jamwal P, Helliwell R (2021) Sustainable water resources through harvesting rainwater and the effectiveness of a low-cost water treatment. J Environ Manag 286:112223. https://doi.org/10.1016/j.jenvman.2021.112223

Rumanti AA, Rizana AF, Achmad F (2023) Exploring the role of organizational creativity and open innovation in enhancing SMEs performance. J Open Innov: Technol Mark Complex 9(2):100045. https://doi.org/10.1016/j.joitmc.2023.100045

Santos RMB et al (2015) Impacts of climate change and land-use scenarios on *Margaritifera margaritifera*, an environmental indicator and endangered species. Sci Total Environ 511:477–488

Scanlon BR, Jolly I, Sophocleous M, Zhang L (2007) Global impacts of conversions from natural to agricultural ecosystems on water resources: quantity versus quality. Water Resour Res 43(3)

Schuster L, Taillardat P, Macreadie PI, Malerba ME (2024) Freshwater wetland restoration and conservation are long-term natural climate solutions. Sci Total Environ 922:171218. https://doi.org/10.1016/j.scitotenv.2024.171218

Semida WM, Beheiry HR, Sétamou M, Simpson CR, Abd El-Mageed TA, Rady MM, Nelson SD (2019) Biochar implications for sustainable agriculture and environment: a review. S Afr J Bot 127:333–347. https://doi.org/10.1016/j.sajb.2019.11.015

Seneviratne, S.I., X. Zhang, M. Adnan, W. Badi, C. Dereczynski, A. Di Luca, S. Ghosh, I. Iskandar, J. Kossin, S. Lewis, F. Otto, I. Pinto, M. Satoh, S.M. Vicente-Serrano, M. Wehner, and B. Zhou, 2021: Weather and climate extreme events in a changing climate. In Climate change 2021: the physical science basis. Contribution of Working Group I to the Sixth Assessment Report of the Intergovernmental Panel on Climate Change [Masson-Delmotte, V., P. Zhai, A. Pirani, S.L. Connors, C. Péan, S. Berger, N. Caud, Y. Chen, L. Goldfarb, M.I. Gomis, M. Huang, K. Leitzell, E. Lonnoy, J.B.R. Matthews, T.K. Maycock, T. Waterfield, O. Yelekçi, R. Yu, and B. Zhou (eds.)]. Cambridge University Press, Cambridge/New York, pp. 1513–1766, doi:10.1017/9781009157896.013

Shawul AA, Chakma S (2020) Trend of extreme precipitation indices and analysis of long-term climate variability in the upper awash basin, Ethiopia. Theor Appl Climatol 140:635–652

Sidhu RK, Kumar R, Rana PS (2021) Chapter five—automation in drip irrigation for enhancing water use efficiency in cereal systems of South Asia: status and prospects. Adv Agron 167:247–300. https://doi.org/10.1016/bs.agron.2021.01.002

Singh S, Nouri A, Singh S et al (2020) Soil organic carbon and aggregation in response to thirty-nine years of tillage management in the southeastern US. Soil Tillage Res 197:104523. https://doi.org/10.1016/j.still.2019.104523

Sinore T, Wang F (2024) Impact of climate change on agriculture and adaptation strategies in Ethiopia: a meta-analysis. Heliyon 10(4):e26103. https://doi.org/10.1016/j.heliyon.2024.e26103

Sinwar D, Dhaka VS, Sharma MK, Rani G (2020) AI-based yield prediction and smart irrigation. In: Pattnaik P, Kumar R, Pal S (eds) Internet of things and analytics for agriculture vol 2. Studies in big data, vol 67. Springer, Singapore. https://doi.org/10.1007/978-981-15-0663-5

Six J, Elliott ET, Paustian K (2000) Soil macroaggregate turnover and microaggregate formation: a mechanism for C sequestration under no-tillage agriculture. Soil Biol Biochem 32:2099–2103

Song X, Li Y, Yue X et al (2019) Effect of cotton straw-derived materials on native soil organic carbon. Sci Total Environ 663:38–44

Sophocleous M (2004) Global and regional water availability and demand: prospects for the future. Nat Resour Res 13:61–75

Ssekyanzi G, Ahmad MJ, Choi K-S (2024) Sustainable solutions for mitigating water scarcity in developing countries: a comprehensive review of innovative rainwater storage systems. WaterSA 16:2394. https://doi.org/10.3390/w16172394
Strudley MW, Green TR, Ascough JC (2008) Tillage effects on soil hydraulic properties in space and time: state of the science. Soil Tillage Res 99:4–48. https://doi.org/10.1016/j.still.2008.01.007
Sukkariyah B, Evanylo G, Zelazny L (2007) Distribution of copper, zinc, and phosphorus in coastal plain soils receiving repeated liquid biosolids applications. J Environ Qual 36:1618–1626
Thakur N, Nigam M, Tewary R, Rajvanshi K, Kumar M, Shukla SK, Mahmoud GA-E, Gupta S (2022) Drivers for the behavioural receptiveness and non-receptiveness of farmers towards organic cultivation system. JOURNAL OF KING SAUD UNIVERSITY SCIENCE 34:102107. https://doi.org/10.1016/j.jksus.2022.102107
Viets FG (1962) Fertilizers and the efficient use of water. Adv Agron 14:223–264
Wahidah N, Isro'ullaili I, Putra BK (2023) The school literacy movement (GLS) and student's interest in reading at SDN 3 Suka Makmur. Jurnal Ilmiah Wahana Pendidikan 9(7):559–564. https://doi.org/10.5281/zenodo.7826963
Wang C, Zhao J, Feng Y et al (2021) Optimizing tillage method and irrigation schedule for greenhouse gas mitigation, yield improvement, and water conservation in wheat–maize cropping systems. Agric Water Manag 248:106762. https://doi.org/10.1016/j.agwat.2021.106762
Wang H, Shao D, Ji B et al (2022) Biochar effects on soil properties, water movement and irrigation water use efficiency of cultivated land in Qinghai-Tibet Plateau. Sci Total Environ 829:154520. https://doi.org/10.1016/j.scitotenv.2022.154520
Williams AP et al (2015) Contribution of anthropogenic warming to California drought during 2012–2014. Geophys Res Lett 42(16):6819–6828
Williams J, Beveridge R, Mayaux P-L (2023) Unconventional waters: a critical understanding of desalination and wastewater reuse. Water Alternat 16(2):429–443
WMO (2024) The Sun's impact on the Earth. https://wmo.int/suns-impact-earth
Xing Y, Wang X (2024) Precision agriculture and water conservation strategies for sustainable crop production in arid regions. Plan Theory 13:3184. https://doi.org/10.3390/plants13223184
Yan S, Wu Y, Fan J et al (2020) A sustainable strategy of managing irrigation based on water productivity and residual soil nitrate in a no-tillage maize system. J Clean Prod 262:121279. https://doi.org/10.1016/j.jclepro.2020.121279
Zeng Y, Chen C, Lin G (2023) Practical application of an intelligent irrigation system to rice paddies in Taiwan. Agric Water Manag 280:108216. https://doi.org/10.1016/j.agwat.2023.108216
Zhang Q, Wang Z, Miao F, Wang G (2017) Dryland maize yield and water-use efficiency responses to mulching and tillage practices. Agron J 109:1196–1209. https://doi.org/10.2134/agronj2016.10.0593
Zhang Y, Ding J, Wang H et al (2020) Biochar addition alleviate the negative effects of drought and salinity stress on soybean productivity and water use efficiency. BMC Plant Biol 20:1–11. https://doi.org/10.1186/s12870-020-02493-2
Zhang Y, Wang J, Feng Y (2021) The effects of biochar addition on soil physicochemical properties: a review. Catena 202:105284. https://doi.org/10.1016/J.CATENA.2021.105284
Zhang W, Dong A, Liu F et al (2022) Effect of film mulching on crop yield and water use efficiency in drip irrigation systems: a meta-analysis. Soil Tillage Res 221:105392. https://doi.org/10.1016/j.still.2022.105392
Zhang X, Shen S, Xue S et al (2023) Long-term tillage and cropping systems affect soil organic carbon components and mineralization in aggregates in semiarid regions. Soil Tillage Res 231:105742. https://doi.org/10.1016/j.still.2023.105742
Zhao M, Boll J (2022) Adaptation of water resources management under climate change. Front Water 4:983228. https://doi.org/10.3389/frwa.2022.983228
Zohary T, Ostrovsky I (2011) Ecological impacts of excessive water level fluctuations in stratified freshwater lakes. Inland Waters 1(1):47–59
Zscheischler J et al (2018) Future climate risk from compound events. Nat Clim Chang 8(6):469–477

Climate Change Effects on Sustainable Field Crops Production in Egypt: Challenges and Prospective

10

Ehab H. El-Harty and Megahed H. Ammar

Abstract

Climate change poses substantial difficulties to development and prosperity by disrupting weather patterns, temperature, precipitation, and sea levels. Throughout the twentieth century, significant climate alterations transpired, including increased temperatures and elevated sea levels. The repercussions of climate change are intricate, encompassing both direct and indirect consequences for human existence. The direct effect encompassed variations in crop yields, contingent upon the crop type, soil conditions, and prevailing environment, exacerbated by salty soil proliferation owing to rising sea levels, unexpected flooding events, and/or drought and heat stress. Indirect impacts encompass food costs and human health. Nevertheless, employing innovative tactics can mitigate losses and enhance profits. Egypt's agriculture is significantly susceptible to climate change owing to its desert environment, water resource constraints, and dependence on food imports. This review analyzes the influence of climate change on agricultural crops in Egypt, encapsulating its impacts on water supplies and crop productivity. We also examine initiatives from the National Egyptian Strategic Plan for climate change mitigation and adaptation. It also underscores the principal problems in implementing adaptation measures to these changes.

E. H. El-Harty (✉)
Food Legumes Research Department, Field Crops Research Institute, Agricultural Research Center (ARC), Giza, Egypt

M. H. Ammar
Rice Research Department, Field Crops Research Institute, Agricultural Research Center (ARC), Giza, Egypt

A. E.-H. Zohry, S. Ouda (eds.), *Resilient Agroecosystems*, Sustainability Sciences in Asia and Africa, https://doi.org/10.1007/978-981-96-9630-7_10

Introduction

Climate change is one of the most significant barriers to sustainable agricultural growth. It is expected to alter weather patterns, influencing temperature, precipitation, sea levels, and the incidence of extreme weather events (IPCC 2007). The United Nations Framework Convention on Climate Change (UNFCCC 2007) characterizes climate change as alterations in climate resulting directly or indirectly from human activities, in conjunction with natural climate variability. The expected temperature increase is estimated to be between 1.2 and 4.1 °C by 2100, rising to 1.9–5.6 °C by 3000 (IPCC 2007). The increases will be greater in winter compared to the monsoon period (Kaini et al. 2020). It is essential to distinguish between global warming and climate change. Global warming denotes the average increase in Earth's surface temperature, representing a critical aspect of climate change that encompasses substantial alterations in climatic variables, such as temperature, precipitation, wind, and humidity (ICTSD-IPC 2009).

According to the IPCC (2007), projected CO_2 levels may be 100% higher by 2100 relative to pre-industrial levels. Furthermore, the IPCC (2023) indicated that human activities during the past decade (2011–2020) have been adequate to elevate the global surface temperature by 1.1 °C above the values documented between 1850 and 1900. Ozone concentrations are rising; even minimal levels near urban areas can result in observable plant tissue damage and physiological alterations (Felzer et al. 2007). During the twentieth century, the global mean surface temperature climbed by 0.6 °C, with nighttime minimum temperatures increasing at twice the pace of daytime maximum temperatures. Additionally, sea levels increased by 10–20 cm, with an average rise of 1–2 mm annually. Intense precipitation events have risen in the mid and high northern latitudes, while tropical countries, particularly in Asia and Africa, have had drier winters. Precipitation increased by 5–10% in the Northern Hemisphere, but it diminished in regions such as North Africa, resulting in more frequent and severe droughts (IPCC 2001).

Extensive research has been undertaken regarding the effects of climate change, indicating that alterations in climate and increasing sea levels will profoundly influence food systems and natural ecosystems. The primary concerns encompass (1) reductions in natural productivity and biodiversity, (2) diminished cereal crop yields in tropical and subtropical areas, (3) exacerbated drought and saline soils resulting from decreased precipitation and heightened evaporation, (4) elevated health risks and growing disparities between affluent and impoverished nations, and (5) an escalation in abrupt climate changes. Although we cannot entirely prevent climate change, we may mitigate its impact by decreasing greenhouse gas emissions or improving the "sinks" that absorb and sequester them (Alcamo et al. 2007). The IPCC (2001) asserts that while complete prevention of climate change is unattainable, its severity can be mitigated by the reduction of greenhouse gases (GHGs) emissions. Human activities markedly intensify these alterations, chiefly through two factors: GHGs including carbon dioxide, methane, nitrous oxide, and chlorofluorocarbons, as depicted in Fig. 10.1, and modifications in land-use patterns resulting from practices such as agricultural changes, deforestation, and

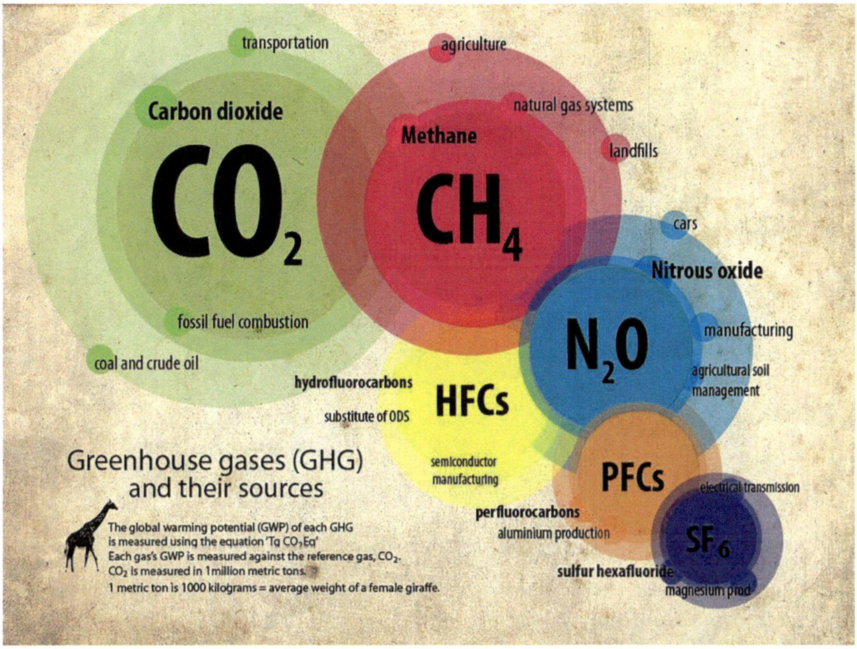

Fig. 10.1 Types of GHGs and factors affecting their levels. (Source: https://www.e-education.psu.edu/ba850/node/774)

reforestation, which have transformed the land's physical and biological attributes. These alterations influence solar radiation absorption, enhance evaporation, and expedite transpiration, diminishing carbon dioxide absorption capacity and amplifying carbon emissions from biomass decomposition (IPCC 2001).

This chapter seeks to examine the effects of climate change on field crop production in Egypt while emphasizing the primary challenges. It evaluates existing tactics and technology for adaptation, including irrigation practices, crop rotation, and the implementation of climate-resilient crop types, while examining future possibilities.

Climate Change Impacts on Agriculture

The Impact of Climate Change on Water Resources

Egypt faces substantial water resource difficulties stemming from insufficient rainfall and a predetermined Nile allocation (Molle et al. 2019). Egypt's water supplies predominantly rely on the Nile River, rendering any alterations in its flow critically important to the nation (Fig. 10.2). The burgeoning population of Egypt is exerting pressure on water supplies, resulting in a decrease in the annual per capita allocation of renewable freshwater, which plummeted to 663 cubic meters in 2018,

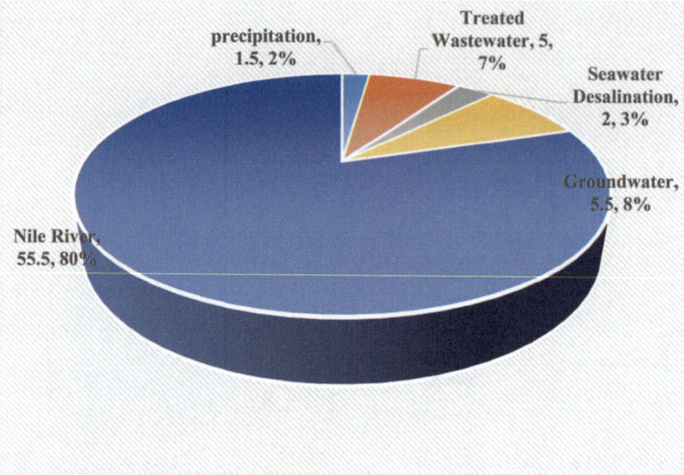

Fig. 10.2 Summary of Egyptian water resources

significantly beneath the crucial threshold. Simultaneously, the cultivated area increased from 8.7 million feddan in 2010 to 9.4 million feddan in 2020, with additional expansion expected. As competition for water escalates from other sectors, agriculture's contribution to national economic growth remains minimal (approximately 9%) from 2000 to 2018. The agriculture sector's allocation of water resources is anticipated to diminish over time (National Strategic Plan 2020). The United Nations anticipates that Egypt could see a severe water crisis by 2025, posing a significant challenge to the agricultural sector, which utilizes more than 80% of the available water (WB 2018).

Khaled and Sherien (2017) observed that the Nile River, extending 3000 km through the arid regions of northern Sudan and Egypt, suffers significant water loss from evaporation. Thus, variations in temperature and precipitation significantly affect water availability. Their research determined that climate change may diminish groundwater availability in the Nile Valley and Delta as a result of heightened water demand and extraction. Dawoud et al. (2024) anticipated a shortfall of almost 20 billion m^3 in 2022, intensified by swift population expansion.

Numerous places encounter reduced surface and groundwater resources due to changes in precipitation, runoff, and consumption patterns. Schilling et al. (2020) examined the susceptibility of North African nations to climate change, emphasizing its socioeconomic ramifications. Their assertion is that the interplay between climate change and accelerated population expansion is expected to exacerbate the existing scarcity of water supplies. Driouech et al. (2020) forecasted that the MENA area will emerge as a global hotspot for droughts by the late twenty-first century. Zittis et al. (2021) estimated that around 600 million individuals in the MENA region may frequently encounter significant heat waves. Furthermore, the per capita freshwater availability has decreased from 990 m^3 in 2005 to 800 m^3 in 2015 (AFED 2017) and is projected to fall below 600 m^3 by 2050, much beneath the poverty level of 1000 m^3 per capita (Abouelnaga 2019). El Agroudy et al. (2014) observed that water retention at the Renaissance Dam may result in a decrease in inflow to Lake Nasser by roughly 25–33 billion m^3 per year. If no compensatory water is extracted from the Dam Lake, around 3–5 million acres of Egypt's agricultural land could be impacted. Research by Siam and Elthahir (2017) linked this variability to heightened climate shifts impacting annual precipitation in the Ethiopian highlands. Alterations in rainfall patterns may lead to a 10–15% average augmentation in the Nile's flow by 2040; nevertheless, there will probably be a decrease in "normal" years alongside an increase in both severe flow years and drought years. Nonetheless, considerable ambiguity exists regarding the effects of climate change on Nile River flows. Research indicates that elevated temperatures may result in increased evaporation, thereby diminishing water availability by as much as 70% (Ashour et al. 2024). In contrast, additional studies suggest that increased precipitation in the Ethiopian highlands and the Blue Nile Basin may elevate flows by 15–25% (USAD 2021). Since the Nile's origins are located beyond Egypt's borders, the nation is significantly susceptible to climate change, both internally and externally. A significant fraction of the population lives adjacent to the Nile, heightening their vulnerability to flooding, especially among the urban impoverished (Climate Risk Profile 2020). Kaini et al. (2021) examined the impact of climate change on the hydrology of the Koshi River basin in India and its consequences for irrigation, analogous to the Nile River, under the RCP4.5 scenario. Their findings suggested that the average annual flow in the Koshi River may rise by 15–30% during the examined periods, with RCP8.5 forecasts indicating potential increases of 20–60%. The study projected elevated flood levels and a temporal shift in peak flow to August and September. Conversely, coastal regions and little islands encounter increased threats from rising sea levels. The coastal area of Egypt's Nile Delta has been recognized as especially susceptible to rising sea levels, influenced by geological and anthropogenic influences (El-Sharkawy et al. 2009).

Research on the potential effects of sea level rise on the Nile Delta revealed that an increase of 0.5 and 1.0 m might submerge 1800 km^2 and 4500 km^2 of agricultural land, respectively, displacing 3.8 and 6.1 million residents at these elevations (Simonett 2002). Research conducted by Elshinnawy and Almaliki (2021) estimated a Mediterranean Sea level rise of 73 cm by the century's end, suggesting that inundated regions will include around 2.16% of the study area (231 km^2). They discovered that major sea wave heights rose by 3.1 cm year from 1999 to 2010, indicating critical implications for future coastal management strategies. Moreover, groundwater levels are projected to increase by 0–0.5 m, leading to saltwater intrusion extending 1 km inland. The study's analysis revealed that around 271 km^2 (60%) of the area will be adversely affected by rising groundwater, hitting almost 70% of the settlements within that region.

Impact of Climate Change on Agricultural Water Demands

Roushdi (2024) employed the CROPWAT model to evaluate reference evapotranspiration and the water needs of principal crops in the Western Nile Delta. This investigation encompassed summer crops (sugar beet, cotton, summer maize, summer tomato, watermelon, and sweet melon), winter crops (wheat, alfalfa, winter potato, winter squash, and broad bean), and permanent crops (orange, apple, grape, and olives). The results indicate a projected rise in agricultural water demands. Makar et al. (2022) forecasted that by 2050, the water requirements for crops in Egypt will increase by 3–3.9% across different crop varieties. Furthermore, Mostafa et al. (2021) projected that the escalation of greenhouse gas emissions will result in a temperature increase of approximately 2.12 °C by 2050 and 3.96 °C by 2100. The irrigation water requirements for wheat crops are projected to rise by 6.2 in 2050 and 11.8 in 2100.

Impact of Climate Change on Crop Growth

Biomass production is significantly dependent on water. All food originates from photosynthesis and biomass; however, water requirements for nutritional sustenance might be approximated using evapotranspiration equations grounded in crop physiology (Roushdi 2024). In Northern Europe, climatic trends may yield beneficial outcomes, increasing productivity and broadening species distributions (Alcamo et al. 2007). In contrast, southern regions such as the Mediterranean have substantial difficulties, including diminished harvestable yields, increased yield variability, and a reduction in areas appropriate for conventional crops (Moriondo et al. 2010). Studies indicate that numerous crops, especially C3 plants, may gain advantages from elevated CO_2 concentrations under ideal settings (Long et al. 2006). C3 crops are anticipated to gain advantages from increased CO_2 levels, owing to their photosynthetic responsiveness to CO_2; yet, this may necessitate enhanced sink capacity (Dingkuhn et al. 2020). Additionally, these advantages may be undermined by many climate change variables, including elevated temperatures, heightened tropospheric ozone levels, and altered precipitation patterns (Da Matta et al. 2010). In Northern Europe, climatic trends may yield beneficial outcomes, increasing productivity and broadening species distributions (Alcamo et al. 2007). In contrast, southern regions such as the Mediterranean have considerable difficulties, including diminished harvestable yields, increased yield variability, and a reduction in areas conducive to conventional crops (Moriondo et al. 2010). Climate change constrains the duration for biomass buildup. Yield variations are contingent upon crop phenology, crop classification (C3 versus C4), and environmental conditions, including water and nutrient accessibility (Moriondo et al. 2010). Current atmospheric CO_2 concentrations are almost 50% elevated compared to pre-industrial levels (400 μmol mol^{-1}) (Ray et al. 2019). Carbon dioxide may restrict the growth and productivity of C3 crops due to sink constraints (Iizumi et al. 2018). If sink capacity is not augmented to sustain balanced source-sink dynamics, C3 crops would face difficulties in harnessing the elevated CO_2 concentrations anticipated in the forthcoming decades (Pardey et al. 2014).

Effects of Climate Change on Field Crops Production

Egypt's agriculture is especially susceptible to climate change owing to its arid and hot environment, anticipated to diminish agricultural yields. The difficulty is exacerbated by the nation's dependence on food imports, swift population expansion, and restricted agricultural acreage, requiring innovative strategies to improve total agricultural production. Should climate change negatively affect crop production as expected, Egypt's reliance on costly food imports is likely to rise. Field crops occupy a significant amount of agricultural land. Among cultivated strategic crops are staple crops such as wheat, rice, maize, and legumes. The nation is deficient in self-sufficiency for certain important crops and is significantly dependent on imports, especially wheat (50% of total consumption), yellow maize, and oilseeds. Conversely, rice and onion attain self-sufficiency. The Field Crops Research Institute (FCRI) in Egypt is instrumental in producing new varieties that exhibit great production potential, diverse resistances to biotic and abiotic challenges, and enhanced resilience to climate change. The institute also develops and updates technical suggestion packages for each variety or hybrid cultivated in every agroecological zone. FCRI also produces foundation seeds for inbreds and hybrids and includes a crucial technology transfer component. The government has demonstrated an increasing dedication to improving food security through efforts designed to augment local production and diminish reliance on imports. Table 10.1 displays the present cultivated area, production, and national strategic objectives for 2030 for several selected field crops.

Elbeltagi et al. (2020) discovered that yield forecasts for wheat and maize fluctuated between −3.21% to 3.47% and −4.93% to 5.88%, respectively, with anticipated precipitation decreases in the western regions surpassing those in the eastern regions of Egypt, ultimately leading to significant reductions in overall production. These substantial variations in precipitation may result in a considerable reduction in the green water footprint (WF) of wheat and maize, ultimately causing a marked decline in total productivity. Gamal et al. (2024) identified a substantial negative association between wheat production and both maximum and minimum air temperatures, in addition to growing degree days. Mostafa et al. (2021) projected that the escalation of greenhouse gas emissions will result in a temperature increase of approximately 2.12 °C by 2050 and 3.96 °C by 2100. Consequently, wheat productivity is projected to decrease by 8.6% in 2050 and 11.1% in 2100.

Additionally, Elsadek et al. (2023) employed the AquaCrop-GIS model (version 2.1) to forecast rice yields in the Nile River Delta (NRD) region of Egypt from 2021 to 2099. The forecasts indicate that the rice industry in the NDR may gain advantages from the anticipated climate, especially in the 2030s and 2050s. Nevertheless, very high temperatures may impede rice production in the 2070s and 2090s, but elevated CO_2 concentrations could mitigate some adverse consequences. Prior

Table 10.1 Expected developments in field crops from 2022 to 2030

Crops	Actual in 2022				Predicted in 2030				Source of increase
	Cultivated area (M fed[a])	Productivity (ton/fed)	Production (M ton)	% Self-sufficiency	Cultivated area (M fed)	Productivity (ton/fed)	Production (M ton)	% Self-sufficiency	
Wheat	3.4	2.7	9.2	55	3.75	3.25	14.2	67	Reduce grain losses and water use by grown-on raised beds
Barley	0.300	1.6	0.360	81	0.450	1.6	0.720	120	Intercropping systems with other crops and cultivation on raised beds.
Corn	2.80	3.22	9.4	5.6	2.9	3.64	10.6	5.6	Reduce grain losses and improve storage efficiency
Sorghum	0.410	2.18	0.940	7.4	0.500	3.0	1.50	104	Reduce grain losses and improve storage efficiency
Faba bean	0.250	1.5	0.400	35	0.400	1.7	0.680	80	Intercropping systems with other crops
Soybean	0.110	1.35	0.143	0.01	0.500	2.0	0.825	9	Intercropping systems with other crops
Sunflower	0.022	1.13	0.0275	0.01	0.400	1.5	0.600	4	Intercropping systems with other crops

Onion	0.207	15.0	3.21	120	0.218	17.0	3.5	150	Use of tolerant cultivars
Rice	1.2	3.8	5.0	96	1.1	5.0	5.5	100	Reduce grain losses and improve water use and storage efficiency
Berseem	1.350	31.25	48	100	1.1	50.0	55	100	
Summer forage	0.230	32.2	7.4	30	0.400	50	20	50	
Sugarcane	0.325	48.0	15.0	77	0.325	55.0	17.9	100	Improve water and fertilizer use
Sugar beet	0.600	18.0	10.0		0.825	28.0	23.1		

[a]One fed = 0.42 ha

Source: National Strategic Plan (2020)

studies utilizing the DSSAT3 and DSSAT3.5 models (Jones et al. 1998) and COTTAM (Jackson et al. 1988) suggested that climate change may diminish national crop production by 2050, with reductions between 11% for rice and 28% for soybean relative to current levels. Nevertheless, cotton production may increase due to altering climatic conditions. Furthermore, water consumption for summer crops is anticipated to rise by 8% for maize and 16% for rice by 2050 (El-Marsafawy and El-Samanody 2009). Furthermore, El-Marsafawy and El-Samanody (2009) evaluated the effects of climate change on sunflower productivity, water utilization, and economic returns, concluding that climate change may decrease sunflower seed yields by 27%, elevate water consumption by 12%, and diminish agricultural water productivity by 34%. Without adaptation, climate change may reduce net returns for landowners by 44% and for renters by 63% while also decreasing economic returns per unit of water by 35%.

Adaptation Strategies to Climate Change

The Intergovernmental Panel on Climate Change (IPCC) defines adaptation to climate change as any modification in natural or human systems in response to present or anticipated climatic stimuli, which reduces harm or capitalizes on advantageous opportunities. The implementation of adaptation measures in Egypt encounters numerous limits and obstacles, including perceptions of knowledge and policy, the inadequate adaptive capacity of rural communities, insufficient scientific data, lack of financial resources, and weak institutional frameworks (Fahim et al. 2013).

Management of Water Resources

Conserving Water Resources

Conserving water resources in Egypt is crucial for maintaining agricultural productivity, accommodating the increasing population, and safeguarding ecosystems. It is essential to use efficient management strategies that guarantee the river's flow remains stable and dependable. Collaboration is essential; interacting with adjacent nations to formulate fair accords on water distribution helps alleviate any disputes while optimizing the advantages of communal resources (WB 2018). Economical solar water desalination serves as a strategic remedy to mitigate these shortages. Egypt operates 97 desalination facilities with a current capacity of 1.3 million cubic meters per day and intends to augment this capacity (Nada et al. 2024). A sustainable development policy must consider the long-term consequences of groundwater extraction. Comprehensive research on groundwater quantity, quality, and availability in dry environments is crucial for formulating efficient aquifer management strategies. The Nubian Sandstone Aquifer System (NSAS) in the Western Desert serves as a viable irrigation source, facilitating land reclamation initiatives and sustaining civilizations in the Oases, including Kharga, Dakhla, Bahariya, Siwa, and Farafra, via shallow and deep wells (Gabr 2023). The National Rainfed Areas

Development Project seeks to augment water harvesting from 3.2 million cubic meters in 2018 to 6.2 million cubic meters by 2030 (National Strategic Plan 2020). Excessive withdrawals from the NSAS have resulted in reduced well capacity and heightened pumping costs, complicating farmers' ability to satisfy crop water needs under present conditions (Gabr 2023). Zelenakova et al. (2022) proposed that utilizing treated wastewater to enhance recharge to the Nile Delta aquifer, along with reducing abstraction from the aquifer, might safeguard it against potential sea level rise.

Minimizing Water Losses and Improving Water Productivity
Water loss is attributed to seepage from canals and drains, evaporation from water surfaces, infiltration from agricultural fields, and the presence of aquatic plants in canals. Mahmoud and El-Bably (2019) assessed Egypt's total irrigation efficiency to range from 40% to 60%, ascribed to evaporation, seepage, deep percolation, and transpiration from aquatic crops in irrigation channels. Moreover, Mohie El Din and Moussa (2016) discovered that, if existing policies persist, water shortages in 2025 could attain 26 billion cubic meters per year. They suggested design options for irrigation canals, scheduling land irrigation, and controlling aquatic weeds in streams and sugarcane regions of former agricultural fields. Furthermore, they advised augmenting the allowable extraction from deep groundwater and the Nubian aquifer, in addition to advancing desalination resources. These approaches would entirely eradicate the anticipated water deficit in 2025. Moreover, advocating for rainwater harvesting and the reclamation of wastewater in metropolitan regions might enhance the availability of water resources. Similarly, to safeguard the northern coastline, Zelenakova et al. (2022) indicated that many scenarios have been implemented to address climate change and sea level rise (SLR). The Precast Diaphragm Wall (PDW) is employed to safeguard Egypt's northern coastlines against sea level rise (SLR). The findings indicated that PDW is both applicable and cost-effective. The expense of constructing PDW along the northern beaches constitutes around 1.0% of the anticipated losses attributable to sea level rise by 2100. Furthermore, the findings indicated that employing treated wastewater to enhance the recharging of the Nile Delta aquifer, alongside reducing abstraction from the aquifer, could aid in safeguarding it against potential sea level rise.

Executing Optimal Agricultural Practices
Effective agricultural methods can markedly minimize water loss. Irrigating during peak temperatures enhances evaporation, and uneven terrain leads to inconsistent water distribution, causing certain regions to be over-irrigated while others are under-irrigated (Srivastava et al. 2016). Since the onset of the century, the domain of agricultural and irrigated agriculture has markedly expanded. Irrigation technologies currently comprise around 75% surface, 20% sprinkler, and 5% micro irrigation. A possible transition from surface irrigation to sprinkler and micro irrigation may diminish the water footprint of irrigated agriculture or augment its land footprint (Al-agele 2020). While some water loss is unavoidable, effective irrigation techniques, such as drip and sprinkler systems, are essential in contemporary

agriculture, reducing waste and evaporation. The application of these methods for rice cultivation is increasing in Egypt owing to their ecological advantages. Amer et al. (2017) assert that broader implementation is essential for water conservation, enhanced rice production, and better circumstances in agricultural regions. Mostafa et al. (2021) determined that shifting from flood to sprinkler irrigation may conserve around 1348 m^3/ha presently, escalating to 1431 m^3/ha by 2050 and 1507 m^3/ha by 2100. Their evaluation of irrigation requirements in the Giza, Beni-Suef, Al-Fayoum, and Minya governorates suggests that anticipated temperature increases of 2.12 °C by 2050 and 3.96 °C by 2100 will elevate irrigation demands, with projected jumps of 6.1–7.3% for winter crops and 4.9–5.8% for summer crops by 2050. This necessitates smart management of water resources in the face of increasing demands and constrained supplies.

Climate change presents additional risks, with Ouda and Zohry (2017) estimating an increase in crop water demands by 2030, alongside a reduction in arable land and food production. To mitigate these effects and enhance land productivity, it is essential to adapt crop management through water-saving techniques and intercropping (Gichangi and Gatheru 2018). Agriculture in Egypt constitutes the predominant consumer of water, representing 80% of freshwater withdrawals (Roushdi 2024). The ridge and raised-bed systems are efficient resource-conserving strategies that improve water use efficiency and boost production. Tang et al. (2024) advised farmers to implement the ridge furrow method and allow a 60% depletion of soil moisture to attain optimal wheat yields. The ridge furrow method outperformed the raised bed method regarding wheat grain production and crop water productivity. Singh et al. (2009) indicated that flat beds are extensively utilized in the agricultural sector of industrialized nations and have demonstrated efficacy for wheat cultivation. Transitioning from traditional basin or narrow furrow agriculture to raised bed cultivation can decrease water usage for crops undersurface irrigation by 20–25% (Karrou et al. 2012). Ismail et al. (2021) demonstrated that the raised bed planting technique efficiently mitigates waterlogging, manages weeds, and decreases water use by 50% while reducing crop loss by 20–25% in comparison to traditional approaches. These enhanced approaches establish ideal moisture levels in the root zone, hence improving seed germination and emergence for maximum production.

Smart Irrigation Technologies

Agriculturalists frequently over-irrigate their fields owing to inadequate understanding of crop water needs. Various models, including AquaCrop, CROPWAT, APSIM, and CropSyst, can determine agricultural water requirements; however, each model possesses uncertainties and necessitates certain input data. CROPWAT necessitates minimal input data and possesses universal applicability (Solangi et al. 2022). The FAO's AquaCrop model is acknowledged for its capability to forecast crop yields during moisture stress (Farahani et al. 2009). Abdelazez et al. (2024) propose the use of CROPWAT and CLIMWAT software, in conjunction with NASA Power resources, to improve irrigation efficiency and crop yield. Enhancing water use efficiency (WUE) of carbon assimilation per unit of water expended can significantly decrease agricultural water usage. Gaddikeri et al. (2024) emphasized the significance of district-specific elements, such as climate and soil characteristics, in formulating successful water management techniques. The employment of CLIMWAT 2.0 and CROPWAT 8.0 can substantially aid in achieving these objectives (Gaddikeri et al. 2024). Utilizing CROPWAT is advisable for accurately evaluating irrigation water needs and formulating irrigation schedules; the findings will aid in enhancing food output and managing water resources effectively. Consequently, effective irrigation methods are essential for alleviating water stress, maximizing crop yields, and attaining sustainable economic returns in agricultural production.

Advancement of Robust Field Crop Varieties/Hybrids

In the last century, crop breeding has progressed markedly, particularly due to the Green Revolution, which brought high-yielding, stress-resistant cultivars. Consequently, there is optimism that this strategy can cultivate types resistant to climate change. Access to technology that reduce crop production losses is essential for producing climate-resilient cultivars (Cabusora 2024). Breeding objectives must prioritize early-maturing varieties that utilize water and nitrogen more efficiently, together with other agricultural inputs, demonstrating durability across diverse conditions (Braun et al. 1997). Breeding practices may significantly contribute to the mitigation of methane and nitrous oxide emissions while improving carbon sequestration (Senapati et al. 2019). Di Filippo et al. (2024) emphasized that selection accuracy has markedly enhanced due to diverse experimental techniques, the implementation of statistical models, and the incorporation of DNA markers for marker-assisted breeding into selection methodologies.

Selective Breeding for Enhanced Water and Nitrogen Utilization Efficiencies

In conjunction with climate change, population increase will lead to a depletion of natural resources. Consequently, it is imperative to prioritize the sustainable utilization of resources like water and nitrogen. Despite recognized variability and heredity, the stability of the water use efficiency (WUE) characteristic remains overlooked. Merchuk and Saranga (2013) emphasized that conventional breeding aimed at enhancing water use efficiency (WUE) in crops is impeded by the difficulties in precisely assessing WUE in extensive field populations. Moreover, the emergence of contemporary genomic methodologies greatly enhanced the detection of QTL and diagnostic DNA markers associated with WUE. Furthermore, EL-Harty (2016) revealed that the volatility in faba bean seed yields may be attributed to its low water use efficiency (WUE). Moreover, EL-Harty et al. (2023) discovered that water use efficiency (WUE) enhanced with diminished irrigation, correlating with fluorescence (Fo) and seed output under both full irrigation and drought conditions. Furthermore, Leakey et al. (2019) revealed that water use efficiency (WUE) rises with increased atmospheric CO_2 concentrations and reduced drought stress;

however, it declines under nitrogen constraints and heightened evaporative demand. Nitrogen use efficiency (NUE) has garnered significant attention recently, particularly due to concerns regarding environmental effect and agricultural sustainability. Enhancing nitrogen use efficiency diminishes water pollution, mitigates algal bloom eutrophication that depletes oxygen in aquatic environments, and lowers greenhouse gas emissions while also decreasing the quantity of nitrogen fertilizer required (Laidig et al. 2024).

Genetically optimizing crops for nitrogen absorption is not the sole approach to improve NUE. Alternative approaches include adjusting fertilizer application rates and timing through the use of slow-release fertilizers. Laidig et al. (2024) determined that the heredity of nitrogen yield is analogous to that of grain yield, suggesting that nitrogen yield should be used as an additional criterion in variety testing to improve nitrogen use efficiency and alleviate adverse environmental impacts.

Developing Climate-Resilient Field Crops

Developing crops with enhanced climate resilience is crucial to alleviate the impacts of climate change on agriculture, especially in light of rising temperatures, unpredictable rainfall, and more frequent extreme weather occurrences. Drought and heat stress can occur simultaneously during crop cycles; localized heat waves or prolonged regional warming can induce drought, significantly affecting plant health by elevating evaporation rates and diminishing soil moisture (Breshears et al. 2021). Wang et al. (2023) emphasized the need of broad adaptability in genetic resource innovation as a breeding objective. Furthermore, cultivating short-duration cultivars can improve crop resilience. EL-Harty et al. (2023) preferred early-maturing quinoa genotypes with moderate yields to late-maturing, high-yield varieties as this strategy conserves irrigation water and facilitates the earlier planting of other crops. A reduced life cycle enables plants to evade elevated summer temperatures (Sabella et al. 2020). Ali et al. (2019) assessed the performance of four bread wheat cultivars: Misr 1, Sakha 93, Giza 168, and Gemmeiza 9, under elevated temperatures. Under a +2 °C scenario, the additional irrigated Gemmeiza 9 cultivar would incur a 9.6% decrease in grain yield, but the average yield reduction for the other cultivars would be 18%. In a +4 °C environment, the yield loss for Gemmeiza 9 would rise to 16.2%, while the average for the other cultivars would be 23.8%. The Gemmeiza 9 cultivar, supplemented with irrigation, is advised to optimize bread wheat productivity and reduce variability caused by increasing temperatures in North Sinai and comparable areas.

Developing and producing heat-tolerant genotypes, along with improving CO_2 fertilization, may provide solutions to future climatic issues (Elsadek et al. 2024). Models that more effectively incorporate adaptation to elevated CO_2 are crucial for forecasting the trait combinations that breeders should prioritize in a CO_2-enriched environment. Dingkuhn et al. (2020) established that e-CO_2 responsiveness is associated with either intrinsic sink capacity or adaptive plasticity, exemplified by enhanced branching. To optimize advantages from e-CO_2, substantial enhancements in vegetative sink strength, inflorescence dimensions or quantity, and root sinks are needed. Moreover, the "feast-famine" sugar signaling pathway may be utilized to augment sink strength in a tissue- and stage-specific manner, hence reinforcing ideotype concepts.

Enhancing resistance to abiotic challenges collectively is anticipated to produce superior outcomes compared to tackling each stress in isolation (Yang et al. 2013). The continual practice of crossing and selecting optimal plants swiftly diminishes genetic diversity, resulting in the rapid fixation of desirable features (Van Ginkel and Ortiz 2018). Therefore, there is a necessity for exotic germplasm in breeding operations due to the enhanced genetic variance it offers as no other convincing explanation exists for these results (Neumann et al. 2023).

Table 10.2 illustrates the accomplishments of the Field Crops Research Institute in Egypt on the development of climate-resilient cultivars.

Utilization of Innovative Breeding Approaches

Biotechnology can significantly contribute to climate change mitigation. In vitro culture, genetic engineering, marker technology, and gene editing can facilitate the development of novel robust genotypes. Genetically modified (GM) technology can enhance sustainable agriculture by augmenting crop yields by 22% and raising farmers' earnings by 68% (Taheri et al. 2017). The planned use of artificial chromosomes for concurrent multigene transformation is unlikely to gain market acceptability (Kan et al. 2022). The scientific community endorses gene-editing technology with caution, treating its regulation independently from GM technologies (Sprink et al. 2022). Worldwide, genetically modified crops such as canola, cotton, soybean, and maize predominate in production, while others, including Golden Rice, occupy a restricted territory (Wu et al. 2012). CRISPR genome editing enhances the

Table 10.2 Climate-resilient Egyptian varieties of selected field crops

Crops	Varieties	Most important characters
Faba bean	Wady 1 Sakha 5 Sakha 1, 4 and Giza 716 Giza 843	Early maturing, low water requirement Resistant for both broomrape and foliar diseases Resistant to foliar diseases Broomrape resistance
Chickpea	Sids 101	Resistant to foliar diseases
Soybean	Misr 6, 10	High yield potential, short grown season
Rice	Giza 179, Sakha S. 300, 301, 302, 303	Early-maturing, tolerant to abiotic stresses, high yield potential
Wheat	Sakha 96 Misr 1, 3,5,6,7	Short season, resistance to three types of wheat rust Tolerant to abiotic stresses, high yield potential
Barely	Giza 2000, 126, 134, 137, 138	Tolerant to abiotic stresses, high yield potential
Sorghum	Dorado, H 305, 306	Tolerant to abiotic stresses, high yield potential

creation of crop varieties, addressing breeding obstacles and advancing sustainable agriculture and food security (Zamani et al. 2024).

Augmented Seed Yield

To improve yield, three genes (OsGW2, OsGW5, and OsTGW6) were concurrently knocked out in rice, resulting in a 30% increase in grain weight (Viana et al. 2019). The ablation of the TaGW2 gene, responsible for encoding a RING E3 ligase, led to the production of longer and broader wheat grains, hence enhancing yield (Zhang et al. 2018). It limits grain size within the maternal tissue prior to anthesis (Simmonds et al. 2016). Regulating cytokinin homeostasis can significantly increase grain yield; silencing TaCKX2-D1, a cytokinin oxidase/dehydrogenase, improves wheat yield (Dahan-Meir et al. 2018). Furthermore, the alteration of the maize gene ARGOS8 using CRISPR/Cas9 enhanced hybrid yield under conditions of drought stress (Shi et al. 2017). Conversely, Shumin et al. (2024) emphasized that the development continues to be elusive. The TaGW2 gene is a crucial regulator of kernel development and drought responses in wheat. Its overexpression enhances drought resistance but leads to diminished yield under full watering. In contrast, the ablation of TaGW2 diminishes drought resistance while markedly enhancing kernel size and weight.

Augment Resilience to Biotic and Abiotic Stressors

CRISPR has been utilized to induce specific mutations in genes associated with plant immunity and defense (Li et al. 2022). Recent CRISPR/Cas9 modifications of chickpea protoplasts focused on drought-tolerance genes 4-coumarate ligase (4CL) and Reveille 7 (RVE7), yielding drought-resistant cultivars (Badhan et al. 2021). In rice, Cas9 mutations in Leaf1 and Leaf2 enhanced protein expression and reactive oxygen species scavenging, hence improving drought tolerance (Liao et al. 2019). Investigations were conducted on mutations in the Drb2a and Drb2b genes that govern drought and salt tolerance in soybeans (Curtin et al. 2018). Moreover, CRISPR/Cas-mediated genome editing has pinpointed many genes to enhance plant salt tolerance. Knockout mutants of ABA-induced transcription repressors (AITRs) were created in soybean to enhance salt stress resistance (Wang et al. 2021). CRISPR-Cas9 base editing and prime editing methods have been employed to improve herbicide tolerance by targeting acetolactate synthase 1 (ZmALS1) and ZmALS2 in Zea mays (Nuccio et al. 2021). In rice, targeted modification of the OsALS gene, succeeded by HDR-mediated repair by primary editing technique, has enhanced herbicide resistance (Ali et al. 2020).

Augmenting Agricultural Diversity and Ecological Robustness

Improving crop diversity and developing stress-resistant cultivars are essential for food security. It allows farmers to adjust to climate change and severe weather conditions (Sarma et al. 2024). Crop variety enhances resilience by augmenting drought and heat tolerance while mitigating pest and disease occurrences (Lin 2011). Underutilized legumes can assimilate atmospheric nitrogen and offer substantial protein content (Cheng et al. 2019). Reliance on a restricted array of staple crops, including wheat, rice, maize, soybean, millet, sorghum, and barley, neglects several underutilized species (Massawe et al. 2015). Crops from the Chenopodiaceae family, including quinoa and amaranth, can markedly enhance production systems in response to changing climates. Crops such as quinoa, maca, psyllium, chia, cassava, and pecan can flourish in many situations, significantly contributing to food security under climate change, despite their restricted regional influence (Georgakopoulos et al. 2016). Moreover, wild edible plants are essential to regional agricultural heritage and local gastronomy. Frequently regarded as weeds, these low-maintenance plants provide significant nutritional and organoleptic advantages (Ladányi et al. 2024).

Braun et al. (1997) emphasized CIMMYT's achievement, indicating that around 70% of the spring wheat acreage in developing nations (excluding China) is cultivated with CIMMYT-derived varieties. Baranski (2022) condemned the Green Revolution for advocating extensive monocultures to the detriment of conventional agricultural practices and marginal ecosystems. Furthermore, it was contended that it disproportionately advantages wealthy farmers through modern technologies and advocated for more inclusive agricultural practices, emphasizing the need for variety in crop research to incorporate traditional methods and marginal areas.

Adjusting Agronomic Practices for Climate-Resilient Agriculture

Enhanced agricultural management increases resource efficiency and diminishes greenhouse gas emissions, with approximately 90% of mitigation resulting from soil carbon sequestration and 10% from emission reductions (Ortiz-Monasterio et al. 2010). Elsadek et al. (2024) advocate for management approaches including the cultivation of heat-tolerant genotypes, modification of planting dates and irrigation techniques, and the enhancement of CO_2 fertilization to mitigate future climate difficulties. Recent research has emphasized ecological sustainability concerning climate change, concentrating on cover crops (Tribouillois et al. 2018), soil organic carbon (Huang et al. 2020), greenhouse gas emissions (Karimi et al. 2021), and chemical leaching (Paff et al. 2022).

Preserving Soil Health

Soil health is crucial for sustainable agriculture and ecological equilibrium. Ditzler et al. (2021) emphasize the need of sustaining permanent soil cover, reducing disturbance, and diversifying plant species. Sarma et al. (2024) assert that proficient soil management necessitates the retention of prior agricultural residues, which mitigates erosion and runoff while augmenting carbon sequestration. Essential practices encompass no-till and minimal tillage, cover cropping, crop rotation, carbon sequestration, soil stabilization, intercropping, enhanced drainage, green manuring, crop weed allelopathy, agroforestry, riparian buffers, windbreaks, shelterbelts, and permanent soil coverage. Wu et al. (2015) discovered that cover crops diminish water evaporation, suppress weed proliferation, and safeguard soil during fallow times, hence reducing erosion. Sarma et al. (2024) further shown that nano-biochar sequesters carbon, reduces greenhouse gas emissions, and addresses climate change. It improves soil structure, water retention, and nutrient accessibility while fostering beneficial microbial activity, resulting in robust and sustainable agriculture. Nano-biochar enhances surface area and reactivity, effectively retaining water and nutrients, which diminishes the necessity for regular watering and fertilization.

Altering the Sowing Date

Research on climate change is ongoing, highlighting the necessity to modify agricultural methods and devise mitigation strategies to reduce disruptions. El-Marsafawy and El-Samanody (2009) suggested that advancing sunflower sowing from June 1–10 to May 1–10 might enhance seed yields by 13–18%, while a 10% reduction in irrigation water usage could conserve resources without markedly impacting yields. Mostafa et al. (2021) determined that early wheat planting can mitigate the adverse effects of climate change on agricultural productivity, identifying November 1st as the ideal sowing date for 2050 and 2100. Augmenting irrigation by 10% in 2050 and 20% in 2100 will enhance wheat yields, leading to diminished output decline.

Modifying Cropping Systems

Noreldin and Zohry (2019) projected a 10% decrease in Egypt's arable land by 2030 as a result of escalating water requirements for agriculture. They propose the adoption of intercropping methods to augment the cultivated area of current cropping patterns by 28%. Crop rotations are advised to alleviate grain production reductions resulting from climate change-induced adverse conditions. Costa et al. (2024) discovered that diversified, functionally robust rotations significantly mitigate yield losses from harsh weather in small grains and maize. Numerous places are streamlining farming systems by implementing reduced rotations (Schaak et al. 2023).

Varied rotations can improve soil fertility and optimize the utilization of water and nutrients (Jian et al. 2020). Species within the same functional category exhibit analogous characteristics, perhaps constraining diversification advantages. In contrast, other functional groupings including cereals, annual legumes, broadleaf crops (e.g., Brassicaceae and Solanaceae), and leys (a combination of perennial grasses and legumes) provide complementing characteristics that improve performance. Functions encompass nitrogen fixation by legumes and soil coverage by leys, which enhance essential ecosystem services such as pathogen management, soil fertility, resource use efficiency, and diverse stress responses (Griffiths et al. 2022).

Precision Agriculture

Precision agriculture enhances the utilization of fertilizers, herbicides, and water to increase crop yield (Gebbers and Adamchuk 2010). Remote sensing technology, such as drones and Unmanned Aerial Vehicles, effectively collect data on crop and soil conditions, assessing growth, water requirements, illnesses, and nutritional status (Karthikeyan et al. 2020). They also enable the administration of foliar fertilizers and the spraying of pesticides in mechanized agriculture (Tellaeche et al. 2008). Matthews et al. (2013) emphasized the significance of models in recognizing effective management strategies for climate adaption. Furthermore, the application of nanotechnology presents prospects for augmenting plant growth by fortifying defense systems, stimulating growth, and minimizing losses through the use of nanoparticles in insecticides, fertilizers, and agrochemicals (Ijaz et al. 2024). Furthermore, Al-agele (2020) delineated innovative technology aimed at enhancing sprinkler and drip systems, specifically highlighting the potential of dynamic nozzle height adjustment in overhead sprinkler systems. An innovation in drip irrigation, referred to as Variable Rate Drip Irrigation (VRDI), has been proposed and assessed. A VRDI emitter, engineered to detect individual water droplets, was constructed and evaluated. This method facilitates the verification of the volume of water dispensed at each outlet. The revolutionary VRDI emitter autonomously checks the total volume of water dispensed and halts flow upon achieving the specified application threshold. Laboratory studies verify that the integrated volume measurements of the VRDI system are unaffected by pressure.

Conclusions

Egypt's agriculture is significantly susceptible to climate change because of its desert climate, water scarcity, and considerable dependence on food imports. The Nile Delta is likewise confronted with rising sea levels, endangering the salinization of its most arable land. Therefore, improving and preserving water resources is crucial for agricultural sustainability, given that irrigation efficiency ranges from 40% to 60% due to variables such as evaporation, seepage, and significant losses from canals and drains. National studies demonstrated that solutions may double water harvesting by 2030; additionally, efficient irrigation techniques such as drip and sprinkler systems can substantially diminish water loss and improve irrigation efficiency by up to 80%. Advanced irrigation technologies and precision agriculture would augment these initiatives. Biotechnology, encompassing genetic engineering and various molecular breeding techniques in conjunction with traditional breeding, is essential for developing resilient crops capable of enduring biotic and abiotic challenges.

The Field Crops Research Institute in Egypt has created high-yield, stress-resistant crop varieties as part of its comprehensive measures to address climatic changes. Expanding crop diversity beyond mainstays such as wheat and rice can enhance resilience, drought tolerance, and insect resistance. Numerous neglected crops, such as quinoa, forage and grain millets, canola, and barley, are receiving increased attention due to their high resilience and capacity to thrive in marginal soils. The Field Crops Research Institute in Egypt is presently developing national projects to breed varieties and hybrids of underused crops. Specifically, when the state aims to recapture 3.5 million acres of new territories that are predominantly low in fertility and experience one or more forms of stress. This chapter also introduces enhanced crop management techniques, including optimum planting schedules and cropping systems. Crop rotation can improve nitrogen utilization efficiency and other agricultural inputs. These integrated management strategies would enhance irrigation efficiency and climatic flexibility.

References

Abdelazez EAM, Shalaby MMH, Ali SM, Omran WM (2024) Impact of climatic change on water consumption in some Egyptian regions. Menoufia J Soil Sci 9:69–72

Abouelnaga M (2019) Why the MENA region needs to better prepare for climate change. Atlantic Council Blogs 7

AFED (2017) Arab environment in 10 years. In: The 2017 annual report of Arab forum for environment and development. Technical Publications, Beirut Lebanon

Al-agele HA (2020) Irrigation innovations to increase efficiency and sustainability. https://ir.library.oregonstate.edu/concern/graduate_thesis_or_dissertations

Alcamo J, Dronin N, Endejan M, Golubev G, Kirilenko A (2007) A new assessment of climate change impacts on food production shortfalls and water availability in Russia. Glob Environ Chang 17:429–444

Ali AM, ElSarag EI, Abdallah MY (2019) Impact of climate change scenarios on some rainfed wheat cultivars under North Sinai region conditions. Sinai J Appl Sci 8:113–122

Ali Z, Shami A, Sedeek K, Kamel R, Alhabsi A, Tehseen M, Mahfouz MM (2020) Fusion of the Cas9 endonuclease and the VirD2 relaxase facilitates homology-directed repair for precise genome engineering in rice. Commun Biol 3(4):44

Amer MH, Abd El Hafez SA, Abd El Ghany MB (2017) Water saving in irrigated agriculture in Egypt. LAP LAMBERT Academic Publishing, Saarbrücken

Ashour MA, El Degwee YA, Hashem RH, Abdou AA, Abu-Zaid TS (2024) The extent to which the available water resources in upper Egypt can be affected by climate change. Limnol Rev 24:164–177

Badhan S, Ball AS, Mantri N (2021) First report of CRISPR/Cas9 mediated DNA-free editing of 4CL and RVE7 genes in chickpea protoplasts. Int J Mol Sci 22:396

Baranski M (2022) The globalization of wheat: a critical history of the green revolution. University of Pittsburgh Press

Braun HJ, Rajaram S, Ginkel M (1997) CIMMYT's approach to breeding for wide adaptation. In: Adaptation in plant breeding: selected papers from the XIVEUCARPIA congress on adaptation in plant breeding held at Jyväskylä Sweden from July 31 to August 4 1995. Springer Netherlands, pp 197–205

Breshears DD, Fontaine JB, Ruthrof KX et al (2021) Underappreciated plant vulnerabilities to heat waves. New Phytol 231:32–39

Cabusora CC (2024) Developing climate-resilient crops: adaptation to abiotic stress-affected areas. Technol Agron 4:e005

Cheng A, Raai MN, Zain NAM, Massawe F, Singh A, Wan-Mohtar WA (2019) In search of alternative proteins: unlocking the potential of underutilized tropical legumes. Food Secur 11:1205–1215

Climate Risk Profile: Egypt (2020) The World Bank Group, World Bank Publications, The World Bank Group Washington, DC USA. Available from www.worldbank.org

Costa NB, de Andrade Bezerra G, de Moraes MG et al (2024) Beneficial rhizobacteria mitigate combined water and phosphorus deficit effects on upland rice. Plant Soil. https://doi.org/10.1007/s11104-024-07158-9

Curtin SJ, Xiong Y, Michno JM (2018) Crispr/cas9 and talen s generate heritable mutations for genes involved in small RNA processing of glycine max and *Medicago truncatula*. Plant Biotechnol J 16:1125–1137

Dahan-Meir T, Filler-Hayut S, Melamed-Bessudo C, Bocobza S, Czosnek H, Aharoni A, Levy AA (2018) Efficient in planta gene targeting in tomato using geminiviral replicons and the CRISPR/Cas9 system. Plant J 95:5–16

DaMatta FM, Grandis A, Arenque BC, Buckeridge MS (2010) Impacts of climate changes on crop physiology and food quality. Food Res Int 43:1814–1823

Dawoud MA, Sallam GR, Abdelrahman MA, Emam M (2024) The performance and feasibility of solar-powered desalination for brackish groundwater in Egypt. Sustainability 16:1630

Di Filippo ES, Chiappalupi S, Falone S, Dolo V, Amicarelli F, Marchianò S, Fulle S (2024) The MyoGravity project to study real microgravity effects on human muscle precursor cells and tissue. npj Microgravity 10:92

Dingkuhn M, Luquet D, Fabre D, Muller B, Yin X, Paul MJ (2020) The case for improving crop carbon sink strength or plasticity for a CO_2-rich future. Curr Opin Plant Biol 56:259–272

Ditzler L, van Apeldoorn DF, Pellegrini F, Antichi D, Bàrberi P, Rossing WAH (2021) Current research on the ecosystem service potential of legume inclusive cropping systems in Europe: a review. Agron Sustain Dev 41(2):26. https://doi.org/10.1007/s13593-021-00678-z

Driouech F, ElRhaz K, Moufouma-Okia W, Arjdal K, Balhane S (2020) Assessing future changes of climate extreme events in the CORDEX-MENA region using regional climate model ALADIN-climate. Earth Syst Environ 4:477–492

El Agroudy N, Shafiq FA, Mokhtar S (2014) The impact of establishing the Ethiopian Dam renaissance on Egypt. J Basic Appl 4:1–5

El-Harty EH (2016) Effect of water deficit on seed yield and proline content in some faba bean genotypes. J Plant Prod 7:653–658

El-Harty EH, Seleiman MF, Afzal M, Alghamdi S (2023) Water use efficiency (WUE) and productivity of promising quinoa (*Chenopodium quinoa* Wild) genotypes grown under three water regimes. Not Bot Horti Agrobot Cluj-Napoca 51:13209

El-Marsafawy SM, El-Samanody MKM (2009) Economic impacts of future climatic changes on sun flower crop in Egypt. In: Proceedings of the fifth international conference of sustainable agricultural development. Faculty of Agriculture Fayoum University Fayoum, pp 21–23

Elbeltagi A, Aslam MR, Malik A, Mehdinejadiani B, Srivastava A, Bhatia AS, Deng J (2020) The impact of climate changes on the water footprint of wheat and maize production in the Nile Delta Egypt. Sci Total Environ 743:14077

Elsadek E, Zhang K, Mousa A, Ezaz GT, Tola TL, Shaghaleh H, Alhaj Hamoud Y (2023) Study on the in-field water balance of direct-seeded rice with various irrigation regimes under arid climatic conditions in Egypt Using the AquaCrop Model. Agronomy 13:609

Elsadek EA, Zhang K, Hamoud YA, Mousa A, Awad A, Abdallah M, Elbeltagi A (2024) Impacts of climate change on rice yields in the Nile River Delta of Egypt: a large-scale projection analysis based on CMIP6. Agric Water Manag 292:108673

Elsharkawy H, Rashed H, Rached I (2009) Climate change: the impacts of sea level rise on Egypt. In: 45th ISOCARP congress, pp 1–11

Elshinnawy IA, Almaliki AH (2021) Vulnerability assessment for sea level rise impacts on coastal systems of Gamasa Ras El Bar Area Nile Delta Egypt. Sustainability 13:3624

Fahim MA, Hassanein MK, Khalil AA, Abou Hadid AF (2013) Climate change adaptation needs for food security in Egypt. Nat Sci 11:68–74

Farahani HJ, Izzi G, Oweis TY (2009) Parameterization and evaluation of the AquaCrop model for full and deficit irrigated cotton. Agron J 101:469–476

Felzer BS, Cronin T, Reilly JM, Melillo JM, Wang X (2007) Impacts of ozone on trees and crops. Compt Rendus Geosci 339:784–798

Gabr ME (2023) Land reclamation projects in the Egyptian Western Desert: management of 1.5 million acres of groundwater irrigation. Water Int. https://doi.org/10.1080/02508060.2023.2185745

Gaddikeri V, Rajput J, Jatav MS, Kumari A, Rana L, Rai A, Gangwar A (2024) Estimating crop water requirement in Madhya Pradesh's agro-climatic regions: a CROPWAT and CLIMWAT software case study. Environ Conserv J 25:308–326

Gamal R, Abou-Hadid AF, Omar MED, Elbana M (2024) Does climate change affect wheat productivity and water demand in arid regions? Case study of Egypt. J Agric Food Res 16:101181

Gebbers R, Adamchuk VI (2010) Precision agriculture and food security. Science 327:828–831

Georgakopoulos P, Travlos IS, Kakabouki I, Kontopoulou CK, Pantelia A, Bilalis DJ (2016) Climate change and chances for the cultivation of new crops. Not Bot Horti Agrobot Cluj-Napoca 44:347–353

Gichangi EM, Gatheru M (2018) Farmers' awareness and perception of climate change and the various adaptation measures they employ in the semi-arid eastern Kenya. Clim Chang 4:112–122

Griffiths M, Wang X, Dhakal K, Guo H, Seethepalli A, Kang Y, York LM (2022) Interactions among rooting traits for deep water and nitrogen uptake in upland and lowland ecotypes of switchgrass (*Panicum virgatum* L.). J Exp Bot 73:967–979

Huang W, González G, Zou X (2020) Earthworm abundance and functional group diversity regulate plant litter decay and soil organic carbon level: a global meta-analysis. Appl Soil Ecol 150:103473

ICTSD-IPC (2009) "International climate change negotiations and agriculture", climate change 320102. Int Centre Trade Sustain Dev (ICTSD). https://doi.org/10.22004/ag.econ.320102

Iizumi T, Shiogama H, Imada Y, Hanasaki N, Takikawa H, Nishimori M (2018) Crop production losses associated with anthropogenic climate change for 1981–2010 compared with preindustrial levels. Int J Climatol 38:5405–5417

Ijaz S et al (2024) Sustainable solutions: nurturing plant growth in a changing climate. In: Fahad S, Saud S, Nawaz T, Gu L, Ahmad M, Zhou R (eds) Environment climate plant and vegetation growth. Springer, Cham. https://doi.org/10.1007/978-3-031-69417-2_20

IPCC (2001) Intergovernmental Panel on Climate Change, Climate change 2001: the scientific basis. Contribution of working group I to the third assessment report of the IPCC. Cambridge University Press

IPCC (2007) Climate change, the physical science basis. Contribution of Working Group I to the Fourth Assessment Report of the Intergovernmental Panel on Climate Change (Solomon SD, Qin M, Manning Z, Chen M, Marquis KB, Averyt M, Tignor HL, Miller (eds)). Cambridge University Press, New York

IPCC (2023) Summary for policymakers. In: Core Writing Team H, Lee, J Romero (eds) Climate change 2023: synthesis report. Contribution of Working Groups I II and III to the Sixth

Assessment Report of the Intergovernmental Panel on Climate Change. IPCC, Geneva. https://doi.org/10.59327/IPCC/AR6-9789291691647.001

Ismail S, Thabet A, El-Al A, Omara AI (2021) Comparative effects of raised bed and traditional flat basin on wheat yield and water productivity under Egyptian conditions. Misr J Agric Eng 38:293–308

Jackson BS, Arkin GF, Hearn AB (1988) The cotton simulation model COTTAM: fruiting model calibration and testing. Trans ASAE 31:846–854

Jian J, Du X, Reiter MS, Stewart RD (2020) A meta-analysis of global cropland soil carbon changes due to cover cropping. Soil Biol Biochem 143:107735

Jones JW, Tsuji GY, Hoogenboom G, Hunt LA, Thornton PK, Wilkens PW, Singh U (1998) Decision support system for agrotechnology transfer: DSSAT v3. In: Understanding options for agricultural production. Springer, pp 157–177

Kaini S, Nepal S, Pradhananga S, Gardner T, Sharma A (2020) Representative general circulation models selection and downscaling of climate data for the transboundary Koshi river basin in China and Nepal. Int J Climatol 40:4131–4149

Kaini S, Nepal S, Pradhananga S, Gardner T, Sharma AK (2021) Impacts of climate change on the flow of the transboundary Koshi River with implications for local irrigation. Int J Water Resour Dev 37:929–954

Kan Z, Kwan MP, Ng MK, Tieben H (2022) The impacts of housing characteristics and built-environment features on mental health. Int J Environ Res Public Health 19:5143

Karimi T, Stöckle CO, Higgins SS, Nelson RL (2021) Impact of climate change on greenhouse gas emissions and water balance in a dryland-cropping region with variable precipitation. J Environ Manag 287:112301

Karrou M, Oweis T, El Enein RA, Sherif M (2012) Yield and water productivity of maize and wheat under deficit and raised bed irrigation practices in Egypt. Afr J Agric Res 7:1755–1760

Karthikeyan L, Chawla I, Mishra AK (2020) A review of remote sensing applications in agriculture for food security: crop growth and yield irrigation and crop losses. J Hydrol 586:124905

Khaled KE, Sherien AZ (2017) Nile basin climate changes impacts and variabilities. The Nile River. In: The handbook of environmental chemistry book series, HEC volume 56. Springer, Cham, pp 533–566

Ladányi M, Divéky-Ertsey A, Csambalik L (2024) Adaptation of traditional crop cultivars to climate change in terms of nutritional aspects. Front Nutr 11:1427068

Laidig F, Feike T, Lichthardt C, Schierholt A, Piepho HP (2024) Breeding progress of nitrogen use efficiency of cereal crops winter oilseed rape and peas in long-term variety trials. Theor Appl Genet 137:45

Leakey ADB, Ferguson JN, Pignon CP, Wu A, Jin Z, Hammer GL, Lobell DB (2019) Water use efficiency as a constraint and target for improving the resilience and productivity of C3 and C4 crops. Annu Rev Plant Biol 70:781–808

Li Y, Wu X, Zhang Y, Zhang Q (2022) CRISPR-Cas genome editing improves abiotic and biotic stress tolerance of crops. Front Genome Ed 4:987817. https://doi.org/10.1111/pbi.13519

Liao S, Qin X, Luo L, Han Y, Wang X, Usman B, Li R (2019) CRISPR/Cas9- induced mutagenesis of semi-rolled leaf1 2 confers curled leaf phenotype and drought tolerance by influencing protein expression patterns and ROS scavenging in rice (*Oryza sativa* L.). Agronomy 9:728

Lin BB (2011) Resilience in agriculture through crop diversification: adaptive management for environmental change. Bioscience 61:183–193

Long SP, Ainsworth EA, Bernacchi CJ et al (2006) Long-term responses of photosynthesis and stomata to elevated [CO2] in managed systems. In: Managed ecosystems and CO_2: case studies, processes and perspectives. Springer, pp 253–270

Mahmoud MA, El-Bably AZ (2019) Crop water requirements and irrigation efficiencies in Egypt. In: Conventional water resources and agriculture in Egypt. Springer, pp 471–487

Makar RS, Shahin SA, El-Nazer M, Wheida A, Abd El-Hady M (2022) Evaluating the impacts of climate change on irrigation water requirements. Sustainability 14:14833

Massawe LN, Massawe ES, Makinde OD (2015) Temporal model for dengue disease with treatment. Adv Inf Dis 5:21–36

Matthews FE, Arthur A, Barnes LE, Bond J, Jagger C, Robinson L, Brayne C (2013) A two-decade comparison of prevalence of dementia in individuals aged 65 years and older from three geographical areas of England: results of the Cognitive Function and Ageing Study I and II. Lancet 382:1405–1412

Merchuk L, Saranga Y (2013) Breeding approaches to increasing water-use efficiency. In: Improving water and nutrient-use efficiency in food production systems. Wiley, pp 145–160

Mohie El Din MO, Moussa AM (2016) Water management in Egypt for facing the future challenges. J Adv Res 7:403–412

Molle F, Rap E, Al-Agha DE, Abou El Hassan W, Freeg M (2019) Irrigation improvement projects in the Nile Delta: promises challenges surprises. Agric Water Manag 216:425–435

Moriondo M, Bindi M, Kundzewicz ZW et al (2010) Impact and adaptation opportunities for European agriculture in response to climatic change and variability. Mitig Adapt Strateg Glob Chang 15:657–679

Mostafa SM, Wahed O, El-Nashar WY, El-Marsafawy SM, Abd-Elhamid HF (2021) Impact of climate change on water resources and crop yield in the Middle Egypt region. Aqua Water Infrastruct Ecosyst Soc 70:1066–1084

Nada A, Ibrahim MG, Elshemy M, Fujii M, Sharaan M (2024) Integrated water quality and performance assessment of seawater desalination plants along two coasts in Egypt. Desalination 586:117844

National Strategic Plan (2020) National strategic plan 2019–2030. MALR

Neumann K, Schulthess AW, Bassi FM, Dhanagond S, Khlestkina E, Börner A, Graner A, Kilian B (2023) Genomic approaches to using diversity for the adaptation of modern varieties of wheat and barley to climate change. In: Ghamkhar K, Williams WM, Dean Brown AH (eds) Plant genetic resources for the 21st century. Academic, pp 47–78. https://doi.org/10.1201/9781003302957

Noreldin T, Zohry AEH (2019) Reduction of the vulnerability of the Egyptian cropping pattern to climate change stress. www.discoveryjournals.org

Nuccio ML, Claeys H, Heyndrickx KS (2021) CRISPR-Cas technology in corn: a new key to unlock genetic knowledge and create novel products. Mol Breed 41:11

Ortiz-Monasterio I, Wassmann R, Govaerts B, Hosen Y, Katayanagi N, Verhulst N (2010) Greenhouse gas mitigation in the main cereal systems: rice wheat and maize. In: Climate change and crop production. CABI, Wallingford, pp 151–176

Ouda SA, Zohry AEH (2017) Crops intensification to reduce wheat gap in Egypt. In: Future of food gaps in Egypt: obstacles and opportunities. Springer, pp 37–56

Paff K, Fleisher D, Timlin D (2022) Changes in the effects of water and nitrogen management for potato under current and future climate conditions in the US. Comput Electron Agric 197:106980

Pardey PG, Beddow JM, Hurley TM, Beatty TK, Eidman VR (2014) A bounds analysis of world food futures: global agriculture through to 2050. Aust J Agric Resour Econ 58:571–589

Ray DK, West PC, Clark M, Gerber JS, Prishchepov AV, Chatterjee S (2019) Climate change has likely already affected global food production. PLoS One 14:e0217148

Roushdi M (2024) Investigation the implications of climate change on crop water requirements in Western Nile Delta Egypt. Water Sci 38:77–91

Sabella E, Aprile A, Negro C (2020) Impact of climate change on durum wheat yield. Agronomy 10:793

Sarma B, Sarmah N, Malakar A (2024) Degradation of water and land resources in the Himalayan mountain ecosystems. In: Addressing the climate crisis in the Indian Himalayas: can traditional ecological knowledge help? Springer, pp 31–52

Schaak H, Bommarco R, Hansson H, Kuns B, Nilsson P (2023) Long-term trends in functional crop diversity across Swedish farms Agric. Ecosyst Environ 343:108269

Schilling J, Hertig E, Tramblay Y, Scheffran J (2020) Climate change vulnerability water resources and social implications in North Africa. Reg Environ Chang 20:1–12

Senapati N, Stratonovitch P, Paul MJ, Semenov MA (2019) Drought tolerance during reproductive development is important for increasing wheat yield potential under climate change in Europe. J Exp Bot 70:2549–2560

Shi J, Gao H, Wang H et al (2017) ARGOS 8 variants generated by CRISPR-Cas9 improve maize grain yield under field drought stress conditions. Plant Biotechnol J 15:207–216

Shumin L, Yifang Z, Yuling L et al (2024) The E3 ligase TaGW2 mediates transcription factor TaARR12 degradation to promote drought resistance in wheat. Plant Cell 36(3):605–625. https://doi.org/10.1093/plcell/koad307

Siam MS, Eltahir EA (2017) Climate change enhances interannual variability of the Nile River flow Nature. Clim Chang 7:350–354

Simmonds JP, Scott J, Brinton TC, Mestre M, Bush A, Del Blanco J, Dubcovsky C (2016) A splice acceptor site mutation in TaGW2-A1 increases thousand grain weight in tetraploid and hexaploid wheat through wider and longer grains. Theor Appl Genet 129:1099–1112

Simonett O (2002) UNEP/GRID-Geneva. Prof. G. Sestini, Florence. Remote Sensing Center, Cairo

Singh Y, Humphreys E, Kukal SS, Singh B, Kaur A, Thaman S, Gajri PR (2009) Crop performance in permanent raised bed rice–wheat cropping system in Punjab India. Field Crop Res 110:1–20

Solangi GS, Shah SA, Alharbi RS, Panhwar S, Keerio HA, Kim TW, Bughio AD (2022) Investigation of irrigation water requirements for major crops using CROPWAT model based on climate data. WaterSA 14(16):2578

Sprink T, Wilhelm R, Hartung F (2022) Genome editing around the globe: an update on policies and perceptions. Plant Physiol 190:1579–1587

Srivastava P, Singh R, Tripathi S, Raghubanshi AS (2016) An urgent need for sustainable thinking in agriculture—an Indian scenario. Ecol Indic 67:611–622

Taheri F, Azadi HD, Haese M (2017) A world without hunger: organic or GM crops? Sustainability 9:580

Tang L, Shaikh IA, Tunio A, Junejo AR, Hao L, Dahri J, Mangrio MA, Soothar RK, Khan ZA (2024) Effect of raised flat bed and ridge planting on wheat crop growth and yield under varying soil moisture depletions. Agronomy 14:1404

Tellaeche A, Burgos Artizzu XP, Pajares G, Ribeiro A, Fernández-Quintanilla C (2008) A new vision-based approach to differential spraying in precision agriculture. Comput Electron Agric 60:144–155

Tribouillois H, Constantin J, Justes E (2018) Cover crops mitigate direct greenhouse gases balance but reduce drainage under climate change scenarios in temperate climate with dry summers. Glob Chang Biol 24:2513–2529

UNFCCC (2007) Investment and financial flows to address climate change. United Nations Framework Convention on Climate Change, Bonn. http://unfccc.int/files/cooperation_and_support/financial_mechanism/application/pdf/background_paper.pdf

USAID (2021) South Sudan Water Resources Profile Overview. Juba, South Sudan: Sustainable Water Partnership

Van Ginkel M, Ortiz R (2018) Cross the best with the best and select the best: HELP in breeding selfing crops. Crop Sci 58:17–30

Viana JS, de Melo JPR, de Almeida Silva JC, Gonçalves EP, Lima JJP, Júnior WRT (2019) Plant biostimulant on the production of biomass and yield grain of soybean: Bioestimulante vegetal e períodos de corte na produção de biomassa erendimentos de grãos de soja. Diversitas J 4:1105–1118

Wang Y, Wang S, Tian Y, Wang Q, Chen S, Li H, Li H (2021) Functional characterization of a sugar beet BvbHLH93 transcription factor in salt stress tolerance. Int J Mol Sci 22:3669

Wang X, Huang H, Lu M, Cui Y, Huang R, Wang X, Ouyang X (2023) Expanding crop adaptability to increase planting area: a promising strategy for enhancing agricultural production. Seed Biol 2(1):20

WB (2018) Egypt: enabling private investments and commercial financing in infrastructure. The World Bank, p 110

Wu J, Guo X, Lu J et al (2012) Effects of continuous straw mulching on soil physical and chemical properties and crop yields in paddy-upland rotation system. J Plant Nutr Fertil 18:587–594

Wu B, Gommes R, Zhang M, Zeng H, Yan N, Zou W, Van Heijden A (2015) Global crop monitoring: a satellite-based hierarchical approach. Remote Sens 7(4):3907–3933

Yang Z, Rao IM, Horst WJ (2013) Interaction of aluminium and drought stress on root growth and crop yield on acid soils. Plant Soil 372:3–25

Zamani K, Mohsenpour M, Shariatpanahi ME (2024) Genome editing based CRISPR system in crop improvement. In: Plant mutagenesis: sustainable agriculture and rural landscapes. Springer, Cham, pp 55–70

Zelenakova M, Abd-Elhamid H, Barańczuk K, Barańczuk J (2022) Assessment and adaptation to climate change and sea level rise impacts on Egypt's northern coasts. IOP Conf Ser Mater Sci Eng 125:012009. IOP Publishing

Zhang YD, Li DB, Zhang XG, Zhao XM, Cao LL, Dong JX, Liu KL, Chen HW, Zhang CX, Gao DW, Wang (2018) Analysis of the functions of TaGW2 homoeologs in wheat grain weight and protein content traits. Plant J 94:857–866

Zittis G, Hadjinicolaou P, Almazroui M et al (2021) Business-as-usual will lead to super and ultra-extreme heat waves in the Middle East and North Africa. NPJ Clim Atmos Sci 4:20

AquaCrop Model for Simulating Maize Productivity with Varying Planting Dates Under Deficit Irrigation in the Context of Climate Change

11

Sameh Mahmoud Mohamed Abdou,
Maher Mohamed Kassab, Ahmed Adel Saad El-Dein,
Ahmed R. Abd El Tawwab,
and Neven Abd Alrazzaq Hassoun

Abstract

In Egypt, maize is a significant summer cereal crop for food and feed utilization. Anticipated climate change is recognized to adversely affect maize productivity in Egypt. Consequently, the appropriate planting date and the precise application of irrigation will significantly influence maize productivity in the context of climate change. In this chapter, the impact of three planting dates and three irrigation levels on maize grain, and biological yields, as well as water productivity was simulated, utilizing the AquaCrop version 7.1 model and data from a 2-year field experiment. Following model validation, data from the CMIP6 mean projection for the SSP5-8.5 future scenario across two time intervals, specifically 2020–2039 and 2040–2059, were used to run the model and estimate anticipated losses in maize grain and biological yields under the examined future scenario.

Keywords

Available soil moisture depletion · Goodness of fit indices · Season actual evapotranspiration · Water productivity · CMIP6 · SSP5-8.5

S. M. M. Abdou (✉) · M. M. Kassab
Soil, Water, and Environment Research Institute, Agricultural Research Center, Giza, Egypt
e-mail: dr.samehabdou2004@arc.sci.eg

A. Abel Saad El-Dein
Department of Soil Science, Faculty of Agriculture, Ain Shams University, Cairo, Egypt

A. R. Abd El Tawwab
Department of Soils and Water, Faculty of Agriculture, Fayoum University, Faiyum, Egypt

N. Abd Alrazzaq Hassoun
Department of Soil Science, Faculty of Agriculture, Tishreen University, Latakia, Syria

A. E.-H. Zohry, S. Ouda (eds.), *Resilient Agroecosystems*, Sustainability Sciences in Asia and Africa, https://doi.org/10.1007/978-981-96-9630-7_11

Introduction

Maize is one of the most extensively used cereal crops globally, with substantial output levels. In the 2022 agricultural year, global production reached roughly 1.2 billion metric tons annually (FAOSTAT 2022). Maize is a significant agricultural product in Egypt for sustenance and livestock feed. The land allocated for maize cultivation in Egypt is around 1.4 million hectares. This region may fluctuate annually, influenced by agricultural policies and meteorological factors. The yearly maize production is approximately 7–8 million metric tons. Local maize productivity fluctuates due to climatic conditions, irrigation methods, and improvements in agricultural techniques, averaging between 5 and 6 tons per hectare (FAOSTAT 2022). Despite Egypt's classification as one of the top 17 maize-producing countries, a significant disparity exists between maize output and consumption owing to rapid population growth. The Egyptian population is expected to exceed 125 million by 2030 (Alkitkat 2017).

Water is categorized as the most limited natural resource, especially in arid and semiarid regions. The River Nile is the primary source of freshwater in Egypt. The river provides around 55.8 billion cubic meters per year, constituting nearly 97% of the nation's total renewable water resources. The mean annual precipitation in Egypt is roughly 12 mm, equivalent to 1.2 billion cubic meters. Approximately 84% of the nation's water resources are allocated for agricultural use. Egypt is designated as a water-scarce nation, with per capita freshwater availability plummeting from 1972 m^3 in 1970 to roughly 570 m^3 in 2018, and forecasts indicating a further reduction to 390 m^3 by 2050 (Rady and Omar 2018).

Minimizing irrigation water wastage in the field and enhancing irrigation efficiency can assist in mitigating the existing water scarcity. Diverse agricultural tactics have been employed to achieve this, including the selection of optimal planting dates, growing on elevated beds, utilization of drought-resistant cultivars, and effective fertilization methods, among others (Ayman 2023). Improving irrigation efficiency is crucial for optimizing output per cubic meter of water. Moreover, improving irrigation systems is essential for attaining the national objective of augmenting irrigation efficiency and ensuring equitable water allocation among farmers to optimize yields. Furthermore, farmers must adapt to the anticipated reduction in water supplies by modifying their excessive irrigation methods (Khatab et al. 2015; Abdelgalil et al. 2022).

The water necessity for maize to get optimal yield varies between 430 and 490 mm each season, contingent upon climatic conditions and the duration of the growing season (Doorenbos et al. 1979). Insufficient soil moisture restricts the metabolic processes of maize, impeding biomass accumulation and diminishing the photosynthetic rate by reducing chlorophyll levels in the leaves, ultimately resulting in decreased maize production (Libing et al. 2019). The limited supply of irrigation water and the irregular moisture distribution throughout the maize growing season resulted in considerable variability in grain and biological yields, seasonal water usage, and maize water productivity.

One method to improve water use efficiency is the implementation of deficit irrigation. Chai et al. (2016) define deficit irrigation as an irrigation method that applies water below the full amounts necessary for optimal growth and yield, with the objective of enhancing plant response to a specific level of water deficit and improving crop water use efficiency. Consequently, deficit irrigation is regarded as a significant factor in water conservation technologies. Numerous research has been conducted in Egypt to evaluate the effects of deficit irrigation on maize productivity (El-Henawy and Elsayed 2018; Abdel-Aal et al. 2020; Abaza et al. 2023). Their findings stated that maize productivity is anticipated to decline with a modest percentage under deficit irrigation application.

The productivity of maize is significantly affected by the planting date, which is indicative of meteorological variables such as air and soil temperature, solar radiation, and relative humidity, all of which also influence the proliferation of insect pests and disease infections (Khalil et al. 2013; Ayman 2023). Delayed maize planting results in diminished vegetative and reproductive growth due to a decrease in solar energy absorption from emergence to the silking stage (Cirilo and Andrade 1994; Shrestha et al. 2018). Postponing planting dates has been shown to significantly reduce the duration from planting to 50% tasseling and silking, as well as diminish plant height, ear length, and final maize grain yield (El-Galfy et al. 2009; Khalil et al. 2013; Hassaan 2018; Maresma et al. 2019; Abaza 2021).

Simulation models are streamlined representations of intricate phenomena that replicate the behavior of the system under examination (Steduto et al. 2009). The principal objective of crop simulation models is to assess yield productivity by accounting for variables such as climatic conditions, agricultural management techniques, and soil properties. Researchers employ several differential equations in these models to calculate variables and rates over time, from planting to harvest (Abdou et al. 2022). In light of the recent increase in demand for agricultural commodities and the consequent food crisis in developing nations, there is an urgent necessity to enhance water productivity efficiency in crop production. As a result, the Food and Agricultural Organization (FAO) developed a simulation model named AquaCrop to enhance water productivity in agricultural output. The model equilibrates precision, resilience, user-friendliness, and straightforwardness. It is optimal for pragmatic users such as extension workers, water specialists, economists, and policymakers who depend on uncomplicated models for planning and scenario analysis (Steduto et al. 2009).

Climate change has been caused by many compounds known as greenhouse gases, which have become more prevalent in the atmosphere, including carbon dioxide (CO_2), nitrous oxide (N_2O), methane (CH_4), ozone (O_3), and chlorofluorocarbons (CFCs) (Blasing 2010). The emission of greenhouse gases presents considerable hurdles for researchers in evaluating its local effects on agricultural systems (Tim 2000; Ulukan 2008).

The global study of maize production and its vulnerability to climate change has been extensively studied. Zhao et al. (2017) projected a 7% reduction in maize productivity for each degree increase in global mean temperature. Kogo et al. (2019) project that maize production may decrease by 8–38% by the conclusion of the twenty-first century under the RCP4.5 and/or RCP8.5 scenarios. A 24% decrease in maize production is anticipated in the RCP8.5 scenario from 2070 to 2099 (Choruma et al. 2022). In Egypt, various simulation models and climate scenarios were employed to project the reduction in maize output due to climate change (Eid et al. 1992; Eid and El-Mowelhi 1998; Ouda et al. 2009, 2012; Ouda and Zohry 2024). No research has utilized the AquaCrop simulation model in Egypt to evaluate the impact of climate change on maize production cultivated under varying planting dates under deficit irrigation conditions.

This chapter employed the AquaCrop model alongside the CMIP6 mean projections for the SSP5-8.5 scenario, analyzing two time intervals, 2020–2039 and 2040–2059, to assess its effects on maize productivity under three planting dates and three irrigation levels.

Materials and Methodology

The Study Site

This section provides an overview of the Tamiea Agricultural Research Station, situated in Fayoum Governorate, Egypt (Fig. 11.1) (latitude: 29.18°N, longitude: 30.51°E, altitude: 30 m above mean sea level). The current research was carried out over the summer seasons of 2022 and 2023. The soil's physical properties were assessed using the methodology described by Klute (1986). The soil at the experimental site is classified as clay loam, with 47.4% clay, 28.2% sand, and 24.3% silt. The chemical characteristics of the soil at the chosen experimental site were examined according to the methods specified by Page et al. (1982). The site had an organic matter level of 1.7% and a $CaCO_3$ content of 7.2%. The soil pH measured 8.23, and the electrical conductivity (ECe) was 3.17 dS m^{-1}. The concentrations of soluble cations were quantified in meq L^{-1}, with the relationship defined as ppm = meq L^{-1} × equivalent weight (mg meq^{-1}). The values are as follows: Mg^{2+} (4.28), Ca^{2+} (8.42), K^{+} (0.50), Na^{+} (18.53), Cl^{-} (20.87), SO_4^{2-} (8.11), and HCO_3^{-} (2.75). The cation exchange capacity (CEC) of the soil was measured at 33.06 meq 100 g^{-1}.

Soil moisture constants were quantified using gravimetric methods from the soil surface to a depth of 60 cm. On an oven-dried basis (%, wt wt^{-1}), soil field capacity was 39.64%, wilting point was 19.29%, and available soil moisture was 20.35%, with a soil bulk density of 1.39 g/cm^3. This results in a moisture content of 178.51 mm inside the upper 60 cm of soil.

The meteorological data were obtained from the Agrometeorological Station at Tamiea, Fayoum Governorate, and presented in Fig. 11.2.

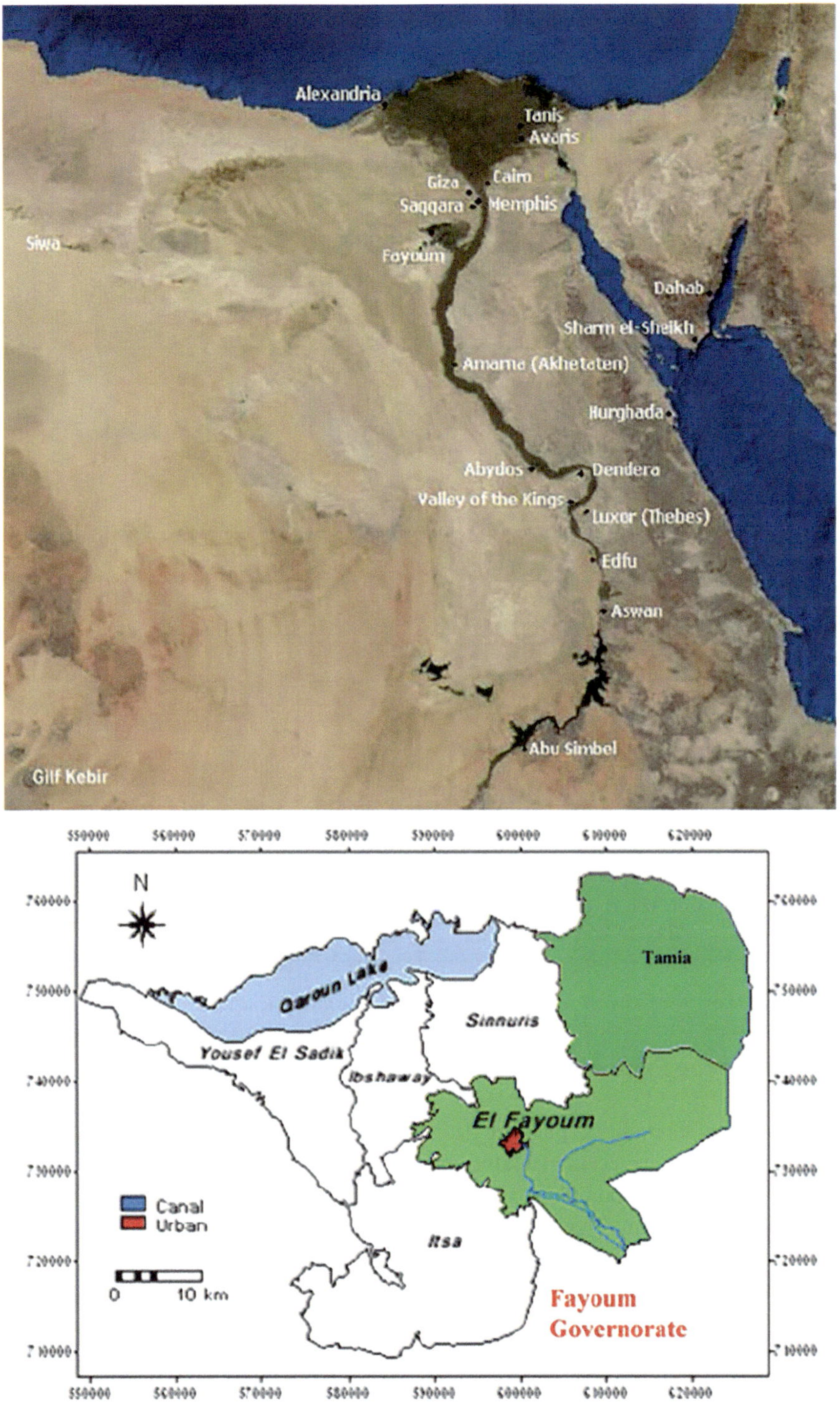

Fig. 11.1 The location of the study at Fayoum Governorate

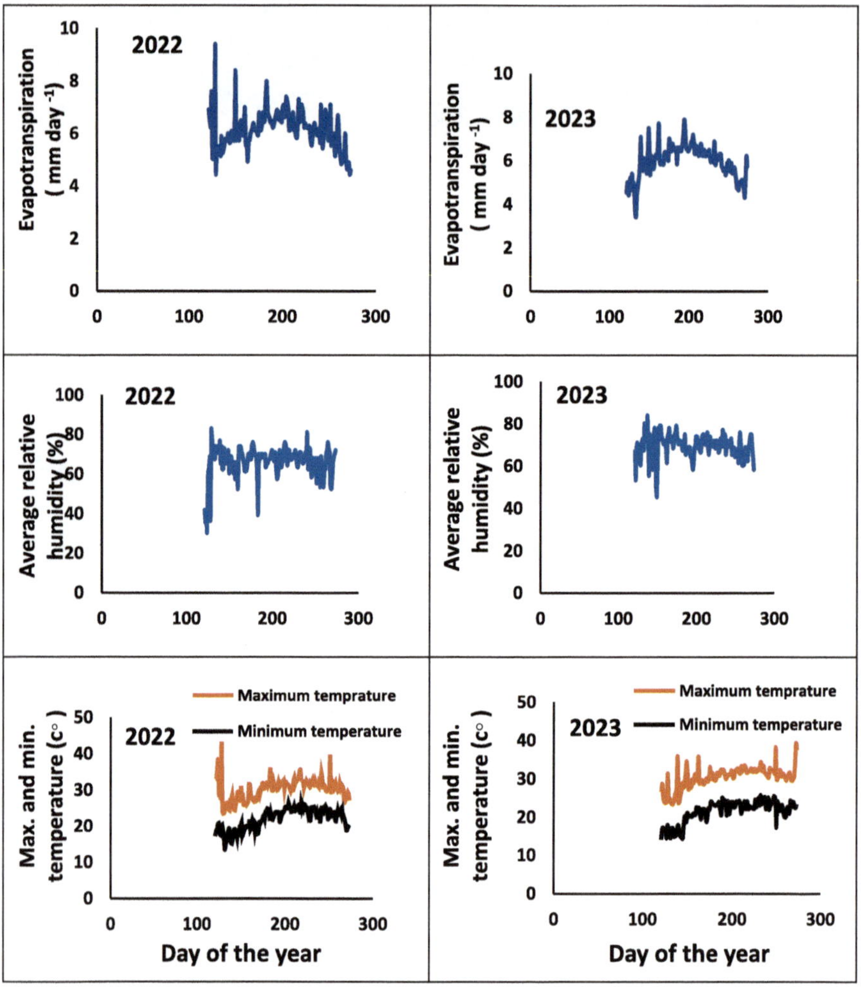

Fig. 11.2 Daily weather during the 2022 and 2023 maize growing seasons

Details of the Experiment

The effects of three maize planting dates (May 1st, 15th, and 30th in both growing seasons) and three levels of available soil moisture depletion (ASMD) (20% ASMD as control, and 40% and 60% ASMD as deficit irrigation treatments) were analyzed on maize grain and biological yield. The TWC310 maize hybrid was sown at a rate of 36 kg ha^{-1} using a split-split plot design with three replications. During land preparation, 360 kg of calcium superphosphate (15.5% P_2O_5) was incorporated. Ammonium nitrate (33.5% N) was utilized as nitrogen fertilization at a rate of 300 kg N ha^{-1}, administered in three equal applications: during planting, during the first irrigation, and during the second irrigation. The plots were divided by 1.5

meter-wide lanes to inhibit lateral water movement. The subplot area was 42.0 m^2, with dimensions of 6.0 m by 7.0 m.

The experiment was executed using a surface irrigation system, with irrigation treatments commencing from the second irrigation application. Dates and irrigation quantities were documented for the evaluated treatments in both growing seasons. Harvest took place on August 28th for the initial planting date and on September 7th and 23rd for the subsequent planting dates in both seasons. The maize grain and biological yields (kg ha^{-1}) for each subplot were quantified and aggregated to determine the yield per hectare.

Hydrological Dynamics in Maize

Maize Seasonal Actual Evapotranspiration (ET_C)

Soil samples were collected before and 48 h after each irrigation event to determine the ET_C, based on the following equation (Israelsen and Hansen 1962).

$$\mathrm{ET}_C - \{(Q_2 - Q_1)/100\} \times Bd \times D \tag{11.1}$$

where:
ETc = crop water evapotranspiration (cm).
Q_2 = soil moisture percentage 48 h after irrigation.
Q_1 = soil moisture just before irrigation.
Bd = soil bulk density (gcm^{-3}).
D = soil layer depth (cm).

Water Productivity of Maize (WP)

Water productivity (WP) was calculated by dividing the maize grain or biological yield (kg ha^{-1}) by the amount of applied water (m^3 ha^{-1}) for each treatment, using the method outlined by Vites (1965):

$$\mathrm{WP} = \text{Maize grain or biological yield}\left(\mathrm{kg\,ha}^{-1}\right) / \text{applied water}\left(\mathrm{m}^3\,\mathrm{ha}^{-1}\right) \tag{11.2}$$

AquaCrop Simulation Model

FAO developed the AquaCrop simulation model to mimic plant development, yield, and agricultural water productivity (Raes et al. 2006; Steduto et al. 2007). AquaCrop is unique among models due to its minimal input parameter requirements for simulating the correlation between yield and water availability. The model simulates variations in soil moisture levels during the growing season. Moreover, the model simulates the moisture levels in the root zone, influencing canopy development and associated crop transpiration, which subsequently contribute to yield formation and biomass production (Raes et al. 2006).

The AquaCrop model comprises three primary components: climate parameters such as precipitation, evaporation, and carbon dioxide content; soil properties and water balance dynamics; and the progression of crop growth and production (Steduto et al. 2007). AquaCrop is adept in evaluating crop water stress in irrigated and rainfed environments, estimating crop yield responses to fluctuating water availability, and calculating crop water productivity (Raes et al. 2006; Steduto et al. 2007).

The AquaCrop model relies on six critical input files for the simulation process: climate, soil profile, groundwater, irrigation, field, and crop. The interaction between the initial planting date and irrigating maize at 20% ASMD yielded the highest average grain and biological outputs, which were utilized to calibrate the model. An error minimization strategy diminishes the disparity between the observed and simulated outcomes. The simulated grain yield, biological yield, and water productivity for both grain and biological yield were compared with the observed data throughout the calibration process. The AquaCrop model was validated using simulated results from the calibration to assess its effectiveness across all combinations of planting dates and irrigation regime treatments. This assessment encompassed the evaluated treatments for the 2022 and 2023 growth seasons.

Assessment of the Performance of AquaCrop Model

The AquaCrop model was assessed by statistical indices to demonstrate the goodness of fit between observed and predicted data, utilizing the following indices:

Coefficient of Determination (R^2)

The R^2 is expressed as the Pearson correlation's square coefficient. It takes a value between 0 and 1.0, with a value of 1.0 representing a perfect fit and a value of 0 indicating no correlation. Additionally, R^2 indicates how much the data points align with the regression line (Jamieson et al. 1998; Moriasi et al. 2007).

$$R^2 = 1 - \frac{\sum_{i=1}^{n} \left(O_i - S_i\right)^2}{\sum_{i=1}^{n} \left(O_i - \bar{O}\right)^2} \tag{11.3}$$

where Oi and Si denote the observed and predicted values, respectively, n represents the total number of observed and predicted values involved in the comparison, and $\bar{O}$ is the average of the observed values.

Root Mean Square Error (RMSE)

RMSE quantifies the standard deviation of the residuals (i.e., prediction errors), which quantifies the distance of data points from the regression line. It is commonly used to evaluate the difference between the values predicted by a model and those observed in the actual experiment being modeled. A lower RMSE indicates higher accuracy in the simulation (Jacovides and Kontoyiannis 1995; Jamieson et al. 1998).

$$\text{RMSE} = \sqrt{\left(\frac{\sum_{i=1}^{n}\left(S_i - O_i\right)^2}{n}\right)} \tag{11.4}$$

Normalized Root Mean Square Error (NRMSE)
Since RMSE is represented in the same units as the variable under analysis, it may not be suitable for evaluating models across diverse climate meteorological conditions (Jacovides and Kontoyiannis 1995). RMSE can be normalized by dividing it by the mean of the observed variable ($\bar{O}$).

$$\text{NRMSE} = \frac{1}{O^{-}}\sqrt{\left(\frac{\sum_{i=1}^{n}\left(S_i - O_i\right)^2}{n}\right)} \tag{11.5}$$

NRMSE, presented as a percentage, offers a relative measure of the discrepancy between the model's predictions and the observed values.

Willmott's Index of Agreement (*d*)
Willmott (1982) introduced the agreement index to assess how closely the predicted data matches the data observed. It reflects the mean squared error ratio to the "prospective error." It is calculated by summing the squared differences between the predicted values and the mean of the observed values, along with the squared differences between the observed values and their mean (Willmott 1984). It addresses the limitations of R^2 and EF in identifying systematic under- or overestimations by the model (Willmott 1984). The index ranges from 0 to 1, where 0 indicates no agreement and 1 denotes perfect agreement between the predicted data and observed.

$$d = 1 - \frac{\Sigma\left(P_i - O_i\right)^2}{\Sigma\left(\left|P_i - \bar{O}\right| + \left\|O_i - \bar{O}\right\|\right)^2} \tag{11.7}$$

A limitation of the index (*d*) is that it can produce relatively high values (above 0.65) even when the model's performance is poor. Furthermore, despite Willmott's (1982) intentions, *d* is insensitive to systematic under- or overestimations (Jacovides and Kontoyiannis 1995).

Future Climate Scenarios

Data from the CMIP6 mean estimates for the SSP5-8.5 future scenario were obtained from The World Bank Group's Climate Change Knowledge Portal (CCKP), accessible at: https://climateknowledgeportal.worldbank.org/. The CCKP provides global climate data, information, and tools to enhance decision-making and deepen our comprehension of future climate conditions and associated hazards. The CCKP was established to provide open access to climate data and associated information, allowing users to better understand future climate scenarios and evaluate potential risks and implications (World Bank Group 2021). The climate projection data on that website is modeled data sourced from the Coupled Model Intercomparison Projects. The CMIP data constitutes the essential information utilized in the IPCC Assessment Reports. CMIP6 underpins the IPCC's Sixth Assessment Report, first published in August 2021. The projection data on that page is offered at a spatial resolution of 1° × 1° (100 km × 100 km).

In CMIP6, future climatic scenarios are delineated by five Shared Socioeconomic Pathways (SSPs): SSP1-1.9, SSP1-2.6, SSP2-4.5, SSP3-7.0, and SSP5-8.5, each embodying distinct societal development trajectories. By 2100, the cumulative radiative forcing, indicative of greenhouse gas emissions from all sources, is presented at the conclusion of each route (e.g., −1.9, −2.6, −4.5, 7.0, 8.5) (World Bank Group 2021). For this evaluation, we chose the SSP5-8.5 climate scenario. SSP5-8.5 illustrates a future marked by heightened exploitation of fossil fuel resources, leading to further global market integration and subsequent innovations and technical progress (World Bank Group 2021).

The anticipated monthly data, encompassing average maximum and minimum temperatures for Egypt and its subregions from 2020 to 2039 and 2040 to 2059, were obtained under the SSP5-8.5 future climate scenario based on a multi-model ensemble. We retrieved the monthly data from this dataset, primarily pertaining to the Fayoum Governorate, where the study was executed.

Findings and Analysis

Hydrological Dynamics in Maize

Maize Seasonal Evapotranspiration (ET_C)

The seasonal water consumption values for maize (ETc) are displayed in Table 11.1. The peak seasonal ETc values occurred with maize planted on May 1st and irrigated at 20% ASMD, whereas the lowest seasonal ETc values were recorded with maize planted on May 30th and irrigated at 60% ASMD. Postponing the planting date from May 1st to May 15th and May 30th resulted in decreases in the seasonal ETc of 6% and 9% during the 2022 growing season and 5% and 8% during the 2023 growing season, respectively. The reductions are probably attributable to reduced

Table 11.1 The seasonal actual evapotranspiration (ET_C) of maize (cm) as influenced by tested treatments

	2022				2023			
Planting dates	20% ASMD	40% ASMD	60% ASMD	Mean	20% ASMD	40% ASMD	60% ASMD	Mean
May 1st	58.10	55.81	52.55	**55.49**	55.88	53.85	49.73	**53.15**
May 15th	55.02	52.25	49.0	**52.09**	53.25	51.22	47.05	**50.51**
May 30th	53.60	51.30	46.80	**50.57**	51.80	49.60	45.40	**48.93**
Mean	**55.57**	**53.12**	**49.45**	**52.71**	**53.64**	**51.56**	**47.39**	**50.86**

evapotranspiration linked to a truncated growth phase (El-Akram et al. 2010; Jibrin et al. 2012).

The reduction in soil moisture from 20% ASMD to deficit irrigation treatments, specifically 40% ASMD and 60% ASMD, resulted with an increase in seasonal ETc by 4% and 11% in the 2022 season and by 4% and 12% in the 2023 season. The results can be ascribed to the regular irrigation at 20% ASMD, which augmented soil moisture availability in the rhizosphere and presumably raised the transpiration rate during the vegetative and reproductive phases.

The Volume of Irrigation-Applied Water to Maize

Table 11.2 illustrates the volume of irrigation water applied to maize across the experimental treatments. The volume of applied water mirrored the seasonal ETc pattern, with peak seasonal ETc values occurring when maize was sown on May 1st and irrigated at 20% ASMD, while the lowest seasonal ETc values were observed when maize was planted on May 30th and irrigated at 60% ASMD during both growing seasons. The findings also indicated that the reduction percentage in applied irrigation water for maize planted on May 30th and irrigated at 60% ASMD was 19%, compared to the value recorded for maize planted on May 1st and irrigated at 20% ASMD, averaged over two growth seasons.

Impact of the Examined Treatments on Maize Grain and Biological Yields

The findings indicated that varying maize planting dates and soil moisture levels significantly influenced grain and biological yields across both subsequent seasons. Table 11.3 indicates that the maximum maize grain and biological yields were attained when maize was planted on May 1st and irrigated with 20% ASMD in both growing seasons. Under this planting date and in the first growing season, maize grain yield decreased by 11% and 20% when irrigated at 40% and 60% ASMD, respectively, while biological yields diminished by 6% and 14% under the same treatments. There was a further decrease in maize grain and biological yields when planted on May 15th and May 30th, particularly under a deficit irrigation of 60% ASMD, with reductions in maize grain yield of 11% and 24%, respectively, compared to planting on May 1st under the same deficit irrigation conditions. A reduction by 11 and 23% in the biological yield of maize under the same respective

Table 11.2 The applied irrigation water amounts (m^3ha^{-1}) as influenced by tested treatments

	2022				2023			
Planting dates	20% ASMD	40% ASMD	60% ASMD	Mean	20% ASMD	40% ASMD	60% ASMD	Mean
May 1st	8938	8586	8085	**8536**	8597	8285	7651	**8178**
May 15th	8465	8038	7538	**8014**	8192	7880	7238	**7770**
May 30th	8246	7892	7200	**7779**	7969	7631	6985	**7528**
Mean	**8550**	**8172**	**7605**	**8110**	**8253**	**7932**	**7291**	**7825**

Table 11.3 Observed and predicted maize grain and biological yields under the studied treatments, as well as statistical indicators in the 2022 and 2023 growing seasons

Treatments		2022 season						2023 season					
		Grain yield (kg ha^{-1})			Biological yield (kg ha^{-1})			Grain yield (kg ha^{-1})			Biological yield (kg ha^{-1})		
Planting date	ASMD %	O	P	Dev (%)	O	P	Dev (%)	O	P	Dev (%)	O	P	Dev (%)
1/5	20%	6491	6450	−0.6	13,665	13,744	0.6	6218	6206	−0.2	13,315	13,452	1.0
	40%	5763	5845	1.4	12,814	13,106	2.3	5431	5445	0.3	12,705	12,861	1.2
	60%	5176	5271	1.8	11,698	11,944	2.1	4805	4937	2.7	11,672	11,796	1.1
15/5	20%	6016	6089	1.2	12,952	13,112	1.2	5741	5812	1.2	12,635	12,389	−1.9
	40%	5096	4990	−2.1	12,281	12,536	2.1	4950	5070	2.4	11,929	12,109	1.5
	60%	4586	4671	1.9	11,132	11,253	1.1	4311	4503	4.5	10,790	10,824	0.3
30/5	20%	5499	5413	−1.6	12,288	12,462	1.4	5114	5127	0.3	11,849	12,082	2.0
	40%	4781	4901	2.5	11,372	11,456	0.7	4428	4684	5.8	10,972	10,809	−1.5
	60%	3942	4078	3.5	10,364	10,575	2.0	3705	3816	3.0	9985	10,216	2.3
Statistical indicators	R^2		0.97			0.98			0.96			0.94	
	RMSE		81.0			200			92.0			193	
	NRMSE		1.57			1.21			1.32			1.41	
	d		0.98			0.98			0.97			0.98	

O measured, *P* predicted, *Dev* deviation between measured and predicted

planting date and irrigation treatment was found, compared to the value obtained under planting in 1st of May and irrigation with 20% ASMD (Table 11.3).

A comparable pattern was noted in the second growing season, where maize grain production decreased by 13% and 23% when planted on May 1st and irrigated with 40% and 60% ASMD, respectively, in contrast to irrigation with 20% ASMD on the same planting date. Additionally, the biological yield decreased by 5% and 12% under the specified planting date and irrigation regimes, respectively. A greater loss in maize grain and biological yield was noted when planted on May 30th and subjected to 60% ASMD deficit irrigation, namely, 40% and 25%, respectively, in contrast to the values recorded for planting on May 1st with 20% ASMD irrigation (Table 11.3).

The yield reductions can be ascribed to the postponement of maize planting, which curtailed the vegetative and reproductive growth phases, resulting in diminished dry matter accumulation in the plant organs and, consequently, a decrease in yield (El-Akram et al. 2010; Jibrin et al. 2012). Moreover, heightened soil moisture depletion will diminish moisture levels in the root zone, thereby inhibiting plant growth and reducing dry matter formation in maize during the grain-filling phase. It also diminishes nutrient absorption from the soil, which the growing plants utilize during their growth phase (El-Akram et al. 2010; Bassou et al. 2012; Khatab et al. 2015; Libing et al. 2019; Ahmed et al. 2020; Abdelgalil et al. 2022).

AquaCrop Prediction of Maize Grain and Biological Yield

The validation of AquaCrop for maize grain and biological yields, as shown in Table 11.3, demonstrates that the model predictions exhibit a high level of accuracy. During the 2022 growing season, the discrepancy between simulated grain production and observed values varied from a minimum of −0.6% for maize planted on May 1st with 20% ASMD irrigation to a maximum of 3.5% for maize planted on May 30th with 60% ASMD irrigation. Regarding the biological yield of maize, the minimal variation recorded was 0.6% for maize planted on May 1st with 20% ASMD irrigation, while the maximal divergence was 2.3% for maize planted on May 1st with 40% ASMD irrigation.

During the following growing season, the minimal departure observed for grain yield was −0.2%, occurring with maize planted on May 1st and irrigated at 20% ASMD. Conversely, the maximum divergence of 4.5% was recorded when maize was planted on May 15th and irrigated at 60% ASMD. The minimal departure in biological yield occurred when maize was sown on May 15th with 60% ASMD, registering at 0.3%, while the maximal deviation was noted when maize was planted on May 30th under the same irrigation conditions, at 2.3% (Table 11.3).

Table 11.3 also indicates that the simulation findings for maize grain and biological yields nearly correspond with the observed values, demonstrating a robust correlation between the predicted and observed data. The results exhibited elevated R^2 values, along with increased *d* values, and diminished RMSE and NRMSE values during both growing seasons.

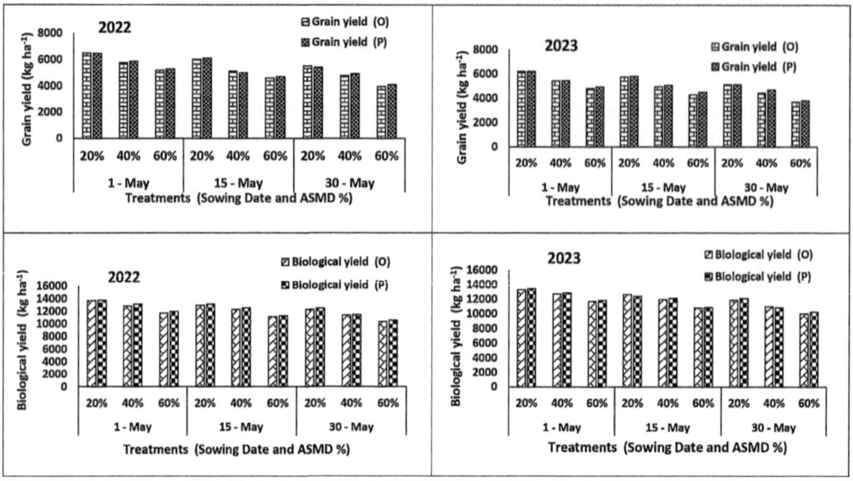

Fig. 11.3 Observed and predicted grain and biological maize yields in 2022 and 2023 growing seasons as influenced by various treatments

The AquaCrop model's efficacy in predicting the yields of various crops has been evaluated by multiple researchers (Kheir and Hassan 2016; Rohini et al. 2019; Shan et al. 2022). Figure 11.3 depicted the actual and predicted grain and biological yields of maize under the evaluated treatments.

Simulation of Water Productivity of Maize by AquaCrop

Table 11.4 indicated that water productivity diminished when the irrigation interval expanded from 20% ASMD to 60% ASMD. The maximum reported water productivity based on grain yield occurred when maize was planted on May 1st with 20% ASMD irrigation, followed by planting on May 15th with 40% ASMD irrigation. The minimum measured water productivity based on grain yield occurred when maize was sown on May 30th and irrigated at 60% ASMD. These results are consistent over both growing seasons. A reduction in water productivity values based on grain yield was noted when planting occurred on May 30th and irrigation was conducted with 60% ASMD, showing declines of 26% and 27% in the first and second growing seasons, respectively, compared to the values recorded for planting on May 1st and irrigation with 20% ASMD (Table 11.4).

Correspondingly, water productivity based on biological yield exhibited a similar tendency to that based on grain yield, with the greatest values recorded when maize was sown on May 1st with 20% ASMD irrigation, followed by maize planted on May 15th with 40% ASMD irrigation. These results were validated in both growing seasons. The reduction in the values of water productivity based on biological yield from planting on May 1st with 20% ASMD irrigation and from planting on May 30th with 60% ASMD irrigation was 8% in both growth seasons (Table 11.4).

Table 11.4 Evaluation statistics for the AquaCrop model assessing maize water productivity (WP) under the studied treatments in the 2022 and 2023 seasons

Treatments		2022 season						2023 season					
		Grain yield WP (kg m^{-1})			Biological yield WP (kg m^{-1})			Grain yield WP (kg m^{-1})			Biological yield WP (kg m^{-1})		
Planting date	ASMD %	O	P	Dev (%)	O	P	Dev (%)	O	P	Dev (%)	O	P	Dev (%)
1/5	20%	0.726	0.720	−0.8	1.53	1.52	−0.7	0.719	0.709	−1.4	1.54	1.53	−0.6
	40%	0.672	0.672	−0.1	1.50	1.51	0.7	0.544	0.634	−1.6	1.51	1.51	0.0
	60%	0.649	0.669	3.1	1.47	1.44	−2.0	0.613	0.623	1.6	1.49	1.49	0.0
15/5	20%	0.702	0.691	−1.6	1.51	1.5	−0.7	0.690	0.67	−2.9	1.52	1.53	0.7
	40%	0.616	0.626	1.6	1.48	1.48	0.0	0.617	0.637	3.2	1.49	1.47	−1.3
	60%	0.597	0.608	1.8	1.45	1.46	0.7	0.583	0.573	−1.7	1.46	1.48	1.4
30/5	20%	0.669	0.647	−3.3	1.50	1.52	1.3	0.651	0.671	3.1	1.51	1.51	0.0
	40%	0.606	0.606	0.0	1.44	1.45	0.7	0.583	0.573	−1.7	1.45	1.46	0.7
	60%	0.536	0.526	−1.9	1.41	1.4	−1.4	0.525	0.545	3.8	1.41	1.42	0.7
Statistical indicators	R^2		0.97			0.98			0.99			0.98	
	RMSE		0.03			0.05			0.04			0.05	
	NRMSE		1.37			1.49			2.12			1.50	
	d		0.99			0.98			0.97			0.99	

O measured, *P* predicted, *Dev* deviation between measured and predicted

The AquaCrop model demonstrated a high degree of accuracy in simulating the growth and production of maize under the evaluated treatments. The minimal variation between the observed and simulated water productivity based on grain yield occurred with planting on May 30th and irrigation at 40% ASMD, while the maximal value was recorded with planting on May 30th and irrigation at 20% ASMD during the first growing season. In the second growing season, the minimal divergence between the observed and simulated values occurred with planting on May 30th and irrigation at 40% ASMD, whereas the maximal deviation was recorded with planting on May 30th and irrigation at 60% ASMD.

Regarding water productivity based on biological yield, the minimal deviation between observed and simulated values occurred with planting on May 1st and irrigation at 40% ASMD and 60% ASMD, while the maximal deviation was observed with planting on May 15th and irrigation at 60% ASMD.

Moreover, the findings indicated elevated values of R^2, NES, and d and reduced values of RMSE and NRMSE over both growing seasons. These results indicated a robust correlation between the observed water productivity and the simulated values. Comparable findings were reported by Linker and Ioslovich (2017), Abdou et al. (2022) and Panek-Chwastyk et al. (2024).

Projection of Climate Change Impacts on Maize Productivity Under the Studied Treatments

The Future Climate Scenario SSP5-8.5 in 2020–2039 Time Interval

Table 11.5 presents the anticipated maize grain and biological yields resulting from the application of three planting dates and three irrigation regimes under the SSP5-8.5 future scenario for 2020–2039 and 2040–2059 time intervals. The findings revealed that the most significant decrease in maize grain and biological yields occurs under the third planting date and a deficit irrigation of 60% ASMD, specifically 31% and 36%, respectively, under the SSP5-8.5 future scenario for the 2020–2039 period. The minimal decrease in maize grain and biological yields is anticipated to occur with the first planting date and irrigation at 20% ASMD, specifically 13% for maize grain and 18% for biological yield, averaged over two growing period seasons within the SSP5-8.5 future scenario for the 2020–2039.

In the SSP5-8.5 scenario for the 2040–2059 period, the most significant anticipated reductions in grain and biological yield for maize are 41% and 42%, respectively, when planted on the third planting date with 60% ASMD irrigation. Conversely, the least expected reductions are 24% and 26%, respectively, for maize planted on the first planting date with 20% ASMD irrigation (Table 11.5).

The results are attributed to the projected temperature rise during the 2040–2059 period, which is projected to exceed that of the 2020–2039 interval under the climate change scenario SSP5-8.5. Moreover, these findings indicated that increasing temperatures in the future scenarios over the two time intervals will substantially diminish moisture in the root zone relative to the levels recorded during the field experiment.

Table 11.5 Observed versus projected maize grain and biological yields under SSP5-8.5 in 2039–2040 and 2040–2059 time intervals

Treatments		2020–2039						2040–2059					
		Grain yield (kg ha^{-1})			Biological yield (kg ha^{-1})			Grain yield (kg ha^{-1})			Biological yield (kg ha^{-1})		
Planting date	ASMD (%)	O	P	Dev (%)	O	P	Dev (%)	O	P	Dev (%)	O	P	Dev (%)
1/5	20%	6491	5647	−13.0	13,665	11,235	−18.0	6491	4933	−24.0	13,665	10,155	−26.0
	40%	5763	4783	−17.0	12,814	9897	−22.0	5763	4092	−29.0	12,814	8918	−30.0
	60%	5176	4089	−21.0	11,698	8814	−25.0	5176	3468	−33.0	11,698	7726	−34.0
15/5	20%	6016	4993	−17.0	12,952	9981	−23.0	6016	4332	−28.0	12,952	9171	−29.0
	40%	5096	4077	−20.0	12,281	9218	−25.0	5096	3465	−32.0	12,281	8180	−0.33.0
	60%	4586	3485	−24.0	11,132	8065	−28.0	4586	2889	−37.0	11,132	6953	−38.0
30/5	20%	5499	4179	−24.0	12,288	8730	−29.0	5499	3684	−33.0	12,288	8100	−34.0
	40%	4781	3442	−28.0	11,372	7621	−33.0	4781	3060	−36.0	11,372	7138	−37.0
	60%	3942	2720	−31.0	10,364	6642	−36.0	3942	2326	−41.0	10,364	6005	−42.0
Statistical indicators	R^2		0.97			0.98			0.96			0.94	
	RMSE		1679			2752			1992			2193	
	NRMSES		0.23			0.31			0.35			0.43	
	d		0.74			0.80			0.57			0.59	

O measured, *P* predicted, *Dev* deviation between measured and predicted

Elevated temperatures accelerate phenological processes such as vegetative and grain-filling phases, frequently diminishing the duration of the grain-filling period, leading to smaller and less developed kernels. Prior studies indicate that each 1 °C rise over the ideal temperature may result in a 3–4% reduction in grain yield (Lobell et al. 2011). Heat stress is particularly detrimental during the initial phases of ear development since it impacts cellular division, starch production, and sugar metabolism in the endosperm (Monjardino et al. 2005). Moreover, extreme water stress resulted in significant decreases in maize's leaf area index, plant height, chlorophyll content, assimilation rate, and gas exchange metrics. These physiological disturbances cumulatively led to a substantial reduction in maize yield (Riar et al. 2022). Moreover, steady soil moisture conditions enhance maize leaf water content, photosynthetic carbon absorption, and grain yields, irrespective of total soil water content. Stable soil moisture conditions promote leaf water-use efficiency and maize output by facilitating improved photosynthesis, in contrast to changing moisture levels (Li et al. 2024). Additionally, water stress during the grain-filling phase adversely impacts the ultimate maize production. Ouda et al. 2015). Heat stress often coincides with water stress, exacerbating its effects by reducing stomatal conductance and transpiration, impeding photosynthesis (Schlenker and Roberts 2009).

Maize Seasonal Actual Evapotranspiration (ET_C)

According to the climate change scenario, the data depicted in Fig. 11.4 indicate an anticipated rise in maize ETc for both time intervals under the SSP5-8.5 circumstances. The rise in ETc is attributed to the anticipated temperature increase of 1.7–2.0 °C by 2039 and 3.2–3.5 °C by 2059 under the future climate scenario SSP5-8.5. The peak seasonal actual evapotranspiration occurs for the first planting date of May 1st, with irrigation at 20% ASMD, recording values of 59.45 cm by 2039 and 63.76 cm by 2059. The minimum values, specifically 54.18 cm by 2039 and 58.69 cm by 2059, are observed for the third planting date with irrigation at

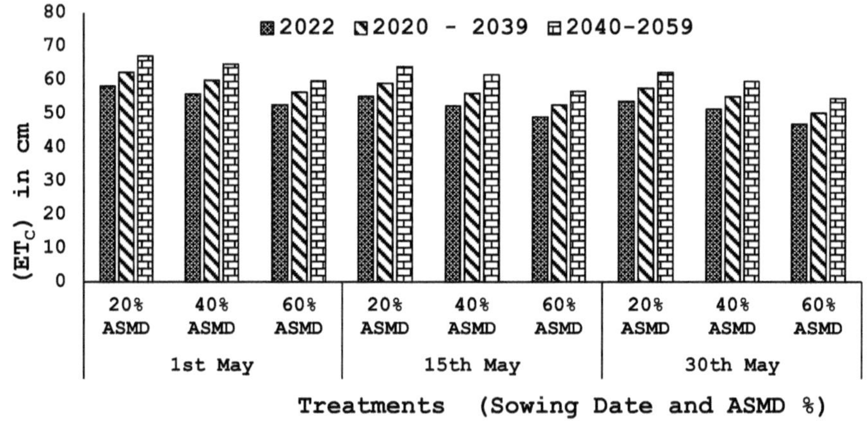

Fig. 11.4 Observed and projected grain maize and biological yields values in 2022, 2020–2039, and 2040–2059 under SSP5-8.5 future scenario

20% ASMD. This difference is due to elevated temperatures throughout growth stages, which are detrimental to plant development and grain production (Ouda et al. 2012).

Maize Water Productivity Under Climate Change

In the SSP5-8.5 future scenario (Table 11.6), the water productivity values based on grain yield will decrease when maize is sown on May 30th and irrigated with 60% ASMD, declining by 29.9% and 54.1% throughout the 2020–2039 and 2040–2059 periods, respectively, compared to measured values under the same treatments. Moreover, the water productivity values based on biological yield are anticipated to decrease by 39.7% and 57.4%, respectively, for maize planted on May 30th and irrigated with 60% ASMD, compared to measured values under the same treatments. The minimal drop in water productivity values based on grain yield is anticipated when maize is sown on May 1st and irrigated with 20% ASMD, resulting in decreases of 9.6% and 49.6% throughout the 2020–2039 and 2040–2059 periods, respectively. Furthermore, the water productivity values based on biological yield are anticipated to decrease by 17.6% and 37.3%, respectively, for maize planted on May 1st and irrigated with 20% ASMD, compared to measured values under the same treatments (Table 11.6).

The anticipated rise in water productivity values under the examined future scenario is attributed to elevated transpiration rates from the maize canopy and increasing evaporative demands from the soil surface due to rising air temperatures (Ouda and Zohry 2024).

Conclusions

Sustainable development involves satisfying current requirements while guaranteeing that future generations can also fulfill their demands. Climate change is expected to impact the social, economic, and environmental dimensions of sustainable development, thereby altering the available avenues for advancement. Agriculture is significantly influenced by climate, and anticipated climate change is likely to modify circumstances for agricultural production, heightening its susceptibility and ultimately impacting global food security.

Therefore, it is essential to create methods that alleviate climate change risks affecting maize productivity. Determining ideal planting dates to endure the abiotic stressors associated with climate change is crucial to these tactics. The efficacy of the maize hybrid TWC 310 was assessed under prevailing climatic conditions with three irrigation levels (20%, 40%, and 60%) of ASMD, alongside three planting dates (May 1st, May 15th, and May 30th). The optimal maize grain yield, biological yield, and water productivity, whether measured on grain or biological yield basis, were achieved by planting maize on May 1st and irrigating with 20% ASMD.

Our findings demonstrated that the AquaCrop model was exceptionally effective in precisely predict maize production under the evaluated treatments. Moreover, AquaCrop predicts a decline in maize grain output, biological yield, and water

Table 11.6 Observed versus projected water productivity on grain and biological bases under SSP5-8.5 in 2039–2040 and 2040–2059 time intervals

Treatments		2040 season						2060 season					
		Grain yield WP (kg m^{-1} water applied)			Biological yield WP (kg m^{-1} water applied)			Grain yield WP (kg m^{-1} water applied)			Biological yield WP (kg m^{-1} water applied)		
Planting date	ASMD %	O 2022	P 2040	Dev.	O 2022	P 2040	Dev.	O 2022	P 2060	Dev.	O 2022	P 2060	Dev.
1/5	20%	0.726	0.656	−9.6	1.53	1.26	−17.6	0.726	0.366	−49.6	1.53	0.96	−37.3
	40%	0.672	0.542	−19.3	1.50	1.19	−20.7	0.672	0.312	−53.6	1.50	0.9	−40.0
	60%	0.649	0.499	−23.1	1.47	1.06	−27.9	0.649	0.309	−52.4	1.47	0.82	−44.2
15/5	20%	0.702	0.582	−17.1	1.51	1.19	−21.2	0.702	0.312	−55.6	1.51	0.9	−40.4
	40%	0.616	0.456	−26.0	1.48	1.1	−25.7	0.616	0.286	−53.6	1.48	0.84	−43.2
	60%	0.597	0.427	−28.5	1.45	0.94	−35.2	0.597	0.257	−57.0	1.45	0.73	−49.7
30/5	20%	0.669	0.569	−14.9	1.50	1.11	−26.0	0.669	0.369	−44.8	1.50	0.87	−42.0
	40%	0.606	0.436	−28.1	1.44	0.96	−33.3	0.606	0.286	−52.8	1.44	0.7	−51.4
	60%	0.536	0.376	−29.9	1.41	0.85	−39.7	0.536	0.246	−54.1	1.41	0.6	−57.4
Statistical indicators	R^2		0.95			0.97			0.98			0.98	
	RMSE		0.18			0.21			0.17			0.25	
	NRMSE		1.69			1.82			1.62			1.73	
	d		0.48			0.46			0.42			0.37	

productivity under the future scenario SSP5-8.5 throughout both evaluated time intervals. Consequently, it is advisable to sow maize in early May and irrigate it with 20% ASMD to alleviate the effects of climate change. In instances of water scarcity, employing deficit irrigation of maize at 40% ASMD can conserve water while minimizing yield losses.

References

Abaza ASD (2021) Effect of irrigation water levels, filter mud cake application and planting dates on maize production under Tushka. https://doi.org/10.1007/s10343-023-00836-1

Abaza ASD, Elshamly AMS, Alwahibi MS, Elshikh MS, Ditta A (2023) Impact of different sowing dates and irrigation levels on NPK absorption, yield and water use efficiency of maize. Sci Rep 13(1):12956. https://doi.org/10.1038/s41598-023-40032-9

Abdel-Aal TS, Ibrahim AM, Abdel-Razek MA, Abd-Allah EGM (2020) Rationalize maize irrigation water using modern irrigation systems and plastic mulching in clayey soils of Fayoum. Fayoum J Agric Res Dev 34(1):48–65

Abdelgalil A, Mustafaa AA, Ali SAM, Yassin OM (2022) Effect of thermal requirements on chemical content of sugar beet and it's reflecting on yield in Upper Egypt. Curr Chem Lett 11:219–226

Abdou SMM, Engel BA, Emam SM, Abd El-Latif KM (2022) AquaCrop model validation for simulation wheat productivity under water stress condition. Commun Soil Sci Plant Anal 53(3):281–292. https://doi.org/10.1080/00103624.2021.2016806

Ahmed MI, Shah AN, Sun J, Song Y (2020) Comparative study on leaf gas exchange, growth, grain yield, and water use efficiency under irrigation regimes for two maize hybrids. Agriculture 10:369. https://doi.org/10.3390/agriculture10090369

Alkitkat H (2017) Projections of the Egyptian population. In: Ouda S et al (eds) Future of food gaps in Egypt, SpringerBriefs in agriculture. Springer. https://doi.org/10.1007/978-3-319-46942-3_3

Ayman MSE (2023) Interaction effects of planting date, irrigation levels, chitosan, and potassium silicate on yield and water use efficiency for maize grown under arid climate. Gesunde Pflanzen 75:1601–1613. https://doi.org/10.1007/s10343-023-00836-1

Bassou B, Dimitri X, Ahmed B, Pierre R, Jean-Claude M (2012) Effect of water stress on growth, water consumption and yield of silage maize under flood irrigation in a semi-arid climate of Tadla (Morocco). Biotechnol Agron Soc Environ 16(4):468–477

Blasing J (2010) Recent greenhouse gas concentrations. Carbon Dioxide Information Analysis Center, Washington

Chai Q, Gan Y, Zhao C, Xu H, Waskom RM, Niu Y, Siddique KHM (2016) Regulated deficit irrigation for crop production under drought stress. A review. Agron Sustain Dev 36(3):2–21

Choruma DJ, Akamagwuna FC, Odume NO (2022) Simulating the impacts of climate change on maize yields using EPIC: a case study in the Eastern Cape Province of South Africa. Agriculture 12:794. https://doi.org/10.3390/agriculture12060794

Cirilo AG, Andrade F (1994) Planting date and maize productivity: I. Crop growth and dry matter partitioning. Crop Sci 34:1039–1043

Doorenbos J, Kassam AH, Bentvelsen CLM, Van Der Wall HK (1979) Yield response to water. Irrigation and drainage paper 33, FAO, Rome, pp 101–104

Eid HM, El-Mowelhi NM (1998) Impact of climate change on field crops and water needs in Egypt. In: African international environmental conference, October 1998. Egypt

Eid HM, Ainer NG, Yousef KMR, Ibrahim MAM, Gad El-Rab GM (1992) Climate change crop modeling study on wheat. In: 5th Egypt botanical conference, Saint Catherine, Sinai, pp 113–130

El-Akram MFI, Abdou SMM, Ashry MRK, Farrag FRM (2010) Water use and yield of maize in relation to irrigation scheduling and Soeing dates. Fayoum J Agric Res Dev 24:2

El-Galfy AMK, Abd El-Azem MEM, El-Mekser Hoda KHA (2009) Response of some maize hybrids to late planting dates. Ann Agric Sci Moshtohor 47(2):121–127

El-Henawy AS, Elsayed FH (2018) Water requirements and irrigation scheduling maize crop by empirical equations using lysemters. Soil Sci Agric Eng Mansoura Univ 9(11):557–560

FAOSTAT (2022) Food and Agriculture Organization of the United Nations. http://www.fao.org/faostat/en/#data/QC

Hassaan MA (2018) Response of some yellow maize hybrids (*Zea mays* L.) to planting date under Toshka conditions. J Plant Prod 9:509–514. https://doi.org/10.14720/aas.2021.117.1.1297

Israelsen OW, Hansen VE (1962) Irrigation principles and practices, 3rd edn. Wiley International Edition, New York

Jacovides CP, Kontoyiannis H (1995) Statistical procedures for the evaluation of evapotranspiration computing models. J Agric Water Manag 27(3–4):365–371. https://doi.org/10.1016/0378-3774(95)01152-9

Jamieson PD, Porter JR, Goudriaan J, Ritchie JT, van Keulen H, Stol W (1998) A comparison of the models AFRCWHEAT2, CERES-Wheat, Sirius, SUCROS2 and SWHEAT with measurements from wheat grown under drought. Field Crop Res 55:23–44. https://doi.org/10.1016/S0378-4290(97)00060-9

Jibrin MJ, Kamara AY, Ekeleme F (2012) Simulating planting date and cultivar effects on dryland maize production using CERES-maize model. Afr J Agric Res 7(40):5530–5536

Khalil MAG, Hassan MMM, El-Makser HKHA, El-Ghonemy MAM (2013) Effect of planting dated on the growth and grain yield of 15 maize hybrids at North and Middle Egypt. Bull Fac Agric Cairo Univ 64:251–258

Khatab AKh, Abd El-Latif KhM, Osman EAM, Abdou SMM (2015) Maize productivity and crop–water relations as affected by irrigation levels and compost rates. J Soil Sci Agric Eng 6(12):1545–1562

Kheir AMS, Hassan MAA (2016) Performance assessment of the FAO AquaCrop model for maize yield, biomass and water productivity along the River Nile, Egypt. J Soil Sci Agric Eng Mansoura Univ 10:721–728

Klute A (1986) Methods of soil analysis. Part-1: physical and mineralogical methods, 2nd edn. American Society of Agronomy, Madison

Kogo BK, Kumar L, Koech R, Langat P (2019) Modelling impacts of climate change on maize (*Zea mays* L.) growth and productivity: a review of models, outputs and limitations. J Geosci Environ Protect 7:76–95

Li G, Long H, Zhang R, Xu A, Niu L (2024) Photosynthetic traits, water use and the yield of maize are influenced by soil water stability. BMC Plant Biol 24:1235. https://doi.org/10.1186/s12870-024-05942-4

Libing S, Jiming J, Jianqiang H (2019) Effects of severe water stress on maize growth processes in the field. Sustain For 11:5086. https://doi.org/10.3390/su11185086

Linker R, Iosilevich I (2017) Assimilation of canopy cover and biomass measurements in the crop model AquaCrop. Biosyst Eng 162:57–66. www.elsevier.com/locate/issn/15375110

Lobell DB, Schlenker W, Costa-Roberts J (2011) Climate trends and global crop production since 1980. Science 333:616–620. https://doi.org/10.1126/science.1204531

Maresma A, Ballesta A, Santiveri F, Lloveras J (2019) Planting date affects maize development and yield in irrigated Mediterranean environments. Agriculture 9(3):67. https://doi.org/10.3390/agriculture9030067

Monjardino P, Smith AG, Jones RJ (2005) Heat stress effects on protein accumulation of maize endosperm. Crop Sci 45:1203–1210

Moriasi DN, Arnold JG, Van Liew MW, Bingner RL, Harmel RD, Veith TL (2007) Model evaluation guidelines for systematic quantification of accuracy in watershed simulations. Trans ASABE 50(3):885–900. https://doi.org/10.13031/2013.23153

Ouda S, Zohry A (2024) Legumes improve wheat and maize productivity grown in different cropping systems under changing climate. In: Integration of legume crops with cereal crops under changing climate: sustainably increases food production. Springer. ISBN: 978-3-031-68101-1

Ouda S, Khalil F, Yousef H (2009) Using adaptation strategies to increase water use efficiency for maize under climate change conditions. In: 13th International conference on water technology, Hurghada, Egypt, 12–15 March

Ouda S, Abdraboh M, Noreldin T (2012) Combating the harmful effect of climate change on wheat using irrigation water management. Egypt J Agric Res 90(4):477–497

Ouda S, Noreldin T, Abd El-Latif K (2015) Water requirements for wheat and maize under climate change in North Nile Delta. Span J Agric Res 13(1):e03-001. https://doi.org/10.5424/sjar/2015131-6412

Page AL, Miller RH, Keeney DR (1982) Methods of soil analysis. Part-2: chemical and microbiological properties, 2nd edn. American Society of Agronomy, Madison

Panek-Chwastyk E, Ozbilge CN, Browska-Zielin KD, Gurdak R (2024) Advancing crop yield predictions: AQUACROP model application in Poland's JECAM fields. Agronomy 14:854. https://doi.org/10.3390/agronomy14040854

Rady R, Omar ME (2018) Background paper: water resources management in Egypt: assessment and recommendations. Alternative policy solutions. American University in Cairo, pp 4–30. http://aps.aucegypt.edu

Raes D, Geerts S, Kipkorir E, Wellens J, Sahli A (2006) Simulation of yield decline as a result of water stress with a robust soil water balance model. Agric Water Manag 81(3):335–357. https://doi.org/10.1016/j.agwat.2005.04.006

Riar M, Loar J, Fasoula V, Holman J (2022) Impact of water stress on physiological traits and yield of maize hybrids under field conditions. Agronomy 12(10):2386. https://doi.org/10.3390/agronomy12102386

Rohini R, Patel SK, Pandey S (2019) Evaluation of a crop simulation model for predicting maize growth and yield under varying environments. Field Crop Res 238:1–10. https://doi.org/10.1016/j.fcr.2019.04.004

Schlenker W, Roberts MJ (2009) Nonlinear temperature effects indicate severe damages to U.S. crop yields under climate change. Proc Natl Acad Sci USA 106:15594–15598. https://doi.org/10.1073/pnas.0906865106

Shan Y, Li G, Su L, Zhang J, Wang Q, Wu J, Mu W, Sun Y (2022) Performance of AquaCrop model for maize growth simulation under different soil conditioners in Shandong Coastal Area, China. Agronomy 12(7):1541. https://doi.org/10.3390/agronomy12071541

Shrestha J, Kandel M, Chaudhary A (2018) Effects of planting time on growth, development and productivity of maize (*Zea mays* L.). J Agric Nat Resour 1(1):43–50

Steduto P, Hsiao TC, Fereres E (2007) On the conservative behavior of biomass water productivity. Irrig Sci 25(3):189–207. https://doi.org/10.1007/s00271-007-0064-1

Steduto P, Hsiao P, Raes TC, Ferengcset DE (2009) AquaCrop-the FAO crop model to simulate yield response to water: I. Concepts and underlying principles. Agron J 101:426e437

Tim H (2000) The impact of climatic variability over the period 1961–1990 on the soil water balance of upland soils in the North East Arid Zone of Nigeria. U.K. Met Office, 2000. Climate change: an update of recent research from the Hadley Centre

Ulukan H (2008) Agronomic adaptation of some field. Crops: a general approach. J Agron Crop Sci 194:169–179

Vites FG (1965) Increasing water use efficiency by soil management in plant environment and efficient water use. J Am Soc Agron 26:537–546

Willmott CJ (1982) Some comments on the evaluation of model performance. Bull Am Meteorol Soc 63(11):1309–1313. https://doi.org/10.1175/1520-0477(1982)063<1309:SCOTEO>2.0.CO;2

Willmott CJ (1984) On the evaluation of model performance in physical geography. In: Gaile GL, Willmott CJ (eds) Spatial statistics and models. D. Reidel, Boston, pp 443–460

World Bank group (2021) Climate change knowledge portal (CCKP) user manual. World Bank Group. 18 pages

Zhao C, Liu B, Piao S, Wang X, Lobell DB, Huang Y, Huang M, Yao Y, Bassu S, Ciais P, Durand JL, Elliott J, Ewert F, Janssens IA, Li T, Lin E, Liu Q, Martre P, Müller C et al (2017) Temperature increase reduces global yields of major crops in four independent estimates. PNAS 114(35):9326–9331. https://doi.org/10.1073/pnas.1701762114

Legume-Based Intercropping Systems Mitigate Climate Change

12

Samiha Ouda and Abd El-Hafeez Zohry

Abstract

Climate change is a critical issue that influences greenhouse gas emissions from soil, with soil texture, pH value, and organic carbon being pivotal elements in regulating these emissions. Climate change is anticipated to elevate global mean surface temperatures, resulting in heightened demand on water supplies and an increased risk of plant infections by pathogens, which would adversely affect crop productivity and stability. Soil degradation is a significant issue resulting from climate change-induced alterations in precipitation, temperature, atmospheric nitrogen deposition, and elevated CO_2 concentrations. This chapter underscores the vital significance of legume-based intercropping systems in enhancing resilience to climate change. Legume-based intercropping systems provide multiple benefits, such as improved soil nutrient cycling, increased water usage efficiency, contribution to soil carbon sequestration, protection of soil and water from erosion, disruption of insect and disease life cycles, and reduction of greenhouse gas emissions from the soil. Thus, legume-based intercropping systems may function as a sophisticated agroecosystem strategy to mitigate the effects of climate change, ultimately enhancing agricultural sustainability and soil security across diverse agroecosystems.

S. Ouda (✉)
Water Requirements and Field Irrigation Research Department, Soils, Water and Environment Research Institute, Agricultural Research Center, Giza, Egypt
e-mail: souda1958@arc.sci.eg

A. E.-H. Zohry
Crop Intensification Research Department, Field Crops Research Institute, Agricultural Research Center, Giza, Egypt

A. E.-H. Zohry, S. Ouda (eds.), *Resilient Agroecosystems*, Sustainability Sciences in Asia and Africa, https://doi.org/10.1007/978-981-96-9630-7_12

Keywords

Nutrients circulation · Water retention · Yield stabilization · Carbon sequestration · Erosion · GHGs

Introduction

The IPCC (2018) contends that alterations in land use resulting from human activities, such as deforestation, cultivation, fertilizer application, biomass combustion, and the removal of crop residuals, contribute to greenhouse gas emissions, which are the principal drivers of climate change. Climate change has affected food system operations, starting from input supply to marketing/distribution and consumption (Singh and Khanna 2021). Ecosystems and food systems are interrelated as food production employs land and water at every level, generating waste that contaminates terrestrial and aquatic habitats, produces greenhouse gases, pollutes water bodies, and degrades soil (Mbow et al. 2019). Food systems account for approximately 19–29% of global greenhouse gas emissions, with agricultural activities responsible for 80–86% of this total (Vermeulen et al. 2012).

Cultivating legume crops enhances soil carbon sequestration and diminishes the requirement for nitrogen fertilizers, decreasing fossil energy inputs (Stagnari et al. 2017). Legumes exhibit multifaceted effects on nutrient accumulation and enhancement of soil fertility (Jhariya et al. 2021). Legumes form symbiotic relationships with rhizobia in the soil, fix atmospheric nitrogen (Quinones et al. 2022), facilitate nutrient cycling, and improve soil water retention (Mrunalini et al. 2022).

Crop diversification in agroecosystems, exemplified by intercropping, often results in greater productivity per unit area than monocropping (Li et al. 2020a, b). Intercropping is a sustainable agricultural technique with significant potential to promote crop yield, increase organic carbon reserves, and ameliorate soil degradation (Li et al. 2022). Intercropping, the practice of cultivating two or more crops in proximity, mitigates the risk of crop failure, enhances yield per unit of land and water, and optimizes resource utilization that would otherwise be monopolized by a single crop (Hossain et al. 2017). Multiple studies have illustrated the impact of crop diversity in intercropping on ecological processes, particularly the correlation between crop production and diversity, carbon sequestration, and carbon emissions (Chai et al. 2021). Intercropping establishes a robust ecological agriculture system by diminishing reliance on external hazardous chemical inputs and optimizing the use of natural resources (Kumawat et al. 2022).

The legume-based intercropping systems offer numerous agroecological benefits: It promotes crop diversification and farm nitrogen self-sufficiency (Tosti et al. 2023), enhances soil health (Ditzler et al. 2021), manages pests and weeds (Daryanto et al. 2020), and achieves yield stability (Koskey et al. 2022). Moreover, legume-based intercropping systems might mitigate the challenges of climate change by curtailing greenhouse gas emissions (Iannetta et al. 2021) and diminishing the carbon footprints of nonlegume cropping systems (Mrunalini et al. 2022). Yang et al.

(2023) discovered that a long-term intercropping system with legumes consistently enhanced productivity and improved soil fertility, chiefly through increased soil nitrogen and carbon resulting from more rhizodeposits and residual inputs. This may enhance net carbon sequestration and promote consistent carbon production by altering the composition and function of soil microbial communities (Wang et al. 2021). Moreover, NH_3 volatilization loss postfertilization was effectively mitigated through intercropping, which enhanced the absorption and utilization of soil inorganic nitrogen, resulting in reduced soil nitrate accumulation and diminished leaching (Long et al. 2021). Legume-based intercropping systems may contribute to achieving the United Nations' sustainable development goals by improving agricultural output, food nutritional quality, and soil health (Kumawat et al. 2022).

This chapter emphasizes the crucial importance of legume-based intercropping systems in increasing the resiliency to climate change. Legume intercropping offers numerous advantages, including enhanced soil nutrient circulation, improved water use efficiency, contribution yield stabilization, enhancing soil carbon sequestration, conservation of soil and water from erosion, disruption of pest and disease life cycles, and reduction of greenhouse gas emissions from the soil.

The Detrimental Effects of Climate Change

Factors Contributing to Climate Change

The IPCC (2014) defines climate change as a "change in the state of the climate that can be identified (e.g., by using statistical tests) through alterations in the mean and variability of its properties, persisting for an extended duration, typically decades or longer." Climate change has been induced by an array of substances termed greenhouse gases (GHGs), which have become increasingly prominent in the atmosphere, including carbon dioxide (CO_2), nitrous oxide (N_2O), methane (CH_4), ozone (O_3), and chlorofluorocarbons (CFCs) (Blasing 2010). The World Resource Institute (2000) reported that CO_2 emissions constitute 77% of GHGs, while NH_4 and N_2O emissions account for 14% and 8%, respectively, with CFCs and other gases comprising 1%. Despite being emitted in lesser quantities than CO_2, CH_4, and N_2O possess a considerably greater global warming potential (Chataut et al. 2023).

Additionally, it has been documented that CO_2 concentration surpassed 400 ppm in 2013 and is anticipated to hit 700 ppm by the century's conclusion (Prentice et al. 2001). Moreover, the combustion of fossil fuels markedly elevates CO_2 concentrations and other GHGs, resulting in the thickening of atmospheric layers that become ensnared, thereby altering the Earth's thermal regulation (Singh and Khanna 2021). The IPCC (2018) has reported that forestry and agricultural land use contribute roughly 6% and 14% of GHGs emissions, respectively, leading to biodiversity loss (Knox et al. 2016).

Climate Change and Global Temperatures

The IPCC's Fifth Assessment Report indicates that global mean surface temperatures are projected to increase by 1.4–5.8 °C by the century's end relative to 1990, cautioning that climate change has been occurring and will persist imminently (IPCC 2014). Global warming is characterized by a general rise in surface air temperatures, alterations in precipitation patterns, and a disturbance of the global hydrologic cycle, leading to an increased occurrence of extreme weather events such as droughts, floods (Mo et al. 2013), and heat or cold waves (van der Wiel and Bintanja 2021). Extreme weather events adversely affect crop productivity in irrigated and rain-fed agriculture (Adamgbe and Ujoh 2013). De Costa (2011) indicated that global climate models forecast a 1–6 °C rise in the average ambient temperature by the conclusion of the twenty-first century.

Climate Change and Hydrological Resources

The IPCC's Fourth Assessment Report indicated that "the magnitude of stress on water resources is anticipated to escalate due to climate change, alongside future population growth and urbanization resulting from land-use change" (IPCC 2012). Teixeira et al. (2013) noted that climate change circumstances are anticipated to exacerbate pressures from increased drought, salinity, waterlogging, and mineral toxicity. Fluctuations in temperature and precipitation may influence runoff and water availability (Cisneros et al. 2014). As much as one-fifth of the global population may face acute freshwater shortages or deteriorating quality in the future (Schiermeier 2014).

Climate Change and Pathogens

Climate change is anticipated to elevate the danger of plant infections by pathogens. It is posited that any anticipated yield enhancements over the next 50 years will be mitigated by disease pressures modified by climate change stemming from both existing and emerging diseases (Chaloner et al. 2021). Climate change is expected to increase plant pathogen infections by altering pathogen evolution, influencing host-pathogen interactions and vector physiology, and facilitating the emergence of new pathogen strains (Velasquez et al. 2018). Climate and ecological alterations, along with modern land management practices marked by monocultures and high-density crops, have likely facilitated the creation and adaptability of plant diseases that can proliferate beyond their usual geographical limits (Cohen and Leach 2020). Motha (2011) also indicated that extreme weather correlated with increased insects, pests, and plant diseases.

Climate Change and Agricultural Production

Climate change adversely influences agriculture by affecting crop productivity and stability (Wheeler and Von Braun 2013). Moreover, the FAO (2012) projected a 50% rise in food consumption and a 30% increase in irrigation water usage by 2030. The yield of main cereal crops (wheat and maize) is anticipated to be influenced by forthcoming climate change. Wheat production fluctuated significantly due to global warming, with anticipated losses reaching up to 6% for each degree Celsius increase in temperature (Miller et al. 2019). Moreover, extreme temperatures negatively impact wheat productivity during essential growth phases of the crop (Talukder et al. 2014). Martre et al. (2015) indicate elevated temperatures diminish spike production and heighten disease susceptibility, potentially reducing the final wheat yield (Barlow et al. 2015). Hernandez-Ochoa et al. (2018) forecast a drop in Mexico's national wheat production by 7% and 8% under the RCP4.5 and RCP8.5 climate scenarios, respectively. Zhao et al. (2017) forecasted a 7% reduction in maize productivity for each degree Celsius increase in global mean temperature. Kogo et al. (2019) indicate that maize production could decline by 8–38% by the conclusion of the twenty-first century under RCP4.5 and RCP8.5 scenarios. Choruma et al. (2022) employed the RCP8.5 climate scenario from the MIROC model to forecast a 24% reduction in maize yield from 2070 to 2099. Ouda and Zohry (2024a, b) reported a reduction in maize productivity in Egypt by 10% under SSP1-2.6 and 25% under SSP5-8.5 over the period 2080–2099. Conversely, wheat productivity will decrease by 16% and 22% under the corresponding scenarios throughout the same timeframe.

Climate Change and Agricultural Soils

The macro- and micronutrient in the soil provided to developing plants are essential for ensuring food security and other ecosystem services. Climate change may lead to soil degradation as climate is the principal determinant of soil formation (Karmakar et al. 2016). Alterations in precipitation, temperature, atmospheric nitrogen deposition, and elevated CO_2 levels due to climate change affect soil's chemical, physical, and biological processes (IPCC 2007). Climate change can impact soil functioning by altering organic carbon conversions and nitrogen cycling due to modified moisture and soil temperature (Hamidov et al. 2018). Coyle et al. (2016) assert that elevated temperatures and diminished soil moisture associated with warming or drought led to decreased carbon storage. Moreover, climate change's impact on soil moisture levels and vegetation cover may promote soil erosion (Borrelli et al. 2020). Climate change is anticipated to influence the quantity and intensity of rainfall and its spatiotemporal distribution, impacting soil erosion (Ziadat and Taimeh 2013).

Climate change is anticipated to increase air temperature, leading to heightened soil evapotranspiration losses and a decline in soil fertility and quality. It also reduces the moisture available to cultivated crops, impacting agricultural production (Mandal and Roy 2024). A rise in evapotranspiration rates may elevate the capillary movement of water and solutes from groundwater to the root zone, increasing salinity (Várallyay 2010).

Climate Change and GHGs Emissions from Soil

A significant quantity of organic carbon is sequestered in the soil, through which approximately 10% of atmospheric CO_2 circulates annually (Tamocai et al. 2009). Multiple factors influence the release of GHGs from the soil: soil texture, pH value, and organic carbon content. Gaillard et al. (2016) assert that soil texture significantly influences N_2O emissions; soils characterized by coarse texture and elevated moisture content are more prone to emit CO_2 compared to clay-textured soils (Sun et al. 2014). Xie et al. (2013) found that sandy soils exhibited more significant fluxes of CH_4 and N_2O compared to clayey soils, whereas silt loam soils exhibit the most significant N_2O emissions compared to sandy loam and loamy sand soils (Gaillard et al. 2016). Serrano-Silva et al. (2011) indicated that using nitrogen fertilizer, particularly urea, elevates soil pH and enhances CO_2 emissions. N_2O emissions may increase when soil pH approaches neutrality but decrease at elevated pH levels (Mørkved et al. 2007).

Previous studies indicate that soil temperature and moisture content influence CO_2 emissions (Wang et al. 2020a, b), with elevated temperatures hastening the decomposition of soil organic matter, thereby increasing the release of CO_2 and CH_4 into the atmosphere due to enhanced soil respiration (Kuzyakov et al. 2019). Cole et al. (2018) demonstrated that elevated temperatures enhance CO_2 emissions from soils into the atmosphere, contributing to further temperature increases. Moreover, inappropriate fertilizer application can markedly elevate GHGs emissions, particularly N_2O (Groenigen 2010). Ventura et al. (2005) indicated that nitrogen fertilizer enhances N_2O generation by supplying a substrate for microbial nitrogen transformation via nitrification and denitrification. The application of urea ($CO(NH_2)_2$) to the soil, in the presence of water and urease enzymes, converts to NH_4^+, hydroxyl ions (OH^-), and bicarbonate (HCO_3^-). The bicarbonate is subsequently decomposed into CO_2 and water (Snyder et al. 2009), with CO_2 being emitted into the atmosphere.

Consequently, as alterations in the Earth's climate have become increasingly evident, developing more resilient agroecosystems is imperative.

Legume-Based Intercropping Systems to Enhance Resilience to Climate Change

Climate change is a complex issue that necessitates developing and implementing climate-resilient methods that optimize the use of available natural resources in food production. Cardoson et al. (2013) assert that climate-resilient strategies enhance crop diversity to augment soil characteristics. Crop diversification techniques enhance net crop yield and improve soil health by fostering interspecific interactions among crop species above and below ground (Rakshit et al. 2018). Aboveground diversification enhances the canopy's ability to catch heat and light, while belowground diversification improves the utilization of water and soil macro- and micronutrients by augmenting soil microbial activity (King and Hofmockel 2017). Reports indicate numerous advantages derived from the implementation of a legume-based intercropping system as follows:

Soil Nutrient Circulation

Nutrient cycling in the soil is directly influenced by climate change via modified moisture and temperature patterns (Hamidov et al. 2018), altered carbon and nitrogen dynamics, reduced soil biodiversity (Benbi and Kaur 2009), and diminished soil fertility. Elevated air temperatures due to climate change affect root zone temperature and moisture conditions, which predominantly dictate nutrient availability to plants, root growth and development, carbon allocation to roots, and nutrient absorption (Mondal 2014). Legume-based intercropping systems can alleviate the impact of climate change by supplying nutrients to nonleguminous plants through root secretions or arbuscular mycorrhizal fungal networks (Adomako et al. 2022).

Soil nitrogen content was reported to increase under legume-based intercropping systems. Ouda et al. (2020) indicated that maize production was enhanced when intercropped with cowpea, attributed to an increase in soil nitrogen content relative to the yield from maize monoculture (Fig. 12.1a). Similarly, incorporating faba bean intercropping with sugar beet systems on sandy soil enhanced soil nitrogen content by 9% relative to the levels observed under sole sugar beet cultivation (Ouda and Zohry 2023) (Fig. 12.1b).

Soil pH is acknowledged as the primary variable that affects biochemical processes in the soil (Philippot et al. 2024). Legumes can alter the soil pH in the rhizosphere, establishing a microenvironment conducive to nutrient availability for

Fig. 12.1 Cowpea intercropping system with maize (**a**) and faba bean intercropped with sugar beet (**b**) in Egypt

intercrops (Goyal and Habtewold 2023). Additionally, hydrogen gas is produced as a by-product during biological nitrogen fixation, facilitating bacterial growth in the rhizosphere (Li et al. 2020a, b). Ferguson et al. (2013) illustrated that legumes can more efficiently extract calcium and magnesium from the soil compared to cereals, substituting these elements with hydrogen, which results in the removal of OH^- ions and an increase in H^+, thereby lowering soil pH and inhibiting N_2O gas emissions (Hénault et al. 2019).

Climate change is anticipated to fundamentally affect the variety of soil microbes and their communities, which are crucial for driving carbon and nutrient cycles, thereby altering the stability of soil carbon stocks and nutrient availability for plants (Yu et al. 2018). Homulle et al. (2022) suggest that intercropped plants may enhance nutrient uptake capability by augmenting the number of absorptive roots or the density of root hairs, hence optimizing the absorption of water and nutrients. The legume-based intercropping systems and root exudates foster a positive rhizobacterial population, enhancing soil quality and the rhizosphere environment for improved resource acquisition (Zhao et al. 2022). Shen et al. (2024) indicated that intercropping leguminous crops with tiger nut markedly enhanced soil organic content and alkali-hydrolyzable nitrogen, attributable to an increase in bacterial populations involved in soil nutrient cycling, including rhizobia, phosphate-solubilizing, and potassium-solubilizing bacteria (Zhang et al. 2018).

Soil enzyme activity is reportedly affected by climate change as these enzymes are highly susceptible to variations in temperature and humidity and their interactions with the substrates they decompose (Luo et al. 2022). The legume-based intercropping systems enhanced the enzymatic activities of soil microorganisms (Kumar et al. 2021), which are crucial mediators of soil biological processes and vital for maintaining soil quality and fertility (Curtright and Tiemann 2021). In intercropping, plant roots exude enzymes, including phosphatases, phytase, and carboxylates (Rojas-Downing et al. 2017), which enhance the mobilization and utilization of nutrients such as phosphorus, iron, manganese, and zinc (Marschner 2012) through the decomposition of organic residues (Xiao et al. 2013). The subterranean direct mutual benefits in legume-cereal intercropping systems include enhanced nitrogen transfer to cereals and improved bioavailability of Fe and Zn to legume crops (Stagnari et al. 2017). Chongloi and Sharma (2018) observed an enhancement in the levels of N, P, K, Zn, and Fe in oats when intercropped with peas, in contrast to sole oat cultivation. Incorporating biological carbon and nitrogen using a legume-based intercropping systems significantly enhanced soil enzyme activity and nutrient cycling, augmenting soil fertility (Li et al. 2020a, b). Liu et al. (2024) report that soil organic matter and total nitrogen in a legume-maize intercropping system increased significantly by 4.4–14.3% compared to monoculture.

Additionally, the mobilization of soil organic P by legume species occurs via the release of carboxylates, the enhancement of phosphatase activity, and alterations in root morphology and physiology, which facilitate phosphorus acquisition by cereals in intercropping systems with legumes (Wang et al. 2020a, b). Shen et al. (2023) documented an elevation in accessible P within the inter-root soils of a soybean-maize intercropping system. Cuartero et al. (2022) similarly observed that in cowpea intercropped with a melon system, there are significantly higher amounts of accessible P than monocropping. Xue et al. (2016) indicated that in a faba bean intercropping system with wheat, the concentrations of carboxylates in the faba bean rhizosphere were 10–20 times greater than those in wheat, demonstrating a significant ability for soil P mobilization and enhancement of P uptake by wheat facilitated by faba bean. Additionally, Ren et al. (2021) indicated that soybeans may release protons and facilitate the mobilization of insoluble soil P due to rhizosphere acidification through the exudation of malate and citrate, hence enhancing soil P availability and its uptake by maize. Guo et al. (2021) assert that intercropping peanuts with maize enhances beneficial soil bacteria, optimizes phosphonate and phosphinate metabolism, and increases P availability for maize. Furthermore, intercropping cowpeas with maize improved soil health by increasing N_2 fixation and facilitating the absorption of soil N by cereals (Boukar et al. 2013).

Enhance Water Retention and Accessibility

The hydrophysical properties of soil are significantly affected by fluctuations in climatic parameters, including seasonal temperatures and precipitation intensities, which influence the soil water regime (Horel et al. 2014). Climate change significantly influences soil water dynamics (Jasper et al. 2006). Intercropping systems that include legumes can markedly enhance soil physic hydraulic parameters and refine soil structure (Shen et al. 2023). Soil water-holding capacity is a fundamental soil function in ecosystems (Werner et al. 2020) and enhances soil resilience to increased climate variability (Abdallah et al. 2021). Crop yield and soil organic matter are substantially connected (Yuan et al. 2022). Moreover, soil organic matter is significantly correlated with water retention ability, absorbing up to 20 times its weight in water (Sun et al. 2021). Shen et al. (2024) reported a substantial increase in soil water content while intercropping tiger nuts with legumes. Shen et al. (2023) similarly observed an elevation in soil water content when maize was intercropped with soybean, attributed to enhanced nitrogen fixation and the production of acidic metabolites from the intercropped legumes, resulting in a reduction of soil pH (Liu et al. 2023). These acids are primary osmotic pressure regulators, enhancing soil water content (Rezig et al. 2013).

Climate change-induced water deficit adversely affects crop development although these impacts can be mitigated through well-designed legume intercropping systems (Gitari et al. 2020). Intercropping can modify specific plant characteristics related to water acquisition, including root distribution (Makumba et al. 2009) and the depth of water sourcing (Sekiya and Yano 2004). Specific investigations indicated that a legume-based intercropping system with maize cultivated under water stress yielded results comparable to or exceeding those of sole maize (Steward et al. 2019).

Mandal and Roy (2024) assert that rising air temperatures due to climate change are anticipated to increase soil evaporation losses, diminishing the moisture available to farmed crops. Intercropping systems have been shown to provide multiple advantages, such as reduced runoff, improved utilization of available soil moisture, and increased productivity per unit of applied irrigation water, thereby enhancing water use efficiency (Blum 2011; Yang et al. 2011; Rahman et al. 2017; Chen et al. 2018). Intercropping cowpeas with maize achieves a compensating soil water effect, as Ouda et al. (2020) noted in Egypt when irrigation was administered to the system in the same quantity as that for pure maize cultivation. Thus, cowpea plants extract water from the superficial soil layers, while maize plants get water from deeper strata. Post-harvest of cowpea, intercropped maize persisted in utilizing soil moisture that cowpea would typically utilize. This led to a cowpea yield comparable to its monoculture and a maize yield exceeding its monoculture. Enhancing water use efficiency in intercropping systems results from water conservation via early foliage cover and expanded leaf area (Maitra et al. 2023), thereby fostering more excellent vegetation cover that diminishes soil evaporation and elevates crop transpiration and soil water content (Rezig et al. 2013). Chimonyo et al. (2016) observed a 55% increase in water use efficiency in a sorghum and cowpea intercropping system compared to single cropping. Land use efficiency improved by 17–53% in the intercropping system relative to mono crops. Choudhary and Kumar (2016) showed a 15–55% enhancement in water use efficiency in maize-legume intercropping systems compared to sole cropping.

Additionally, legume-based intercropping systems enable deep-rooted legumes to extract water and nutrients from deeper soil strata. At the same time, shallow-rooted intercrops can access resources from upper soil layers, as demonstrated in the pigeon pea intercropping system with finger millet (Singh et al. 2020a, b). Legumes can enhance soil micro- and macropores, improving water infiltration to greater root zone depths (Kumar and Goh 2016).

Chen et al. (2018) observed that when shorter-season peas were intercropped in alternating strips with longer-season maize, the peas predominantly extracted soil water from the uppermost soil layers. The maize plants extracted water from the deepest strata of the neighboring pea strips. The intercropped maize acquired compensatory soil moisture from the pea strips post-harvest, completely recovering from the growth inhibition experienced during the concurrent growth phase (Li et al. 2014). Furthermore, Chai et al. (2013) documented a nearly 95% enhancement in water use efficiency in a maize intercropping system with pea compared to a pure pea stand. Chen et al. (2015) also showed a 22% enhancement in water use efficiency in a maize intercropping system with pea compared to sole maize and pea cultivation.

Intercropping tall plants with shorter plants has been shown to mitigate the effects of drought and heat stress by providing partial shading and decreasing wind speed (Raza et al. 2020), reducing soil water loss. The intercropping system of potatoes and legumes led to a notable reduction in soil temperature by 7.3 °C and an increase in soil water content by 38% (Nyawade et al. 2019).

Contribution to Yield Stabilization

Wheeler and Von Braun (2013) predict that climate change will negatively affect crop productivity and stability. Several research assessed the impact of prospective climate scenarios on the productivity of legume-based intercropping systems. Aminata et al. (2023) conducted a study in Burkina Faso assessing the impact of cowpea intercropping with sorghum on climate change resilience, utilizing the STICS model, two CO_2 emission scenarios, and five potential climatic models for the mid-century. Their findings indicated that intercropping cowpea with sorghum stabilized grain production, exhibiting lower coefficients of variation compared to monoculture systems, suggesting that intercropping enhances crop resistance to climate change. In a separate study conducted in Germany, Yu et al. (2025) employed the agroecosystem model MONICA to project the impact of two emission scenarios, RCP2.6 and RCP8.5, on the yield of a winter wheat-soybean relay intercropping system. Their findings indicated a 9% production enhancement in the intercropping system relative to sole cropping in the future. In Egypt, Ouda and Zohry (2024a, b) employed the SALTMED model to simulate maize yield in its intercropping system with cowpea, utilizing CMIP6 mean projections for two future scenarios (SSP1-2.6 and SSP5-8.5) across two time intervals (2060–2079 and 2080–2099). Their findings demonstrated that the yield losses of intercropped maize were inferior to those of sole maize under the two future scenarios over both time periods. The table's results indicated that intercropping cowpea with maize serves as an adaptation strategy to climate change, enhancing the resilience of nonlegume crops to climate-related stress.

Contribution to Carbon Sequestration

Climate change, particularly global warming, can influence organic carbon decomposition by regulating microorganisms, enzyme activity, and soil respiration (Conant et al. 2011). The stability of soil organic carbon mostly depends on temperature and land use (Kurganova et al. 2019). Wang et al. (2022) discovered that warming led to more significant proportionate losses of soil organic carbon in the topsoil compared to the subsoil, especially in high-latitude, carbon-rich ecosystems. Agricultural management practices, such as legume-based intercropping systems, might alleviate the impacts of climate change by enhancing the sequestration of soil organic carbon (Haddaway et al. 2015). Intercropping is a conventional diversified cropping system with significant carbon sequestration potential (Liang and Zhu 2021) owing to the comprehensive usage of agroecosystem resources (Stomph et al. 2020). Liu et al. (2024) indicated that intercropping legumes with maize enhanced carbon input by 17–79% at the maize V9 stage and by 81–140% at the R6 stage relative to maize monoculture. Improving soil fertility in legumes intercropping systems with cereals is chiefly linked to augmented soil organic carbon storage and nutrient turnover (Abera and Gerkabo 2021). Excessive biomass production in intercropping systems enhances carbon input to the soil, which contributes to the creation of soil organic matter (Wang et al. 2022). Stagnari et al. (2017) demonstrated enhanced soil organic carbon and humus levels in a legume-based intercropping system, wherein the legume crop can elevate soil organic carbon and nitrogen (Li et al. 2020a, b). The soil organic carbon content in a soybean intercropping system with maize at a 3:2 row ratio was 4% more than that of sole maize in the upper 20 cm soil layer (Bichel et al. 2016). Likewise, the inter-root soil organic content in the soybean intercropping system with sugarcane is 6% greater than that in sole sugarcane cultivation (Lian et al. 2019).

The legume-based intercropping systems can enhance soil organic matter (Li et al. 2020a, b). Abbady et al. (2016) reported in Egypt that a 17% increase in soil organic matter was observed under a pea intercropping system with wheat, in contrast to the levels recorded with pure wheat cultivation. Zohry et al. (2020) demonstrated that soil organic matter rose by 11% when the identical intercropping method was employed, in contrast to the levels observed with mono wheat cultivation.

Mitigating Erosion and Runoff

Mitigating soil degradation in the context of climate change through technology and agricultural practices that preserve ground cover is crucial for soil conservation (Karmakar et al. 2016). Intercropping maximizes ground coverage, minimizing the likelihood of runoff, soil erosion, and nutrient depletion (Nyawade et al. 2019). Implementing a legume-based intercropping system contributes to soil erosion control by mitigating the impact of precipitation on the soil surface (Maitra et al. 2021). Certain legume crops are superior at sustaining crop cover compared to others. Kariaga (2004) compared cowpeas intercropped with maize to beans intercropped with maize regarding soil erosion and concluded that cowpea was a superior cover crop to beans in mitigating soil erosion. Lithourgidis (2011) observed that legume-based intercropping systems resulted in diminished runoff and soil erosion, with cowpea intercropped with sorghum reducing runoff by 20–30% relative to solo sorghum planting and by 45–55% compared to sole cowpea planting. Furthermore, soil erosion was diminished by over 50% under the identical intercropping system compared to monocultures of sorghum and cowpea.

According to Dass and Sudhishir (2010), the intercropping of finger millet with black gram yielded the least runoff and minimized losses of soil, N, P, and K due to erosion, in contrast to sole cropping when planted in contour since black gram effectively covered the ground area. Furthermore, Singh et al. (2020a, b) noted a 24% reduction in soil loss with an intercropping system compared to monoculture. Singh et al. (2014) also showed a reduction in runoff of 24% and 19% under the intercropping systems of soybean with maize and pigeon pea with maize, respectively, compared to the values obtained from pure maize cultivation, attributed to the high canopy cover.

Disrupting Insect and Disease Cycles and Weed Proliferation

Climate change may exacerbate the distribution and intensity of pest outbreaks, impacting both the quality and quantity of agricultural yields (Maitra et al. 2019). The legume-based intercropping system is an ecosystem-oriented strategy that mitigates agricultural chemical pollution by decreasing the need for chemical pesticides (Singh and Adjeigbe 2002). Legume-based intercropping systems improve biodiversity and disrupt the life cycle of detrimental insect pests and diseases (Gontijo 2019). Maitra et al. (2019) asserted that diversifying cropping systems with legume-based intercropping enhances the prevalence of beneficial pest parasites and helps keep harmful pest populations below the economic threshold.

Loko et al. (2021) observed a significant decrease in termite infestation in maize when intercropped with peanut, soybean, and common bean, in contrast to maize monoculture. Moreover, Seran and Brintha (2010) demonstrated that intercropping maize with leguminous crops resulted in a reduction in detrimental maize pests, specifically the corn borer, budworm, stalk borer, and leafhopper, while simultaneously enhancing the population of beneficial insects. Hailu et al. (2018) discovered that intercropping maize with leguminous crops reduces S. frugiperda damage, which occurs primarily during the early growth stages of maize till tasseling, as certain legume plants offer nectar, pollen, shelter, and chemical protection for resident parasitoids (Dannon et al. 2010). Saritha and Visalakshi (2024) observed a reduction in the average percentage of maize plants affected by autumn armyworm in an intercropping system with cowpea relative to mono cultivation of maize. Hamd-Alla et al. (2014) in Egypt suggested that cowpea, when intercropped with maize, attracts beneficial insects, such as C. septempunctata, C. undecimpunctata, and Chrysopa pallens. These insects prey on aphids that infest maize tassels, resulting in significant output losses.

Furthermore, the increased canopy cover of legume-based intercropping diminishes weed biomass due to their suppressive impact. Banik et al. (2006) showed that the chickpea intercropping system with wheat led to a 70% reduction in weed biomass and population compared to sole wheat cultivation. Similarly, intercropping faba bean and lentil with winter oilseed rape diminished weed issues and decreased the prevalence of stem weevil (Cadoux et al. 2015). Moreover, Maitra and Ray (2019) stated that legume-based intercropping systems and their associated Rhizobium species impede the proliferation of soilborne diseases and create a complex environment that limits the reproductive capabilities of pathogens such as Fusarium spp. and Phytophthora spp.

Mitigation of GHGs Emissions

Approximately 5–20% of anthropogenic GHGs emissions can be mitigated globally through enhanced soil and crop management practices (Olsson et al. 2019). Crop rotation and intercropping systems can significantly reduce GHGs emissions from the soil by optimizing carbon and nitrogen management in agricultural ecosystems

(Cerrie et al. 2004). The legumes-based intercropping system may significantly contribute by delivering various agroecosystem services while adhering to sustainability principles (Kumar et al. 2019). Research indicates that GHGs emissions from legume crops per unit area are 5–7 times lower than those from other crops (Jeuffroy et al. 2013). Implementing legume-based intercropping systems can provide numerous advantages to agroecosystems, including reducing GHGs emissions, particularly N_2O and CO_2 (Kakraliya et al. 2018). The emission of N_2O is often greater from soils cultivated with nonleguminous crops than those with leguminous crops (Jensen et al. 2012). Senbayram et al. (2016) documented a 31% reduction in cumulative N_2O emissions in a faba bean intercropping system with wheat relative to N-fertilized sole wheat. Additionally, a 3-year monitoring study of GHGs emissions in a soybean intercropping system with maize demonstrated a 16–36% reduction in soil N_2O emissions compared to maize monocropping (Shen et al. 2018). Furthermore, a reduction in N_2O emissions has been documented for barley intercropping systems with field beans (Mariotti et al. 2015) and peas (Pappa et al. 2011) relative to the sole crop system.

Conclusions

Human activities contribute to GHGs emissions, which are the principal driver of climate change. Legume-based intercropping systems can mitigate GHGs emissions and diminish the carbon footprints of nonlegume cropping systems. Reports indicate that climate change is anticipated to exacerbate soil degradation, reduce water use efficiency, heighten yield losses, diminish soil carbon sequestration, escalate soil and water erosion, promote pest and disease infestations, and elevate GHGs emissions from the soil.

This chapter addressed the vital significance of legume-based intercropping systems in enhancing resilience to climate change and mitigating the aforementioned adverse effects.

Research indicates that legume-based intercropping systems promote soil nutrient cycling, evidenced by increased levels of N, P, K, Zn, and Fe in oats when intercropped with peas. Moreover, research indicates that augmenting soil organic matter in legume-based intercropping systems is strongly associated with enhanced water retention capacity. A significant rise in soil water content was noted during the intercropping of tiger nuts with legumes. Additionally, an increase in soil water content has been documented when maize is intercropped with soybean.

Simulation studies indicate that legume-based intercropping systems, specifically cowpea with sorghum, mitigates yield losses due to climate change, stabilizing grain production and enhancing crop resilience. A 9% increase in production was observed in a winter wheat-soybean relay intercropping system compared to sole cropping in the future. Moreover, in a cowpea-maize intercropping system, the yield losses of intercropped maize were less severe than those of sole maize in the context of future climate change.

It has been reported that legume-based intercropping system can enhance soil carbon sequestration. A 17% improvement in soil organic matter has been reported under a pea intercropping system with wheat, compared to the levels observed with sole wheat farming, contributing to soil carbon sequestration. Intercropping legumes with maize increased carbon input by 17–79% at the maize V9 stage and by 81–140% at the R6 stage compared to maize monoculture.

Legume-based intercropping systems aid to reducing runoff and soil erosion. Intercropping cowpeas and beans with maize helps mitigate soil erosion by establishing ground cover early in the season. Moreover, cowpea intercropped with sorghum decreases runoff by 20–30% compared to sole sorghum cultivation and by 45–55% relative to sole cowpea cultivation while also reducing soil erosion by over 50%.

Legume-based intercropping systems interrupt the life cycles of pests, diseases, and weeds. Intercropping maize with peanut, soybean, or common bean significantly reduced termite infestation compared to maize monoculture. Moreover, cowpea, when intercropped with maize, draws beneficial insects that target aphids infesting maize tassels, leading to considerable production reductions. The augmented canopy cover of legume-based intercropping systems can reduce weed biomass due to their inhibitory effects. The chickpea-wheat intercropping system resulted in a 70% decrease in weed biomass and population relative to monoculture wheat agriculture.

Legume-based intercropping systems have demonstrated a reduction in greenhouse gas emissions from the soil. A 31% decrease in cumulative N_2O emissions in a faba bean-wheat intercropping system was found. A reduction of 16–36% in soil N_2O emissions in a soybean-maize intercropping system was also documented. Additionally, a decrease in N_2O emissions has been seen for barley intercropping systems with field beans and peas.

Thus, a legume-based intercropping system may function as a sophisticated agroecosystem strategy to mitigate the effects of climate change, consequently enhancing agricultural sustainability and soil security across diverse agroecosystems.

References

Abbady K, El-Maaz E, Ahmed H, Zohry A (2016) Carbon sequestration as a function of intercropping management practice and different nitrogenous fertilizer types. J Soil Sci Agric Eng 7(8):565–586

Abdallah AM, Jat HS, Choudhary M, Abdelaty EF, Sharma PC, Jat ML (2021) Conservation agriculture effects on soil water holding capacity and water-saving varied with management practices and agroecological conditions: a review. Agronomy 11(9):1681. https://doi.org/10.3390/agronomy11091681

Abera G, Gerkabo H (2021) Effects of green manure legumes and their termination time on yield of maize and soil chemical properties. Arch Agron Soil Sci 67:397–409. https://doi.org/10.1080/03650340.2020.1733536

Adamgbe EM, Ujoh F (2013) Effect of variability in rainfall characteristics on maize yield in Gboko. Nigeria J Environ Prot 4:881–887

Adomako MO, Roiloa S, Yu FH (2022) Potential roles of soil microorganisms in regulating the effect of soil nutrient heterogeneity on plant performance. Microorganisms 10(12):2399. https://doi.org/10.3390/microorganisms10122399

Aminata G, Florian L, Salifou T, Myriam A (2023) Potential of cereal-legume intercropping systems in future climates: a modelling study in Burkina Faso. In: Sorghum in the 21th century: resiliency and sustainability in the face of climate change. Book of abstracts. CIRAD, Kansas State University, Sorghum International Development, IRD, CERAAS. Montpellier: CIRAD-Kansas State University, Résumé, p 289. Sorghum in the 21st century, global sorghum conference, Montpellier, France, 5 June 2023/9 June 2023

Banik P, Midya A, Sarkar BK, Ghose SS (2006) Wheat and chickpea intercropping systems in an additive series experiment: advantages and weed smothering. Eur J Agron 24:325–332

Barlow KM, Christy BP, O'leary GJ, Riffkin PA, Nuttall JG (2015) Simulating the impact of extreme heat and frost events on wheat crop production: a review. Field Crop Res 171:109–119

Benbi DK, Kaur R (2009) Modeling soil processes in relation to climate change. J Indian Soc Soil Sci 57:433–444

Bichel A, Oelbermann M, Voroney P, Echarte L (2016) Sequestration of native soil organic carbon and residue carbon in complex agroecosystems. Carbon Manag 7(5–6):261–270. https://doi.org/10.1080/17583004.2016.1230441

Blasing J (2010) Recent greenhouse gas concentrations. Carbon Dioxide Information Analysis Center, Washington, DC

Blum A (2011) Plant breeding for water-limited environment. Springer, New York. ISBN 978-1-4419-7490-7

Borrelli P, Robinson DA, Panagos P et al (2020) Land use and climate change impacts on global soil erosion by water (2015–2070). Proc Natl Acad Sci 117(36):21994–22001

Boukar O, Bhattacharjee R, Fatokun C, Kumar PL, Gueye B (2013) Cowpea. In: Singh M, Upadhyaya HD, Bisht IS (eds) Genetic and genomic resources of grain legume improvement. Academic, pp 137–155

Cadoux S, Sauzet G, Morison MV, Pontet C, Champolivier L, Robert C, Lieven J, Flénet F, Mangenot O, Fauvin P, Landé N (2015) Intercropping frost-sensitive legume crops with winter oilseed rape reduces weed competition, insect damage, and improves nitrogen use efficiency. OCL 22(3):D302

Cardoson EJBN, Vasconcellos RLF, Bini D, Miyauchi MYH, Santos CAD, Alves PRL, Paula AMD, Nakatani AS, Pereira JDM, Nogueira MA (2013) Soil health: looking for suitable indicators. What should be considered to assess the effects of use and management on soil health? Sci Agric 70(4):274–289

Cerrie CC, Bernoux M, Cerrie CEP, Feller C (2004) Carbon cycling and sequestration opportunities in South America: the case of Brazil. Soil Use Manag 20:248–254

Chai Q, Qin A, Gan Y, Yu A (2013) Higher yield and lower carbon emission by intercropping maize with rape, pea, and wheat in arid irrigation areas. Agron Sustain Dev 34:535–543. https://doi.org/10.1007/s13593-013-0161-x

Chai Q, Nemecek T, Liang C, Zhao C, Yu A, Coulter JA et al (2021) Integrated farming with intercropping increases food production while reducing environmental footprint. Proc Natl Acad Sci USA 118:e2106382118

Chaloner TM, Gurr SJ, Bebber DP (2021) Plant pathogen infection risk tracks global crop yields under climate change. Nat Clim Chang 11:710–715

Chataut G, Bhatta B, Joshi D, Subedi K, Kafle K (2023) Greenhouse gases emission from agricultural soil: a review. J Agric Food Res 11:100533. https://doi.org/10.1016/j.jafr.2023.100533

Chen G, Chai Q, Huang G, Yu A, Feng F, Mu Y, Kong X, Huang P (2015) Belowground interspecies interaction enhances productivity and water use efficiency in maize-pea intercropping systems. Crop Sci 55:420–428

Chen G, Kong X, Gan Y et al (2018) Enhancing the systems productivity and water use efficiency through coordinated soil water sharing and compensation in strip-intercropping. Sci Rep 8:10494. https://doi.org/10.1038/s41598-018-28612-6

Chimonyo VGP, Modi AT, Mabhaudhi T (2016) Water use and productivity of a sorghum cowpea–bottle gourd intercrop system. Agric Water Manag 165:82–91
Chongloi KL, Sharma KK (2018) Quality attributes of pea and oat in an intercropping system in rice fallows as influenced by integrated nutrient management. JPP SP1:1663–1667
Choruma DJ, Akamagwuna FC, Odume NO (2022) Simulating the impacts of climate change on maize yields using epic: a case study in the eastern cape province of South Africa. Agriculture 12(6):794. https://doi.org/10.3390/agriculture12060794
Choudhary VK, Kumar PS (2016) Productivity, water use and energy profitability of staggered maize–legume inter-cropping in the Eastern Himalayan Region of India. Proc Natl Acad Sci India Sect B Biol Sci 86(3):547–557
Cisneros JBE, Oki T, Arnell NW, Benito G, Cogley JG, Doll P, Jiang T, Mwakalila SS (2014) Freshwater resources. In: Field CB, Barros VR, Dokken DJ, Mach KJ, Mastrandrea MD, Bilir TE, Chatterjee M, Ebi KL, Estrada YO, Genova RC, Girma B, Kissel ES, Levy AN, MacCracken S, Mastrandrea PR, White LL (eds) Climate change 2014: impacts, adaptation, and vulnerability. Part A: global and sectoral aspects. Contribution of Working Group II to the Fifth Assessment Report of the Intergovernmental Panel on Climate Change. Cambridge University Press, Cambridge/New York, pp 229–269
Cohen SP, Leach JE (2020) High temperature-induced plant disease susceptibility: more than the sum of its parts. Curr Opin Plant Biol 56:235–241
Cole MB, Augustin MA, Robertson MJ, Manners JM (2018) The science of food security. npj Sci Food 2:14. https://doi.org/10.1038/s41538-018-0021-9
Conant RT, Ryan MG, Ågren GI, Birge HE, Davidson EA, Eliasson PE et al (2011) Temperature and soil organic matter decomposition rates—synthesis of current knowledge and a way forward. Glob Chang Biol 17(11):3392–3404
Coyle C, Creamer RE, Schulte RPO, O'Sullivan L, Jordan P (2016) A functional land management conceptual framework under soil drainage and land use scenarios. Environ Sci Pol 56:39–48. https://doi.org/10.1016/j.envsci.2015.10.012
Cuartero J, Pascual JA, Vivo JM, Özbolat O, Sánchez-Navarro V, Egea Cortines M et al (2022) A first-year melon/cowpea intercropping system improves soil nutrients and changes the soil microbial community. Agric Ecosyst Environ 328:107856. https://doi.org/10.1016/j.agee.2022.107856
Curtright AJ, Tiemann LK (2021) Intercropping increases soil extracellular enzyme activity: a meta-analysis. Agric Ecosyst Environ 319:Article 107489. https://doi.org/10.1016/j.agee.2021.107489
Dannon EA, Tamò M, Van Huis A, Dicke M (2010) Effects of volatiles from *Maruca vitrata* larvae and caterpillar-infested flowers of their host plant *Vigna unguiculata* on the foraging behavior of the parasitoid *Apanteles taragamae*. J Chem Ecol 36(10):1083–1091. https://doi.org/10.1007/s10886-010-9859-2
Daryanto S, Fu B, Zhao W, Wang S, Jacinthe P, Wang L (2020) Ecosystem service provision of grain legume and cereal intercropping in Africa. Agric Syst 178:102761. https://doi.org/10.1016/j.agsy.2019.102761
Dass A, Sudhishir S (2010) Intercropping in finger millet (Eleusine coracana) with pulses for enhanced productivity, resource conservation and soil fertility in uplands of southern Orissa. Indian J Agron 55:89–94
De Costa WAJM (2011) A review of the possible impacts of climate change on forests in the humid tropics. J Nat Sci Found Sri 39:281–302. https://doi.org/10.4038/jnsfsr.v39i4.3879
Ditzler L, van Apeldoorn DF, Pellegrini F, Antichi D, Bàrberi P, Rossing WAH (2021) Current research on the ecosystem service potential of legume inclusive cropping systems in Europe: a review. Agron Sustain Dev 41(2):26. https://doi.org/10.1007/s13593-021-00678-z
FAO (2012) Developing a climate-smart agriculture strategy at the country level: lessons from recent experience. Background paper for the second global conference on agriculture, food security and climate change. Food and Agriculture Organization of the United Nations, Rome
Ferguson BJ, Lin MH, Gresshoff PM (2013) Regulation of legume nodulation by acidic growth conditions. Plant Signal Behav 8(3):e23426

Gaillard R, Duval BD, Osterholz WR, Kucharik CJ (2016) Simulated effects of soil texture on nitrous oxide emission factors from corn and soybean agroecosystems in Wisconsin. J Environ Qual 45:1540–1548. https://doi.org/10.2134/jeq2016.03.0112

Gitari HI, Nyawade S, Kamau S, Raza MA, Karanja NN, Gachene CKK (2020) Revisiting intercropping indices with respect to potato-legume intercropping systems. Field Crop Res 258:Article 107957

Gontijo LM (2019) Engineering natural enemy shelters to enhance conservation biological control in field crops. Biol Control 130(4):155–163. https://doi.org/10.1016/j.biocontrol.2018.10.014

Goyal RK, Habtewold JZ (2023) Evaluation of legume-rhizobial symbiotic interactions beyond nitrogen fixation that help the host survival and diversification in hostile environments. Microorganisms 11(6):1454. https://doi.org/10.3390/microorganisms11061454

Groenigen WV (2010) Towards an agronomic assessment of N_2O emissions: a case study for arable crops. Eur J Soil Sci 61(6):903–913. https://doi.org/10.1111/j.1365-2389.2009.01217.x

Guo F, Wang M, Si T, Wang Y, Zhao H, Zhang X, Yu X, Wan S, Zou X (2021) Maize-peanut intercropping led to an optimization of soil from the perspective of soil microorganism. Arch Agron Soil Sci 67:1986–1999

Haddaway NR, Hedlund K, Jackson LE, Kätterer T, Lugato E, Thomsen IK, Söderström B (2015) What are the effects of agricultural management on soil organic carbon in boreo-temperate systems? Environ Evid 4:23. https://doi.org/10.1186/s13750-015-0049-0

Hailu G, Niassy S, Zeyaur KR, Ochatum N, Subramanian S (2018) Maize-legume intercropping and push-pull for management of fall armyworm, stemborers, and striga in Uganda. Agron J 110(6):1–10. https://doi.org/10.2134/agronj2018.02.0110

Hamd-Alla WA, Shalaby EM, Dawood RA, Zohry AA (2014) Effect of cowpea (*Vigan sinensis* L.) with maize (*Zea mays* L.) intercropping on yield and its components. Int Scholar Sci Res Innov 8(11):1170–1176

Hamidov A, Helming K, Bellocchi G, Bojar W et al (2018) Impacts of climate change adaptation options on soil functions: a review of European case-studies. Land Degrad Dev 29(8):2378–2389. https://doi.org/10.1002/ldr.3006

Hénault C, Bourennane H, Ayzac A et al (2019) Management of soil pH promotes nitrous oxide reduction and thus mitigates soil emissions of this greenhouse gas. Sci Rep 9:20182. https://doi.org/10.1038/s41598-019-56694-3

Hernandez-Ochoa IM, Asseng S, Kassie BT, Xiong W, Robertson R, Pequeno DNL, Sonder K, Reynolds M, Babar MA, Milan AM, Hoogenboom G (2018) Climate change impact on Mexico wheat production. Agric For Meteorol 263:373–387. https://doi.org/10.1016/j.agrformet.2018.09.008

Homulle Z, George TS, Karley AJ (2022) Root traits with team benefits: understanding belowground interactions in intercropping systems. Plant Soil 471:1–26. https://doi.org/10.1007/s11104-021-05165-8

Horel A, Lichner L, Alaoui A, Czachor H, Nagy V, Tóth E (2014) Transport of iodide in structured clay–loam soil under maize during irrigation experiments analysed using HYDRUS model. Biologia 69:1531–1538

Hossain MS, Kader MA, Islam N (2017) Multiple cropping for sustainable and exaggerated agricultural production system. J Biosci Agric Res 14(02):1202–1209

Iannetta P, Hawes C, Begg G, Maab H, Ntatsi G, Savvas D et al (2021) A multifunctional solution for wicked problems: value-chain wide facilitation of legumes cultivated at bioregional scales is necessary to address the climate-biodiversity nutrition nexus. Front Sustain Food Syst 5:692137. https://doi.org/10.3389/fsufs.2021.692137

IPCC (2012) Managing the risks of extreme events and disasters to advance climate change adaptation. A special report of Working Groups I and II of the Intergovernmental Panel on Climate Change. (Field CB, Barros V, Stocker TF, Qin D, Dokken DJ, Ebi KL, Mastrandrea MD, Mach KJ, Plattner G-K, Allen SK, Tignor M, Midgley PM (eds)). Cambridge University Press, Cambridge/New York, 582 pp

IPCC (2014) Climate change 2014: impacts, adaptation, and vulnerability. Part A: global and sectoral aspects. Summary for policymakers. Contribution of Working Group II to the Fifth

Assessment Report of the Intergovernmental Panel on Climate Change. Cambridge University Press, Cambridge/New York, pp 1–32
IPCC (2018) Summary for policymakers of IPCC special report on global warming of 1.5 C approved by governments. Available at Ipcc.ch. https://www.ipcc.ch/2018/10/08/summary-for-policymakers-of-ipcc-special-report-on-lobal-warming-of-1-5c-approved-bygovernments/
Jasper K, Calanca P, Fuhrer J (2006) Changes in summertime soil water patterns in complex terrain due to climatic change. J Hydrol 327:550–563. https://doi.org/10.1016/j.jhydrol.2005.11.061
Jensen ES, Peoples MB, Boddey RM, Gresshoff PM, Hauggaard Nielsen H, Alves BJ, Morrison MJ (2012) Legumes for mitigation of climate change and the provision of feedstock for biofuels and biorefineries. A review. Agron Sustain Dev 32(2):329–364
Jeuffroy MH, Baranger E, Carrouée B, Chezelles ED, Gosme M, Hénault C, Schneide A, Cellier P (2013) Nitrous oxide emissions from crop rotations including wheat, oilseed rape and dry peas. Biogeosciences 10(3):1787–1797
Jhariya MK, Banerjee A, Meena RS, Kumar S, Raj A (2021) Sustainable intensification for agroecosystem services and management: an overview. Springer, Singapore. https://doi.org/10.1007/978-981-16-3207-5_1
Kakraliya SK, Singh U, Bohra A, Choudhary KK, Kumar S, Meena RS, Jat ML (2018) N and legumes: a meta-analysis. In: Meena et al (eds) Legumes for soil health and sustainable management. Springer, Singapore
Kariaga BM (2004) Intercropping maize with cowpeas and beans for soil and water management in Western Kenya. In: Proceedings of the 13th international soil conservation organization conference on conserving soil and water for society, Brisbane, July, pp 1–5
Karmakar R, Das I, Dutta D, Rakshit A (2016) Potential effects of climate change on soil properties: a review. Sci Int 4:51–73
King AE, Hofmockel KS (2017) Diversified cropping systems support greater microbial cycling and retention of carbon and nitrogen. Agric Ecosyst Environ 240:66–76
Knox J, Daccache A, Hess T, Haro D (2016) Meta-analysis of climate impacts and uncertainty on crop yields in Europe. Environ Res Lett 11:113004
Kogo B, Kumar L, Koech R, Langat P (2019) Modelling impacts of climate change on maize (*Zea mays* L.) growth and productivity: a review of models, outputs and limitations. J Geosci Environ Protect 7:76–95. https://doi.org/10.4236/gep.2019.78006
Koskey G, Leoni F, Carlesi S, Avio L, Bàrberi P (2022) Exploiting plant functional diversity in durum wheat–lentil relay intercropping to stabilize crop yields under contrasting climatic conditions. Agronomy 12(1):210. https://doi.org/10.3390/agronomy12010210
Kumar K, Goh KM (2016) Crop residues and management practices: effects on soil quality, soil nitrogen dynamics, crop yield and nitrogen recovery. Adv Agron 68:198–279
Kumar S, Meena RS, Jakhar SR, Jangir CK, Gupta A, Meena BL (2019) Adaptation strategies for enhancing agricultural and environmental sustainability under current climate. In: Meena RS (ed) Sustainable agriculture. Scientific publisher, Jodhpur, pp 226–274
Kumar A, Blagodaskaya E, Dippold MA, Temperton VM (2021) Positive intercropping effects on biomass production are species-specific and involve rhizosphere enzyme activities: evidence from a field study. Soil Ecol Lett 4(4):444–453. https://doi.org/10.1007/s42832-021-0108-0
Kumawat A, Bamboriya SD, Meena RS, Yadav D, Kumar A, Kumar S, Raj A, Pradhan G (2022) Legume-based inter-cropping to achieve the crop, soil, and environmental health security. In: Ram SM, Kumar S (eds) Advances in legumes for sustainable intensification. Academic, pp 307–328
Kurganova I, Semenov VM, Kudeyarov VN (2019) Climate and land use as key factors of the stability of organic matter in soils. Dokl Biol Sci 489:189–192
Kuzyakov Y, Horwath WR, Dorodnikov M, Blagodatskaya E (2019) Review and synthesis of the effects of elevated atmospheric CO_2 on soil processes: no changes in pools, but increased fluxes and accelerated cycles. Soil Biol Biochem 128:66–78
Li L, Tilman D, Lambers H, Zhang FS (2014) Plant diversity and overyielding: insights from belowground facilitation of intercropping in agriculture. New Phytol 203:63–69. https://doi.org/10.1111/nph.12778

Li C, Hoffland E, Kuyper TW, Yu Y, Zhang C, Li H et al (2020a) Syndromes of production in intercropping impact yield gains. Nat Plant 6:653–660

Li C, Hoffland E, Kuyper TW, Yu Y, Li H, Zhang C, Zhang F, Wopkevan der W (2020b) Yield gain, complementarity and competitive dominance in intercropping in China: a meta-analysis of drivers of yield gain using additive partitioning. Eur J Agron 113:Article 125987. https://doi.org/10.1016/j.eja.2019.125987

Li H, Guan Q, Sun Y, Wang Q, Liang L, Ma Y, Du Q (2022) Spatiotemporal analysis of the quantitative attribution of soil water erosion in the upper reaches of the Yellow River Basin based on the RUSLE-TLSD model. Catena 212:106081

Lian T, Mu Y, Jin J, Ma Q, Cheng Y, Cai Z, Nian H (2019) Impact of intercropping on the coupling between soil microbial community structure, activity, and nutrient-use efficiencies. PeerJ 7:Article e6412. https://doi.org/10.7717/peerj.6412

Liang C, Zhu X (2021) The soil Microbial Carbon Pump as a new concept for terrestrial carbon sequestration. Sci China Earth Sci 64(4):545–558. https://doi.org/10.1007/s11430-020-9705-9

Lithourgidis A, Dordas C, Damalas C, Vlachostergios D (2011) Annual intercrops: an alternative pathway for sustainable agriculture. Aust J Crop Sci 5:396–410

Liu HY, Huang N, Zhao CM, Li JH (2023) Responses of carbon cycling and soil organic carbon content to nitrogen addition in grasslands globally. Soil Biol Biochem 186:109164. https://doi.org/10.1016/j.soilbio.2023.109164

Liu R, Yang L, Zhang J, Zhou G, Chang D, Chai Q, Cao W (2024) Maize and legume intercropping enhanced crop growth and soil carbon and nutrient cycling through regulating soil enzyme activities. Eur J Agron 159:127237. https://doi.org/10.1016/j.eja.2024.127237

Loko YLE, Toffa J, Orobiyi A, Gavoedo Dé M, Dansi A, Tamò M, Roisin Y (2021) Termites and maize crops: assemblage composition, damage level, and varietal sensitivity in contrasting agro-ecological zones of the Republic of Benin. Int J Pest Manag 5(2):1–18. https://doi.org/10.1080/09670874.2021.1969472

Long G, Li L, Wang D, Zhao P, Tang L, Zhou Y et al (2021) Nitrogen levels regulate intercropping-related mitigation of potential nitrate leaching. Agric Ecosyst Environ 319:107540

Luo L, Guo M, Wang E, Yin C, Wang Y, He H, Zhao C (2022) Effects of mycorrhiza and hyphae on the response of soil microbial community to warming in eastern Tibetan plateau. Sci Total Environ 837:Article 155498

Maitra S, Ray DP (2019) Enrichment of biodiversity, influence in microbial population dynamics of soil and nutrient utilization in cereal-legume inter-cropping systems: a review. Int J Bioresour Sci 6:11–19

Maitra S, Palai JB, Manasa P, Kumar DP (2019) Potential of inter-cropping system in sustaining crop productivity. Int J Agric Environ Biotechnol 12(1):39–45. https://doi.org/10.30954/0974-1712.03.2019.7

Maitra S, Hossain A, Brestic M, Skalicky M, Ondrisik P, Gitari HI, Brahmachari K, Shankar T, Bhadra P, Palai JB, Jana J, Bhattacharya U, Duvvada SK, Lalichetti S, Sairam M (2021) Intercropping—a low input agricultural strategy for food and environmental security. Agronomy 11:343. https://doi.org/10.3390/agronomy11020343

Maitra S, Sahoo U, Sairam M, Gitari HI, Rezaei-Chiyaneh E, Battaglia ML, Hossain A (2023) Cultivating sustainability: a comprehensive review on intercropping in a changing climate. Res Crop 24(4):702–715. https://doi.org/10.31830/2348-7542.2023.ROC1020

Makumba W, Akinnifesi FK, Janssen BH (2009) Spatial rooting patterns of gliricidia, pigeon pea and maize intercrops and effect on profile soil N and P distribution in southern Malawi. Afr J Agric Res 4:278–288. https://doi.org/10.5897/AJAR.9000652

Mandal D, Roy T (2024) Climate change impact on soil erosion and land degradation. In: Pathak H, Chatterjee D, Saha S, Das B (eds) Climate change impacts on soil-plant-atmosphere continuum. Advances in global change research, vol 78. Springer, Singapore. https://doi.org/10.1007/978-981-99-7935-6

Mariotti M, Masoni A, Ercoli L, Arduini I (2015) Nitrogen leaching and residual effect of barley/field bean intercropping. Plant Soil Environ 61:60–65

Marschner P (2012) Mineral nutrition of higher plants (Marschner P (ed)), 3rd edn. Elsevier Ltd

Martre P, Wallach D, Asseng S, Ewert F, Jones JW, Rötter RP et al (2015) Multi model ensembles of wheat growth: many models are better than one. Glob Chang Biol 21(2):911–925

Mbow C, Rosenzweig C, Barioni LG, Benton TG, Herrero M, Krishnapillai M, Liwenga E, Pradhan P, Rivera-Ferre MG, Sapkota T, Tubiello FN, Xu Y (2019) Food security. In: Shukla PR, Skea J, Buendia EC, Masson-Delmotte V, Pörtner O, Roberts DC, Zhai P, Slade R, Connors S, van Diemen R, Ferrat M, Haughey HE, Luz S, Neogi S, Pathak M, Petzold J, Portugal Pereira J, Vyas P, Huntley E, Kissick K, Belkacemi M, Malley J (eds) Climate change and land: an IPCC special report on climate change, desertification, land degradation, sustainable land management, food security, and greenhouse gas fluxes in terrestrial ecosystems. Cambridge University Press

Miller O, Helman D, Svoray T, Morin E, Bonfil DJ (2019) Explicit wheat production model adjusted for semi-arid environments. Field Crop Res 231:93–104

Mo X-G, Guo R-P, Liu S-X et al (2013) Impacts of climate change on crop evapotranspiration with ensemble GCM projections in the North China Plain. Clim Chang 120:299–312

Mondal S (2014) Impact of climate change on soil fertility. In: Choudhary DK et al (eds) Climate change and the microbiome, soil biology. Springer. https://doi.org/10.1007/978-3-030-76863-8_28

Mørkved PT, Dörsch P, Bakken LR (2007) The N_2O product ratio of nitrification and its dependence on long-term changes in soil pH. Soil Biol Biochem 39:2048–2057. https://doi.org/10.1016/j.soilbio.2007.03.006

Motha RP (2011) The impact of extreme weather events on agriculture in the United States. In: Challenges and opportunities in agrometeorology. https://doi.org/10.1007/978-3-642-19360-6

Mrunalini K, Behera B, Chandana C, Patnaik GP, Modi RU, Saraswat A, Rathi N, Kumar N (2022) Legumes to reduce ecological footprints for climate-smart cropping systems. In: Advances in legumes for sustainable intensification. Academic, pp 403–420. https://doi.org/10.1016/B978-0-323-85797-0.00032-X

Nyawade SO, Karanja NN, Gachene CKK et al (2019) Intercropping optimizes soil temperature and increases crop water productivity and radiation use efficiency of rainfed potato. Am J Potato Res 96:457–471. https://doi.org/10.1007/s12230-019-09737-4

Olsson L, Barbosa H, Bhadwal S, Cowie A, Delusca K, Flores-Renteria D, Hermans K, Jobbagy E, Kurz W, Li D, Sonwa DJ, Stringer L (2019) Land degradation. In: Shukla et al (eds) Climate change and land: an IPCC special report on climate change, desertification, land degradation, sustainable land management, food security, and greenhouse gas fluxes in terrestrial ecosystems. Lund University Centre for Sustainability Studies, pp 345–436

Ouda S, Zohry A (2023) Increasing sustainability by intercropping legume crops with sugar beet under imposed water stress. Irrig Drain:1–26. https://doi.org/10.1002/ird.2884

Ouda S, Zohry A (2024a) Legumes improve wheat and maize productivity grown in different cropping systems under changing climate. In: Integration of legume crops with cereal crops under changing climate: sustainably increases food production. Springer

Ouda S, Zohry A (2024b) Climate variability and change disturbs maize production. In: Integration of legume crops with cereal crops under changing climate: sustainably increases food production. Springer. ISBN: 978-3-031-68101-1

Ouda S, Zohry A, Abdel-Wahab S (2020) Sustainable intensive cropping to reduce irrigation-induced erosion: II. Changing cropping sequence under sprinkler irrigation practice. Mor J Agric Sci 1(1):26–33

Pappa VA, Rees RM, Walker RL, Baddeley JA, Watson CA (2011) Nitrous oxide emissions and nitrate leaching in an arable rotation resulting from the presence of an intercrop. Agric Ecosyst Environ 141:153–161

Philippot L, Chenu C, Kappler A, Rillig MC, Fierer N (2024) The interplay between microbial communities and soil properties. Nat Rev Microbiol 22:226–239. https://doi.org/10.1038/s41579-023-00980-5

Prentice IC, Farquhar GD, Fasham MJR, Goulden ML, Heimann M, Jaramillo VJ, Kheshgi HS, Le Quéré C, Scholes RJ, Wallace DWR (2001) The carbon cycle and atmospheric carbon dioxide. In: Climate change 2001: the scientific basis. Cambridge University Press, Cambridge, pp 183–237

Quinones M, Lucas M, Pueyo J (2022) Adaptive mechanisms make lupin a choice crop for acidic soils affected by aluminum toxicity. Front Plant Sci 12:810692. https://doi.org/10.3389/fpls.2021.810692

Rahman T, Liu X, Hussain S, Ahmed S, Chen G, Yang F (2017) Water use efficiency and evapotranspiration in maize-soybean relay strip intercrop systems as affected by planting geometries. PLoS One 12(6):e0178332. https://doi.org/10.1371/journal.pone.0178332

Rakshit A, Sarkar B, Abhilash PC (2018) Soil amendments for sustainability: challenges and perspectives. CRC Press, Boca Raton

Raza A, Asghar M, Ahmad B, Bin C, Iftikhar M, Hussain W, Li T et al (2020) Agro-techniques for lodging stress management in maize-soybean intercropping system—a review. Plan Theory 9(11):1592

Ren Y, Zhang L, Yan M, Zhang Y, Chen Y, Palta JA, Zhang S (2021) Effect of sowing proportion on above- and below-ground competition in maize–soybean intercrops. Sci Rep 11:15760. https://doi.org/10.1038/s41598-021-95242-w

Rezig M, Sahli A, Benjeddi F, Harbaoui Y (2013) Potato (*Solanum tuberosum* L.) and sulla (*Hedysarum coronarium* L.) intercropping in Tunisia: effects in water consumption and water use efficiency. J Agric Sci 5:123. https://doi.org/10.5539/jas.v5n10p123

Rojas-downing MM, Nejadhashemi AP, Harrigan T, Woznicki SA (2017) Climate risk management climate change and livestock: impacts, adaptation, and mitigation. Clim Risk Manag xxx:1–19. https://doi.org/10.1016/j.crm.2017.02.001

Saritha R, Visalakshi M (2024) Legume based profitable intercropping system for management of fall armyworm in maize. Legum Res 47(9):1583–1587. https://doi.org/10.18805/LR-5179

Schiermeier Q (2014) The parched planet: water on tap. Nature 510:326–328. https://doi.org/10.1038/510326a

Sekiya N, Yano K (2004) Do pigeon pea and sesbania supply groundwater to intercropped maize through hydraulic lift?—hydrogen stable isotope investigation of xylem waters. Field Crop Res 86:167–173. https://doi.org/10.1016/j.fcr.2003.08.007

Senbayram M, Wenthe C, Lingner A, Isselstein J, Steinmann H, Kaya C, Köbke S (2016) Legume-based mixed inter-cropping systems may lower agricultural born N_2O emissions. Energy Sustain Soc 6:2

Seran TH, Brintha I (2010) Review on maize based inter-cropping. J Agron 9(3):135–145. https://doi.org/10.3923/ja.2010.135.145

Serrano-Silva N, Luna-Guido M, Fernández-Luqueño F, Marsch R, Dendooven L (2011) Emission of greenhouse gases from an agricultural soil amended with urea: a laboratory study. Appl Soil Ecol 47:92–97

Shen Y, Sui P, Huang J, Wang D, Whalen JK, Chen Y (2018) Greenhouse gas emissions from soil under maize–soybean intercrop in the North China Plain. Nutr Cycl Agroecosyst 110:451–465

Shen L, Wang X, Liu T, Wei W, Zhang S, Keyhani A et al (2023) Border row effects on the distribution of root and soil resources in maize–soybean strip intercropping systems. Soil Tillage Res 233:105812. https://doi.org/10.1016/j.still.2023.105812

Shen X, Liu Y, Li X, Li L (2024) Effect of intercropping with legumes at different rates on the yield and soil physicochemical properties of *Cyperus esculentus* L. in arid land. Front Plant Sci 15:1351843. https://doi.org/10.3389/fpls.2024.1351843

Singh BB, Adjeigbe HA (2002) Improving cowpea cereals-based cropping systems in the dry savannas of West Africa. In: Fatokun CA et al (eds) Challenges and opportunities for enhancing sustainable cowpea production. IITA, Ibadan, pp 276–284

Singh R, Khanna V (2021) Climate change and food systems: implications on food security. In: Hebsale Mallappa VK, Shirur M (eds) Climate change and resilient food systems. Springer Nature Singapore Pte Ltd. https://doi.org/10.1007/978-981-33-4538-6_3

Singh RK, Chaudhary RS, Somasundaram J, Rashmi I, Hati KM, Sinha NK, Rao AS (2014) Impact of crop covers on soil properties, runoff, soil-nutrients losses and crop productivity in Vertisols of central India. Indian J Soil Conserv 42(3):268–275

Singh D, Mathimaran N, Boller T, Kahmen A (2020a) Deep-rooted pigeon pea promotes the water relations and survival of shallow-rooted finger millet during drought—despite strong competitive interactions at ambient water availability. PLoS One 15:e0228993. https://doi.org/10.1371/journal.pone.0228993

Singh RK, Chaudhary RS, Somasundaram J, Sinha NK, Mohanty M, Hati KM, Rashmi I, Patra AK, Chaudhari SK, Lal R (2020b) Soil and nutrients losses under different crop covers in vertisols of Central India. J Soils Sediments 20(2):609–620

Snyder CS, Bruulsema TW, Jensen TL, Fixen PE (2009) Review of greenhouse gas emissions from crop production systems and fertilizer management effects. Agric Ecosyst Environ 133(3–4):247–266. https://doi.org/10.1016/j.agee.2009.04.021

Stagnari F, Maggio A, Galieni A, Pisante M (2017) Multiple benefits of legumes for agriculture sustainability: an overview. Chem Biol Technol Agric 4(2):1–13

Steward PR, Thierfelder C, Dougill AJ, Ligowe I (2019) Conservation agriculture enhances resistance of maize to climate stress in a Malawian medium-term trial. Agric Ecosyst Environ 277:95–104. https://doi.org/10.1016/j.agee.2018.07.009

Stomph T, Dordas C, Baranger A, de Rijk J, Dong B, Evers J, Gu C et al (2020) Designing intercrops for high yield, yield stability and efficient use of resources: are there principles. In: Sparks DL (ed) Advances in agronomy, vol 160. Academic, pp 1–50. https://doi.org/10.1016/bs.agron.2019.10.002

Sun L, Li L, Chen Z, Wang J, Xiong Z (2014) Combined effects of nitrogen deposition and biochar application on emissions of N_2O, CO_2 and NH_3 from agricultural and forest soils. Soil Sci Plant Nutr 60:254–265. https://doi.org/10.1080/00380768.2014.885386

Sun F, Xiao B, Li S, Kidron GJ (2021) Towards moss biocrust effects on surface soil water holding capacity: soil water retention curve analysis and modeling. Geoderma 399:Article 115120. https://doi.org/10.1016/j.geoderma.2021.115120

Talukder AS, McDonald GK, Gill GS (2014) Effect of short-term heat stress prior to flowering and early grain set on the grain yield of wheat. Field Crop Res 160:54–63

Tamocai C, Canadell JG, Schuur EAG, Kuhry P, Mazhitova G, Zimov S (2009) Soil organic carbon pools in the northern circumpolar permafrost region. Glob Biogeochem Cycles 23(2):1–11. https://doi.org/10.1029/2008GB003327

Teixeira EI, Fischer G, Velthuizen HV, Walter C, Ewert F (2013) Global hot spots of heat stress on agricultural crops due to climate change. Agric For Meteorol 170:206–215. https://doi.org/10.1016/j.agrformet.2011.09.002

Tosti G, Falcinelli B, Guiducci M (2023) Lentil–cereal intercropping in a Mediterranean area: yield, pests and weeds. Agron J 115:2570–2578

van der Wiel K, Bintanja R (2021) Contribution of climatic changes in mean and variability to monthly temperature and precipitation extremes. Commun Earth Environ. https://doi.org/10.1038/s43247-020-00077-4

Várallyay G (2010) The impact of climate change on soils and on their water management. Agron Res 8(Special Issue II):385–396

Velasquez AC, Castroverde CDM, He SY (2018) Plant–pathogen warfare under changing climate conditions. Curr Biol 28:R619–R634

Ventura M, Opstad N, Zanotelli D, Scandellari F, Quartieri M, Tagliavini M (2005) Monitoring soil nitrogen losses through leaching in a pear orchard. Riv Frutticolt Ortofloricolt 67:40–44

Vermeulen SJ, Campbell BM, Ingram JSI (2012) Climate change and food systems. Annu Rev Environ Resour 37:195–222. https://doi.org/10.1146/annurev-environ-020411-130608

Wang H, Köbke S, Dittert K (2020a) Use of urease and nitrification inhibitors to reduce gaseous nitrogen emissions from fertilizers containing ammonium nitrate and urea. Glob Ecol Conserv 22:e00933. https://doi.org/10.1016/j.gecco.2020.e00933

Wang L, Hou B, Zhang D, Li H, Rengel Z, Shen J (2020b) The niche complementarity driven by rhizosphere interactions enhances phosphorus-use efficiency in maize/alfalfa mixture. Food Energy Secur 9(e252):1–14

Wang X, Chen Y, Yang K, Duan F, Liu P, Wang Z et al (2021) Effects of legume intercropping and nitrogen input on net greenhouse gas balances, intensity, carbon footprint and crop productivity in sweet maize cropland in South China. J Clean Prod 314:127997

Wang Z, Ma S, Hu Y, Chen Y, Jiang H, Duan B, Lu X (2022) Links between chemical composition of soil organic matter and soil enzyme activity in alpine grassland ecosystems of the Tibetan Plateau. Catena 218:Article 106565. https://doi.org/10.1016/j.catena.2022.106565

Werner WJ, Sanderman J, Melillo JM (2020) Decreased soil organic matter in a long-term soil warming experiment lowers soil water holding capacity and affects soil thermal and hydrological buffering. J Geophys Res Bio Geosci 125:e2019JG005158. https://doi.org/10.1029/2019JG005158

Wheeler T, Von Braun J (2013) Climate change impacts on global food security. Science 341:508–513. https://doi.org/10.1126/science.1239402

World Resource Institute (2000) World resources 2000–2001—people and ecosystems: the fraying web of life. World Resource Institute. www.wri.org/wr2000/

Xiao X, Cheng Z, Meng H, Liu L, Li H, Dong Y (2013) Intercropping of green garlic (*Allium sativum* L.) induces nutrient concentration changes in the soil and plants in continuously cropped cucumber (*Cucumis sativus* L.) in a plastic tunnel. PLoS One 8(4). https://doi.org/10.1371/journal.pone.0062173

Xie Z, Xu Y, Liu G, Liu Q, Zhu J, Tu C, Amonette JE, Cadisch G, Yong JWH, Hu S (2013) Impact of biochar application on nitrogen nutrition of rice, greenhouse-gas emissions and soil organic carbon dynamics in two paddy soils of China. Plant Soil 370:527–540. https://doi.org/10.1007/s11104-013-1636-x

Xue Y, Xia H, Christie P, Zhang Z, Li L, Tang C (2016) Crop acquisition of phosphorus, iron, and zinc from the soil in cereal/legume intercropping systems: a critical review. Ann Bot 117:363–377. https://doi.org/10.1093/aob/mcv182

Yang C, Huang G, Chai Q, Luo Z (2011) Water use and yield of wheat/maize intercropping under alternate irrigation in the oasis field of northwest China. Field Crop Res 124:426–432

Yang L, Luo Y, Lu B, Zhou G, Chang D, Gao S et al (2023) Long-term maize and pea intercropping improved subsoil carbon storage while reduced greenhouse gas emissions. Agric Ecosyst Environ 349:108444

Yu H, Deng Y, He Z, Van Nostrand JD (2018) Elevated CO_2 and warming altered grassland microbial communities in soil top–layers. Front Microbiol 9:1790. https://doi.org/10.3389/fmicb.2018.01790

Yu J, Rezaei EE, Reckling M, Nendel C (2025) Winter wheat–soybean relay intercropping in conjunction with a shift in sowing dates as a climate change adaptation and mitigation strategy for crop production in Germany. Field Crop Res 322. https://doi.org/10.1016/j.fcr.2024.109695

Yuan Q, Feng XG, Tian JY, Song X, Li GY, Fang C (2022) Controlling factors of soil organic carbon and nitrogen in lucerne grasslands in a semiarid environment. Catena 211:Article 105983. https://doi.org/10.1016/j.catena.2021.105983

Zhang MM, Wang N, Hu YB, Sun GY (2018) Changes in soil physicochemical properties and soil bacterial community in mulberry (*Morus alba* L.)/alfalfa (*Medicago sativa* L.) intercropping system. Microbiol Open 7(2):e00555. https://doi.org/10.1002/mbo3.555

Zhao C, Liu B, Piao S et al (2017) Temperature increase reduces global yields of major crops in four independent estimates. PNAS 114:9326–9331

Zhao J, Chen J, Beillouin D, Lambers H, Yang Y, Smith P et al (2022) Global systematic review with meta-analysis reveals yield advantage of legume-based rotations and its drivers. Nat Commun 13:4926. https://doi.org/10.1038/s41467-022-32464-0

Ziadat FM, Taimeh AY (2013) Effect of rainfall intensity, slope, land use and antecedent soil moisture on soil erosion in an arid environment. Land Degrad Dev 24(6):582–590

Zohry A, Ouda S, Abdel-Wahab T (2020) Sustainable intensive cropping to reduce irrigation-induced erosion: intercropping systems under surface irrigation practice. Mor J Agric Sci 1(2):63–71

Evaluating Soil Degradation Processes and Their Exacerbation by Climate Change in the Nile Delta of Egypt

13

A. A. M. Mohamedin and Ebtsam M. Morsy

Abstract

The Nile Delta is a vital region for Egypt's agricultural production, and climate change speeds up a number of degradation processes, such as erosion, waterlogging, land subsidence, salinization, and loss of soil organic matter. In this chapter, we examined some of the problems related to soil degradation in the Nile Delta. The effect of increasing salinization and land subsidence, especially in coastal areas, due to rising sea levels brought on by global warming was also reviewed. As a consequence of melting of polar ice caps, saltwater seeps into agricultural areas severely reducing soil fertility and crop productivity. Furthermore, the natural balance of sediment deposition is disrupted by changes in precipitation patterns and sediment flow caused by upstream dam construction, which accelerates land subsidence. Other issues affecting land degradation were also discussed, namely, drought, loss of soil organic matter, soil compaction, and desertification. Analysis of long-term temperature and precipitation in the Nile Delta governorates was done to examine its past trend and expect how it will contribute in land degradation in the future. A variety of adaptive strategies were proposed in the study, such as breeding for tolerant crop varieties to drought and salt stress, enhanced irrigation techniques and increasing its efficiency, and conserving soil moisture. In addition, reducing the effects of salinization and waterlogging requires the application of sustainable agricultural practices, and better drainage systems were also presented.

A. A. M. Mohamedin (✉)
Department of Field Drainage, Soils, Water and Environment Research Institute, Agricultural Research Center (ARC), Giza, Egypt

E. M. Morsy
Department of Agricultural Microbiology, Soils, Water and Environment Research Institute, Agricultural Research Center (ARC), Giza, Egypt

A. E.-H. Zohry, S. Ouda (eds.), *Resilient Agroecosystems*, Sustainability Sciences in Asia and Africa, https://doi.org/10.1007/978-981-96-9630-7_13

Keywords

Sea level rise · Land subsidence · Waterlogging · Salinization · Erosion · Loss of soil organic matter

Introduction

Currently, 25% of the Earth's land and 40% of its agricultural land are affected by degradation. The environmental consequences of soil degradation are extensive, leading to increased soil erosion, deterioration of water quality, loss of biodiversity, and degradation of ecological resources and their associated values, particularly in areas where land use conflicts with environmental sustainability (Sahoo and Majid 2020). Variations in temperature, wind speed, and precipitation can influence plant biomass production, land use, land cover, soil moisture, infiltration rates, runoff, and crop management and ultimately contribute to land degradation (Telo da Gama 2023). In recent decades, a significant relationship has emerged between global climate change and land loss.

The ongoing global warming affects rain patterns, causing agricultural yields to slump. Rain, heat, and temperature are three critical climatic variables for farming and can induce variations in both crop yield and quality (Rezaei et al. 2023). The impacts of these changes are especially significant in developing countries, where the poorest farmers are mainly located and agricultural production is already severely weather dependent due to the great cost involved in commercial irrigation systems (Mueller et al. 2012). The majority of the people in the developing world have no access to any insurance and are therefore forced to look after themselves at their own risk (Malhi et al. 2021). With ongoing warming, the vulnerability of an already water-scarce region such as the Mediterranean Basin increases day by day.

The Nile Delta, located in the northeastern region, is the Egypt's fertile agricultural region and receives sediment from the Ethiopian Plateau and Rift Valley. The Nile flood records show significant hydroclimatic variability due to climate changes, including droughts (Badawy et al. 2024). This has led to changes in sediment delivery, influenced by paleoclimate and hydraulic infrastructure (Trenberth 2011). The Nile Delta plains have been shaped by flood sediments, creating dynamic floodplain environments (Olson and Kome 2024). These plains are dominated by young, loosely compacted silt, promoting the establishment of vegetative roots and organic-rich bio platforms (Abdel-Kawy and Ali 2012).

The environmental history of the Nile Delta is complex, involving sediment delivery, living organism impact, underwater expansion, and sediment burial processes (Olson and Kome 2024). This creates a fertile soil base and habitat for fish and waterfowl, supporting ecosystems for millennia (Badawy et al. 2024). Human intervention has influenced agricultural soils in the Nile Delta to preserve fertility, which is sustained by sediment-rich waters from the Nile River. However, over the last 150 years, soil management has not fully addressed the negative effects of insufficient sediment, rising sea levels, increasing temperatures, drought, and suboptimal land use (Ribbe and Ahmed 2006). This has led to low-porosity, moderately saline, and sodic soil environments (Abdel-Kawy and Ali 2012). The long-term prospects for soil resources in the Nile Delta is uncertain as rising sea levels and land subsidence pose a threat of flooding to soil layers (Ducassou et al. 2009). Understanding the impact of global climate change on soil degradation processes is crucial to forecast future soil degradation, identify the influence of various factors on soil degradation, and propose the production capacity for cultivable land and its changes under applied scenarios.

Recently, the Nile Delta is experiencing a decline in soil quality due to soil degradation processes worsened by climate change and other factors (Badawy et al. 2024). Elhini et al. (2024) reported that this degradation is expected to significantly impact crop yields, farmers' profitability, and food security in the Delta. The long-term climate change projections, including rising temperatures, extreme rainfall, and increased evapotranspiration-rainfall ratios, may worsen soil degradation effects (Balwan and Kou 2021). To address this issue, it is crucial to understand other contributing factors such as irrigation, cropping systems, land preparation, fertilization, subsoil cultivation, and secondary soil compaction, which often exacerbate climate change's impacts.

In this chapter, we examined some of the problems related to soil degradation in the Nile Delta of Egypt. The expected impact of rising temperature and rainfall variability on soil degradation in the Nile Delta governorates was presented. Furthermore, agricultural practices that can contribute in reducing soil degradation were discussed.

Soil Degradation Processes in the Nile Delta

Soil degradation is one of the most saline environmental threats facing developing countries. The current and potential pernicious consequences of this trend for poverty, low agricultural productivity, water scarcity, and carbon emissions are significant (Kraamwinkel et al. 2021). It is estimated that the reductions in present and future global gross domestic product caused by soil nutrient depletion would amount to hundreds of billions of dollars by 2050 (FAO 2018). Climate change, population pressure, short-sighted government policies, and deteriorating soil are the main factors behind soil degradation (El-Ramady et al. 2024). Land degradation boosts the poor's vulnerability vis-à-vis other disasters, such as floods, economic shocks, and the impact of demographic pressures. These are just some of the many negative effects that should encourage the adoption of effective international countermeasures.

Soil degradation in the Nile Delta, a critical agricultural hub in Egypt, has intensified due to climate change and unsustainable human practices. Abou-Kota et al. (2024) emphasized the primary degradation processes include sea level rise, land subsidence, salinization, waterlogging, erosion, soil compaction, loss of soil organic matter, and desertification. The following diagram summarizes the impact of main climate change factors and their interactions on soil degradation (Fig. 13.1).

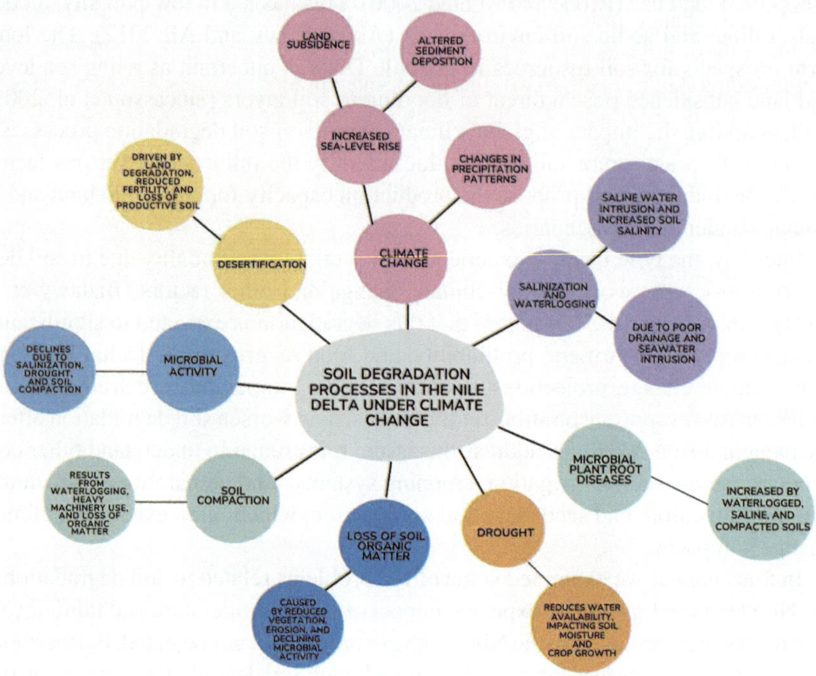

Fig. 13.1 The impact of main climate change factors and their interactions on soil degradation

Sea Level Rise

The Nile Delta experiences exacerbated salinization and subsidence due to climate change-induced sea level rise. As the sea penetrates further inland, coastal soils become saltier, diminishing their agricultural productivity and making them more susceptible to erosion and subsidence (El-Raey 2010). A key impact of climate change is the elevation of global sea levels caused by polar ice caps melting and seawater expansion due to temperature increases (Elshinnawy and Almaliki 2021). The northern Nile Delta, particularly its low-lying coastal areas, is extremely susceptible to rising sea levels. As sea levels increase, the pressure on coastal lands intensifies, resulting in more land submersion and exacerbating subsidence. Additionally, El-Raey (2010) indicates that the Mediterranean Sea has been rising at a rate of 1.7–3.3 mm annually, with expectations of an accelerated rise under future climate change scenarios. This sea level rise amplifies land subsidence by increasing the flooding of coastal regions, reducing their natural ability to accumulate sediments that could counteract subsidence. Specifically, the Rosetta and Damietta branches of the Nile Delta face the risk of subsidence and potential significant land loss by 2100 if current trends persist (Elshinnawy and Almaliki 2021).

Eman-Mounir (2024) provided evidence that the Nile Delta may lose its suitability for habitation, farming, or investment on thousands of acres. Egypt's sea levels have risen by 3.2 mm per year since 2012, which will cause the northern shore of the Nile Delta to continue to flood and erode.

Land Subsidence

Land subsidence in the northern Nile Delta is a significant concern, particularly as it relates to climate change, which exacerbates the issue through sea level rise and the alteration of sediment dynamics. Abou El-Magd et al. (2024) explained that this phenomenon is mainly driven by a combination of natural and human-induced factors, with climate change playing an increasingly important role. A detailed discussion of the impact of climate change on land subsidence in the northern Nile Delta and strategies for mitigating this problem is explained by Eelnahry and Doluschitz (2009), and they concluded that if this problem is not addressed promptly, a significant part of this densely populated and agriculturally vital area could be lost to subsidence. It is worth mentioning that climate change affects the phenomenon of land subsidence by several processes including sea level rise, altered sediment deposition, saline water intrusion and increased groundwater extraction.

Land subsidence has historically been maintained by the sediment deposition of the Nile River; however, the equilibrium is being upset by climate change and the construction of dams (Abou El-Magd et al. 2024). Rapid sinking has been caused by the Aswan High Dam's reduction of sediment flow to the delta. Flooding and storm surges are examples of extreme weather occurrences that interfere with sediment deposition (Sharaan et al. 2024). The Nile Delta in Egypt, which is 1% above sea level, is unevenly submerging as a result of decreasing land levels, rising sea levels, and decreased sediment replenishment. Stanley and Clemente (2017) indicated that, as a result of Ethiopia's Grand Renaissance Dam building, the lower Nile Basin's fresh water is predicted to increase the issue of reduced Blue Nile water and sediment distribution to the lower Nile Basin and Egypt's Delta coast. In this concern, a study by Stanley and Clemente in 2014 investigates the effect of clay on the northern Nile Delta's Holocene sedimentary series' sinking. They discovered that the thickness and amount of clay varies greatly, with the Manzala lagoon having the greatest proportion of clay. The yearly stream of Nile water and sediment has diminished due to climate change and human activities. This has resulted in compaction of the silt, an increase in sea level in the region, and occasional readjustment, which has reduced the coastal edge of the delta.

Rising sea levels due to climate change leads to saline water intrusion in coastal aquifers. The northern Nile Delta is especially vulnerable to this phenomenon as the increased salinity in groundwater speeds up the deterioration of subsurface soils, leading to further subsidence (Saqr and Abd-Elmaboud 2024). This degradation weakens the structural integrity of the land, making it more susceptible to sinking. However, Sefelnasr and Sherif (2013) pointed out that seawater intrusion is a significant issue in coastal aquifers, particularly in flat, low-land coastal alluvial plans. Studies have shown that seawater level rise can cause landward movement of the distribution zone and alter boundary conditions at the seaside and domain geometry

(Bear et al. 1999). The effects of seawater level rise in the Mediterranean Sea on the Nile Delta aquifer are considered, including a 0.5 m seawater rise, a 1.0 m rise, and a modified shoreline. The study concludes that large areas in the coastal zone of the Nile Delta will be submerged by seawater, leading to landward shifts in the eastern and western sides of the Delta (Sefelnasr and Sherif 2013).

Additionally, Negm et al. (2019) stated that due to the decreasing availability of fresh water and shifting precipitation patterns associated with climate change, there has been a rise in the extraction of groundwater in the northern Nile Delta. Extracting too much groundwater reduces the pressure in aquifers, leading to the sinking of the land above them. This is a major factor contributing to human-induced subsidence in the Nile Delta (Saqr 2023).

Nevertheless, implementing strategies like through Integrated Coastal Zone Management (ICZM) better water management and the construction of adaptable infrastructure can help minimize the effects and build resilience against the evolving climate. Effective management and mitigation of land subsidence in the Nile Delta will require coordinated efforts among local governments, international organizations, and scientific communities (Tolba et al. 2018).

Salinization and Waterlogging

Salinization in the Nile Delta is a significant soil degradation affecting plant development and agricultural productivity. Salts remain after water evaporates, exacerbated by insufficient drainage and excessive irrigation. The issue is further exacerbated by Mediterranean Sea saltwater intrusion (Stanley and Clemente 2017). Tawfik et al. (2024) indicated that salinization poses a major threat to soil quality in the Nile Delta, leading to the buildup of soluble salts in the soil, which can hinder plant growth and diminish agricultural output. In the Nile Delta, the primary causes of salinization are elevated groundwater levels and inadequate drainage systems, allowing salts to surface (Hammam and Mohamed 2020). Excessive irrigation, coupled with ineffective drainage, results in water evaporation from the soil, leaving behind salts (Mohamed et al. 2019). Furthermore, saltwater intrusion from the Mediterranean Sea, influenced by natural and human-induced factors, compounds the salinization issue (Stanley and Clement 2014). The escalation of sea levels due to global climate change is intensifying saltwater intrusion in the Nile Delta. Climate forecasts indicate that the Mediterranean may rise by 0.5–1 m by 2100, significantly heightening the risk of salinization (El-Raey 2010). As sea levels increase, saltwater permeates deeper into coastal aquifers and soils, elevating their salinity. This process is worsened by diminished freshwater flow in the Nile, caused by upstream water withdrawals and altering precipitation patterns, further compromising the Delta's ability to remove salts from the soil (Conway et al. 2019).

Whereas waterlogging in the Nile Delta is a major issue due to level topography, insufficient drainage systems, and flood irrigation, climate changes may exacerbate waterlogging, affecting soil aeration and root development and reducing agricultural productivity (Attaher et al. 2009). Waterlogging condition where soil is saturated with water, inadequate drainage infrastructure, and high groundwater levels are significant issues in the Nile Delta (Kaiser et al. 2013). This deficiency in oxygen in the root zone hinders plant growth and agricultural yield. Factors contributing to waterlogging include over-irrigation, elevated groundwater levels, and insufficient drainage systems (El-Marsafawy 2007). Waterlogged soils hinder air infiltration, leading to oxygen deficiency in the root zone, which obstructs plant development and reduces agricultural productivity. Anticipated changes in precipitation patterns due to climate change may exacerbate waterlogging in the Nile Delta (Kaiser et al. 2013). Furthermore, increasing sea levels may result in elevated groundwater levels, exacerbating the risk of waterlogging, particularly in low-lying coastal areas next to the Mediterranean (Conway et al. 2019). Elevated sea levels also increase groundwater levels, increasing waterlogging risk in low-lying coastal areas (El-Raey 2010).

Drought

Rising temperatures and reduced rainfall contribute to drought conditions, directly impacting plant growth and making them more susceptible to diseases. Siam and Eltahir (2017) suggest that climate change will enhance the interannual variability of Nile River flow, further contributing to unpredictable water availability and increasing the frequency and severity of droughts. Moreover, 80% of the Nile Delta use surface and groundwater for food production that can be negatively affected by climate-induced drought causing salinization of the soil by seawater intrusion and groundwater discharge (Elsayed et al. 2024). Furthermore, it can alter the microbial community's habitat in the soil and/or rhizosphere by introducing fusarium or other soil-pathogenic fungi that harm the crops' roots, brood, and crop plants.

Loss of Soil Organic Matter

The soil of the Nile Delta is experiencing accelerated decomposition of soil organic matter (SOM) due to rising temperatures directly resulting from climate change. With increasing temperatures, organic materials in the soil break down quickly, resulting in a decrease in SOM. This reduces soil fertility and hampers the soil's capacity to hold moisture, worsening the decline in agricultural productivity (Abdel Kawy and Ali 2012).

Soil Compaction

Soil compaction, a significant issue in the Nile Delta, is exacerbated by climate change due to rising temperatures and altered rainfall patterns. This results in thicker soil structures with reduced air and water flow, hindering root growth and restricting water infiltration and drainage (Soane et al. 2012). Batey (2009) emphasized that soil compaction in the Nile Delta is primarily caused by the compression of soil particles, which reduces pore space and has detrimental effects on various soil processes. This can lead to decreased porosity, weakened structure, and poor permeability, making them more vulnerable to soil degradation. Farmers in the region rely heavily on irrigation, which can reduce the soil's natural capacity to recover by settling and compacting soil particles under water pressure. Heat stress from climate change can also deplete the soil's organic content, increasing its vulnerability to compaction (Hamza and Anderson 2005). Soil compaction can significantly impair agricultural yields, plant development, and soil health, particularly in clay-rich soils. It can also cause nutritional deficits and water stress, which can hinder plant growth. For crops that depend on water, dense soils can cause hypoxia, or oxygen shortages in the root region, leading to stunted growth, yellowing leaves, and decreased vigor (Soane et al. 2012). Climate change also increases soil density and compaction due to changing rainfall patterns and increased temperatures.

Desertification

The building of dams upstream and the effects of climate change are causing a decrease in water flow from the Nile, leading to desertification on the edges of the Nile Delta. Processes of desertification include a decrease in soil moisture, salinization, and erosion, all of which pose a threat to the long-term sustainability of farming in the area (Hussein 2011). The reduced flow of fresh water restricts the natural irrigation provided by the Nile, making the Nile Delta more susceptible to land degradation (Abou-Kota et al. 2024). Rising sea level, which is an undeniable impact of climate change, will drastically exacerbate the detrimental effects of desertification in the Nile Delta. Elbasiouny (2018) indicated that it is essential to update timely measurements of land degradation and identify its causes for the security of populations living on and near the Nile Delta.

Microbial Activity

Increases in temperature, CO_2 levels, soil moisture content, salinity, and drought are examples of climate changes that can have an impact on soil microbial activity and, in turn, the biogeochemical cycles of nutrients in agricultural soils (Frouz (2018). Determining the importance of these cycles and climatic feedbacks requires an understanding of these microbial activities. Significant microbial activity is occurring in Egyptian soils, which is important for soil fertility and nutrient recycling. These behaviors are sensitive markers of the biological condition of the soil that are impacted by environmental stress (Hossain and Majumde 2018). It is crucial to do research on how climate change affects the microbiological activity of Nile Delta soil, including microbial populations, biomass, enzymes, beneficial bacteria, and carbon sequestration (Abd El-Azeem 2020).

The Expected Effect of Rising Temperatures on the Soil of the Nile Delta

Climate change is expected to lead to increased temperatures and shifts in precipitation patterns in the Nile Delta (Abou-Hadid 2006). These changes will have far-reaching consequences for agriculture, water resources, and overall environmental health (Abou-Kota et al. 2024). Higher temperatures increase evaporation rates, which in turn heighten the risk of soil desiccation and salinization. The combination of high heat and poor water management practices can exacerbate water scarcity, making soils more susceptible to degradation (Conway et al. 2019).

Egypt's average temperatures have already risen by about 1.5 °C over the past century, and projections suggest that they could increase by an additional 2–3 °C by 2100 (Conway et al. 2019). Higher temperatures accelerate the evaporation of surface water, exacerbate water scarcity, and stress crops, reducing agricultural productivity (Abou-Hadid 2006). Crops like wheat and maize, which are sensitive to heat stress, are particularly vulnerable to yield reductions as temperatures rise (Abuzaid 2018). In addition, World Bank (2014) reported that important aspect of the Nile Delta's changing climate is rising temperatures. Egypt's average temperature may increase by 1.5–2.5 °C by 2050 according to projections. Elevated temperatures may cause a decrease in soil moisture, a rise in evaporation rates, and a rise in the need for irrigation water (Abou-Hadid 2006). Egypt already depends largely on the Nile River for its freshwater needs, and upstream developments and decreased precipitation in the river's source regions could reduce the river's flow, making the already severe water stress worse (El-Raey 2010). The rise in temperature has profound implications for soil degradation in the coastal and inland governorates of the Nile Delta, exacerbating the already fragile environmental conditions in this critical agricultural region of Egypt (El-Raey 2010). Soil degradation manifests through various processes such as salinization, erosion, nutrient depletion, and organic matter loss. As temperatures rise due to climate change, these processes are intensified, with distinct impacts in both coastal and inland areas of the Nile Delta.

Increased heat stress also affects crop yields, particularly for temperature-sensitive crops like wheat and maize, which are staple food crops in Egypt. According to pervious research, wheat yields could decrease by 15–20% by 2050 due to increased temperatures, while maize yields may fall down by 14% (Abou-Hadid 2006). These changes could jeopardize Egypt's food security, especially as the population continues to grow.

It illustrates the average changes in temperature through 23 years from year 2000 to 2023 among eight governorates representing the Nile Delta. The temperature data was downloaded from the following website: https://power.larc.nasa.gov/data-access-viewer/. The annual average temperature trends for the various governorates in the Nile Delta (Alexandria, Dakahlia, Damietta, Gharbia, Kafr El-Sheikh, Monufia, Port Said, and Sharkia) in the studied period exhibit clear patterns of rising temperatures over the years. The following is an in-depth examination of the trends in each governorate.

In Alexandria governorate, the average annual temperature trend from 2000 to 2023 varies between 26.5 and 28.0 °C. There was a notable increase in peak temperature and fluctuations in 2010, reaching 28.5 °C, followed by a decrease to 26.5 °C in 2011. Subsequently, there were fluctuations around 27.0 °C with a peak of 28.1 °C in 2023. Despite fluctuations from year to year, the overall trend shows a gradual rise in average temperatures over the 23 years. The average annual temperature trend in Dakahlia governorate in 2000 stood at 29.7 °C. By 2010, the temperature had risen to 31.8 °C, marking the highest recorded temperature during the study period. Following this peak, temperature experienced a slight dip but remained elevated, with 30.8 °C recorded in 2018 and 31.3 °C in 2023. In general, Dakahlia displays a noticeable warming trend, with average temperatures increasing consistently from year to year.

Regarding the annual average temperature trend in Damietta governorate in 2000, it has reached 23.4 °C. By 2010, there was a gradual rise in temperature to 24.9 °C, followed by smaller fluctuations in the subsequent years. In 2023, the average temperature was 25.0 °C, indicating a consistent increase from the early 2000s. Damietta has shown a steady warming trend, with temperatures gradually rising over the 23-year period. Similar to Dakahlia, in the year 2000, Gharbia governorate had an average annual temperature of 29.7 °C. By 2010, temperatures in Gharbia had risen, reaching a peak of 31.8 °C. At the end of the studied period, the annual temperature was 31.3 °C, indicating a consistent warming trend in Gharbia, particularly after 2010. There is a clear pattern of increasing temperatures, with peaks in 2010 and consistently high temperatures thereafter. This trend indicates a steady rise in temperatures over time, with noticeable peak.

In 2000, the average yearly temperature in Kafr El-Sheikh governorate was 29.0 °C. The annual average of temperature in Kafr El-Sheikh experienced changes, reaching a peak of 31.0 °C in 2010 and fluctuating in the following years. By the end of the study in 2023, the average temperature had risen to 30.4 °C, surpassing the majority of early 2000s readings. The region of Kafr El-Sheikh displays a consistent trend of warming, marked by significant peaks and continual increases in recent year. Regarding Sharkia governorate in the year 2000, it had an average annual temperature of 28.9 °C. The highest temperature in Sharkia was recorded in 2010 at 31.0 °C, and subsequent years showed consistent increases. By 2023, the average temperature in Sharkia had risen to 30.6 °C, indicating a continuous warming trend similar to that seen in other governorates, especially in recent years (Fig. 13.2).

Based on the above results, it can be inferred that there is a consistent trend of increasing temperatures in all governorates of the Nile Delta over 23 years, with the highest temperatures occurring around 2010, making it a scorching year throughout the region. The temperature increase is particularly noticeable in Menofia and Gharbia, where temperatures surpassed 31 °C in the 2020s. Even coastal areas such as Alexandria and Port Said, which typically have milder climates, experienced significant temperature rises over time, indicating that the effects of climate change are

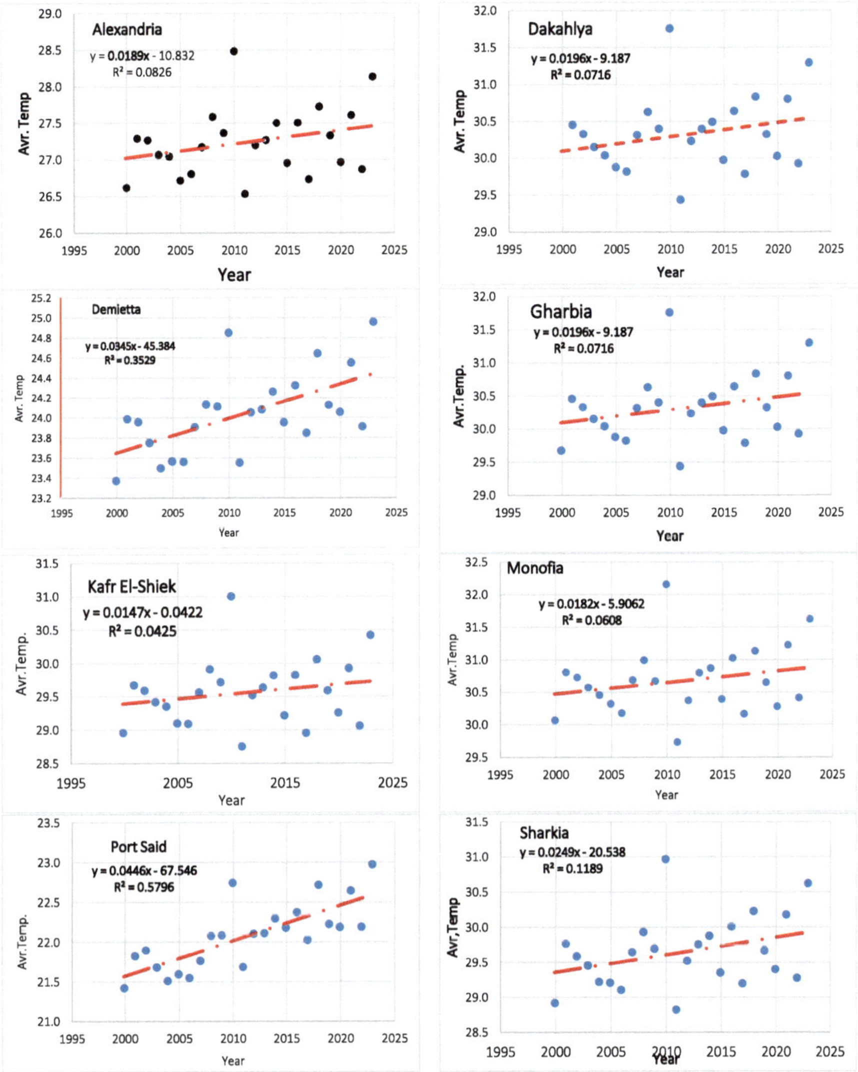

Fig. 13.2 The annual average values of temperature, regression equation, and the R^2 represent the key indicators of rising temperatures for eight governorates representing the Nile Delta over 23 years

widespread across the Nile Delta. These results reflect global trends of escalating temperatures, and the continual warming in the Nile Delta could have significant implications for agriculture, water resources, and the health of the region's inhabitants. Temperature rises in general could resulted in several climatic impacts on these areas such as the following:

Increased Soil Salinity

Soil salinity poses a significant problem in the coastal governorates of the Nile Delta, especially close to the Mediterranean Sea. The region's low-lying areas are susceptible to saltwater intrusion, and the escalating temperatures worsen this issue. Elevated temperatures result in increased evaporation rates, causing the accumulation of salts in the soil as water evaporates and leaves behind salts (Saqr and Abd-Elmaboud 2024). This phenomenon is particularly pronounced in coastal regions like Alexandria, Damietta, Kafr El-Sheikh, Port Said, and Beheira, where agricultural groundwater extraction is prevalent. With more groundwater being extracted, the intrusion of saltwater into freshwater aquifers intensifies, leading to further soil salinization (Tarolli et al. 2024). Additionally, in inland governorates like Menofia, Dakahlia, Gharbia, and Sharkia, the effects of rising temperatures on salinity are more indirect but still significant. Increasing temperature leads to greater evaporation from irrigation waters, elevating the salt concentration in the soil even if the water source is not saline (Abdel-Kawy and Ali 2012). Over time, this process diminishes soil productivity, particularly in intensive agricultural areas such as Menofia and Dakahlia.

Accelerated Soil Erosion

The increase in temperatures results in a more severe soil erosion in both coastal and inland areas of the Nile Delta. Soil erosion along the coast is driven by the rising sea levels, which are directly correlated with the temperature rise. As the sea levels increase, the delta's coastline recedes, causing the erosion of the fertile agricultural lands (El-Raey 2010). The elevated temperatures also contribute to more intense and unpredictable weather patterns, including heavier storms, which worsen the coastal erosion (Hussein 2011). Conversely, in the inland areas, soil erosion is a result of the combined impact of higher temperatures and changes in precipitation patterns (Negm et al. 2019). The drier soils, due to increased evaporation rates, become more prone to wind erosion. Additionally, when rainfall does occur, it often takes the form of heavy storms, leading to water erosion. In governorates like Gharbia and Sharkia, this has resulted in the deterioration of agricultural topsoil, further decreasing crop yields (Elham and Abou El-Magd 2016).

Loss of Soil Organic Matter

The presence of soil organic matter is essential for preserving soil fertility and structure, and its breakdown is closely linked to increasing temperatures. Elevated temperatures speed up the breakdown of organic matter in the soil, resulting in a decrease in its quality (Abou-Kota et al. 2024). This poses a significant issue in both coastal and inland regions of the Nile Delta, where organic matter levels are already low due to intensive farming methods. In coastal areas, the combination of rising temperatures and saltwater intrusion can accelerate the decomposition of organic material, leading to further soil deterioration. In inland regions, where organic matter is often replenished through crop residues and manure, higher temperatures cause quicker decomposition, reducing the soil's capacity to retain water and nutrients. This renders the soil more susceptible to other forms of degradation, such as erosion and salinization. In this concern, Mondal (2021) revealed that since preindustrial times, the average global temperature has increased by 1.1 °C, and because of growing quantities of greenhouse gases, it may reach 4 °C by the twenty-first century. The physical, chemical, and biological characteristics of soil fertility are anticipated to be impacted by this shift because of rising temperatures, altered precipitation patterns, and elevated greenhouse gas concentrations (Abdel Kawy and Ali 2012). Measures for adaptation and mitigation are required to reduce these consequences. The effects of climate change on soil characteristics and fertility processes are examined in this research.

Reduction in Agricultural Productivity

Abdullahi et al. (2023) demonstrated that the increasing temperatures is causing soil degradation, which is directly impacting the productivity of agriculture in the Nile Delta. Coastal governorates, which are already experiencing saltwater intrusion and erosion, are experiencing reduced yields due to declining soil quality. Inland governorates are also facing a decrease in crop productivity because of severe salinization, nutrient depletion, and loss of organic matter in the soil (Tawfik et al. 2024). With rising temperatures, the demand for water for crops is increasing, putting additional stress on water resources and contributing to soil degradation through over-irrigation and the accumulation of salts in the soil. Furthermore, the results mentioned earlier by Abdel Rahman (2023) indicated that the increase in temperature significantly affects soil degradation in both coastal and inland governorates of the Nile Delta. Coastal areas are more affected by salinization and erosion, while inland regions witness accelerated loss of organic matter, depletion of nutrients, and erosion. With these factors combining, the agricultural productivity of the Nile Delta, which is one of Egypt's most fertile regions, is facing significant challenges (Attaher et al. 2009). If adequate mitigation measures are not implemented, the soils in the region will continue to degrade, posing a threat to food security and livelihoods.

Changes in Precipitation Patterns in the Nile Delta

The fluctuating precipitation patterns associated with climate change are projected to result in more unpredictable rainfall, encompassing periods of both drought and intense storms. Prolonged dry spells have the potential to compromise the stability of soil, rendering it more susceptible to erosion when heavy precipitation eventually occurs (Negm et al. 2019). Simultaneously, heightened rainfall intensity raises the likelihood of both erosion and waterlogging as soils struggle to rapidly absorb substantial amounts of water (Sowers and Weinthal 2021). Despite the Delta region receiving comparatively low levels of rainfall, alterations in the timing and strength of precipitation are anticipated to have substantial repercussions (Abdel Kawy and Ali 2012). Increased variability in rainfall may lead to more frequent occurrences of droughts and sudden floods. Droughts would worsen existing water scarcity challenges, while floods could amplify soil erosion and degrade agricultural land (Conway et al. 2019). Additionally, shifts in rainfall patterns within the Nile Basin could impact the flow of the Nile River, upon which the Delta relies for irrigation.

The information presented in Table 13.1 provides a breakdown of the variations in annual precipitation totals in eight governorates located in the Nile Delta (Alexandria, Dakahlia, Damietta, Gharbia, Kafr El-Sheikh, Menofia, Port Said, and Sharkia) over a span of 23 years (2000–2023). The following is an in-depth examination of the patterns and notable shifts identified in each governorate.

Alexandria

During the initial years from 2000 to 2010, Alexandria experienced varying levels of precipitation, with a relatively stable average of 106.1 mm. However, there was an increase in 2002 (109.7 mm) and sharp decline in 2010 (22.9 mm). In the mid period from 2011 to 2019, there was a small recovery in precipitation, peaking in 2015 (172.0 mm) and 2016 (173.2 mm). Moving on to the later years from 2020 to 2023, there was a significant increase in 2020, with precipitation reaching a high of 312.1 mm. This was followed by a sharp drop in 2023 (158.0 mm), although still higher than the averages in the early 2000s. This drastic change in the last period suggests a potential anomaly that could be linked to climate variability (Table 13.1).

Dakahlia

From 2000 to 2010, Dakahlia experienced moderate precipitation, with 64.1 mm in 2000 and values fluctuating between 31.8 mm (2005) and 67.8 mm (2002). The year 2010 marked the lowest precipitation at 18.5 mm, reflecting a general decrease across many governorates. During the mid period (2011–2019), there was a consistent increase in precipitation, reaching 122.1 mm in 2015 and peaking at 150.1 mm in 2017. In the later years (2020–2023), precipitation sharply rose to 286.9 mm in 2020 but declined in subsequent years, with 102.6 mm recorded in 2023, still higher than the early years such as 2001 and 2003 (Table 13.1).

Damietta

From the early 2000s, Damietta experienced some fluctuations in precipitation, starting at a relatively high rate of 114.3 mm and it was increased to 116.8 mm in 2004. However, in 2010, like other regions, it experienced its driest year, with total

Table 13.1 Changes in total yearly precipitation among the eight governorates representing the Nile Delta over 23 years

Year	Alexandria	Dakahlia	Damietta	Gharbia	Kafr El-Sheikh	Menofia	Port Said	Sharkia
2000	106.1	64.1	114.3	64.1	93.6	51.8	188.3	69.6
2001	74.0	44.7	82.9	44.7	67.1	35.3	121.0	47.2
2002	109.7	67.8	103.7	67.8	102.4	51.8	144.2	67.2
2003	71.2	37.6	77.5	37.6	61.9	23.3	113.4	37.7
2004	91.4	55.7	116.8	55.7	81.1	33.0	178.7	56.6
2005	58.4	31.8	67.8	31.8	50.7	17.7	84.3	30.8
2006	85.6	60.4	93.6	60.4	80.7	41.6	156.0	62.3
2007	71.3	44.6	79.9	44.6	63.3	27.0	124.4	44.1
2008	80.2	49.8	84.4	49.8	73.6	33.7	148.8	52.3
2009	58.8	44.3	86.4	44.3	55.0	21.2	139.4	45.3
2010	22.9	18.5	33.9	18.5	21.2	16.6	97.7	20.4
2011	87.6	65.8	99.5	65.8	83.7	48.2	196.3	65.6
2012	78.9	55.6	112.4	55.6	73.3	36.7	175.9	57.3
2013	70.9	46.0	113.9	46.0	63.2	26.8	182.9	47.6
2014	58.8	59.8	84.4	59.8	59.5	45.8	165.4	62.1
2015	172.0	122.1	160.1	122.1	164.1	79.2	196.1	118.7
2016	173.2	109.1	157.6	109.1	171.9	93.4	166.2	99.6
2017	192.2	150.1	136.6	150.1	185.9	122.7	103.2	145.4
2018	122.2	80.8	150.9	80.8	114.2	71.9	221.0	83.2
2019	105.5	75.1	119.9	75.1	100.6	54.4	136.2	74.4
2020	312.1	286.9	311.5	286.9	312.8	278.3	277.1	288.0
2021	367.4	171.7	205.3	171.7	344.7	122.3	215.7	157.3
2022	202.5	136.1	225.3	136.1	199.9	84.5	255.1	133.8
2023	158.0	102.6	140.4	102.6	151.5	103.2	161.6	105.1

precipitation dropping to an extremely low value, namely, 33.9 mm. Throughout the mid period from 2011 to 2019, there was a general increase in precipitation, with notable peaks in 2015 (160.1 mm) and in 2016 (157.6 mm). In the later years, from 2020 to 2023, precipitation increases dramatically to 311.5 mm in 2020, which is one of the highest records. However, in 2023, a decrease in the precipitation to 140.4 mm occurred, which still remained higher than the levels in the early 2000s, continuing the overall pattern of increasing variability (Table 13.1).

Gharbia

During the early 2000s, Gharbia received low value of rainfall, namely, 64.1 mm. From 2000 to 2009, the precipitation varied from 31.8 mm in 2005 to 44.3 mm in 2009. In 2010, the precipitation dropped to 18.5 mm, following a similar pattern in other regions. Between 2011 and 2019, there was a gradual rise in precipitation, reaching its peak in 2017 at 150.1 mm. In the later years, from 2020 to 2023, Gharbia experienced a significant increase in rainfall in 2020 with 286.9 mm, continuing the trend of heavy rainfall seen in other regions. This was followed by a decrease to 102.6 mm in 2023 (Table 13.1).

Kafr El-Sheikh

In the year 2000, Kafr El-Sheikh experienced a relatively high total yearly precipitation of 93.6 mm. From 2001 to 2010, there were moderate fluctuations in precipitation, with the lowest recorded in 2010 at 21.2 mm, indicating a significant decrease. Throughout the mid period from 2011 to 2019, the total yearly precipitation in Kafr El-Sheikh steadily increased, reaching its highest points in 2015 at 164.1 mm and in 2017 at 185.9 mm. The total yearly precipitation experienced a remarkable increase in 2020, reaching 312.8 mm, which marked the highest recorded for this region. In the later years from 2020 to 2023, similar to other governorates, Kafr El-Sheikh experienced a spike in 2020 with 312.8 mm, which decreased to 151.5 mm by 2023, still significantly higher compared to earlier years (Table 13.1).

Menofia

The annual precipitation in Menofia governorate from 2000 to 2010 was at its highest in 2000, with 51.8 mm, and reached its lowest point in 2005, recording only 17.7 mm. In 2010, there was a significant decrease in total yearly precipitation, with a recorded value of 16.6 mm, the lowest among all governorates for that year. During the period from 2011 to 2019, there was a noticeable increase in the total yearly precipitation, with peaks in 2015 (79.2 mm) and 2016 (93.4 mm). The total annual precipitation reached a high of 278.3 mm in 2020, reflecting regional patterns, followed by a reduction to 103.2 mm in 2023, showing a decrease after 2020 but still higher than the previous decade (Table 13.1).

Port Said

In the year 2000, Port Said received 188.3 mm of precipitation, which was one of the highest levels among the governorates. From 2000 to 2010, Port Said had higher precipitation compared to other governorates, with amounts ranging from 188.3 mm in 2000 to 97.7 mm in 2010, which was the lowest recorded precipitation in the series for Port Said although it still remained higher than in other regions. During the mid period (2011–2019), the precipitation remained relatively high, reaching peaks of 196.3 mm in 2011 and 196.1 mm in 2015. In the later years (2020–2023), Port Said experienced a sharp increase in precipitation, with 277.1 mm in 2020 and 215.7 mm in 2021, before dropping to 161.6 mm in 2023 (Table 13.1).

Sharkia

In 2000, Sharkia governorate received a total annual precipitation of 69.6 mm, which was considered moderate compared to other governorates. In the subsequent years (2000–2010), Sharkia experienced low and fluctuating precipitation, with values ranging from 20.4 mm in 2010 to 69.6 mm in 2000, following the general decline in 2010 across the Nile Delta. From 2011 to 2019, there was a moderate increase in precipitation, reaching a peak of 118.7 mm in 2015 and rising to 145.4 mm in 2017. In the later years (2020–2023), similar to other governorates, Sharkia witnessed a significant increase in 2020 (288.0 mm), followed by a decrease to 105.1 mm in 2023 (Table 13.1).

To sum, throughout 2020, there was a noticeable increase in rainfall in several governorates of the Nile Delta, such as Alexandria, Dakahlia, Gharbia, Kafr El-Sheikh, and Port Said, indicating an unusual weather occurrence or a departure from typical climate patterns. Furthermore, from 2011 to 2019, there was a consistent upward trend in annual precipitation in several governorates, including Gharbia, Kafr El-Sheikh, and Port Said. Following the peak in 2020, most governorates experienced a general decline in precipitation levels. By 2023, there was a significant decrease in rainfall although it remained higher than in previous years, signaling ongoing variability in precipitation patterns.

Accordingly, in the governorates of the Nile Delta, there has been a clear uptick in the variability of precipitation patterns over the last two decades. The notable increase in 2020, succeeded by a decline in precipitation levels in the subsequent years, suggests a potential shift in local climate trends, possibly associated with worldwide climate change. This fluctuation underscores the importance of monitoring and adapting agricultural and water resource management methods to cope with erratic precipitation patterns.

Agricultural Practices to Combat Soil Degradation in the Nile Delta

As an important agricultural area of Egypt, the Nile Delta is more vulnerable to the effects of climate change, including drought, salinization, waterlogging, and sea level rise. The region's already precarious hydrological balance, which has been further strained by human activities like excessive irrigation and inadequate water management, exacerbates these difficulties. Many adaptive strategies are being used in this region to guarantee the sustainability of agriculture. The purposes of these methods are to protect crops from environmental stresses, increase soil resilience, and optimize water usage efficiency. The impacts of salinization and drought on farming are of great importance as these two factors pose the greatest challenges to the global food supply security (Al-Mailam et al. 2023).

Combating Drought

Drought conditions in the Nile Delta are expected to worsen due to climate change. Reduced rainfall and higher temperatures result in decreased water availability for irrigation, leading to significant stress on crops. To address these challenges, several drought-adaptation practices are being implemented:

Drought-Resistant Crops/Varieties

One of the best ways to deal with water constraint is to introduce crop kinds that can withstand drought. Due to their improved root systems, decreased water consumption, and increased water absorption efficiency, these varieties are genetically adapted to endure protracted droughts (Seleiman et al. 2021). To sustain crop yield in increasingly dry circumstances, drought-tolerant wheat and maize varieties are being developed and marketed in the Nile Delta by the Ministry of Agriculture.

Efficient Irrigation Techniques

Adopting water-efficient irrigation methods is crucial in areas where water is limited. Techniques such as drip irrigation and sprinkler systems have been introduced to replace traditional flood irrigation in Egypt, which is highly inefficient and leads to water wastage. Drip irrigation, in particular, delivers water directly to the plant roots, reducing evaporation and improving water use efficiency by up to 40–50% (Abou-El-Magd et al. 2016). This method is increasingly being adopted in the Nile Delta to conserve water while maintaining crop yields.

Soil Moisture Conservation

Conserving soil moisture through practices such as mulching, reduced tillage, and cover crops is also being promoted to combat drought. Mulching, which involves covering the soil with organic or synthetic materials, reduces water evaporation, moderates soil temperature, and improves soil moisture retention. Cover crops, such as legumes, are grown during off-seasons to protect the soil from erosion and improve its water-holding capacity (Abd El-Wahed et al. 2019). Cover crops and organic amendments can enhance soil health and structure, as well as strengthen plants' resistance to drought and salinity.

Using microbes that promote plant development is one possible strategy. By a number of processes, including osmolyte synthesis, ion homeostasis modulation, and antioxidant system stimulation, these bacteria can increase plant tolerance to salt and drought (Gupta and Huang 2014).

Addressing Salinization

One of the biggest issues affecting agriculture in the Nile Delta is salinization. When soil salt concentrations rise to a point where they might stunt plant development and lower agricultural output, this is known as salinization (Sakadevan and Nguyen 2010). The incursion of salty water from the Mediterranean Sea, coupled with the use of saline groundwater for irrigation, has led to increasing salt concentrations in the soil, lowering its fertility and production.

More efficient irrigation techniques and improved drainage systems can help manage soil salinity (Hayat et al. 2020). Furthermore, the use of plant growth-promoting bacteria can help mitigate the negative effects of salinity on plants (Kumar et al. 2020). Furthermore, researchers have identified a number of halophytic (salt-tolerant) plant species that can be cultivated in saline environments (Mukhopadhyay et al. 2020). These plants not only tolerate high levels of salt but can also actively sequester salt in their tissues, potentially reducing soil salinity over time (Mukhopadhyay et al. 2020; Sakadevan and Nguyen 2010). In addition, salinity problems require sustainable management strategies, including identifying and further developing crop cultivation practices adapted to saline conditions, enhanced drainage systems, using salt-tolerant varieties/crops, and exchanging knowledge and transferring practical and adaptive solutions (Kotb et al. 2000). In the end, to maintain sustainable agricultural output in the face of these escalating environmental concerns, a combination of genetic, agronomic, and technological techniques will probably be required (Qadir et al. 2008; Türkan and Demiral 2009; Mukhopadhyay et al. 2020). Saline land management has also demonstrated potential for agroforestry systems that incorporate salt-tolerant plants and shrubs. In general, to sustainably increase agricultural productivity in the face of these escalating environmental constraints, a multidimensional approach integrating technological, ecological, and management-based techniques will be essential (Qadir et al. 2008; Sakadevan and Nguyen 2010). A number of measures are being taken into consideration to lessen the consequences of salinization:

Salt-Tolerant Crop Varieties

To preserve agricultural production in saline soils, crop types that are tolerant to salt stress must be developed and adopted. In the Nile Delta, salt-tolerant wheat, barley, and rice cultivars have made possible for farmers in regions that were previously thought to be too salty for conventional crops (Hassanein et al. 2016). These cultivars can withstand and are even produced in highly salinized environments because of their physiological modifications.

Improved Drainage Systems

One of the best strategies to stop or lessen salinization is to improve soil drainage, in which poor drainage causes salt to accumulate in the root zone. According to Qadir et al. (2001), improving drainage aids in the removal of surplus water, which removes dissolved salts and lowers their concentration in the soil. This lessens the detrimental effects of excessive salinity on plants by preventing salt from building up at the surface where plant roots collect water. Additionally, good drainage promotes better soil conditions and sustainable farming methods by allowing salts to flow away from the root zone (Mukhopadhyay et al. 2020). To eliminate surplus water and salts from the soil, drainage systems, including underground drainage networks, are being built throughout the Nile Delta's impacted areas. These devices serve to wash salts out of the root zone, decreasing the negative effects of salinity on crops (Ismail et al. 2020).

Leaching and Soil Amendments

Salinity can be also being reduced via leaching, which is the act of flooding the soil with a lot of water to wash off salts. However, this approach requires a significant supply of fresh water, which is becoming limited (Ismail et al. 2020). Furthermore, adding soil amendments, such gypsum, can assist to enhance soil structure and displace sodium ions, reducing the likelihood of salinization (El-Ramady et al. 2020). Moreover, Mohamedin (2012) conducted a study on the effectiveness of leaching processes combined with gypsum or compost in improving salt-affected soil, rice growth, and nutrient uptake in the Nile Delta's El-Hamoul area. The results showed that leachate volumes increased with increasing gypsum application rates, causing faster water percolation through the gypsum-treated soil. Moreover, compost combined with gypsum treatment was more effective than compost alone in leaching salt-affected soil and was entirely safe, as well as more profitable than leaching with water alone.

Managing Waterlogging

Abdrabo and Hassaan (2014) put spotlight on another major economic problem dominated in the Nile Delta, namely, waterlogging, which is brought primarily by inadequate drainage and excessive irrigation. It lowers the amount of oxygen present in the soil, resulting in anaerobic conditions that damage root systems and lower crop yield. Several adaptive methods are being used to alleviate waterlogging:

Raised Beds Planting

Raised beds planting is a process where crops are cultivated on elevated rows or beds, allowing excess water to drain away from the root zone. This technique is appropriate for locations that frequently experience excessive rainfall because it increases soil aeration and decreases waterlogging (Ahmed et al. 2017). It has been demonstrated that planting on raised beds boosts agricultural yields and enhances the Nile Delta's water usage efficiency.

Improved Irrigation Scheduling

Waterlogging may be avoided by carefully planning of irrigation schedules based on crop water requirements and soil moisture content, which helps in minimizing over-irrigation. Farmers may lower their risk of waterlogging by applying the appropriate quantity of water at the correct time with the use of weather-based irrigation systems and soil moisture sensors (Moursy et al. 2018).

Subsurface Drainage

By eliminating extra water from the root zone, subsurface drainage systems are also employed to reduce waterlogging. In order to keep water from building up in the soil, these systems are made up of subterranean pipes that gather and redirect surplus water. Subsurface drainage is a frequently used technique in the Nile Delta to increase soil health and lessen waterlogging (Abu-Zeid et al. 2014).

Adapting to Sea Level Rise

A long-term concern to the Nile Delta is sea level rise, which can cause saltwater intrusion into freshwater systems and agricultural areas. Arable land is lost and salinity rises as a result. Tarolli et al. (2024) highlighted on several measures that are being implemented to adapt to the consequences of sea level rise:

Coastal Protection Measures

Coastal protection structures, such as seawalls, levees, and dikes, are being constructed to prevent seawater intrusion into agricultural areas. These structures act as physical barriers, protecting farmland from saltwater inundation and reducing the risk of salinization (El-Raey 2016). In addition, soft engineering solutions, such as the restoration of coastal wetlands and mangroves, are being explored as natural barriers to sea level rise.

Land Use Zoning and Relocation

In areas most vulnerable to sea level rise, land use zoning is being implemented to restrict agricultural activities and promote alternative land uses, such as aquaculture. In extreme cases, farmers are being encouraged to relocate to less vulnerable areas to protect their livelihoods and reduce the risk of crop loss due to rising sea levels (Fahim et al. 2017). Furthermore, Roy et al. (2022) pointed that as major danger to coastal populations and environments worldwide, sea level rise is one of the most catastrophic consequences of climate change and global warming. Around the planet, it would seriously damage coastal wetlands and pollute groundwater and surface water with salt. Rice fields and freshwater aquaculture facilities would become salinized, low-lying coastal areas would flood, and an increase in salinity-tolerant insects that cause various diseases would force people to migrate to poor developing countries, especially those in East, Southeast, and South Asia (Ducassou et al. 2009). At the same time, it can be damage the coastal infrastructure and assets, such as highways, seaports, airports, schools, hospitals, and many other.

Desalination and Freshwater Management

To combat the intrusion of saline water, the use of desalination technologies for irrigation is being explored. Desalination plants can convert seawater into fresh water, providing an alternative water source for agriculture in coastal areas. In addition, improved freshwater management practices, such as the construction of reservoirs and the efficient allocation of freshwater resources, are being promoted to ensure a reliable water supply for agriculture in the Nile Delta, particularly in areas affected by sea level rise and salinity (Kariman et al. 2023).

Conclusions

Based on the previously discussed issues, it can be inferred that the Nile Delta is confronted with several climate change-related problems, such as sea level rise, waterlogging, salinization, and drought. To counter these dangers, a variety of adaptive agricultural techniques are being used. Improved irrigation methods, soil moisture conservation strategies, the installation of appropriate drainage systems, and the introduction of crops resistant to salt and drought are some examples of these activities. The future of agriculture in this area also depends on initiatives to manage freshwater supplies and shield coastal communities from sea level rise.

In conclusion, a comprehensive strategy that incorporates the combination of physical and bioremediation technologies is required to successfully address the problem of soil degradation in the Nile Delta. Egypt can prevent soil deterioration, improve agricultural output, save the environment, and promote sustainable socio-economic development in the area by effectively deploying these technologies and making sure they are maintained on a regular basis.

In addition to raising the general standard of living for the local population, the effective use of these tactics will support Egypt's larger objectives of sustainable development and food security. The agricultural industry in the Nile Delta will need to remain resilient in the face of continuous climate change, which will need sustained innovation and investment in sustainable farming methods.

References

Abd El-Azeem SAM (2020) Impacts of climate change on microbial activity in agricultural Egyptian soils. Springer Water Book Ser (SPWA) 4:97–114

Abd El-Wahed MS, Farahat M, Atta-Alla M (2019) Mulching and cover cropping for improved soil moisture retention in arid regions. J Arid Environ 162:27–33

Abdel Kawy WAM, Ali RR (2012) Assessment of soil degradation and resilience at Northeast Nile Delta, Egypt: the impact on soil productivity. Egypt J Remote Sens Space Sci 15:19–30

Abdel Rahman MAE (2023) An overview of land degradation, desertification and sustainable land management using GIS and remote sensing applications. Rend Lincei Sci Fis Nat 34:767–780. https://doi.org/10.1007/s12210-023-01155-3

Abdrabo MA, Hassaan MA (2014) Economic valuation of sea level rise impacts on agricultural sector: Damietta Governorate, Egypt. J Environ Prot 5:87–95

Abdullahi B, El-Naggar A, Omar M, Murtala AI, Lawal M, Mosa A (2023) Land degradation, causes, implications, and sustainable management in arid and semi-arid regions: a case study of Egypt. Egypt J Soil Sci 63:659–676

Abou El-Magd I, Zakzouk M, Ali EM, Foumelis M, Blasco JMD (2024) Exploring the potentiality of InSAR data to estimate land subsidence of the Nile Delta. Egypt J Remote Sens Space Sci 27:342–355. http://creativecommons.org/licenses/by-nc-nd/4.0/

Abou-El-Magd MM, El-Sherif M, Hegab S (2016) Water management in the Nile Delta: a case for efficient irrigation systems. Agric Water Manag 178:186–193

Abou-Hadid AF (2006) Chapter 5: impact of climate change: vulnerability and adaptation food production. In: Tolba MK, Saab NW (eds) Arab Environment climate change—impact of climate change on Arab countries. Edition: 2009 report. The Arab Forum For Environment and Development, pp 63–74

Abou-Kota ME, Hassaballa H, Elhini M, Ganzou SK (2024) Land degradation, desertification and environmental sensitivity to climate change in Alexandria and Beheira, Egypt. Egypt J Soil Sci 64:167–180

Abuzaid AS (2018) Assessing degradation of flood plain soils in North East Nile Delta, Egypt. Egypt J Soil Sci 58:135–146

Abu-Zeid M, El-Zawahry AM, El-Nahry AH (2014) Subsurface drainage systems for improving agricultural productivity in waterlogged areas of the Nile Delta. Irrig Drain 63:382–391

Ahmed MM, El-Sherif H, Darwish R (2017) Raised bed planting as an adaptation to waterlogging in Egypt's Nile Delta. Int J Agric Res 9:156–164

Al-Mailam M, Arkeh J, Hamzawy A (2023) Climate change in Egypt: opportunities and obstacles. CARNEGIE Environment for International Peace. https://carnegie-production-assets.s3.amazonaws.com/static/files/Al-Mailam_et_al_Egypt_Climate_2.pdf

Attaher SM, Medany MA, Abou-Hadid AF (2009) Possible adaptation measures of agriculture sector in the Nile Delta to climate change impacts. Adv Sci Res 3:123–126

Badawy A, Sultan M, Abdelmohsen K et al (2024) Floods of Egypt's Nile in the 21st century. Sci Rep 14:27031. https://doi.org/10.1038/s41598-024-77002-8

Balwan WK, Kou S (2021) Wetland—an ecological boon for the environment. East African Scholars J Agric Life Sci 4:38–48

Batey T (2009) Soil compaction and soil management—a review. Soil Use Manag 25:335–345

Bear J, Sorek AH, Sorek S, Quazar D, Herrera I (1999) Seawater intrusion in coastal aquifers, concepts, methods and practices. Kluwer Academic Publishers, Dordrecht. ISBN 0-7923-5573-3

Conway D, Nicholls RJ, Brown S, Tebboth MGL, Adger WN, Ahmad B, Biemans H, Crick F, Lutz AF, De Campos RS, Said M, Singh C, Zaroug MAH, Ludi E, New M, Wester P (2019) The need for bottom-up assessments of climate risks and adaptation in climate-sensitive regions. Nat Clim Chang 9:503–511

Ducassou E, Migeon S, Mulder T, Murat A, Capotondi L, Bernasconi S, Zurich E, Mascle J (2009) Evolution of the Nile deep sea turbidite system during the Late Quaternary: influence of climate change on fan sedimentation. Sedimentology 56:2061–2090

Eelnahry A, Doluschitz R (2009) Climate change and its impacts on the coastal zone of the Nile Delta, Egypt. Environ Earth Sci 59:1497–1506

El-Marsafawy SM (2007) Impact of climate change on sunflower crop production in Egypt. Egypt J Agric Res Porto Portugal 85:1547–1563

El-Raey M (2010) Impacts and implications of climate change for the coastal zones of Egypt. Coast Zone Clim Change 7:31–50

El-Raey M (2016) Adaptation to climate change in the coastal zones of Egypt. Coast Res J 25:189–204

El-Ramady H, Domokos-Szabolcsy E, Fári M (2020) Soil amendments to combat salinization in the Nile Delta: a review. J Soil Sci Plant Nutr 20:1–16

El-Ramady H, Brevik EC, Abowaly ME, Ali RA, Moghanm FS, Gharib MS, Mansour H, Fawzy ZF, Prokisch J (2024) Soil degradation under a changing climate: management from traditional to nano-approaches. Egypt J Soil Sci 64:287–298

Elbasiouny H (2018) Assessment of environmental sensitivity to desertification, soil quality and sustainability in an area of the North Nile Delta, Egypt. Egypt J Soil Sci 58(4):399–415. https://ejss.journals.ekb.eg/article_21553.html

Elham MA, El-Magd IA (2016) Impact of human interventions and coastal processes along the Nile Delta coast, Egypt during the past twenty-five years. Egypt J Aquat Res 42:1–10

Elhini M, Hassaballa H, Simpson NP, Balbaa M, Ibrahim R, Mansour S, Abou-Kota M, Ganzour S (2024) The land degradation and desertification-socioeconomic nexus in Egypt's delta region: a case study on Alexandria and Beheira. Heliyon 10:e31165

Elsayed AH, Nashwan MS, Eltahan MH, Shahid S (2024) Millions more Egyptians will be exposed to drought by 2100 under the goals of the Paris climate agreement. Int J Disaster Risk Reduct 101. https://doi.org/10.1016/j.ijdrr.2024.104257

Elshinnawy IA, Almaliki AH (2021) Vulnerability assessment for sea level rise impacts on coastal systems of Gamasa Ras El Bar Area, Nile Delta, Egypt. Sustain For 13(7):3624. https://doi.org/10.3390/su13073624

Eman-Mounir (2024) Egypt's Nile Delta under threat, part 1: the sea Engulfs Kafr El-Sheikh. Earth Journalism Network

Fahim YA, Shalaby A, El-Raey M (2017) Land use zoning and coastal protection in response to sea level rise in the Nile Delta. Environ Monit Assess 189:96

FAO (2018) The future of food and agriculture—alternative pathways to 2050. FAO, Rome, 224 pp. Licence: CC BY-NC-SA 3.0 IGO

Frouz J (2018) Effects of soil macro- and mesofauna on litter decomposition and soil organic matter stabilization. Geoderma 332:161–172

Gupta B, Huang B (2014) Mechanism of salinity tolerance in plants: physiological, biochemical, and molecular characterization. Int J Genomics (Hindawi Publishing Corporation) 1:1–18. https://doi.org/10.1155/2014/701596

Hammam AA, Mohamed ES (2020) Mapping soil salinity in the East Nile Delta using several methodological approaches of salinity assessment. Egypt J Remote Sens Space Sci 23(2):125–131. https://doi.org/10.1016/j.ejrs.2018.11.002

Hamza MA, Anderson WK (2005) Soil compaction in cropping systems. A review of the nature, causes and possible solutions. Soil Tillage Res 82:121–145

Hassanein HS, El-Sherif H, Abdel-Rahman M (2016) Salt-tolerant crops as a solution to salinization in Egypt's Nile Delta. Plant Stress J 12:265–275

Hayat K, Zhou Y, Menhas S, Bundschuh J, Hayat S, Ullah A, Wang J, Chen X, Zhang D, Zhou P (2020) Pennisetum giganteum: an emerging salt accumulating/tolerant non-conventional crop for sustainable saline agriculture and simultaneous phytoremediation. Environ Pollut (Elsevier BV) 265:114876. https://doi.org/10.1016/j.envpol.2020.114876

Hossain MS, Majumde AK (2018) Impact of climate change on agricultural production and food security: a review on coastal regions of Bangladesh. Int J Agric Res Innov Tech 8(1):62–69

Hussein GIAE (2011) Desertification process in Egypt. In: Brauch H et al (eds) Coping with global environmental change, disasters and security. Hexagon series on human and environmental security and peace, vol 5. Springer, Berlin/Heidelberg. https://doi.org/10.1007/978-3-642-17776-7

Ismail A, Saleh M, Dawood M (2020) Managing salinity in irrigated agriculture: case studies from the Nile Delta. J Water Resour Manag 34:2143–2155

Kaiser MF, El Rayes A, Ghodeif K, Geriesh B (2013) GIS data integration to manage waterlogging problem on the eastern Nile Delta of Egypt. Int J Geosci 4:680–687

Kariman H, Shafieian A, Khiadani M (2023) Small scale desalination technologies: a comprehensive review. Desalination 567:116985. https://doi.org/10.1016/j.desal.2023.116985

Kotb THS, Watanabe T, Ogino Y, Kenneth K (2000) Soil salinization in the Nile Delta and related policy issues in Egypt. Agric Water Manag 43:239–261

Kraamwinkel CT, Beaulieu A, Dias T et al (2021) Planetary limits to soil degradation. Commun Earth Environ 2:249. https://doi.org/10.1038/s43247-021-00323-3

Kumar A, Singh S, Gaurav AK, Srivastava S, Verma JP (2020) Plant growth-promoting bacteria: biological tools for the mitigation of salinity stress in plants. Front Med 11. https://doi.org/10.3389/fmicb.2020.01216

Malhi GS, Kaur M, Kaushik P (2021) Impact of climate change on agriculture and its mitigation strategies. Sustainability 13:1318. https://doi.org/10.3390/su1303131

Mohamed E, Belal RRA, Ali AS, Hendawy EA (2019) Land degradation. In: The soils of Egypt. Springer, Cham, pp 159–174

Mohamedin AAM (2012) Effect of leaching with adding gypsum and rice straw compost on improving salt affected soil and rice yield. In: Minia international conference for agriculture and irrigation in the Nile Basin countries, 26–29 March 2012, El-Minia, Egypt

Mondal S (2021) Chapter 28: impact of climate change on soil fertility. In: Climate change and the microbiome, sustenance of the ecosphere. Springer

Moursy M, Youssef A, Khalil M (2018) Improved irrigation scheduling using soil moisture sensors for water conservation in Egypt's Nile Delta. Irrig Sci 36:417–427

Mueller ND et al (2012) Closing yield gaps through nutrient and water management. Nature 490:254–257

Mukhopadhyay R, Sarkar B, Jat H, Sharma PC, Bolan N (2020) Soil salinity under climate change: challenges for sustainable agriculture and food security. J Environ Manage (Elsevier BV) 280:111736–111736. https://doi.org/10.1016/j.jenvman.2020.111736

Negm AM, Sakr S, Abd-Elaty I, Abd-Elhamid HF (2019) An overview of groundwater resources in Nile Delta aquifer. https://www.researchgate.net/publication/324370442

Olson K, Kome G (2024) Review and analysis of Africa's lifelines: the Nile River and the Aswan High Dam. Open J Soil Sci 14:741–764. https://doi.org/10.4236/ojss.2024.1412036

Qadir M, Schubert S, Ghafoor A (2001) Amelioration strategies for saline soils. Adv Agron 70:197–246

Qadir M, Tubeileh A, Akhtar J, Larbi A, Minhas P, Khan MA (2008) Productivity enhancement of salt-affected environments through crop diversification. Land Degrad Dev (Wiley) 19:429–453. https://doi.org/10.1002/ldr.853

Rezaei EE, Webber H, Asseng S et al (2023) Climate change impacts on crop yields. Nat Rev Earth Environ 4:831–846. https://doi.org/10.1038/s43017-023-00491-0

Ribbe L, Ahmed S (2006) Transboundary water management in the Nile River Basin. Technol Resour Manag Dev 4:13–27

Roy B, Penha-Lopes GP, Uddin MS, Kabir MH, Lourenço TC, To A (2022) Sea level rise induced impacts on coastal areas of Bangladesh and local-led community-based adaptation. Int J Disaster Risk Reduct 73:102905

Sahoo G, Majid WA (2020) Effect of climate change on land degradation. Int J Innov Eng Manage Res 9:483–494

Sakadevan K, Nguyen ML (2010) Extent, impact, and response to soil and water salinity in arid and semiarid regions. Adv Agron (Elsevier BV):55–74. https://doi.org/10.1016/b978-0-12-385040-9.00002-5

Saqr AM (2023) Groundwater resource management: a quantitative-based study for meeting sustainable development in Southwestern Nile Delta, Egypt. PhD thesis, Egypt-Japan University of Science and Technology, Alexandria, Egypt

Saqr MA, Abd-Elmaboud M (2024) Management of saltwater intrusion in coastal aquifers: a review and case studies from Egypt. Online J Eng Sci 3(1):1–17

Sefelnasr A, Sherif M (2013) Impacts of seawater rise on seawater intrusion in the Nile Delta aquifer. Egypt Ground Water 52:1–13

Seleiman MF, Al-Suhaibani N, Ali N, Akmal M, Alotaibi M, Refay Y, Dindaroglu T, Abdul-Wajid HH, Battaglia ML (2021) Drought stress impacts on plants and different approaches to alleviate its adverse effects. Plan Theory 10:259–284

Sharaan M, Ibrahim MG, Moubarak H, ElKut AE, Romya AA, Hamouda M, Soliman A, Iskander M (2024) A qualitative analysis of climate impacts on Egyptian ports. Sustain For 16(3):1015. https://doi.org/10.3390/su16031015

Siam MS, Eltahir EAB (2017) Climate change enhances interannual variability of the Nile river flow. Nat Clim Chang 7(5):350–354

Soane BD, Ball BC, Arvidsson J, Basch G, Moreno F, Roger-Estrade J (2012) No-till in northern, western and south-western Europe: a review of problems and opportunities for crop production and the environment. Soil Tillage Res 118:66–87. https://doi.org/10.1016/j.still.2011.10.015

Sowers J, Weinthal E (2021) Humanitarian challenges and the targeting of civilian infrastructure in the Yemen war. Int Aff 97:157–177

Stanley JD, Clemente PL (2014) Clay distributions, grain sizes, sediment thicknesses, and compaction rates to interpret subsidence in Egypt's Northern Nile Delta. J Coast Res 30:88–101

Stanley JD, Clemente PL (2017) Increased land subsidence and sea-level rise are submerging Egypt's Nile Delta coastal margin. GSA Today 27. https://doi.org/10.1130/GSATG312A.1

Tarolli P, Luo J, Park E, Barcaccia G, Masin R (2024) Soil salinization in agriculture: mitigation and adaptation strategies combining nature-based solutions and bioengineering. iScience 27:108830. http://creativecommons.org/licenses/by/4.0/

Tawfik M, Hoogestege J, Moussa M, Hellegers P (2024) 'Squeezing out' the Nile Delta's drainage water to irrigate Egypt's desert land. WaterSA 16:157. https://doi.org/10.3390/w16010157

Telo da Gama J (2023) The role of soils in sustainability, climate change, and ecosystem services: challenges and opportunities. Ecologies 4:552–567

The World Bank Annual Report (2014) Main report (English). Report number 91155

Tolba E, Selim T, Salem A (2018) Integrated coastal zone management in the North of Egypt a case study: natural gas treatment plant west of Port Said. Port-Said Eng Res J 22(1):34–51. https://doi.org/10.21608/pserj.2018.32287

Trenberth K (2011) Changes in precipitation with climate change. Clim Res 47:123–138. https://doi.org/10.3354/cr00953

Türkan İ, Demiral T (2009) Recent developments in understanding salinity tolerance. Environ Exp Bot (Elsevier BV) 67:2–9. https://doi.org/10.1016/j.envexpbot.2009.05.008

Impact of Climate Change Drivers on Soil Microbial Communities and Nutrient Cycling in Agricultural Soils in the Nile Delta of Egypt

14

Ebtsam M. Morsy and A. A. M. Mohamedin

Abstract

Soil microbial populations and nutrient cycling are significantly impacted by climate change, especially in areas with intense agriculture like the Nile Delta. Soil fertility, organic matter decomposition, and plant-microbe interactions have been impacted by changes in microbial composition brought about by rising temperatures, changing precipitation patterns, elevated atmospheric CO_2 levels, and soil salinization. Furthermore, a rise in the incidence of soilborne illnesses and disturbances in the nitrogen cycle resulted from climate change. In this chapter, the ways in which climate change drivers affect microbial diversity and function were investigated. The chapter also highlights how soil microbial populations may help mitigate climate change by fixing nitrogen, sequestering carbon, and controlling plant diseases biologically, with an emphasis on their effects on nutrient cycling and greenhouse gases emissions. Furthermore, viable methods for improving soil resilience were examined. In order to lessen the negative consequences of climate change on soil ecosystems, the chapter highlights the significance of incorporating microbial-based remedies into agricultural strategies.

Keywords

Climate change drivers · Soil microbial · Nutrient cycling · Greenhouse gases emission · Carbon sequestration

E. M. Morsy (✉)
Agricultural Microbiology Department, Soils, Water and Environment Research Institute, Agricultural Research Center, Giza, Egypt

A. A. M. Mohamedin
Field Drainage Department, Soils, Water and Environment Research Institute, Agricultural Research Center, Giza, Egypt

A. E.-H. Zohry, S. Ouda (eds.), *Resilient Agroecosystems*, Sustainability Sciences in Asia and Africa, https://doi.org/10.1007/978-981-96-9630-7_14

Introduction

The emission of atmospheric greenhouse gases (GHGs) has a substantial influence on the global climate, affecting ecosystems, agriculture, and food security (Bolan et al. 2024). The drivers of global climate change, such as rising temperatures, changed precipitation patterns, increasing carbon dioxide, and changed nitrogen deposition, have a significant impact on soils (Telo da Gama 2023). Monocropping, frequent tillage, and shallow plowing are examples of intensive agricultural techniques that greatly contribute to soil degradation and increased atmospheric CO_2 (Sahoo and Majid 2020). According to Ferreira et al. (2022), continuous human-driven disruptions of the world's soil-climate systems pose a threat to our ability to adapt to or adequately mitigate some of the most detrimental effects of climate change due to changes in soil microbial metabolic outputs and changes in soil properties like aggregate stability, erosion, and leaching. Furthermore, soil biological activities that are essential to ecosystem health face increased challenges due to the expected 2 °C increase in global temperatures, uneven precipitation, and severe droughts. This highlights the urgent need for sustainable soil management (El-Ramady et al. 2019).

Nutrient dynamics, which include the conversion of minerals, organic molecules, and other energy sources necessary for ecosystem functioning, are what soil organisms depend on (Muscolo and Sidari 2008). Climate change has a significant influence on these dynamics, which are susceptible to shifts in soil characteristics and biotic/abiotic variables (Yuan et al. 2024). Extreme weather, changed moisture levels, increased CO_2, and increasing temperatures are important factors (Wu et al. 2024). The soil microbiota's metabolic activities liberate a variety of nutrients from organic debris and transform them into forms that plants can absorb, but they may also absorb a wide range of contaminants (Telo da Gama 2023). Because soil bacteria are susceptible to changes in abiotic environmental conditions, Suman et al. (2022) show how these climatic variations, in conjunction with several governance and socioeconomic factors, have caused changes in soil nutrient availability and contamination risk. They stated that the primary drivers of climate change are local emergencies that are more heavily reliant on rainfall and interventions will be impacted the uncertainty of precipitation reductions and increases, while heat-induced vulnerability detections are likely to be directly observed from the exposure of soil microbes to higher temperatures during system robustness checks. Indirect suppressions between the blocks linked to increased accumulation and decreases of pollutants and other contaminant-related relationships, as well as disturbances by root-associated microbes with the recently introduced activities, can result from increased concentrations or changed rhizosphere interactions.

One of the most significant biotic elements affecting soil health is the soil microbial community, which transforms pollutants from harmful substances to innocuous ones and lessens their levels and harmful impacts (Wang et al. 2024). Many scientists have been working extensively to investigate how the health of soil microbial communities changes as a result of climate change. Nevertheless, little is known about how microbial communities react to the primary causes of climate change during the growth season and off-season, especially when looking at the world as a whole. The processes behind these changes are controversial and probably context dependent despite the general consensus that climate change will have a significant impact on microbial composition and nutrient cycle rates. The microbial groups that are most active in various agricultural sectors and complicating variables like substrate availability will determine how big these predicted impacts are (Angon et al. 2024).

The rich and fertile soils of the Nile Delta, which have been preserved for thousands of years by the yearly natural fertilization with debris from the periodic inundation of the Nile flood pulse, are the principal source of the region's considerable agricultural productivity. Since the natural Nile organic pulse has been replaced by synthetic fertilizers and manure, the Nile Delta's agricultural production has continuously depended on the ongoing and growing usage of artificial nitrogen fertilizers to offset the loss of soil nutrients. In addition to intensifying soil nutrient flow, the growing use of inorganic fertilizers also indirectly contaminates water supplies and increases GHGs emissions (El-Ramady et al. 2019).

Over the past 40 years, Nile Delta agriculture has seen substantial changes due to climate change. Sea level rise, increases in air and sea surface temperatures, and changes in precipitation patterns are the most commonly documented aspects of climate change. These changes may affect local temperature, precipitation, and evaporation rates, resulting in frequent droughts and floods disasters. The architecture, behaviors, and functional characteristics of soil microbial communities are both directly and indirectly impacted by these climatic changes (Wang et al. 2024). The information regarding the adaptation of Nile Delta soil microbiomes to climate change will help in assessing and formulating strategies that will preserve and restore the natural productivity of these agriculturally significant ecosystems in the future as the microbial communities and enzyme activities in the soil adapt to various conditions.

This chapter investigates the ways in which climate change drivers affect microbial diversity and function. It also highlights how soil microbial populations may help mitigate climate change by fixing nitrogen, sequestering carbon, and controlling plant diseases biologically, with an emphasis on their effects on nutrient cycling and GHGs emissions. Furthermore, viable methods for improving soil resilience were examined. In order to lessen the negative consequences of climate change on soil ecosystems, the chapter highlights the significance of incorporating microbial-based remedies into agricultural strategies.

The Role of Soil Microbial Biodiversity to Mitigate Climate Change in the Nile Delta

Abd El-Azeem (2020) further highlights the vulnerability of Egyptian soils, including those in the Nile Delta, to climate change and the potential impacts on soil microbial activity. Soil microbial biodiversity, particularly in the Nile Delta, can play a crucial role in mitigating climate change. Hossain and Majumde (2018) and

Frouz (2018) emphasize the importance of soil microflora in reducing the adverse effects of climate change and in maintaining key ecosystem functions. Furthermore, Kassem et al. (2019) underscores the need for "microbiome conservation" as a strategy to mitigate the ecological impacts of climate change, which could be particularly relevant in the Nile Delta. In the same context, Delgado-Baquerizo et al. (2016) confirmed that soil microbial biodiversity plays a crucial role in mitigating climate change effects, particularly in sensitive ecosystems like the Nile Delta. These studies collectively suggested that preserving and enhancing soil microbial biodiversity in the Nile Delta can contribute to climate change mitigation efforts.

The relationship between microbial diversity and ecosystem functions has been documented by Niu et al. (2016), indicating that diverse microbial communities enhance soil health, nutrient cycling, and carbon storage, which are essential for resilience against climate change impacts. The study has shown that higher microbial diversity contributes to the stability and functionality of soil ecosystems. Moreover, diverse microbial communities can enhance soil nitrogen cycling, which is vital for plant productivity and ecosystem health. This is particularly important in the Nile Delta where agricultural practices are heavily influenced by soil health. Furthermore, the presence of a rich microbial community can help suppress soil-borne diseases, thereby improving crop resilience and productivity (Wang et al. 2017). Other study indicated that soils with higher microbial diversity are more resistant to stressors, including those induced by climate change, such as drought and flooding (Banerjee and Heijden 2022).

Moreover, the interaction between plant diversity and soil microbial communities is significant. Diverse plant communities can promote microbial diversity by providing various litter types and root exudates that support different microbial populations (Lange et al. 2015). This relationship could be critical in the Nile Delta where agricultural practices often lead to monocultures that can diminish microbial diversity and, consequently, soil health (Salimi et al. 2021). Similarly, Tiemann et al. (2015) revealed that enhancing plant diversity through crop rotation and cover cropping can foster a more resilient soil microbial community, which in turn can improve soil carbon storage and mitigate GHGs emissions.

Climate variability, particularly changes in precipitation and temperature, can significantly impact soil microbial diversity. For example, drought conditions can lead to reduced microbial diversity and function, which negatively affects soil health and productivity (Philippot et al. 2013). In the context of the Nile Delta, where climate change is expected to exacerbate water scarcity, maintaining microbial diversity becomes increasingly important for sustaining agricultural productivity and ecosystem resilience (Yetgin 2023).

Furthermore, Bebber and Richards (2020) spotlight on the essential of managing soil health through practices that promote microbial diversity. Strategies such as reducing tillage, minimizing chemical inputs, and implementing organic farming practices can enhance soil microbial communities. These practices not only improve soil structure and fertility but also contribute to carbon sequestration, which is vital for mitigating climate change (Suman et al. 2022). Wagg et al. (2014) concluded that promoting soil microbial biodiversity in the Nile Delta is a critical strategy for

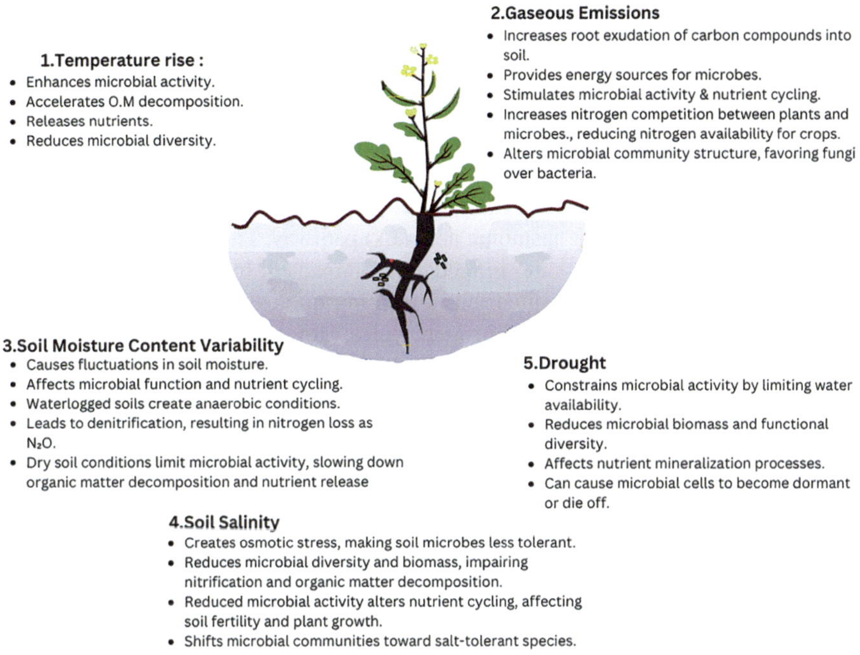

Fig. 14.1 Influence of environmental factors on soil microbes

enhancing soil health and resilience against climate change. By fostering diverse microbial communities through sustainable agricultural practices, it is possible to improve ecosystem functions, support agricultural productivity, and contribute to climate change mitigation efforts.

Figure 14.1 presents information about how various environmental factors influence soil microbes. It is divided into five key sections. The image also includes an illustration of a plant with visible soil layers, showing microbial interactions underground.

Impact of Temperature Changes on Microbial Activity

Microorganisms are divided into three main groups based on their optimal growth temperatures: mesophiles, which thrive between 20 and 45 °C; psychrophiles, which grow in cold environments with optimal temperatures ranging from 15 (or lower) to 20 °C; and thermophiles, which prefer higher temperatures of 50 °C or more (Keenleyside 2019). Temperature is a critical factor in microbial activity, enzyme functionality, disease progression, and plant-fungal interactions, with optimal ranges and effects varied among different organisms and soil processes (Barnard et al. 2013).

The activity of important enzymes in nitrogen-fixing bacteria is strongly impacted by temperature fluctuations. According to Dubey et al. (2019), vanadium nitrogenase is the most efficient enzyme for N_2 fixation at lower temperatures (around 5 °C). Molybdenum nitrogenase, on the other hand, is more effective at warmer temperatures (around 30 °C) because of its greater affinity for nitrogen.

According to Goicoechea (2020), elevated CO_2 levels raise soil temperature, which increases plant defenses against *V. dahliae* by promoting the salicylic acid (SA) route and reducing the jasmonic acid (JA) pathway. The diversity of soil bacteria and metabolic sports are immediately impacted by rising temperatures. Temperature typically causes microbial breathing quotes to increase, most likely due to changes in the carbon and nitrogen cycles. However, prolonged exposure to high temperatures might result in a loss of sensitive microbial species, which reduces the range and usefulness of microorganisms (Singh et al. 2010).

Furthermore, soil thermals can affect the microbial population directly or indirectly through their impacts on plant growth, nutrient availability, and soil moisture. As the temperature rises, soil microbial communities will have to either adapt or die. Accordingly, the 5 °C temperature increase changed the soil bacteria and increased the bacterial to fungal ratio (B/F ratio) in the community (De Angelis et al. 2015). However, studies have shown that the composition of microbial communities has changed, with a decrease in fungal-to-bacterial ratios and an increase in Gram-positive bacteria. The decrease in fungus with warming may be caused by the traits that bacteria have that give them a competitive advantage at higher temperatures, such as faster growth rates and better nutrition competition (Barnard et al. 2013). Soil microorganisms are essential to the fertility and productivity of soil ecosystems because they produce mucus compounds that help maintain soil stability and structure. In addition to managing the C and N cycles, litter decomposition, C mineralization, nutrient release, and N nitrification and denitrification, they establish symbiotic partnerships with plants (Andronov et al. 2012).

Higher temperatures have been shown in several studies to have a direct impact on arbuscular mycorrhizal (AM) fungal colonization, community structure, and function. For example, warmer conditions encourage the growth of AM fungi (Heinemeyer et al. 2006; Compant et al. 2010). In particular, Hawkes et al. (2008) looked at the effects of soil temperature on AM fungal communities linked to Plantago lanceolata. They suggested that raising the soil temperature from 14 to 26 °C drastically changes the distribution and makeup of AM fungal hyphal networks, causing them to grow more actively in warmer soils and produce more vesicles. In a similar vein, studies by Heinemeyer and Fitter (2004) and Gavito et al. (2005) found that temperatures between 24 and 30 °C boost glucose absorption, encourage the growth of vast extraradical mycelium, and improve AM fungal root penetration. However, various AM fungal species have different ideal temperatures. These results were explained by the fact that greater temperatures caused extraradical mycelium to respire at higher rates and allocate carbon to the rhizosphere more quickly. Such temperature-induced impacts on AM fungal biomass have important ecological ramifications because of the crucial role that AM fungi play in plant nutrition, community dynamics, and ecosystem responses to climate change (Heinemeyer et al. 2006).

The colonization of arbuscular mycorrhizal fungi (AMF) is also influenced by temperature. Higher temperatures promote root elongation and plant-microbe interactions, which in turn improve AMF colonization, according to studies by Frater et al. (2018) and Jerbi et al. (2020). In contrast, AMF nutrition uptake declines at lower temperatures, which results in less colonization. Due to regional soil properties like the carbon-to-nitrogen (C:N) ratio, different AMF species have different ideal temperatures for growth and development (Jerbi et al. 2020).

Excessive temperatures can either positively or negatively affect the overall performance of symbiotic and asymbiotic bacteria in the Egyptian soil. Several bacteria cannot develop at 38 °C, and the ideal temperature range for growth inside the medium is 28–30 °C. However, certain rhizobia lines may be cultivated well at 40 °C or 44 °C. Furthermore, whereas R. leguminosarum bv. Phaseoli have superior warmth tolerance and may form strong nodules with their host legumes, increased soil temperature results in worthless nodules (Zahran 2001). When soil temperatures rose to around 30 °C, nitrogen-fixing *Azotobacter* and *Clostridium* showed the most interest in Egyptian soils; nevertheless, Clostridium is more suited to higher soil temperatures than *Azotobacter* (Abd-El-Malek 1971).

Furthermore, soil microorganisms react differently to global temperature fluctuations or temperature increases. For instance, dwindling substrate supplies may be the reason of microbe populations growing with rising temperatures. However, at higher temperatures, indicators of fungus and actinobacteria decreased (Zogg et al. 1997). The nitrification sports in the Egyptian clay loam soil have been shown to increase with temperature, with the highest charge being verified at 30 °C. Activation and population duration of ammonium-oxidizing or nitrite-oxidizing microorganisms, nitrate-decreasing microorganisms, and maximum likely numbers are best achieved around 25 or 30 °C. Some nitrifiers and denitrifiers, however, have also been shown to grow at or above 50 °C (Saad and Conrad 1993).

In contrast, frost is a major wintertime weather phenomena in Northern Egypt, especially in the Nile Delta. Plant development and soil microbial activity are negatively impacted by frost, which happens on nights when temperatures fall below freezing (Torres and Miarnau 2024). Low temperatures have a significant impact on soil microbial communities by delaying important soil processes, changing community composition, and decreasing metabolic activity especially when coupled with frost occurrences (Torres and Miarnau 2024). In this regard, frost episodes frequently result in sudden temperature shifts in the soil, which can harm or inhibit microbial communities that are not acclimated to such harsh circumstances. This leads to the emergence of psychrophilic microorganisms, which are able to endure and proliferate in cold and icy conditions (Drepper et al. 2021). Frost also decreases soil enzymatic activity, such as cellulases and proteases, which are critical for the breakdown of organic materials. Frost-related cycles of freezing and thawing also physically disturb the structure of the soil, which has an impact on microbial communities and nutrient availability (Henry 2007).

Moreover, frost directly affects soil properties, causing biological and physical alterations that might influence microbial activity and ecosystem functions. Through freeze-thaw cycles, frost events cause soil to expand and contract, physically breaking apart soil aggregates (Drepper et al. 2021). In addition to dislodging microorganisms and upsetting nutrient routes, this procedure might enhance soil porosity, which may improve oxygen availability (Sharratt and Glenn 2006). Frost releases organic materials into the soil by breaking down microbial cells. Although this offers a brief nutritional rush, the cold inhibits microbial activity, which limits the nutrients' rapid use (Schimel and Clein 1996).

Despite the negative consequences of frost episodes, nematodes may be greatly impacted by them as their activity, survival, and interactions within the soil-food relationship are disrupted. Nematodes, particularly those that are not acclimated to cold conditions, are directly harmed by freezing temperatures or frost episodes. Frost can change the balance of soil fauna by killing sensitive nematode species (Forge and Simard 2001). Nematodes become immobile due to frost-induced soil freezing, which limits their capacity to consume root systems or microbial populations. As a result, their function in microbial population control and nutrient cycling is diminished. However, cold conditions have distinct effects on fungi that cause root rot, such as Fusarium and Phytophthora, which can alter their pathogenicity and interactions within the soil ecosystem (Raymond and Wharton 2016).

Generally speaking, freezing temperatures lessen the pathogenic fungi's immediate impact on plants by slowing their proliferation and sporulation. But when circumstances improve, spores or resting structures, such as chlamydospores, might reappear because they remain latent in the soil (Pautasso et al. 2012). Additionally, Henry (2007) highlighted that cold inhibits fungus, and it also lowers the activity of rival microorganisms such as antagonistic bacteria. When temperatures rise again, fungus may unintentionally profit from this decreased competition. Frost's combined impacts on nematodes, root rot fungus, and microbial populations cause a number of significant alterations in soil processes. Decomposition rates are slowed by frost, which causes organic matter to accumulate. In the near run, this decreases plant nutrient availability even as it improves soil carbon storage. Continuous cycles of freezing and thawing also diminish microbial diversity, which may make soil ecosystems less resilient to future environmental shifts. Last but not least, alterations in nematode populations and microbial activity brought on by cold might modify crop development patterns, thereby delaying planting seasons or lowering yields (Raymond and Wharton 2016).

Impact of Gaseous Emissions on Microbial Activity

Because they take part in the nitrogen and carbon cycles, microbial communities are essential to reducing the effects of climate change. Controlling microbial processes, such as carbon fixation and methane oxidation, can affect GHGs emissions and sequestration (Crowther et al. 2015). By controlling GHGs including carbon dioxide (CO_2), methane (CH_4), and nitrous oxide (N_2O), microbial actions are essential to reducing the effects of climate change. The carbon and nitrogen cycles, among other biogeochemical cycles, depend on these microbes and have a direct effect on the release and sequestration of these gases (Singh et al. 2010). By fixing atmospheric carbon, promoting plant development, and decomposing organic materials, bacteria play a role in the carbon cycle. Microbial activity has a major impact on the stability of stored carbon in ecosystems that function as carbon sinks, such as permafrost, grasslands, and tropical forests. This stability determines whether the carbon is released into the atmosphere as CO_2 or stays sequestered in the soil (Weiman 2015).

Particularly through the fluxes of CO_2 and CH_4, soil microorganisms are essential in controlling carbon exchanges between the land and atmosphere. Climate variables including temperature, precipitation, and extreme weather occurrences have an impact on their metabolic processes both directly and indirectly (Singh et al. 2010). For example, increasing temperatures speed up microbial activity, which raises CO_2 emissions and feeds a vicious cycle that makes climate change worse (Crowther et al. 2015). Mostly through the emission of CO_2 during the breakdown of organic matter, a process that gets more intense with warming, microbes contribute to climate feedbacks. The breakdown rates of organic carbon molecules vary based on their substrate quality and temperature sensitivity, making this feedback process extremely complicated (Davidson and Janssens 2006). Additionally, Zhang et al. (2017) and Johnson et al. (2005) found that higher CO_2 levels promote plant development and enhance carbon allocation to roots, which in turn affects greenhouse gas emissions and microbial activity.

By aiding in the breakdown of organic materials, microbial enzyme systems help in obtaining carbon from the atmosphere into the soil. These systems can aid in lowering CO_2 levels through respiration and other metabolic activities, which helps sequester carbon and control atmospheric carbon concentrations (Zimmerman and Labonte 2015). Plants benefit from increased atmospheric CO_2 because it promotes photosynthesis, which increases root biomass and produces more root exudates. This extra carbon in the soil may provide soil microorganisms with more food sources and affect the organization of microbial communities, favoring fungal populations over bacterial communities because of the increase in carbon availability (Treseder and Allen 2000). Through improving carbon transport to plant roots, mycorrhizal fungi have a major impact on the makeup and activity of microbial communities. Moreover, by boosting nutrient intake and providing protection against root pathogens, mycorrhizal fungi's improved proliferation under elevated CO_2 might help plants and indirectly lower the frequency of root rot (Pritchard et al. 2008). However, because more root biomass means more entry sites for nematodes and fungi to infect, increasing CO_2 can also encourage the growth of certain root infections (Thornton and Gerber 2010). Furthermore, changes in the makeup of microbial communities under increased CO_2 may change the ratio of helpful to detrimental species, which might lower microbial diversity and resistance to infections (Cheng et al. 2014).

Methanotrophic bacteria are essential for reducing atmospheric CH_4 concentrations because they use CH_4 as an energy source. These bacteria work especially well in places that release a lot of methane, such as landfills and marshes (Shindell et al. 2012). However, the anaerobic environments, methanogens transforming CO_2 into CH_4 demonstrate the cyclical nature of methane in ecosystems (Charu et al. 2014).

Given that nitrogen is essential to the biosphere and may exist in a variety of gaseous forms, such as NH_4, NO, and N_2O, the nitrogen cycle and climate change are also intimately related. The nitrogen cycle in soil is driven by microbial activities such as nitrogen fixation, nitrification, assimilation, ammonification, and denitrification. Plants can receive bioavailable nitrogen from rhizobium bacteria through symbiotic nitrogen fixation. Furthermore, nitrogen is converted by bacteria into ammonia, which is then nitrified to produce nitrates, allowing plants to take nitrogen effectively (Zimmerman and Labonte 2015) as followed:

- *Nitrogen Fixation:* Using either free-living or symbiotic bacteria, microbes transform atmospheric nitrogen into ammonium, which is crucial for the cycling of nutrients (Zimmerman and Labonte 2015).
- *Nitrification:* Nitrosomonas and Nitrobacter, two soil bacteria, transform ammonia into nitrates that plants may take. This procedure promotes soil fertility and avoids ammonia toxicity (Singh and Kundu 2014).
- *Denitrification:* The cycle is completed when bacteria like Clostridium and Pseudomonas convert nitrates back to inert nitrogen gas, which enables nitrogen to return to the atmosphere. Waterlogged soils and other low-oxygen situations are frequent places for this process to occur (Zimmer 2010).

Emissions of GHGs have a major impact on soil microbial populations and the roles they play. The increase in GHGs emissions and their impact on soil ecosystems provide difficulties to agricultural and environmental sustainability in Egypt, a country that is particularly sensitive to climate change because of its dry and semi-arid climate.

Impact of Soil Moisture on Microbial Activity

According to Schimel et al. (2007), soil moisture is crucial for controlling microbial activity, which in turn influences microbial diversity, metabolic activities, and vital ecological processes including the breakdown of organic matter and the cycling of nutrients. Water availability has a significant impact on microbial activity although extreme moisture levels, either too high or too low, inhibit microbial operations (Smith et al. 2003). Ideal moisture conditions promote respiration, enzyme activity, and biomass generation. The lack of oxygen in wet soils promotes anaerobic microorganisms including denitrifies and methanogens, which raises emissions of N_2O and CH_4 (Linn and Doran 1984). On the other hand, because of osmotic stress and limited nutrient transport, dryness causes microbial dormancy (Manzoni et al. 2012). Important nutrient cycles are driven by soil microorganisms, and changes in their activity brought on by variations in moisture can affect crop yield and soil fertility, especially in agricultural systems that rely on consistent microbial contributions to the cycling of carbon and nitrogen (Smith et al. 2003).

Variations in soil moisture alter the makeup of microbial populations. While sandy soils dry up more rapidly, reducing microbial survival, fine-textured soils, like clay, hold water longer, promoting microbial activity during dry seasons (Six et al. 2002). The dynamics of soil moisture also affect GHGs emissions; drier circumstances boost the release of CO_2, while higher moisture encourages the creation of CH_4 (Smith et al. 2003). Furthermore, the "Birch effect" describes how abrupt increases in moisture after dry spells can cause a strong jump in microbial activity and a subsequent significant increase in CO_2 emissions (Birch 1958). Because they affect the equilibrium of GHGs in the atmosphere and play a role in global climate feedback systems, such variations in microbial activity can have a substantial impact on the environment (Adviento-Borbe and Linquist 2016).

Rising sea levels and unpredictable precipitation patterns brought forth by climate change have increased the unpredictability of soil moisture. According to Steinmuller and Chambers (2018), seawater intrusion in coastal locations changes microbial activities, which raises GHGs emissions and nutrient leaching. Microbial diversity changes in coastal soils exposed to seawater, with sulfate-reducing bacteria, become more prevalent in nitrogen cycling and decomposition (Chambers et al. 2013). More frequent flooding in low-lying agricultural regions is another long-term impact of sea level rise. This causes alternating anaerobic and aerobic conditions, which disturbs the structure and function of soil microbes (Moomaw et al. 2018). According to Grzyb et al. (2020), flooding compacts soil, restricts gas exchange, lowers oxygen availability, and causes microbial communities to move toward anaerobic dominance. Long-term flooding alters soil microbial communities by reducing fungal populations and boosting anaerobic bacteria (Bai 2000). This change has an effect on soil carbon sequestration because it slows down the release of stored carbon by decreasing the activity of some microbial groups that aid in the breakdown of organic matter (Sjögersten et al. 2014).

Osmoregulation, dormancy, and enzyme synthesis are some of the adaptive strategies that microbial communities have evolved to deal with moisture stress (Barnard et al. 2013). Actinobacteria are one type of microorganism that survives drought by saving water and going into hibernation (Bouskill et al. 2016). According to Guhr et al. (2015), certain fungi create hyphal networks that help distribute water across the soil environment, enhancing drought resistance (Naylor et al. 2017). Other bacteria, on the other hand, show a boom-and-bust cycle, with populations quickly growing when moisture becomes available and declining when circumstances dry up (Smercina et al. 2019). The abundance of microbial species that can adapt to harsh environments, plant microbe interactions, and soil texture all affect the microbial communities' capacity to withstand moisture changes (Treseder et al. 2018). Predicting ecosystem resilience under climate change and creating efficient soil management plans need an understanding of these microbial responses (Jerbi et al. 2020).

Irrigation techniques are essential in agricultural contexts for preserving the equilibrium of soil moisture and promoting microbial activity. Microbial-driven CH_4 and N_2O emissions have been demonstrated to be influenced by techniques such as alternating wetting and drying irrigation where drier circumstances improve denitrification and wetter conditions encourage methanogenesis (Zhang et al. 2012; Yao et al. 2013). By controlling soil moisture using cover crops, organic amendments, and efficient irrigation, microbial stability may be improved, promoting sustainable agricultural output and reducing the consequences of climate change (Adviento-Borbe and Linquist 2016). In order to ensure that natural ecosystems and agriculture can adjust to the growing uncertainty of the climate, future research should concentrate on incorporating microbial dynamics into predictive models for soil health management (Wieder et al. 2013).

Impact of Soil Salinity on Microbial Activity

One of the main environmental variables lowering agricultural yield in dry and semiarid regions is soil salinity, which is made worse by climate change. In the twenty-first century, it presents a serious obstacle to the world's food production. The magnitude of this problem is demonstrated by the fact that in Egypt, salty, saline sodic, and sodic soils make up around 37% of the agricultural soils in the Nile Delta (Pitman and Läuchli 2002).

Due to irrigation methods and the overuse of chemical fertilizers, Egypt's soils and groundwater are becoming more salinized, which has become a serious problem (Omar et al. 1994). Elevated salinity increases the osmotic potential of the soil, which limits aeration, decreases water permeability, and decreases plant absorption of water and nutrients. These elements diminish the sustainability of advantageous rhizosphere bacteria and have a detrimental effect on soil microbial activities, including enzyme functions. Unsuitable circumstances for microbial development are produced by toxic salts and the degradation of soil structure (Lu et al. 2015).

Because of their large area-to-volume ratio, high permeability of cell membrane, and rapid turnover rate, it is assumed that salinity reduces the biomass, activity, and community structure of microbes. This is primarily because osmotic pressure causes cells to shrink, which causes water to efflux from the microbial cells and slows their growth (Narsing et al. 2019; Mokrani et al. 2020).

Halophiles, or organisms that flourish in high-salt settings and need salt for growth, and halotolerant organisms, or organisms that can survive in saline conditions but do not require high salt levels to proliferate, are the two primary categories of bacteria that can withstand salinity. These microbes frequently create special enzymes known as halozymes that work in salty environments. Cellulases, xylanases, proteases, amylases, lipases, and gelatinases are a few examples of these enzymes (Delgado-Garcia et al. 2012; Mukhtar et al. 2019). Halozymes can function well in saline conditions because of unique characteristics that make them dependent on or tolerant of salt (Gupta et al. 2016).

According to Mukhtar et al. (2019), halophilic microorganisms may create important biological molecules including phytohormones and exopolysaccharides, which are essential for plant microbiome interaction and also help to stabilize soil structures and retain water in soil particles. In salty environments, they are also helpful for bioremediation of these contaminants (Bergi and Trivedi 2020; Dastgheib et al. 2011). Large amounts of beneficial microbiomes in the rhizospheres of salt-tolerant plants allow them to develop and withstand severe salinity and drought (Mahmud et al. 2021). In order to adjust to low osmotic pressure, microorganisms and plants may store osmolytes. Numerous studies have demonstrated that salinity changes the form of microbial communities and reduces microbial activity and relative abundance (Mukhtar et al. 2019; Imhoff et al. 2020). Andronov et al. (2012) also discovered that when salinity gradients increase, the taxonomic makeup of bacterial and fungal communities changes.

The bacterium/fungi ratio may be higher in saline soils because fungi are more susceptible to salt stress than bacteria (Peiffer et al. 2013; Pérez-Jaramillo et al. 2016). The notion that salinity might be a crucial indication of microbial diversity and composition at the community level stems from the fact that different microorganisms have different resistance to salt, which alters the makeup of microbial communities (Gros et al. 2003). Proline and glycine betaine are the main organic osmolytes, but potassium ions are the most common inorganic materials used as osmolytes in bacteria that can withstand salt (Csonka 1989).

Beneficial microbiomes are found in the rhizospheres of salt-tolerant plants, which help them withstand drought and excessive salinity (Wichern et al. 2006). Although it takes a lot of energy, microbes and plants create osmolytes like proline and glycine betaine to control osmotic pressure. On the other hand, excessive salinity affects soil health, changes community makeup, and decreases microbial activity. Compared to bacteria and actinobacteria, fungi are especially vulnerable to salt stress; higher salinity inhibits spore germination and hyphal development and changes the shape of fungi (de Souza Silva and Fay 2012).

The majority of soil organisms are stressed by soil salinity, which is frequently caused by irrigation techniques and sea level rise. This lowers microbial activity and biodiversity. Particularly at risk are microorganisms and nematodes that are sensitive to salt, which can disturb soil ecosystems and reduce soil fertility. However, if beneficial bacteria are unable to compete, pathogenic nematodes and root rot pathogens may adapt to salty conditions and flourish (Singh et al. 2014). According to Rath and Rousk (2015), salinity tends to decrease the overall number of beneficial microorganisms such as Bacillus and Pseudomonas species, which are renowned for their ability to inhibit pathogens and promote plant development. This change raises the possibility of root rot by allowing salt-tolerant pathogenic organisms to take over. Although their ability to inhibit pathogens is often less than that of their non-salt-tolerant counterparts, some salt-tolerant beneficial bacteria, such as *Halomonas* spp., show promise in salty soils (Setia et al. 2011).

Impact of Drought on Microbial Activity

One important climatic occurrence that has a detrimental effect on food security and agricultural output is drought. The frequency and intensity of droughts are predicted to rise over the next 10 years, which will have a substantial impact on the soil microbiome, and it is controlled by temperature and moisture. According to Schimel et al.

(2007), during drought, the microbiomes of plant roots shift, preferring Actinobacteria and other Gram-positive species over the more prevalent Gram-negative microbes.

Different tactics are used by soil microorganisms to endure drought. They increase the production of extracellular polymeric substances (EPS), which act as an efficient response to drought stress, create suitable osmolytes as a defense mechanism, and collect organic chemicals to reduce solute potential without interfering with cellular metabolism. Due to their large hyphal networks, which enable them to absorb water and nutrients more efficiently than other microbes, fungi in particular are more drought resistant (Timmusk et al. 2014).

The growth and root colonization of mycorrhizae were either adversely or favorably impacted by the drought, in which it generally decreased mycorrhizal colonization. Davies et al. (2002) showed that drought increased the production of arbuscular and hyphal development of Glomus sp. Strain ZAC-19 while decreasing the colonization of roots by *Glomus fasciculatum*, which varies depending on the strain. Conversely, the findings of other research revealed that some mycorrhizal strains have adapted to drought circumstances (Augé 2001).

However, in Mediterranean regions, drought has a negative impact on soil enzyme activities, which either directly or indirectly affect photosynthetic output potential, water usage efficiency, and nutrient availability. According to Sardans and Peñuelas (2005), drought circumstances reduced the activity of soil enzymes such as urease, protease, β-glucosidase, and acid phosphatase but had no discernible effect on alkaline phosphatase activity.

Reduced soil moisture due to drought has a major effect on creatures that live in the soil, including as nematodes, root rot pathogens, and soil microorganisms. Nematodes may become less mobile due to low water availability, which would decrease their capacity to infect plant roots (Rillig et al. 2002). On the other hand, extended drought damages plants, weakening their root systems and increasing their vulnerability to root rot infections when they rehydrate (Schimel et al. 2007). Additionally, drought reduces microbial activity, which results in a decrease in the good bacteria that control pathogens and cycle nutrients. Beneficial bacteria that improve nitrogen fixation and promote plant development, such as *Azospirillum* and *Rhizobium* species, frequently show decreased activity in dry soils, which lessens their ability to defend against root diseases (Rillig et al. 2002). Root rot infections spread quickly because drought-adapted pathogens may take advantage of weaker plant defenses and become more virulent when water stress is reduced (Pritchard et al. 2008).

Specific Challenges in Egyptian Soils

For a number of reasons, Egyptian soils especially those in the Nile Delta and reclaimed desert areas are extremely susceptible to the effects of warming temperatures. According to Abou-Kota et al. (2024), Egypt's dry environment already causes severe heat stress on soil microbial communities, which lowers microbial diversity and functioning. The consequences of high temperatures are further exacerbated by the poor organic matter content of many Egyptian soils, which reduces soil water retention capacity and impairs microbial resilience to heat stress (Zhou et al. 2020). Although organic matter is necessary for improving soil structure, cushioning temperature extremes, and supplying microbial communities with vital nutrients, its lack in Egyptian soils renders these ecosystems more vulnerable to warming (Luo et al. 2020).

Additionally, the negative impacts of rising temperatures on soil microbial communities are made worse by water shortages, which is a recurring problem in Egypt because of a lack of freshwater supplies and erratic precipitation (Abou-kota et al. 2024). Reduced microbial biomass and enzymatic activity brought on by drought have a detrimental effect on soil fertility by lowering the availability of essential nutrients like phosphorus and nitrogen (Schimel et al. 2007). Degradation of the soil and a rise in GHGs emissions result from limited water availability, which also hinders microbially mediated activities including carbon cycling and organic matter breakdown (Manzoni et al. 2012). Furthermore, microbial communities are further stressed by excessive evaporation and soil salinization, which reduces their ability to maintain soil health (Rath and Rousk 2015).

By improving plant resistance to heat stress, maintaining soil ecosystems, and encouraging carbon sequestration, soil microorganisms can significantly reduce the consequences of climate change despite these obstacles (Jansson and Hofmockel 2020). Through improving root water absorption and enabling nutrient exchange, heat-tolerant microbes like some strains of Actinobacteria and mycorrhizal fungi can increase plant resilience to drought (Bouskill et al. 2016; Treseder et al. 2018). By means of enhancing soil moisture retention, raising carbon availability, and lowering temperature swings, the use of organic amendments, such as compost and biochar, can also support the maintenance of microbial diversity (Lehmann et al. 2011). Additionally, via maximizing water availability without overextracting finite freshwater resources, sustainable water management techniques like drip irrigation and deficit irrigation can promote microbial activity (Adviento-Borbe and Linquist 2016).

Despite rising temperatures, Egypt's agricultural productivity may be maintained and soil health preserved by incorporating certain soil management strategies. In order to improve climate adaption tactics, future research should concentrate on discovering robust microbial strains and enhancing soil amendment techniques. Furthermore, in light of continuing climate change, Egypt's food security may be protected by integrating microbial-based soil restoration techniques into national agricultural plans (Wieder et al. 2013).

Case Studies from Egyptian Soils

Egypt's agriculture and economy depend heavily on the Nile Delta, which is regarded as one of the world's most fertile areas. Nonetheless, this low-lying coastal region is especially vulnerable to the adverse consequences of climate change. The region's soil quality, water availability, and general biodiversity are all under risk due to rising sea levels, warming temperatures, and erratic weather patterns (El-Ramady et al. 2022). Known as the invisible builders of ecosystems, soil microbes play a crucial role in maintaining soil health, promoting nutrient cycling, and boosting plant production (Jansson and Hofmockel 2020).

Through changing important environmental parameters including salinity, moisture content, temperature, and organic matter content, climate change has a substantial influence on the form and function of microbial communities. Seawater intrusion brought on by sea level rise, for instance, raises soil salinity and favors salt-tolerant microbial species while decreasing the number of helpful bacteria that are essential for plant growth (Abd-Elgawad et al. 2023). Furthermore, higher temperatures have the ability to increase microbial respiration rates, which may hasten the breakdown of soil organic carbon and thus reduce soil fertility (Luo et al. 2020).

Knowing how soil microbial populations react to climate change is crucial to maintaining agricultural output in the Nile Delta. These negative consequences can be lessened by putting adaptation measures into practice, such as adopting improved soil management practices and introducing stress-tolerant microbial inoculants (Schimel et al. 2007). In the face of climatic difficulties, sustainable measures like as conservation tillage, organic amendments, and optimal irrigation can help maintain soil health over the long term and enhance microbial resilience (Adviento-Borbe and Linquist 2016).

Table 14.1 shows the effects of climate change on the distribution and diversity of microorganisms in Egypt's Nile Delta during the last 20-year period.

- *Temperature Rise:* Over the past 20 years, the Nile Delta's temperature has increased by around 1.2 °C, which has had a major impact on soil microbial dynamics (IPCC 2021). While heat-tolerant microorganisms like *Bacillus* and *Streptomyces* species have grown more prevalent, many temperature-sensitive bacteria and fungus have decreased (El-Ramady et al. 2017). This change in microbial communities, which favors species that do well at warmer temperatures, is visible both regionally and within soil layers. Furthermore, higher temperatures speed up the decomposition of organic matter by increasing soil respiration rates, which may ultimately result in nutrient depletion (Hassan et al. 2024).
- *Drought and Water Scarcity:* The region's soil ecosystems have suffered from decreased rainfall and an increased occurrence of drought episodes (Wassef and Schüttrumpf 2016). While drought-resistant organisms, such as spore-forming *Aspergillus* species, have become more common, microbial communities that depend on moisture have decreased (Abdul Rahman et al. 2021). Microbial biomass and enzyme activity, which are crucial for soil production, have declined as a result of these circumstances. Additionally, the availability of nitrogen for crops has decreased due to the negative impact on drought stress-sensitive nitrogen-fixing microorganisms such as *Rhizobium* (El-Sayed et al. 2018).
- *Sea Level Rise:* According to Wassef and Schüttrumpf (2016), the northern Nile Delta's soil salinity levels have increased due to the incursion of saline water from the Mediterranean Sea. As a result, halophilic bacteria, such as those in the

Table 14.1 Impact of climate changes on microbial diversity and distribution in the Nile Delta, Egypt

Climate factor	Observed changes in the Nile Delta	Impact on microbial diversity	Impact on soil microbial distribution
Temperature rise	Increase in annual temperature (~1.2 °C over 20 years)	Decline in temperature-sensitive species; increase in thermophilic microbes	Redistribution of microbial populations toward heat-tolerant species
Drought and water scarcity	Reduced annual precipitation and more frequent droughts	Decrease in moisture-dependent microbes; rise of drought-tolerant fungi and bacteria	Reduced microbial biomass and enzymatic activity
Sea level rise	Increased salinity from Mediterranean Sea intrusion into northern soils	Decline in freshwater microbes; growth of halophilic species	Uneven microbial distribution due to salinity gradients
Extreme flooding events	More frequent and intense flooding	Short-term increase in microbial diversity due to nutrient influx	Proliferation of anaerobic microbes in waterlogged soils
Agricultural adaptations	Increased use of fertilizers and pesticides to combat climate stress	Loss of native microbes; dominance of pesticide-resistant species	Shift toward pesticide-degrading microbial communities
Urbanization and land use changes	Expansion of urban areas reducing agricultural land	Decline in microbial diversity due to soil compaction and habitat loss	Lower microbial biomass in urban soils with reduced organic matter

Halomonas genus, have increased while freshwater-adapted microbial species have decreased (Abdul Rahman et al. 2021). Uneven microbial distribution brought on by salinity gradients has disrupted important nutrient cycle mechanisms and created microhabitats dominated by salt-tolerant species (Wassef and Schüttrumpf 2016).

- *Extreme Flooding Events:* As a result of altered rainfall patterns, flooding has become more often and severe. Long-term soggy conditions eventually decrease microbial diversity even if these occurrences initially increase it by importing nutrient-rich sediments (Hassan et al. 2024). Microbes such as methanogens, which produce methane, take over in anaerobic environments (El-Ramady et al. 2017).
- *Agricultural Adaptations:* Farmers have used more chemical pesticides and fertilizers to lessen the effects of climatic stress, which has changed the balance of soil microbial populations (El-Sayed et al. 2018). This change decreases total microbial richness and benefits bacteria resistant to pesticides. Microbial communities that can break down manmade substances gradually take over, which might result in imbalances in the health of the soil (Abdul Rahman et al. 2021).
- *Land Use Changes and Urbanization:* As urban areas have grown, agricultural lands have shrunk, which has a detrimental effect on soil structure and limits the amount of organic matter that can be added (Wassef and Schüttrumpf 2016).

Microbial diversity and biomass have decreased as a result of soil compaction and habitat loss, which has hampered the natural nutrient cycling processes and decreased soil fertility (El-Ramady et al. 2017).

In order to lessen the negative effects of climate change on microbial ecosystems, the Nile Delta urgently needs sustainable soil and water management techniques.

Impact of Climate Change on Soilborne Diseases in the Nile Delta

Egypt's reliance on agriculture in the Nile Valley and Delta makes it especially vulnerable to the severe effects of climate change on agricultural systems. Crop production is being severely impacted by drought, rising temperatures, erratic rainfall patterns, and sea level rise-induced salinization. These changes in the climate make it easier for soilborne illnesses to proliferate and worsen, which is a serious risk to plant health and agricultural productivity. Root rot pathogens (Pritchard et al. 2008) and plant-parasitic nematodes (Rillig et al. 2002) are two of the most important problems because they both flourish in environments that are constantly changing. These changes have the potential to upset the delicate balance between helpful soil microbes and dangerous pathogens, endangering plant health and harvests.

Effect of Rising Temperatures

Temperature increases are one of the main effects of climate change, and they have a direct impact on the activity of nematodes and soilborne diseases. The metabolic rates and reproduction of plant parasitic nematodes, especially those that impact commercially significant crops like cotton, tomatoes, and potatoes, are accelerated by warmer temperatures (Rillig et al. 2002). This exacerbates root damage by causing quicker life cycles, greater rates of reproduction, and larger population densities. Additionally, diseases that like warm soils, including Fusarium and Rhizoctonia, can erode root systems even more, making them more vulnerable to infections (Back et al. 2002). Higher temperatures, however, can also benefit helpful microorganisms like *Trichoderma* species, which inhibit root rot pathogens by producing antifungal compounds and competing for nutrients (Harman 2000). This dynamic interplay may cause changes in the microbial population when temperatures increase more, changing the ratio of diseases to beneficial species. Accordingly, it is anticipated that nematode infestations would become more widespread and severe throughout Egypt's Nile Delta, leading to large crop losses and decreased agricultural output (El-Ramady et al. 2013).

Changes in Precipitation Patterns

Rising temperatures and altered rainfall patterns lead to both drought-prone and waterlogged situations, which have varying effects on soil organisms. Anaerobic diseases such as *Fusarium*, *Pythium*, and *Rhizoctonia*, which flourish in low oxygen settings and inflict substantial root damage, are favored in wet soils because oxygen levels drop there (Stromberg et al. 2001; Hussain et al. 2020). Saturated soils, on the other hand, may hinder the establishment of beneficial aerobic bacteria that contribute to disease suppression (Schimel et al. 2007). However, because there is less moisture, nematode activity is often decreased during drought circumstances. Because weaker roots are more likely to get infected, plants under drought stress are more susceptible to root rot pathogens, especially when rehydration takes place (McElrone et al. 2013). Furthermore, changes in soil moisture can upset the microbial population, making it harder for good bacteria to successfully inhibit dangerous diseases.

Effect of Soil Salinity

Soil organisms are stressed by soil salinity, a problem that is becoming worse in the Nile Delta of Egypt as a result of sea level rise and unsustainable irrigation methods. Beneficial microorganisms and salt-sensitive nematodes are especially at risk, which lowers soil fertility and microbial diversity. However, harmful nematodes and root rot diseases frequently adjust to salty conditions and flourish in areas where helpful microorganisms cannot compete (Singh et al. 2023). Beneficial microorganisms like Bacillus and Pseudomonas species, which are recognized for their ability to control pathogens and promote plant development, tend to have lower numbers in salinity (Rath and Rousk 2015). This raises the danger of root rot diseases by allowing salt-tolerant pathogens to take control of the soil ecosystem. Halomonas species and other salt-tolerant beneficial bacteria may live in salty soils although they are usually less successful at suppressing pathogens than their non-salt-tolerant counterparts (Setia et al. 2011).

The effects of climate change on soilborne illnesses, which have gotten worse over the last 20 years, are especially dangerous in the Nile Delta. Major agricultural losses have been caused by the growth of plant parasitic nematodes and root rot diseases, which have been exacerbated by rising temperatures, changing soil moisture, increasing salt, and elevated CO_2 levels (Hassan et al. 2024). Heat-tolerant species continue to flourish in spite of intense heat, whereas nematodes like *Meloidogyne* spp. and fungus like Fusarium spp. benefit from higher temperatures for reproduction (Dutta and Phani 2023). While droughts push nematodes into dormancy, allowing them to survive until favorable circumstances return, changes in rainfall patterns, notably excessive wetness, promote nematode movement and the spread of diseases like Pythium spp. (El-Sayed and Kamel 2020). In addition to suppressing many diseases, salinity, which is made worse by irrigation techniques and saltwater intrusion, damages plants and increases their vulnerability to halotolerant organisms like Tylenchulus semipenetrans, which is the cause of decline in the yield of citrus trees (Mahfouz and Abd-Elgawad 2021). Additionally, elevated CO_2 levels enhance root growth, providing feeding sites for nematodes and encouraging fungal sporulation, which increases outbreaks of pathogens such as *Verticillium* spp. (Pathak et al. 2018; Wang et al. 2020). These elements highlight how urgently sustainable management techniques are needed to lessen the impact of climate change on the area's agriculture.

The influence of climate change on soilborne illnesses during the past 20 years in the northern region of Egypt (Nile Delta Soils) is shown in Table 14.2.

Economic and Agricultural Consequences

Nematode infestations and root rot diseases result in significant output losses in important staple crops including cotton, rice, and wheat according to Perry and Moens (2011). While root rot diseases completely kill roots, causing stunted development and crop loss (Pritchard et al. 2008), nematodes harm root systems, affecting the intake of water and nutrients (Rillig et al. 2002). Such losses pose significant economic concerns, endangering farmers' livelihoods and the stability of the national food supply in Egypt. Egyptian farmers frequently use chemical treatments to control soilborne illnesses in order to address these problems. But in addition to increasing production costs, this reliance also speeds up the development of pest and disease resistance and degrades the environment (Hafez and El-Ghanam 2010). Furthermore, extended chemical usage degrades soil over time and reduces microbial diversity, which is detrimental to soil health and endangers sustainable agriculture (Tahat et al. 2020).

Mitigation Strategies of Climate Change Drivers on Microorganisms

Microorganisms are essential for reducing the effects of climate change in a number of ways as followed:

- *Carbon Sequestration:* Applying biochar and agroforestry techniques improves soil organic matter and creates favorable conditions for helpful microorganisms, which improves carbon storage (Lehmann et al. 2006).
- *GHGs Reduction:* Nitrification inhibitors and a reduction in nitrogen-based fertilizers may both significantly reduce N_2O emissions from soils, which helps mitigate climate change (Lam et al. 2017).
- *Microbial Engineering:* A sustainable strategy for tackling climate issues is the introduction or selection of microbial strains with enhanced carbon sequestration capacity or lower GHGs emissions (Singh et al. 2010).
- *Microbial Inoculants:* Under stressful environmental circumstances, the use of microbial inoculants that are resistant to salt and drought can help restore soil microbial processes and preserve nitrogen cycling (Elnahal et al. 2022).

Table 14.2 Impact of climate changes in the Nile Delta on soilborne diseases (nematodes and root rots) over the past 20 years

Factor	Examples of soilborne diseases	Impact of climate changes	References
Temperature rise	*Meloidogyne* spp. (root-knot nematodes) *Fusarium* spp. (root rot)	Higher temperatures accelerate the life cycle of *Meloidogyne* spp., increasing their population and host damage. Warmer conditions favor the growth of *Fusarium* spp. and other pathogenic fungi, leading to more severe root rot outbreaks. Extreme heat (>35 °C) may reduce some nematodes but stress crops, making them more susceptible to root diseases.	Dutta and Phani (2023) and Hassan et al. (2024)
Changes in rainfall and soil moisture	*Pratylenchus* spp. (lesion nematodes) *Pythium* spp. (damping off and root rot)	Increased rainfall and poor drainage promote nematode mobility and the proliferation of root rot pathogens like *Pythium* spp. Prolonged droughts force nematodes into dormancy (anhydrobiosis) but also weaken plant defenses, making them vulnerable to infections once moisture returns.	El-Sayed and Kamel (2020)
Salinity	*Tylenchulus semipenetrans* (citrus nematode) *Rhizoctonia solani* (root rot)	Increased salinity (>5 dS/m) reduces the activity of many nematodes and pathogens but stresses crops, weakening their resistance to diseases like *Rhizoctonia solani*. Halotolerant nematodes and fungi may adapt and thrive in saline soils, exacerbating crop losses.	Mahfouz and Abd-Elgawad (2021) and Patankar et al. (2021)
CO_2 levels	*Meloidogyne* spp. (root-knot nematodes) *Verticillium* spp. (vascular wilt)	Elevated CO_2 levels may enhance nematode reproduction and fungal sporulation by boosting plant root biomass, which provides more feeding sites. Increased susceptibility of plants due to shifts in nutrient uptake and water use efficiency.	Pathak et al. (2018) and Wang et al. (2020)

– *Soil Amendments:* Organic inputs, such as compost and biochar, enhance soil structure, provide microbial habitats, and help buffer against salinity and drought stress (Lehmann et al. 2006).
– *Efficient Water Management:* Implementing strategies such as drip irrigation and controlled flooding minimizes water stress while optimizing microbial activity in the soil (Adviento-Borbe and Linquist 2016).

- *Crop Rotation and Cover Crops:* These practices foster microbial diversity, increase soil organic matter, and enhance soil resilience to climate variations (Tiemann et al. 2015).
- *Integrated Pest Management (IPM):* This practice is a comprehensive strategy for controlling nematodes and root rot diseases. IPM combines biological, physical, cultural, and chemical approaches. Techniques such as crop rotation, resistant crop varieties, and organic soil amendments like compost and biochar help in managing nematodes. Similarly, root rot diseases can be mitigated by improving soil drainage, cultivating resistant varieties, and avoiding excessive irrigation (Goot and Lee 2007).
- *Biological Control Alternatives:* The use of beneficial microorganisms, including nematode-trapping fungi provides an environmentally friendly method for managing nematodes and root rot pathogens. Unlike chemical treatments, these biological agents support soil microbial balance (Rahman et al. 2024).
- *Soil Health Management:* Organic farming, compost application, and reduced tillage practices improve soil structure, promote microbial diversity, and enhance resilience against climate-related stressors (Pimentel 2006).
- *Climate-Resilient Crop Varieties:* Developing and cultivating crops with resistance to nematodes, root rot, and environmental stressors such as drought, salinity, and high temperatures can boost agricultural productivity while minimizing dependence on chemical controls (Singh et al. 2007).
- *Monitoring and Predictive Tools:* Advanced soil and crop monitoring technologies enable early detection of nematode infestations and disease outbreaks, allowing timely intervention. Additionally, predictive models that integrate climate data help anticipate potential disease risks and facilitating proactive management (Brown et al. 2022).

Conclusions

Climate change's impact on soil microbial communities and nutrient cycling is becoming increasingly apparent, especially in agricultural regions like the Nile Delta where soil health directly impacts food security and environmental sustainability. Rising temperatures, unpredictable precipitation patterns, and increasing soil salinity are causing microbial dynamics to shift, biogeochemical cycles to be upset, and soilborne illnesses to spread. These problems reduce agricultural output, jeopardize long-term soil fertility, and increase the demand for chemical fertilizers and pesticides, all of which deteriorate soil health.

Soil bacteria provide practical strategies for climate change adaptation and mitigation in spite of these challenges. Microbially driven processes including carbon sequestration, nitrogen fixation, organic matter decomposition, and the avoidance of soilborne diseases are essential for preserving soil function and resilience. Soil fertility, structure, and crop resilience to climate stressors can all be improved by increasing soil microbial diversity through the use of microbial inoculants, biochar additions, and sustainable water management practices. Advanced soil monitoring techniques and predictive models can also aid in the early detection of environmental changes, allowing for swift intervention to mitigate their detrimental effects on agriculture.

Future research should focus on optimizing microbial-based treatments to enhance plant-microbe interactions, promote soil carbon sequestration, and develop climate-resilient crop types. The use of microbial technologies to conventional farming practices and sustainable soil management strategies might potentially be encouraged by policy frameworks. Cooperation between researchers, legislators, and farmers will be essential to developing all-encompassing plans to mitigate the effects of climate change on soil ecosystems.

Lastly, even though soil microbial populations and nutrient cycling face major challenges due to climate change, employing microbial-based solutions offers a viable path toward sustainable agriculture. The Nile Delta, an agricultural region vulnerable to climate change, is an important case study highlighting the potential and hazards of microbially driven climate adaptation. By encouraging microbial biodiversity and using innovative soil management strategies, agricultural systems can become more resilient to the ongoing effects of climate change, ensuring both environmental sustainability and long-term output.

References

Abd El-Azeem SAM (2020) Impacts of climate change on microbial activity in agricultural Egyptian soils. Springer Water Book Ser (SPWA) 4:97–114

Abd-El-Malek Y (1971) Free-living nitrogen-fixing bacteria in Egyptian soils and their possible contribution to soil fertility. Plant Soil 35(1):423–442

Abd-Elgawad H, Zinta G, Hornbacher J, Papenbrock J, Markakis MN, Asard H, Beemster GTS (2023) Elevated CO_2 mitigates the impact of drought stress by upregulating glucosinolate metabolism in Arabidopsis thaliana. Plant Cell Environ 46:812–830. https://doi.org/10.1111/pce.14521

Abdul Rahman NSN, Abdul Hamid NW, Nadarajah K (2021) Effects of abiotic stress on soil microbiome. Int J Mol Sci 22(16):9036. https://doi.org/10.3390/ijms22169036. PMID: 34445742; PMCID: PMC8396473

Abou-Kota ME, Hassaballa H, Elhini M, Ganzou SK (2024) Land degradation, desertification and environmental sensitivity to climate change in Alexandria and Beheira, Egypt. Egypt J Soil Sci 64:167–180

Adviento-Borbe MA, Linquist B (2016) Assessing fertilizer N placement on CH4 and N_2O emissions in irrigated rice systems. Geoderma 266:40–45

Andronov EE, Petrova SN, Pinaev AG, Pershina EV, Rakhimgaliyeva S, Akhmedenov KM, Gorobets AV, Sergaliev NKH (2012) Analysis of the structure of microbial community in soils with different degrees of salinization using T-RFLP and real-time PCR techniques. Soil Biol 45:147–156

Angon PB, Shafiul Islam M, Shreejana KC, Das A, Anjum N, Poudel A, Suchi SA (2024) Sources, effects and present perspectives of heavy metals contamination: soil, plants and human food chain. Heliyon 10:e28357

Augé RM (2001) Water relations, drought and vesicular-arbuscular mycorrhizal symbiosis. Mycorrhiza 11:3–42. https://doi.org/10.1007/s005720100097

Back MA, Haydock PPJ, Jenkinson P (2002) Disease complexes involving plant parasitic nematodes and soil borne pathogens. Plant Pathol 51:683–697. https://doi.org/10.1046/j.1365-3059.2002.00785.x

Bai GZ (2000) Characterization of microbial consortia in paddy rice soil by phospholipid analysis. Microb Ecol 39:273–281

Banerjee S, Heijden M (2022) Soil microbiomes and one health. Nat Rev Microbiol 21(1):6–20. https://doi.org/10.1038/s41579-022-00779-w

Barnard RL, Osborne CA, Firestone MK (2013) Responses of soil bacterial and fungal communities to extreme desiccation and rewetting. ISME J 7:2229. https://doi.org/10.1038/ismej.2013.104

Bebber D, Richards V (2020) A meta-analysis of the effect of organic and mineral fertilizers on soil microbial diversity. Appl Soil Ecol 175:104450. https://doi.org/10.1101/2020.10.04.325373

Bergi J, Trivedi R (2020) Bioremediation of saline soil by cyanobacteria. In: Microbial bioremediation & biodegradation. Springer Singapore, Singapore, pp 447–465

Birch HF (1958) The effect of soil drying on humus decomposition and nitrogen availability. Plant Soil 10(1):9–31

Bolan S, Padhye LP, Jasemizad T, Govarthanan M, Karmegam N, Wijesekara H, Amarasiri D, Hou D, Zhou P, Biswal BK, Balasubramanian R, Wang H, Siddique KHM, Rinklebe J, Kirkham MB, Bolan N (2024) Impacts of climate change on the fate of contaminants through extreme weather events. Sci Total Environ 909:168388

Bouskill NJ, Wood TE, Baran R, Hao Z, Ye Z, Bowen BP, Lim HC, Nico PS, Holman HY, Gilbert B, Silver WL, Northen TR, Brodie EL (2016) Belowground response to drought in a tropic forest soil. II. Change in microbial function impacts carbon composition. Front Microbiol 7:323

Brown ME, Mugo S, Petersen S, Klauser D (2022) Designing a pest and disease outbreak warning system for farmers, agronomists and agricultural input distributors in East Africa. Insects 13:232. https://doi.org/10.3390/insects13030232

Chambers LG, Osborne TZ, Reddy KR (2013) Effect of salinity-altering pulsing events on soil organic carbon loss along an intertidal wetland gradient: a laboratory experiment. Biogeochemistry 115:363–383. https://doi.org/10.1007/s10533-013-9841-5

Charu G, Prakash D, Sneh G (2014) Role of microbes in combating global warming. Int J Pharm Sci Lett 4:359–363

Cheng L, Agha Kouchak A, Gilleland E, Katz RW (2014) Non-stationary extreme value analysis in a changing climate. Climate Change 127:353–369. https://doi.org/10.1007/s10584-014-1254-5

Compant S, van der Heijden MGA, Sessitsch A (2010) Climate change effects on beneficial plant-microorganism interactions. FEMS Microbiol Ecol 73(2):197–214. https://doi.org/10.1111/j.1574-6941.2010.00900.x. Epub 2010 May 4

Crowther TW, Thomas SM, Maynard DS, Baldrian P, Covey K, Frey SD, van Diepen LTA, Bradford MA (2015) Biotic interactions mediate soil microbial feedbacks to climate change. PNAS 112:7033–7038

Csonka LN (1989) Physiological and genetic responses of bacteria to osmotic stress. Microbiol Rev 53:121–147

Dastgheib SMM, Amoozegar MA, Khajeh K, Ventosa A (2011) A halotolerant *Alcanivorax* sp. strain with potential application in saline soil remediation. Appl Microbiol Biotechnol 90:305–312

Davidson EA, Janssens IA (2006) Temperature sensitivity of soil carbon decomposition and feedbacks to climate change. Nature 440:165–173

Davies T Jr, Olalde-Portugal V, Aguilera-Gomez L, Alvarado MJ, Ferrera-Cerrato RC, Boutton TW (2002) Alleviation of drought stress of Chile ancho pepper (*Capsicum annuum* L. cv. San Luis) with arbuscular mycorrhiza indigenous to Mexico. Sci Hortic 92:347–359

De Angelis KM, Pold G, Topçuoglu BD, van Diepen LT, Varney RM, Blanchard JL, Melillo J, Frey SD (2015) Long-term forest soil warming alters microbial communities in temperate forest soils. Front Microbiol 104:1–13. https://doi.org/10.3389/fmicb.2015.00104

de Souza Silva CMM, Fay EF (2012) Effect of salinity on soil microorganisms. In: Hernandez-Soriano MC (ed) Soil health and land use management. Intech Open

Delgado-Baquerizo M, Maestre F, Reich P, Jeffries T, Gaitán J, Encinar D, Singh B (2016) Microbial diversity drives multifunctionality in terrestrial ecosystems. Nat Commun 7(1):10541. https://doi.org/10.1038/ncomms10541
Delgado-Garcia M, Valdivia-Urdiales B, Aguilar-Gonzalez C (2012) Halophilic hydrolase as a new tool for the biotechnological industries. Sci Food Agric 92:2575–2580. https://doi.org/10.1007/s10658-012-9936-1
Drepper B, Bamps B, Gobin A, Van Orshoven J (2021) Strategies for managing spring frost risks in orchards: effectiveness and conditionality—a systematic review protocol. Environ Evid 10:32
Dubey A, Malla MA, Khan F, Chowdhary K, Yadav S, Kumar A, Sharma S, Khare PK, Khan ML (2019) Soil microbiome: a key player for conservation of soil health under changing climate. Biodivers Conserv 28:2405–2429
Dutta TK, Phani V (2023) The pervasive impact of global climate change on plant- nematode interaction continuum. Front Plant Sci 14:1143889. https://doi.org/10.3389/fpls.2023.1143889
El-Ramady HR, El-Marsafawy SM, Lewis LN (2013) Sustainable agriculture and climate changes in Egypt. In: Lichtfouse E (ed) Sustainable agriculture reviews, sustainable agriculture reviews, vol 12. Springer, Dordrecht, pp 41–95. https://doi.org/10.1007/978-94-007-5961-9_2
El-Ramady H, Alshaal T, Abowaly M, Abdalla N, Taha HS, Al-Saeedi AH, Shalaby T, Amer M, Fári M, Domokos-Szabolcsy É, Sztrik A, Prokisch J, Selmar D, Pilon-Smits EAH, Pilon M (2017) Nanoremediation for sustainable crop production. In: Ranjan S et al (eds) Nanoscience in food and agriculture, sustainable agriculture reviews, vol 26. Springer, pp 335–363. https://doi.org/10.1007/978-3-319-58496-6_12
El-Ramady H, Abowaly M, Elbehiry F, Omara AE, Elsakhawy T, Mohamed ES, Belal AA, Elbasiouny H, Fawzy ZF (2019) Stressful environments and sustainable soil management: a case study of Kafr El-Sheikh, Egypt. Environ Biodiv Soil Secur 3:193–213
El-Ramady H, Abdalla N, Badgar K, Llanaj X, Töros G, Hajdú P, Eid Y, Prokisch J (2022) Edible mushrooms for sustainable and healthy human food: nutritional and medicinal attributes. Sustainability 14:4941. https://doi.org/10.3390/su14094941
El-Sayed A, Kamel M (2020) Climatic changes and their role in emergence and re-emergence of diseases. Environ Sci Pollut Res 27:22336–22352. https://doi.org/10.1007/s11356-020-08896-w
El-Sayed AA, El-Leithy AS, Bazraa WM, Abdel-Latef MS (2018) Effect of growing media, bio and organic fertilization on the flowering and chemical constituents of *Calendula officinalis* L. Plant Biosci Res 15(3):2029–2040
Elnahal ASM, El-Saadony MT, Saad AM et al (2022) The use of microbial inoculants for biological control, plant growth promotion, and sustainable agriculture: a review. Eur J Plant Pathol 162:759–792. https://doi.org/10.1007/s10658-021-02393-7
Ferreira CSS, Seifollahi-Aghmiuni S, Destouni G, Ghajarnia N, Kalantari Z (2022) Soil degradation in the European Mediterranean region: processes, status and consequences. Sci Total Environ 805:150106
Forge TA, Simard SW (2001) Effects of soil freezing and thawing on nematodes in forest soils. Can J Soil Sci 81(1):49–54. https://doi.org/10.4141/S00-035
Frater PN, Borer ET, Fay PA, Jin V, Knaeble B, Seabloom E, Sullivan L, Wedin DA, Harpole WS (2018) Nutrients and environment influence arbuscular mycorrhizal colonization both independently and interactively in *Schizachyrium scoparium*. Plant Soil 425:493–406
Frouz J (2018) Effects of soil macro- and mesofauna on litter decomposition and soil organic matter stabilization. Geoderma 332:161–172
Gavito ME, Olsson PA, Rouhier H, Medina-Peñafiel A, Jakobsen I, Bago A, Azcón-Aguilar C (2005) Temperature constraints on the growth and functioning of root organ cultures with arbuscular mycorrhizal fungi. New Phytol 168:179–188. https://doi.org/10.1111/j.1469-8137.2005.01481.x
Goicoechea N (2020) Mycorrhizal fungi as bioprotectors of crops against verticillium wilt a hypothetical scenario under changing environmental conditions. Plan Theory 9:1–15
Goot KA, Lee S (2007) Integrated management of plant-parasitic nematodes. Nematology 9(5):691–705

Gros R, Poly F, Monrozier LJ, Faivre P (2003) Plant and soil microbial community responses to solid waste leachates diffusion on grassland. Plant Soil 255:445–455

Grzyb A, Wolna-Maruwka A, Niewiadomska A (2020) Environmental factors affecting the mineralization of crop residues. Agronomy 10:1951. https://doi.org/10.3390/agronomy10121951

Guhr A, Borken W, Spohn M, Matzner E (2015) Redistribution of soil water by a saprotrophic fungus enhances carbon mineralization. Proc Natl Acad Sci 112:14647. https://doi.org/10.1073/pnas.1514435112

Gupta S, Sharma P, Dev K, Sourirajan A (2016) Halophilic bacteria of Lunsu produce an array of industrially important enzymes with salt tolerant activity. Biochem Res Int 2016:9237418

Hafez SM, El-Ghanam H (2010) Integrated management of root rot diseases in Egypt. Phytopathology 100(8):1035–1042

Harman GE (2000) Myths and dogmas of biocontrol changes in perceptions derived on *Trichoderma harzianum* T-22. Plant Dis 84:377–393. https://doi.org/10.1094/PDIS.2000.84.4.377

Hassan A, Khattab SA, Gohar IMA, El-Sharnoby HM (2024) Ecofriendly management approaches for sugar beet root-knot nematodes Meloidogyne ssp. by seed treatment with bio-origin compounds and Oxamyl Mohamed. Egypt Sugar J 22:19–33. https://doi.org/10.21608/esugj.2024283987.1062

Hawkes CV, Hartley IP, Ineson P, Fitter AH (2008) Soil temperature affects carbon allocation within arbuscular mycorrhizal networks and carbon transport from plant to fungus. Glob Chang Biol 14(5):1181–1190

Heinemeyer A, Fitter AH (2004) Impact of temperature on the arbuscular mycorrhizal symbiosis: growth responses of the host plant and its AM fungal partner. J Exp Bot 55:525–534

Heinemeyer A, Ineson P, Ostle N, Fitter AH (2006) Respiration of the external mycelium in the arbuscular mycorrhizal symbiosis shows strong dependence on recent photosynthates and acclimation to temperature. New Phytol 171:159–170

Henry HAL (2007) Soil freeze thaw cycle experiments: trends, methodological weaknesses and suggested improvements. Soil Biol Biochem 39(5):977–986. https://doi.org/10.1016/j.soilbio.2006.11.017

Hossain MS, Majumde AK (2018) Impact of climate change on agricultural production and food security: a review on coastal regions of Bangladesh. Int J Agric Res Innov Tech 8(1):62–69

Hussain T, Akhtar N, Aminedi R, Danish M, Nishat Y, Patel S (2020) Role of the potent microbial based bioagents and their emerging strategies for the ecofriendly management of agricultural phytopathogens. In: Natural bioactive products in sustainable agriculture. Springer, Berlin, pp 45–66. https://doi.org/10.1007/978-981-15-3024-1_4

Imhoff JF, Rahn T, Künzel S, Keller A, Neulinger SC (2020) Osmotic adaptation and compatible solute biosynthesis of phototrophic bacteria as revealed from genome analyses. Microorganisms 9:46

IPCC (2021) Climate change 2021: the physical science basis. Contribution of Working Group I to the Sixth Assessment Report of the Intergovernmental Panel on Climate Change. Cambridge University Press

Jansson JK, Hofmockel KS (2020) Soil microbiomes and climate change. Nat Rev Microbiol 18:35–46

Jerbi M, Labidi S, Lounès-Hadj Sahraoui A, Chaar H, Jeddi F (2020) Ben Higher temperatures and lower annual rainfall do not restrict, directly or indirectly, the mycorrhizal colonization of barley (*Hordeum vulgare* L.) under rainfed conditions. PLoS One 15:e0241794

Johnson DW, Murphy JF, Susfalk RB, Caldwell TG, Miller WW, Walker RF, Powers RF (2005) The effects of wildfire, salvage logging, and post-fire N-fixation on the nutrient budgets of a Sierran forest. For Ecol Manag 220:155–165

Kassem HS, Bello AS, Alotaibi BM, Aldosri FO, Straquadine GS (2019) Climate change adaptation in the delta Nile region of Egypt: implications for agricultural extension. Sustainability 11:685. http://www.mdpi.com/journal/sustainability

Keenleyside W (2019) Temperature and microbial growth. In: Microbiology: Canadian edition. Pressbooks, Toronto, p 636

Lam SH, Suter H, Mosier A, Chin D (2017) Using nitrification inhibitors to mitigate agricultural N2O emission: a double-edged sword? Glob Chang Biol 23:485–489. https://doi.org/10.1111/gcb.13338

Lange M, Eisenhauer N, Sierra C, Beßler H, Engels C, Griffiths R, Gleixner G (2015) Plant diversity increases soil microbial activity and soil carbon storage. Nat Commun 6(1):6707. https://doi.org/10.1038/ncomms7707

Lehmann J, Gaunt J, Rondon M (2006) Bio-char sequestration in terrestrial ecosystems—a review. Mitig Adapt Strateg Glob Chang 11(2):403–427

Lehmann J, Rillig MC, Thies J, Masiello CA, Hockaday WC, Crowley D (2011) Biochar effects on soil biota—a review. Soil Biol Biochem 43(9):1812–1836. https://doi.org/10.1016/j.soilbio.2011.04.022

Linn DM, Doran JW (1984) Aerobic and anaerobic microbial populations in no-till and plowed soils. Soil Sci Soc Am J 48(4):794–799

Lu H, Lashari MS, Liu X, Ji H, Li L, Zheng J, Kibue GW, Joseph S, Pan G (2015) Changes in soil microbial community structure and enzyme activity with amendment of biochar-manure compost and pyroligneous solution in a saline soil from Central China. Eur J Soil Biol 70:67–76

Luo N, Wang X, Hou J, Wang Y, Wang P, Meng Q (2020) Agronomic optimal plant density for yield improvement in the major maize regions of China. Crop Sci 60:1580–1590

Mahfouz MM, Abd-Elgawad (2021) Optimizing safe approaches to manage plant-parasitic nematodes. Plan Theory 10(9):1911. https://doi.org/10.3390/plants10091911

Mahmud K, Missaoui A, Lee K, Ghimire B, Presley HW, Makaju S (2021) Rhizosphere microbiome manipulation for sustainable crop production. Curr Plant Biol 27:100210

Manzoni S, Schimel JP, Porporato A (2012) Responses of soil microbial communities to water stress: results from a meta-analysis. Ecology 93(4):930–938

McElrone AJ, Choat B, Gambetta GA, Brodersen CR (2013) Water uptake and transport in vascular plants. Nat Educ Knowl 4(5):6

Mokrani S, Nabti E, Cruz C (2020) Current advances in plant growth promoting bacteria alleviating salt stress for sustainable agriculture. Appl Sci 10:7025

Moomaw WR, Chmura GL, Davies GT, Finlayson CM, Middleton BA et al (2018) Wetlands in a changing climate: science, policy and management. Wetlands 38(2):183–205. https://doi.org/10.1007/s13157-018-1023-8

Mukhtar S, Abdulla Malik K, Mehnaz S (2019) Microbiome of halophytes: diversity and importance for plant health and productivity. Microbiol Biotechnol Lett 47:1–10

Muscolo A, Sidari M (2008) Soil ecosystem: dynamic and sustainability. In: Chen J, Guo C (eds) Ecosystem ecology research trends. Nova Science Publishers, Inc, pp 23–63. ISBN: 978-1-60456-183-8

Narsing Rao MP, Dong ZY, Xiao M, Li WJ (2019) Effect of salt stress on plants and role of microbes in promoting plant growth under salt stress. In: Microorganism in saline environments. Springer, Cham, pp 423–435

Naylor D, DeGraaf S, Purdom E, Coleman-Derr D (2017) Drought and host selection influence bacterial community dynamics in the grass root microbiome. ISME J 11:2691–2704. https://doi.org/10.1038/ismej.2017.118

Niu J, Rang Z, Zhang C, Wu C, Tian F, Yin H, Linjian D (2016) The succession pattern of soil microbial communities and its relationship with tobacco bacterial wilt. BMC Microbiol 16(1):233. https://doi.org/10.1186/s12866-016-0845-x

Omar SA, Abdel-Sater MA, Khallil AM, Abd-Alla MH (1994) Growth and enzyme activities of fungi and bacteria in soil salinized with sodium chloride. Folia Microbiol 39(1):23–28. https://doi.org/10.1007/bf02814524

Patankar RS, Zambare VP, Ponraj M (2021) Physiological aspects of the halophilic and halotolerant fungi, and their potential applications. Nov Res Microbiol J 5(5):1371–1391

Pathak R, Singh SK, Tak A, Gehlot P (2018) Impact of climate change on host, pathogen and plant disease adaptation regime: a review. Biosci Biotechnol Res Asia 15(3) Available from: https://www.biotech-asia.org/?p=31049

Pautasso M, Döring TF, Garbelotto M, Pellis L, Jeger MJ (2012) Impacts of climate change on plant diseases opinions and trends. Eur J Plant Pathol 133:295–313

Peiffer JA, Spor A, Koren O, Jin Z, Tringe SG, Dangl JL, Buckler ES, Ley RE (2013) Diversity and heritability of the maize rhizosphere microbiome under field conditions. Proc Natl Acad Sci 110:6548–6553

Pérez-Jaramillo JE, Mendes R, Raaijmakers JM (2016) Impact of plant domestication on rhizosphere microbiome assembly and functions. Plant Mol Biol 90:635–644

Perry RN, Moens M (2011) Introduction to plant-parasitic nematodes; modes of parasitism. In: Jones J, Gheysen G, Fenoll C (eds) Genomics and molecular genetics of plant-nematode interactions. Springer, Dordrecht. https://doi.org/10.1007/978-94-007-0434-3

Philippot L, Spor A, Hénault C, Bru D, Bizouard F, Jones C, Maron P (2013) Loss in microbial diversity affects nitrogen cycling in soil. ISME J 7(8):1609–1619. https://doi.org/10.1038/ismej.2013.34

Pimentel D (2006) Soil health and environmental sustainability. Soil Tillage Res 89(1):1–5

Pitman MG, Läuchli A (2002) Global impact of salinity and agricultural ecosystems. In: Salinity: environment-plants-molecules. Springer, Berlin, pp 3–20

Pritchard RD, Harrell MM, Diaz-Granados D, Guzman MJ (2008) The productivity measurement and enhancement system: a meta-analysis. J Appl Psychol 93(3):540–567. https://doi.org/10.1037/0021-9010.93.3.540

Rahman MU, Zhong X, Uzair M, Ben Fan B (2024) Application of fungi as biological control strategies for nematode management in horticultural crops. Phytopathol Res 6:38. https://doi.org/10.1186/s42483-024-00257-6

Rath KM, Rousk J (2015) Salt effects on the soil microbial decomposer community and their role in organic carbon cycling: a review. Soil Biol Biochem 81:108–123

Raymond MR, Wharton DA (2016) The ability to survive intracellular freezing in nematodes is related to the pattern and distribution of ice formed. J Exp Biol 219(Pt 13):2060–2065. https://doi.org/10.1242/jeb.137190

Rillig MC, Wright SF, Shaw MR, Field CB (2002) Artificial climate warming positively affects arbuscular mycorrhizae but decreases soil aggregate stability in an annual grassland. Oikos 97:52–58

Saad OALO, Conrad R (1993) Temperature dependence of nitrification, denitrification, and turnover of nitric oxide in different soils. Biol Fertil Soils 15:21–27. https://doi.org/10.1007/bf00336283

Sahoo G, Majid WA (2020) Effect of climate change on land degradation. Int J Innov Eng Manag Res 9:483–494

Salimi S, Almuktar SAAAN, Scholz M (2021) Impact of climate change on wetland ecosystems: a critical review of experimental wetlands. J Environ Manag 286:112160

Sardans J, Peñuelas J (2005) Drought decreases soil enzyme activity in a Mediterranean Quercus ilex L. forest. Soil Biol Biochem 37(3):455–461. https://doi.org/10.1016/j.soilbio.2004.08.004

Schimel JP, Clein JS (1996) Microbial response to freeze-thaw cycles in tundra and taiga soils. Soil Biol Biochem 28(8):1061–1066

Schimel J, Balser TC, Wallenstein M (2007) Microbial stress-response physiology and its implications for ecosystem function. Ecology 88(6):386–1394

Setia R, Smith P, Marschner P, Baldock J, Chittleborough D, Smith J (2011) Introducing a decomposition rate modifier in the Rothamsted Carbon Model to predict soil organic carbon stocks in saline soils. Environ Sci Technol 45(15):6396–6403. https://doi.org/10.1021/es200515d

Sharratt B, Glenn DM (2006) Freeze-thaw and soil structure: an overview with implications for soil management. Agron J 98(1):1–6. https://doi.org/10.2134/agronj2005.0001

Shindell D, Johan CI, Kuylenstierna, Elisabetta V, Rita VD (2012) Simultaneously mitigating near-term climate change and improving human health and food security. Science 335:183–189

Singh RK, Shwetang Kundu S (2014) Review on changing natural nitrogen cycle: special reference to kingdom of Saudi Arabia. Int J Eng Sci Invent Res Dev I:2349–6185

Singh B et al (2007) Climate change and agriculture: adaptation strategies. Agric Syst 94(1):3–12

Singh BK, Bardgett RD, Smith P, Dave SR (2010) Microorganisms and climate change: terrestrial feedbacks and mitigation options. Nat Rev Microbiol 8:779–790
Singh P, Nedumaran S, Traore PCS, Boote KJ, Rattunde HFW, Prasad PVV, Singh NP, Srinivas K, Bantilan MCS (2014) Quantifying potential benefits of drought and heat tolerance in rainy season sorghum for adapting to climate change. Agric For Meteorol 185:37–48
Singh BK, Delgado-Baquerizo M, Egidi E, Guirado E, Leach JE, Liu H, Trivedi P (2023) Climate change impacts on plant pathogens, food security and paths forward. Nat Rev Microbiol 21:640–656
Six J, Bossuyt H, Degryze S, Denef K (2002) A history of research on the link between (micro) aggregates, soil biota, and soil organic matter dynamics. Soil Tillage Res 79(1):7–31
Sjögersten S, Black CR, Evers S, Hoyos-Santillan J, Wright EL, Turner BL (2014) Tropical wetlands: a missing link in the global carbon cycle? Glob Biogeochem Cycles 28(12):1371–1386
Smercina DN, Evans SE, Friesen ML, Tiemann LK (2019) To fix or not to fix: controls on free-living nitrogen fixation in the rhizosphere. Appl Environ Microbiol 85(6):e02546-18. https://doi.org/10.1128/AEM.02546-18
Smith KA, Ball T, Conen F, Dobbie KE, Massheder J, Rey A (2003) Exchange of greenhouse gases between soil and atmosphere: interactions of soil physical factors and biological processes. Eur J Soil Sci 54(4):779–791
Steinmuller HE, Chambers LG (2018) Can saltwater intrusion accelerate nutrient export from freshwater wetland soils? An experimental approach. Soil Sci Soc Am J 82:283–292
Stromberg M, Kephar TP, Vern Y (2001) Composition, invisibility, and diversity in coastal California grasslands. Madrono 48(4):236–252
Suman J, Rakshit A, Ogireddy SD, Singh S, Gupta C, Chandrakala J (2022) Microbiome as a key player in sustainable agriculture and human health. Front Soil Sci 2:821589
Tahat MM, Alananbeh KM, Othman YA, Leskovar DI (2020) Soil health and sustainable agriculture. Sustain For 12:4859. https://doi.org/10.3390/su12124859
Telo da Gama J (2023) The role of soils in sustainability, climate change, and ecosystem services: challenges and opportunities. Ecologies 4:552–567
Thornton PK, Gerber P (2010) Climate change and the growth of the livestock sector developing countries. Mitig Adapt Strateg Glob Chang 15(2):169–184
Tiemann L, Grandy A, Atkinson E, Marín-Spiotta E, McDaniel M (2015) Crop rotational diversity enhances belowground communities and functions in an agroecosystem. Ecol Lett 18(8):761–771. https://doi.org/10.1111/ele.12453
Timmusk S, Abd El-Daim IA, Copolovici L, Tanilas T, Kannaste A, Behers L, Nevo E, Seisenbaeva G, Stenstrom E, Niinemets U (2014) Drought-tolerance of wheat improved by rhizosphere bacteria from harsh environments: enhanced biomass production and reduced emissions of stress volatiles. PLoS One 9(5):e96086
Torres E, Miarnau X (2024) Frost damage mitigation in flowers and fruitlets of peach and almond from the application of a multi-attribute approach biostimulant. Plan Theory 13:1603. https://doi.org/10.3390/plants13121603
Treseder KK, Allen MF (2000) Mycorrhizal fungi have a potential role in soil carbon storage under elevated CO2 and nitrogen deposition. New Phytol 147(1):189–200
Treseder KK, Berlemont R, Allison SD, Martiny AC (2018) Drought increases the frequencies of fungal functional genes related to carbon and nitrogen acquisition. PLoS One 13:e0206441
Wagg C, Bender S, Widmer F, Heijden M (2014) Soil biodiversity and soil community composition determine ecosystem multifunctionality. Proc Natl Acad Sci 111(14):5266–5270. https://doi.org/10.1073/pnas.1320054111
Wang R, Zhang H, Sun L, Qi G, Shu C, Zhao X (2017) Microbial community composition is related to soil biological and chemical properties and bacterial wilt outbreak. Sci Rep 7(1):343. https://doi.org/10.1038/s41598-017-00472-6
Wang H, Bier R, Zgleszewski L, Peipoch M, Omondi E, Mukherjee A, Chen F, Zhang C, Kan J (2020) Distinct distribution of archaea from soil to freshwater to estuary: implications of archaeal composition and function in different environments. Front Microbiol 11:576661

Wang X, Chi Y, Song S (2024) Important soil microbiota's effects on plants and soils: a comprehensive 30-year systematic literature review. Front Microbiol 15:134774

Wassef R, Schüttrumpf H (2016) Impact of sea-level rise on groundwater salinity at the development area western delta, Egypt. Groundw Sustain Dev 2–3:85–103

Weiman S (2015) Microbes help to drive global carbon cycling and climate change. Microbe Mag 10:233–238

Wichern J, Wichern F, Joergensen RG (2006) Impact of salinity on soil microbial communities and the decomposition of maize in acidic soils. Geoderma 137(1–2):100–108

Wieder WR, Bonan GB, Allison SD (2013) Global soil carbon projections are improved by modeling microbial processes. Nat Clim Chang 3(10):909–912

Wu H, Cui H, Fu C, Li R, Qi F, Liu Z, Yang G, Xiao K, Qiao M (2024) Unveiling the crucial role of soil microorganisms in carbon cycling: a review. Sci Total Environ 909:168627

Yao Z, Zheng X, Wang R, Xie B, Butterbach-Bahl K, Zhu J (2013) Nitrous oxide and methane fluxes from a rice-wheat crop rotation under wheat residue incorporation and no-tillage practices. Atmos Environ 79:641–646

Yetgin A (2023) Exploring the link between soil microbial diversity and nutritional deficiencies. J Agric Prod 4(2):81–90. https://doi.org/10.56430/japro.1279830

Yuan X, Li S, Chen J, Yu H, Yang T, Wang C, Huang S, Chen H, Ao X (2024) Impacts of global climate change on agricultural production: a comprehensive review. Agronomy 14:1360

Zahran HH (2001) Rhizobia from wild legumes: diversity, taxonomy, ecology, nitrogen fixation and biotechnology. J Biotechnol 91:143–153. https://doi.org/10.1016/S0168-1656(01)00342-X

Zhang W, Xu M, Wang X, Huang Q, Nie J, Li Z, Li S, Hwang SW, Lee KB (2012) Effects of organic amendments on soil carbon sequestration in paddy fields of subtropical China. J Soils Sediments 12:457–470

Zhang Z, Zimmermann NE, Stenke A, Li X, Hodson EL, Zhug G, Huangc C, Poulter B (2017) Emerging role of wetland methane emissions in driving 21st century climate change. PNAS 114(36):9647–9652

Zhou M, Liu C, Wang J, Meng Q, Yuan YE, Ma X, Liu X, Zhu Y, Ding G, Zhang J, Zeng X, Du W (2020) Soil aggregates stability and storage of soil organic carbon respond to cropping systems on Black Soils of Northeast China. Sci Rep 10(1):265. https://doi.org/10.1038/s41598-019-57193-1

Zimmer C (2010) The microbe factor and its role in our climate future, Yale Environment 360. Published at the Yale School of Forestry & Environmental Studies

Zimmerman L, Labonte B (2015) Climate change and the microbial methane banquet. Clim Alert 27:1–6

Zogg GP, Zak DR, Ringelberg DB, MacDonald MW, Pregitzer KS, White DC (1997) Compositional and functional shifts in microbial communities due to soil warming. Soil Sci Soc Am J 61:475–481

Part V

Machin Learning Use in Crop Production

Climate Change Impacts on Crop Yield Prediction: A Machine Learning Perspective

15

Dina Elreedy and Amir F. Atiya

Abstract

Machine learning plays an emerging role in agricultural systems to further enhance crop production and efficiently manage water and soil resources. For example, machine learning algorithms are applied to crop management, water management and smart irrigation systems, and soil management. Recently, climate change has significant impacts on agriculture. In particular, the climate change influences numerous agricultural aspects including soil properties, fertilization, irrigation mechanisms, and crop yield. Specifically, climate change has substantial impacts on crop production and food security. Thus, this work focused on the impacts of climate change on crop yield. In particular, this work essentially reviews the main machine learning approaches employed for studying and analyzing the impacts of the climate change on crop yield prediction.

Keywords

Classification methods · Regression methods · Deep learning approaches · Statistical measures

Introduction

The IPCC's Fifth Assessment Report indicates that global mean surface temperatures are projected to increase by 1.4–5.8 °C by the century's end relative to 1990, cautioning that climate change has been occurring and will persist imminently (Field et al. 2014). Global warming is characterized by a general rise in surface air

D. Elreedy (✉) · A. F. Atiya
Computer Engineering Department, Cairo University, Giza, Egypt
e-mail: dinaelreedy@eng.cu.edu.eg

A. E.-H. Zohry, S. Ouda (eds.), *Resilient Agroecosystems*, Sustainability Sciences in Asia and Africa, https://doi.org/10.1007/978-981-96-9630-7_15

temperatures, alterations in precipitation patterns, and a disturbance of the global hydrologic cycle, leading to an increased occurrence of extreme weather events such as droughts, floods (Mo et al. 2013), and heat or cold waves (van der Wiel and Bintanja 2021), in which crop productivity and stability are adversely affected in irrigated and rain-fed agriculture (Adamgbe and Ujoh 2013).

The yield of main cereal crops (wheat and maize) is anticipated to be influenced by forthcoming climate change. Wheat production fluctuated significantly due to global warming, with anticipated losses reaching up to 6% for each degree Celsius increase in temperature (Miller et al. 2019), whereas Zhao et al. (2017) projected a 7% reduction in maize productivity for each degree Celsius increase in global mean temperature. Ouda and Zohry (2024) reported a reduction in maize productivity in Egypt by 10% under SSP1-2.6 and 25% under SSP5-8.5 over the period 2080–2099. Conversely, wheat productivity will decrease by 16% and 22% under the corresponding scenarios throughout the same timeframe.

Recently, applying machine learning to agriculture has attained a growing interest (Liakos et al. 2018; Benos et al. 2021; Meshram et al. 2021; Pathan et al. 2020). Machine learning approaches have been employed in various aspects of agriculture (Araujo et al. 2023) such as smart irrigation and water management (Saggi and Jain 2022; Younes et al. 2024), crop management (Gautron et al. 2022; Attri et al. 2024), crop disease detection (Shruthi et al. 2019; Sujatha et al. 2021), and soil management (Huang et al. 2023; Padmapriya and Sasilatha 2023).

As indicated by Van Klompenburg et al. (2020), estimating crop yield is a complicated task since it involves several factors including crop data, climate variables, fertilizers, soil, and seeds (Xu et al. 2019). Crop yield prediction could help decision makers to plan for their crops to know which crops could be more profitable, and for efficient resource utilization to decide which crops suit the environmental and climate conditions (Van Klompenburg et al. 2020; Paude-l et al. 2021). Furthermore, deciding the suitable crops and having good estimates for crop yields could promote food security especially in essential crops like wheat and rice (Tilman et al. 2011).

Specifically, several machine learning paradigms have been adopted for crop yield prediction (Van Klompenburg et al. 2020). For example, some studies have employed regression models for crop yield prediction (Gopal and Bhargavi 2019; Iniyan et al. 2023), and other works have adopted classification models (Sujatha and Isakki 2016). In this work, we consider basic machine learning approaches for crop yield prediction considering climate changes.

Sidhu et al. (2022) performed a study investigating the influence of climatic versus non-climatic features on crop yield prediction in India. They considered three crops: rice, wheat, and pearl millet. The study scrutinized various climate factors such as mean daily average, minimum, and maximum temperatures during the growing season, total seasonal precipitation, total seasonal precipitation days, subseasonal precipitation, and sub-seasonal precipitation days over the four crop growing stages (Allen et al. 1998). For non-climatic features, they considered geography and time. They adopted the ordinary least squares (OLS) linear regression model for crop yield prediction (Merriman 1877; Craven and Islam 2011). They designed nine OLS regression models using different set of climate and non-climate variables. The

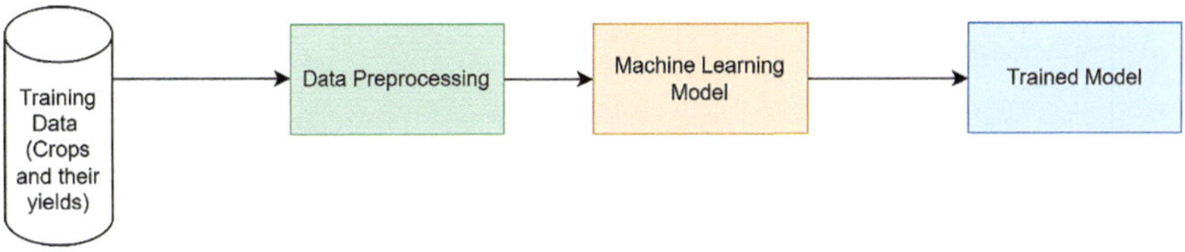

Fig. 15.1 Training crop yield prediction using machine learning

authors evaluated their regression models using the root mean square error and adjusted R^2 and assessed relative performance of climatic versus non-climatic variables using total variance and R^2. Their findings demonstrated that the relative importance of climate features outweigh other non-climatic ones for all the considered three crops. Furthermore, the authors analyzed the impact of climate change since 1966 on crop yield prediction for the three considered crops using the nine developed statistical models. Their results depicted that for rice and pearl millet, crop yield prediction levels were reduced. However, for wheat, average yield prediction levels increased as more climate variables are incorporated in the statistical learning models. Figures 15.1 and 15.2 demonstrate the basic training and testing pipelines for crop yield prediction.

This work focuses on the impacts of climate change on crop yield. In particular, this work essentially reviews the main machine learning approaches employed for studying and analyzing the impacts of the climate change on crop yield prediction.

The Main Machine Learning Approaches

Classification Methods

Classification method is a supervised learning technique that involves categorizing data into distinct classes (Bansal 2023). A hybrid machine learning model was developed by Crane-Droesch (2018), combining deep learning and statistical methods. Specifically, the author designed semiparametric neural networks (SNN). Moreover, the author applied bagging (Breiman 1996) to improve crop yield prediction performance. Furthermore, the author adopted Bayesian hyperparameter optimization (BHO) for tuning the model hyperparameters. The results highlighted a negative influence of climate change on corn yield, whereas Hrytsiuk et al. (2023) adopted machine learning to quantify the effect of climate change on wheat yield prediction. The authors formalized yield prediction into a classification problem by having two classes: high and low according to whether the yield is higher or lower than the average yield, respectively. Precisely, the authors employed five different machine learning classifiers such as logistic regression (LR) model (Wright 1995), linear discriminant analysis (LDA) (Balakrishnama and Ganapathiraju 1998; Izenman 2013), decision tree (DT) (Von Winterfeldt and Edwards 1986; Hssina et al. 2014), support vector machines (SVM) with linear and Radial basis function

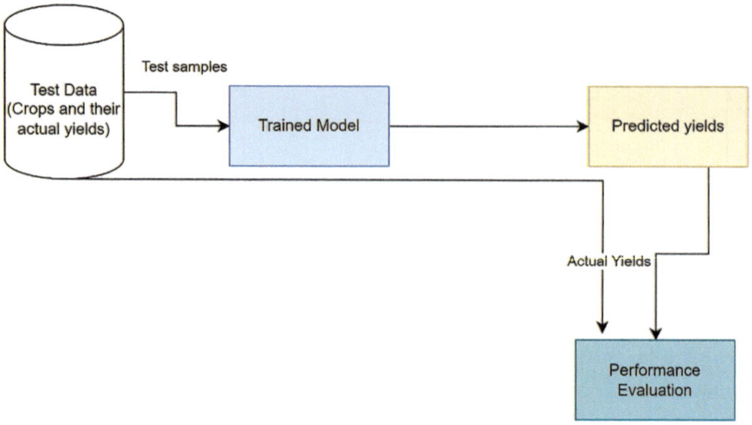

Fig. 15.2 Crop yield prediction using machine learning

(RBF) kernels (Cortes and Vapnik 1995), and random forest (RF) (Ho 1998; Biau and Scornet 2016). Their best adopted classification model achieved around 87% classification accuracy. They used a dataset of wheat yields for around 22 years at five regions of Ukraine. The presented analysis depicted a linear correlation between average monthly temperature values and wheat yield.

An ensemble model for crop yield prediction was employed by Jayanthi et al. (2024) composed of three classifiers: random forest (Ho 1995), XGBoost (Chen and Guestrin 2016), and AdaBoost (Freund and Schapire 1997). They adopted different aggregation mechanisms: model averaging, stacking, and voting regression. Their model achieved a coefficient of determination (R^2) score of 98.99%. Moreover, the authors adopted several Explainable Artificial Intelligence (XAI) methods including SHapley Additive exPlanations (SHAP) (Lundberg 2017), Local Interpretable Model-Agnostic Explanations (LIME) (Ribeiro et al. 2016), and Explain Like I'm 5 (ELI5) (Fan et al. 2019). Their presented results indicated that according to LIME explanation for the random forest model, the average temperature has a moderate impact on crop yield prediction. However, the LIME explanation for the XGBoost model depicts a negative influence on crop yield, and it highlights that average rainfall is the most significant feature affecting crop yield.

A former work developed by Veenadhari et al. (2014) applied machine learning for crop yields. The authors developed a software application for the effects of several climatic parameters on crop yields using machine learning. Specifically, they have adopted the C4.5 decision tree algorithm (Salzberg 1994) for detecting the confounding factors on crop yields. In their experiments, the authors applied their model to four crops: soybean, paddy, maize, and wheat. The classification results ranged from 76% to 90% for different crops.

Furthermore, the study developed by Cedric et al. (2022) applied machine learning models for crop yield prediction of six crops in West Africa. In particular, the authors employed three classifiers: decision tree (Von Winterfeldt and Edwards

1986; Hssina et al. 2014), multivariate logistic regression (MLR) (Walker and Duncan 1967; Glonek and McCullagh 1995), and K-nearest neighbors (KNN) (Fix and Hodges 1989; Peterson 2009). Their adopted decision tree model attained the best performance of R^2 score of 95.3%. The authors studied six crops: rice, maize, cassava, seed cotton, yams, and bananas. They considered nine West African countries: Burkina Faso, the Gambia, Ghana, Guinea, Mali, Mauritania, Niger, Senegal, and Togo.

Regression Methods

Regression is a term used in statistics that aims to understand the relationship between a dependent variable (response variable) and one or more independent variables (predictors) (Theodoridis 2015). The Crop Yield Prediction Algorithm (CYPA) was proposed by Talaat (2023) using machine learning methods and statistical methods. The author employed three machine learning classifiers: DecisionTreeRegressor (Maimon and Rokach 2014), RandomForestRegressor (Breiman 2001), and ExtraTreeRegressor (Geurts et al. 2006) for crop yield prediction. The aforementioned classifiers achieved accuracy measures as follows: 98.14% for DecisionTreeRegressor, 99.03% for RandomForestRegressor, and 99.33% ExtraTreeRegressor. Moreover, the author adopted two statistical measures: Pearson correlation coefficient (PCC) (Costa de Oliveira et al. 2014; Pearson 1895) and multiple regression (MR) (Sen et al. 1990) to assess the climate change impact on crop yield. In the conducted experiments, the author considered the following crops: cassava, maize, plantains, potatoes, rice, paddy, sorghum, soybeans, sweet potatoes, wheat, and yams. The presented results demonstrated a negative correlation between the average temperature and crop yield.

The impacts of climate change on wheat yield prediction were investigated by Iqbal et al. (2024), where two climate variables: rainfall and temperature values, were considered. The authors employed several machine learning regression models including XGBoost (Chen and Guestrin 2016), artificial neural networks (ANN) (Rosenblatt 1958; Bishop 1995), random forest regression (Ho 1995), and multiple linear regression (Freedman 2009). Moreover, the authors proposed a heterogeneous ensemble model composed of multilayer perceptron (Cybenko 1989), random forest, and boosting tree. Their proposed ensemble model achieved R^2 value of 0.953 and RMSE of 0.107. Their results indicated that the temperature values are compelling aspects for wheat yield. On the other hand, the rainfall feature has a minor influence on wheat yield.

The study presented in Sidhu et al. (2023) employed boosted regression trees (BRTs) (Elith et al. 2008) for crop yield prediction. They compared their model with least squares regression (Charnes et al. 1976; Watson 1967) with quadratic and piecewise linear functions. They applied both models to investigate the climate change impact on crop yield of three different crops in India: rice, wheat, and pearl millet. Specifically, the authors considered the model without climate variables to feature sets of climate variables: the first one includes average seasonal temperature and total seasonal precipitation. The second climate features were average seasonal temperature, total seasonal precipitation, and total precipitation days over the growing season. The BRT model surpasses linear regression models at crop yield prediction performance. The presented results demonstrated that incorporating climate features into the considered regression models enhanced the prediction performance for two crops: rice and pearl millet. On the other hand, the prediction performance of wheat was not influenced.

Several ensemble learning models for crop yield prediction by Gharakhanlou and Perez (2024) have been adopted in Canada. The authors developed five ensemble learning models: adaptive boosting (AdaBoost) (Freund and Schapire 1995), gradient boosting machine (GBM) (Friedman 2002), XGBoost (Chen and Guestrin 2016), LightGBM (Ke et al. 2017), and random forest (RF) (Ho 1995). The datasets include climate variables, satellite-derived vegetation, soil characteristics, and honeybee colonies on soybeans and canola crops. The considered climate variables include total precipitation, temperature, minimum and maximum monthly temperature, and mean temperature. They utilized different evaluation metrics including root mean square error (RMSE), R^2, and mean absolute error (MAE). The XGBoost classifier achieved the best prediction performance for canola and soybeans. The XGBoost model attained RMSE values of 119.48 for canola and 102.39 for soybeans and R^2 values of 95% for canola and 96% for soybeans. The results demonstrated negative impacts of climate change on canola and soybean yields. Moreover, the experimental results depicted the significant impact of honeybee colonies on the crop yields of canola and soybeans.

Furthermore, the study presented by Aparecido et al. (2022) aimed to quantify the climate change impact on cotton yield in Brazil. Several climatic conditions were included in that work such as temperature, precipitation, evapotranspiration, water storage, water deficit, and water surplus. The authors worked on data from 1983 to 2018 in different regions of Brazil. They applied several machine learning regression models for crop yield prediction such as: multiple linear regression (Freedman 2009), K-nearest neighbors (KNN) regressor (Fix and Hodges 1989; Peterson 2009), random forest regressor (RFT) (Ho 1998; Biau and Scornet 2016), artificial neural networks—multilayer perceptron (MLP) (Cybenko, 1989), gradient boosting for regression (BOO) (Friedman 2002), and extra-trees regressor (TREE) (Quinlan 1986). The authors assessed the regression models using three performance metrics: Pearson correlation coefficient (PCC), mean absolute percentage error (MAPE), and adjusted R^2. The experimental results indicated that the TREE regressor achieved top performance for cotton yield prediction. In addition, the results implied that the potential evapotranspiration and soil water storage exhibit a positive correlation with crop yield. Moreover, the authors found that these studied variables have significant impact on cotton crop yield in the main production regions in Brazil.

Another study investigating wheat yield prediction under climate change was performed by Ashfaq et al. (2024). The authors worked on a yield prediction dataset of a province in Pakistan from 2017 to 2022. The dataset combined several aspects using Google Earth Engine service including climate, soil, satellite, and spatial data.

In their experiments, the authors employed three machine learning models: SVM (Cortes and Vapnik 1995), random forest (Ho 1998; Biau and Scornet 2016), and Least Absolute Shrinkage and Selection Operator (LASSO) regression (Santosa and Symes 1986; Tibshirani 1996). The authors adopted three performance evaluation metrics: RMSE, MAE, and R^2. The random forest model achieved the best predictive performance of RMSE of 0.05 and R^2 value of 0.88.

Deep Learning Approaches

Recently, deep learning has been increasingly applied to various smart agriculture (Zhu et al. 2018; Altalak et al. 2022; Fiehn et al. 2018; Dhanya et al. 2022). For example, deep learning algorithms have been utilized in: plant disease detection (Zhu et al. 2018), crop identification (Altalak et al. 2022), water management (Altalak et al. 2022; Wai et al. 2022), and fruit counting (Rahnemoonfar and Sheppard 2017; Altalak et al. 2022).

Several works have leveraged deep learning approaches for crop yield prediction. For example, Alibabaei et al. (2021) developed a deep learning approach for estimating crop yield under climate changes in Portugal. Specifically, they applied long short-term memory (LSTM) (Hochreiter and Schmidhuber 1997; Smagulova and James 2019) and gated recurrent unit (GRU) (Cho et al. 2014) architectures. They worked on time series datasets. Their experiments considered two crops: potato and tomato. The datasets contained climate variables, irrigation scheduling, and soil water content. In particular, the climate features included minimum, maximum, and average temperature; minimum, maximum, and average relative humidity; average solar radiation; minimum and average wind speed; and precipitation.

The authors evaluated their model using MSE and R^2 and compared the adopted deep learning methods with several conventional machine learning regression models such as a multilayer perceptron model (Cybenko 1989) and random forest regression (Ho 1998; Biau and Scornet 2016). However, the LSTM deep learning model outperformed them, where it achieved a MSE value of 0.017–0.039 and an R^2 score of 0.97–0.99.

Another work was introduced by Demirhan (2024), in which a deep learning model was developed for crop yield prediction in Australia. The author considered climate change impacts on the yield prediction of several crops, namely, oats, corn, rice, and wheat. The proposed deep learning approach consisted of a fully connected feedforward deep neural network. They worked on a dataset including temperature, rainfall, carbon dioxide emissions, fertilizer use, and crop area. The deep learning model achieved RMSE scores of 0.156, 0.154, 0.039, and 0.129 for oats, corn, rice, and wheat, respectively. The experimental results depicted that the deep learning approach outperformed other statistical and machine learning methods for three crops: oats, corn, rice. However, for the wheat crop, the Autoregressive Integrated Moving Average with eXogenous variables (ARIMAX) statistical forecasting model outperformed the deep learning model. In addition, the author evaluated crop yield forecasting performance using MASE. Specifically, the deep learning model attained MASE scores of 0.360, 0.514, 0.335, and 0.315 for oats, corn, rice, and wheat, respectively. Furthermore, this study investigated six different climate change conditions and studied crop yield prediction.

Evaluation Metrics

Classification Models

There are several performance metrics to evaluate classifier's performance (Ferri et al. 2009), for example, classification accuracy which evaluates the percentage of correctly classified labels (Ferri et al. 2009). For a binary classifier having two classes: a positive and a negative one, the classification accuracy is evaluated as follows:

$$\text{Accuracy} = \frac{TP + TN}{N} \tag{15.1}$$

where *TP* is the number of true positive samples correctly classified as the positive class, *TN* is the number of true negative samples correctly classified as the negative class, and *N* is the number of test samples.

Another performance metric is the area under the Receiver operating characteristic curve (ROC) curve (AUC) which is defined as the probability that a randomly chosen example of the positive class has a higher rank than that of the negative class (Ferri et al. 2009; Hand 2012; Hrytsiuk et al. 2023).

Regression Models

For regression models, there are several evaluation metrics used in the literature to assess the prediction accuracy of crop yield (Iqbal et al. 2024; Sidhu et al. 2023; Shastry et al. 2017; Demirhan 2024). Some of the evaluation measures adopted in crop yield prediction are categorized into statistical measures and error measures, and they are described as follows:

Statistical Measures

Some statistical measures are employed to measure the correlation between the variable to be predicted (dependent variable) and the independent variables (Carpenter 1960; Wright 1921; Chicco et al. 2021). According to Wright (1921) and Chicco et al. (2021), the R^2 evaluates the proportion of the variance in the dependent variable (target) that could be predicted from the independent variables (features). The R^2 is calculated as:

$$R^2 = 1 - \frac{\sum_{i=1}^{N} (X_i - y_i)^2}{\sum_{i=1}^{N} (\overline{y} - y_i)^2} \tag{15.2}$$

where y_i denotes the target variable y for sample i, $\overline{y}$ denotes the mean of the target variable y, and X_i defines the feature vector for sample i.

Another useful statistical measure is the adjusted R^2, which is used in several studies of crop yield prediction (Leach and Henson 2007; Aparecido et al. 2022; Sidhu et al. 2022). The adjusted R^2 penalized the number of included predictor features (Sidhu et al. 2022). Conversely, R^2 keeps increasing while adding more features even if they are not correlated (James et al. 2013; Sidhu et al. 2022). The adjusted R^2 is computed as:

$$\text{Adjusted}\,R^2 = 1 - \frac{\left(1 - R^2\right)\left(N - 1\right)}{N - d - 1} \tag{15.3}$$

where d denotes the number of features in the regression task.

Pearson correlation coefficient (PCC) (Sidhu et al. 2022) is another measure that quantifies the linear correlation between two variables. PCC ranges between −1 and 1, where the sign of PCC indicates whether the correlation is positive or negative. The PCC measure the relation between two variables X and Y as follows:

$$\rho\left(X,Y\right) - \frac{\text{cov}\left(X,Y\right)}{\sigma_X \sigma_Y} \tag{15.4}$$

where Cov(X,Y) denotes the covariance between two variables X and Y. In addition, σ_X and σ_Y define the standard deviations of variables X and Y, respectively.

Error Measures

The root mean square error (RMSE) is one of the widely adopted performance metrics for regression tasks (Chicco et al. 2021). RMSE is computed as follows:

$$\text{RMSE} = \frac{1}{N} \sum_{N}^{i=1} \sqrt{\left(y_i - z_i\right)^2} \tag{15.5}$$

where N is the number of samples, y_i is the true value, and z_i is the predicted value.

The mean squared error (MSE) is calculated as:

$$\text{MSE} = \frac{1}{N} \sum_{N}^{i=1} \left(y_i - z_i\right)^2 \tag{15.6}$$

The mean absolute error (MAE) (Cedric et al. 2022; Demirhan 2024) is calculated as follows:

$$\text{MAE} = \frac{1}{N} \sum_{N}^{i=1} \left|y_i - z_i\right| \tag{15.7}$$

The MAE penalizes outliers or noisy data samples less than MSE and RMSE which square errors (Chicco et al. 2021).

The mean absolute percentage error (MAPE) measure is adopted in Aparecido et al. (2022) and computed as:

$$\text{MAPE} = \frac{\sum_{i=1}^{N}\left|\frac{yi - zi}{yi}\right| \times 100}{N} \tag{15.8}$$

The mean absolute scaled error (MASE) (Demirhan 2024) is a scale-independent measure defined as:

$$\text{MASE} = \frac{1}{N}\sum_{N}^{i=1}\left|q_i\right| \tag{15.9}$$

where qi is defined as:

$$q_i = \frac{y_i - z_i}{\frac{1}{N-1}\sum_{i=2}^{N}\left|y_i - y_{i-1}\right|} \tag{15.10}$$

The Summary of the Selected Studies

Table 15.1 indicates the summary of the reviewed studies in section "The Main Machine Learning Approaches" indicating study year, task type, the included climate parameters, the employed machine learning algorithms, the performance metrics used for measuring performance of crop yield prediction models, the investigated crops, and the region of the study.

Furthermore, Table 15.2 summarizes non-climatic variables utilized in the reviewed studies. The adopted non-climatic variable set includes soil data, crop data, fertilizer data, and pesticide data.

Figure 15.3 depicts the percentage of each climate variable adopted in the reviewed studies. It could be observed from the figure that temperature, rainfall, and precipitation are the most prevalent climate factors in the reviewed works.

Conclusions

This chapter reviews the basic works of crop yield prediction using machine learning while considering climate change impacts. This work presents classification models, regression models, and deep learning models adopted for crop yield prediction under climate changes. Moreover, the adopted performance measures for evaluating the machine learning models are presented.

For future work directions, probabilistic machine learning models could be explored as they provide uncertainty measures which could be beneficial for decision making in smart agriculture. In addition, more research could be devoted to interpretable machine learning models for crop yield prediction so that useful insights could be gained. Furthermore, the presented studies could be expanded and applied to more crops.

Table 15.1 Summary of studies adopting machine learning algorithms for crop yield prediction under climate changes

Study	Task	ML methods	Climate variables	Performance metrics	Crops	Region
Hrytsiuk et al. (2023)	Regression	LR LDA DT SVM-linear SVM-RBF	Average. temperature Total precipitation	Accuracy	Wheat	Ukraine
Ashfaq et al. (2024)	Regression	SVM RF LASSO	Daily weather data	RMSE R^2 MAE	Wheat	Pakistan
Jayanthi et al. (2024)	Classification	Ensemble of RF, XGBoost, and AdaBoost	Average Temperature Average rainfall	MSE R^2 MAE Explained variance	Many crops	Many regions
Veenadhari et al. (2014)	Classification	DT (C4.5)	Cloud cover Rainfall Minimum temperature	Accuracy	Soybean Paddy Maize Wheat	Dewas in India
Cedric et al. (2022)	Classification	DT MLR KNN	Temperature Precipitation	R^2 MAE	Rice Maize Cassava Seed cotton Yams Bananas	Burkina Faso The Gambia Ghana Guinea Mali Mauritania Niger Senegal Togo

(continued)

Table 15.1 (continued)

Study	Task	ML methods	Climate variables	Performance metrics	Crops	Region
Talaat (2023)	Regression	DT RF Extra tree	Rainfall Average temperature	PCC MR MAE RMSE	Cassava Maize Plantains Potatoes Rice Paddy Sorghum Soybeans Sweet potatoes Wheat Yams	Many regions
Iqbal et al. (2024)	Regression	XGBoost ANN RF MLR Ensemble of MLP, RF, and boosting tree	Rainfall Maximum. and minimum temperature Total precipitation Maximum and minimum. near-surface air temperature Average air temperature Surface downward motion shortwave heat Ave. wind speed	R^2 RMSE	Wheat	Punjab in Pakistan

Sidhu et al. (2023)	Regression	BRTs Linear regression-quadratic function Linear regression-piecewise linear function	Averageseasonal temperature Total seasonal precipitation Total precipitation days over the growing season	RMSE	Rice Wheat Pearl millet	India
Gharakhanlou and Perez (2024)	Regression	AdaBoost GBM XGBoost LightGBM RF	Total precipitation Minimum, maximum, and average monthly temperature	RMSE R^2 MAE	Soybeans Canola	Canada
Aparecido et al. (2022)	Regression	KNN RF ANN BOO TREE	Temperature Precipitation	PCC MAPE Adjusted R^2	Cotton	Brazil
Alibabaei et al. (2021)	Regression	LSTM GRU	Minimum., maximum., and average temperature Precipitation Maximum humidity Maximum humidity Averagerelative humidity Average solar radiation Minimum. and average. wind speed	MSE R^2	Tomato Potato	Portugal

(continued)

Table 15.1 (continued)

Study	Task	ML methods	Climate variables	Performance metrics	Crops	Region
Demirhan et al. (2024)	Regression	Feedforward Deep NN	Temperature Rainfall Carbon dioxide emissions	RMSE MASE	Oats Corn Rice Wheat	Australia

Table 15.2 Summary of non-climatic factors adopted in the reviewed studies

Study	Non-climatic variables
Hrytsiuk et al. (2023)	–
Ashfaq et al. (2024)	Satellite data Soil data Crop data
Jayanthi et al. (2024)	Year Season Crop data Area Production
Veenadhari et al. (2014)	–
Cedric et al. (2022)	Pesticide data Chemical data NO_2
Talaat (2023)	Pesticide data Chemical data
Iqbal et al. (2024)	–
Sidhu et al. (2023)	District data Year
Gharakhanlou and Perez (2024)	Soil data Honeybee census Vegetation indices Cropland land cover
Aparecido et al. (2022)	Water storage Water deficit Water surplus
Alibabaei et al. (2021)	Evapotranspiration Total soil profile Irrigation amount
Demirhan et al. (2024)	Fertilizer Crop area

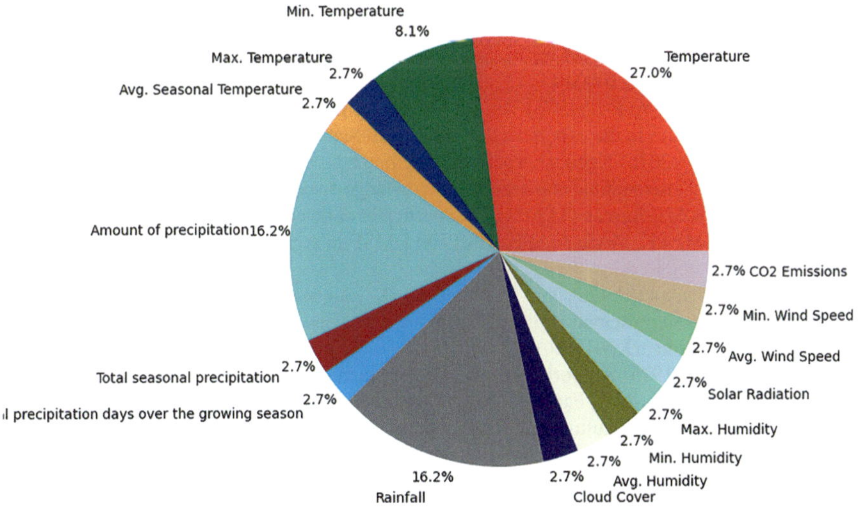

Fig. 15.3 Percentage of each adopted climate variable in the reviewed studies

References

Adamgbe EM, Ujoh F (2013) Effect of variability in rainfall characteristics on maize yield in Gboko. Nigeria J Environ Prot 4:881–887

Alibabaei K, Gaspar PD, Lima TM (2021) Crop yield estimation using deep learning based on climate big data and irrigation scheduling. Energies 14(11):3004

Allen RG, Pereira LS, Raes D, Smith M et al (1998) Crop evapotranspiration-guidelines for computing crop water requirements-FAO irrigation and drainage paper 56. FAO, Rome 300(9):D05109

Altalak M, Ammaduddin M, Alajmi A, Rizg A (2022) Smart agriculture applications using deep learning technologies: a survey. Appl Sci 12(12):5919

Aparecido LEDO, Meneses KCD, Rolim de Souza G, Carvalho MJN, Pereira WBS, da Silva PA et al (2022) Algorithms for forecasting cotton yield based on climatic parameters in Brazil. Arch Agron Soil Sci 68(7):984–1001

Araújo SO, Peres RS, Ramalho JC, Lidon F, Barata J (2023) Machine learning applications in agriculture: current trends, challenges, and future perspectives. Agronomy 13(12):2976

Ashfaq M, Khan I, Alzahrani A, Tariq MU, Khan H, Ghani A (2024) Accurate wheat yield prediction using machine learning and climate-NDVI data fusion. IEEE Access 12:40947–40961

Attri I, Awasthi LK, Sharma TP (2024) Machine learning in agriculture: a review of crop management applications. Multimed Tools Appl 83(5):12875–12915

Balakrishnama S, Ganapathiraju A (1998) Linear discriminant analysis—a brief tutorial. Inst Signal Inform Process 18(1998):1–8

Bansal S (2023) Analytixlabs, machine learning. What is classification algorithm in machine learning? With examples, 25 January. [Online]. Available https://www.analytixlabs.co.in/blog/classification-in-machine-learning/

Benos L, Tagarakis AC, Dolias G, Berruto R, Kateris D, Bochtis D (2021) Machine learning in agriculture: a comprehensive updated review. Sensors 21(11):3758

Biau G, Scornet E (2016) A random forest guided tour. TEST 25(2):197–227

Bishop CM (1995) Neural networks for pattern recognition. Oxford university Press

Breiman L (1996) Bagging predictors. Mach Learn 24:123–140

Breiman L (2001) Random forests. Mach Learn 45:5–32

Carpenter R (1960) Principles and procedures of statistics, with special reference to the biological sciences. Eugen Rev 52(3):172

Cedric LS, Adoni WYH, Aworka R, Zoueu JT, Mutombo FK, Krichen M, Kimpolo CLM (2022) Crops yield prediction based on machine learning models: case of West African countries. Smart Agric Technol 2:100049

Charnes A, Frome EL, Yu P-L (1976) The equivalence of generalized least squares and maximum likelihood estimates in the exponential family. J Am Stat Assoc 71(353):169–171

Chen T, Guestrin C (2016) Xgboost: a scalable tree boosting system. In: Proceedings of the 22nd ACM SIGKDD international conference on knowledge discovery and data mining, pp 785–794

Chicco D, Warrens MJ, Jurman G (2021) The coefficient of determination r-squared is more informative than SMAPE, MAE, MAPE, MSE and RMSE in regression analysis evaluation. PeerJ Comput Sci 7:e623

Cho K, Van Merrienboer B, Gulcehre C, Bahdanau D, Bougares F, Schwenk H, Bengio Y (2014) Learning phrase representations using RNN encoder decoder for statistical machine translation. arXiv preprint arXiv:1406.1078

Cortes C, Vapnik V (1995) Support-vector networks. Mach Learn 20(3):273–297

Costa de Oliveira A, Marini N, Farias D (2014) Climate change: new breeding pressures and goals. In: Encyclopedia of agriculture and food systems. Elsevier, pp 284–293

Crane-Droesch A (2018) Machine learning methods for crop yield prediction and climate change impact assessment in agriculture. Environ Res Lett 13(11):114003

Craven B, Islam SM (2011) Ordinary least-squares regression. In: The SAGE dictionary of quantitative management research, vol 1. SAGE Publications, Thousand Oaks, pp 224–228

Cybenko G (1989) Approximation by superpositions of a sigmoidal function. Math Control Signals Syst 2(4):303–314
Demirhan H (2024) A deep learning framework for prediction of crop yield in Australia under the impact of climate change. Inf Process Agric 12(1):125–138
Demirhan, H. (2025). A deep learning framework for prediction of crop yield in Australia under the impact of climate change. Information Processing in Agriculture, 12(1):125–138
Dhanya V, Subeesh A, Kushwaha N, Vishwakarma DK, Kumar TN, Ritika G, Singh A (2022) Deep learning based computer vision approaches for smart agricultural applications. Artif Intell Agric 6:211–229
Elith J, Leathwick JR, Hastie T (2008) A working guide to boosted regression trees. J Anim Ecol 77(4):802–813
Fan A, Jernite Y, Perez E, Grangier D, Weston J, Auli M (2019) Eli5: long form question answering. arXiv preprint arXiv:1907.09190
Ferri C, Hernandez-Orallo J, Modroiu R (2009) An experimental comparison of performance measures for classification. Pattern Recogn Lett 30(1):27–38
Fiehn HB, Schiebel L, Avila AF, Miller B, Mickelson A (2018) Smart agriculture system based on deep learning. In: Proceedings of the 2nd international conference on smart digital environment, pp 158–165
Field CB, Van Aalst M, Adger WN, Arent D, Barnett J, Betts R et al (2014) Part A: global and sectoral aspects: volume 1, global and sectoral aspects: Working Group II Contribution to the Fifth Assessment Report of the Intergovernmental Panel on Climate Change. In: Climate change 2014: impacts, adaptation, and vulnerability. IPCC, pp 1–1101
Fix E, Hodges JL (1989) Discriminatory analysis. nonparametric discrimination: consistency properties. Int Stat Rev 57(3):238–247
Freedman DA (2009) Statistical models: theory and practice. Cambridge University Press
Freund Y, Schapire RE (1995) A decision-theoretic generalization of online learning and an application to boosting. In: European conference on computational learning theory, pp 23–37
Freund Y, Schapire RE (1997) A decision-theoretic generalization of online learning and an application to boosting. J Comput Syst Sci 55(1):119–139
Friedman JH (2002) Stochastic gradient boosting. Comput Statist Data Anal 38(4):367–378
Gautron R, Maillard O-A, Preux P, Corbeels M, Sabbadin R (2022) Reinforcement learning for crop management support: review, prospects and challenges. Comput Electron Agric 200:107182
Geurts P, Ernst D, Wehenkel L (2006) Extremely randomized trees. Mach Learn 63:3–42
Gharakhanlou NM, Perez L (2024) From data to harvest: leveraging ensemble machine learning for enhanced crop yield predictions across Canada amidst climate change. Sci Total Environ 951:175764
Glonek GF, McCullagh P (1995) Multivariate logistic models. J R Stat Soc B Stat Methodol 57(3):533–546
Gopal PM, Bhargavi R (2019) A novel approach for efficient crop yield prediction. Comput Electron Agric 165:104968
Hand DJ (2012) Assessing the performance of classification methods. Int Stat Rev 80(3):400–414
Ho TK (1995) Random decision forests. In: Proceedings of 3rd international conference on document analysis and recognition, vol 1, pp 278–282
Ho TK (1998) The random subspace method for constructing decision forests. IEEE Trans Pattern Anal Mach Intell 20(8):832–844
Hochreiter S, Schmidhuber J (1997) Long short-term memory. Neural Comput 9(8):1735–1780
Hrytsiuk P, Babych T, Baranovsky S, Havryliuk M (2023) Assessing of climate impact on wheat yield using machine learning techniques. In: ICST, pp 314–329
Hssina B, Merbouha A, Ezzikouri H, Erritali M (2014) A comparative study of decision tree id3 and c4. 5. Int J Adv Comput Sci Appl 4(2):13–19
Huang Y, Srivastava R, Ngo C, Gao J, Wu J, Chiao S (2023) Data driven soil analysis and evaluation for smart farming using machine learning approaches. Agriculture 13(9):1777
Iniyan S, Varma VA, Naidu CT (2023) Crop yield prediction using machine learning techniques. Adv Eng Softw 175:103326

Iqbal N, Shahzad MU, Sherif E-SM, Tariq MU, Rashid J, Le T-V, Ghani A (2024) Analysis of wheat-yield prediction using machine learning models under climate change scenarios. Sustainability 16(16):6976

Izenman AJ (2013) Linear discriminant analysis. In: Modern multivariate statistical techniques. Springer, pp 237–280

James G, Witten D, Hastie T, Tibshirani R et al (2013) An introduction to statistical learning, vol 112. Springer

Jayanthi S, Indraneel K, Vivekanandan M, Srinivasan J et al (2024) Investigation of explainable crop yield prediction: leveraging ensemble learning and a novel blend model approach. Springer

Ke G, Meng Q, Finley T, Wang T, Chen W, Ma W et al (2017) Lightgbm: a highly efficient gradient boosting decision tree. In: Advances in neural information processing systems, 30

Leach LF, Henson RK (2007) The use and impact of adjusted r2 effects in published regression research. Multiple Linear Regression Viewpoints 33(1):1–11

Liakos KG, Busato P, Moshou D, Pearson S, Bochtis D (2018) Machine learning in agriculture: a review. Sensors 18(8):2674

Lundberg S (2017) A unified approach to interpreting model predictions. arXiv preprint arXiv:1705.07874

Maimon OZ, Rokach L (2014) Data mining with decision trees: theory and applications, vol 81. World scientific

Merriman M (1877) A list of writings relating to the method of least squares: with historical and critical notes, vol 4. Academic

Meshram V, Patil K, Meshram V, Hanchate D, Ramkteke S (2021) Machine learning in agriculture domain: a state-of-art survey. Artif Intell Life Sci 1:100010

Miller O, Helman D, Svoray T, Morin E, Bonfil DJ (2019) Explicit wheat production model adjusted for semi-arid environments. Field Crop Res 231:93–104

Mo X-G, Guo R-P, Liu S-X et al (2013) Impacts of climate change on crop evapotranspiration with ensemble GCM projections in the North China Plain. Clim Chang 120:299–312

Ouda S, Zohry A (2024) Climate variability and change disturbs maize production. In: Integration of legume crops with cereal crops under changing climate: sustainably increases food production. Springer. ISBN: 978-3-031-68101-1

Padmapriya J, Sasilatha T (2023) Deep learning based multi-labelled soil classification and empirical estimation toward sustainable agriculture. Eng Appl Artif Intell 119:105690

Pathan M, Patel N, Yagnik H, Shah M (2020) Artificial cognition for applications in smart agriculture: a comprehensive review. Artif Intell Agric 4:81–95

Paudel, D., Boogaard, H., de Wit, A., Janssen, S., Osinga, S., Pylianidis, C., & Athanasiadis, I. N. (2021). Machine learning for large-scale crop yield forecasting. Agricultural Systems, 187:103016.

Pearson K (1895) Note on regression and inheritance in the case of two parents. Proc R Soc Lond 58(347–352):240–242

Peterson LE (2009) K-nearest neighbor. Scholarpedia 4(2):1883

Quinlan JR (1986) Induction of decision trees. Mach Learn 1:81–106

Rahnemoonfar M, Sheppard C (2017) Deep count: fruit counting based on deep simulated learning. Sensors 17(4):905

Ribeiro MT, Singh S, Guestrin C (2016) "Why should I trust you?" Explaining the predictions of any classifier. In: Proceedings of the 22nd ACM SIGKDD international conference on knowledge discovery and data mining. Association for Computing Machinery, pp 1135–1144

Rosenblatt F (1958) The perceptron: a probabilistic model for information storage and organization in the brain. Psychol Rev 65(6):386

Saggi MK, Jain S (2022) A survey towards decision support system on smart irrigation scheduling using machine learning approaches. Arch Comput Methods Eng 29(6):4455–4478

Salzberg SL (1994) C4.5: programs for machine learning by J. Ross Quinlan. Morgan Kaufmann Publishers, Inc., 1993. Kluwer Academic Publishers

Santosa F, Symes WW (1986) Linear inversion of band-limited reflection seismograms. SIAM J Sci Stat Comput 7(4):1307–1330

Sen A, Srivastava M, Sen A, Srivastava M (1990) Multiple regression. In: Regression analysis: theory, methods and applications. Springer, pp 28–59
Shastry A, Sanjay H, Bhanusree E (2017) Prediction of crop yield using regression techniques. Int J Soft Comput 12(2):96–102
Shruthi U, Nagaveni V, Raghavendra B (2019) A review on machine learning classification techniques for plant disease detection. In: 2019 5th international conference on advanced computing & communication systems (ICACCS). IEEE, pp 281–284
Sidhu BS, Mehrabi Z, Kandlikar M, Ramankutty N (2022) On the relative importance of climatic and non-climatic factors in crop yield models. Clim Chang 173(1):8
Sidhu BS, Mehrabi Z, Ramankutty N, Kandlikar M (2023) How can machine learning help in understanding the impact of climate change on crop yields? Environ Res Lett 18(2):024008
Smagulova K, James AP (2019) A survey on LSTM memristive neural network architectures and applications. Eur Phys J Spec Top 228(10):2313–2324
Sujatha R, Isakki P (2016) A study on crop yield forecasting using classification techniques. In: 2016 international conference on computing technologies and intelligent data engineering (ICCTIDE'16). IEEE, pp 1–4
Sujatha R, Chatterjee JM, Jhanjhi N, Brohi SN (2021) Performance of deep learning vs machine learning in plant leaf disease detection. Microprocess Microsyst 80:103615
Talaat FM (2023) Crop yield prediction algorithm (CYPA) in precision agriculture based on IOT techniques and climate changes. Neural Comput & Applic 35(23):17281–17292
Theodoridis S (2015) Machine learning: a Bayesian and optimization perspective. Academic
Tibshirani R (1996) Regression shrinkage and selection via the lasso. J R Stat Soc Ser B Stat Methodol 58(1):267–288
Tilman D, Balzer C, Hill J, Befort BL (2011) Global food demand and the sustainable intensification of agriculture. Proc Natl Acad Sci 108(50):20260–20264
Van der Wiel K, Bintanja R (2021) Contribution of climatic changes in mean and variability to monthly temperature and precipitation extremes. Commun Earth Environ. https://doi.org/10.1038/s43247-020-00077-4
Van Klompenburg T, Kassahun A, Catal C (2020) Crop yield prediction using machine learning: a systematic literature review. Comput Electron Agric 177:105709
Veenadhari S, Misra B, Singh C (2014) Machine learning approach for forecasting crop yield based on climatic parameters. In: 2014 international conference on computer communication and informatics. IEEE, pp 1–5
Von Winterfeldt D, Edwards W (1986) Decision analysis and behavioral research. Cambridge University Press, Cambridge
Wai KP, Chia MY, Koo CH, Huang YF, Chong WC (2022) Applications of deep learning in water quality management: a state-of-the-art review. J Hydrol 613:128332
Walker SH, Duncan DB (1967) Estimation of the probability of an event as a function of several independent variables. Biometrika 54(1–2):167–179
Watson GS (1967) Linear least squares regression. Ann Math Stat 38:1679–1699
Wright S (1921) Correlation and causation. J Agric Res 20(7):557
Wright RE (1995) Logistic regression. In: Grimm LG, Yarnold PR (eds) Reading and understanding multivariate statistics. American Psychological Association, pp 217–244
Xu X, Gao P, Zhu X, Guo W, Ding J, Li C et al (2019) Design of an integrated climatic assessment indicator (ICAI) for wheat production: a case study in Jiangsu Province, China. Ecol Indic 101:943–953
Younes A, Abou Elassad ZE, El Meslouhi O, Abou Elassad DE, Majid E-DA (2024) The application of machine learning techniques for smart irrigation systems: a systematic literature review. Smart Agric Technol 7:100425
Zhao C, Liu B, Piao S et al (2017) Temperature increase reduces global yields of major crops in four independent estimates. PNAS 114:9326–9331
Zhu N, Liu X, Liu Z, Hu K, Wang Y, Tan J et al (2018) Deep learning for smart agriculture: concepts, tools, applications, and opportunities. Int J Agric Biol Eng 11(4):32–44

Coupling Process-Based Models with Machine Learning for Robust Predictions of Soil, Water, and Crop Dynamics

16

Ahmed M. S. Kheir, Mahmoud M. A. Shabana, Ahmed Attia, Marwa G. M. Ali, Mohamed A. Abd El-Aziz, Vinothkumar Kolluru, and Til Feike

Abstract

Process-based mechanistic models incorporate physical, chemical, and biological knowledge, providing a comprehensive understanding of climate-agriculture-management interactions. They are instrumental in addressing global environmental challenges such as climate change, food security, biodiversity conservation, and natural resource management. However, despite their utility, these models face significant limitations, including uncertainties due to model structure and parameterization, computational challenges, and the need for large-scale, high-resolution data. Machine learning, on the other hand, excels in processing large datasets and uncovering patterns, yet lacks the mechanistic interpretability of process-based models. By combining the mechanistic understanding from process-based models with the data-driven capabilities of machine

A. M. S. Kheir (✉)
Julius Kühn Institute (JKI)—Federal Research Centre for Cultivated Plants, Institute for Strategies and Technology Assessment, Kleinmachnow, Germany

Soils, Water and Environment Research Institute, Agricultural Research Center, Giza, 12112, Egypt
e-mail: Ahmed.Kheir@julius-kuehn.de

M. M. A. Shabana · M. G. M. Ali · M. A. Abd El-Aziz
Soils, Water and Environment Research Institute, Agricultural Research Center, Giza, 12112, Egypt

A. Attia · T. Feike
Julius Kühn Institute (JKI)—Federal Research Centre for Cultivated Plants, Institute for Strategies and Technology Assessment, 14532, Kleinmachnow, Germany
e-mail: Til.Feike@julius-kuehn.de

V. Kolluru
Department of Data Science, Stevens Institute of Technology, Hoboken, NJ, USA

A. E.-H. Zohry, S. Ouda (eds.), *Resilient Agroecosystems*, Sustainability Sciences in Asia and Africa, https://doi.org/10.1007/978-981-96-9630-7_16

learning algorithms, this chapter presents a novel approach aimed at improving forecasting precision and robustness in agricultural systems. This chapter delves into the innovative integration of process-based models with machine learning techniques to enhance the accuracy and reliability of predictions concerning soil, water, and crop dynamics. It demonstrates the efficacy of coupling process-based biophysical models with machine learning to facilitate regional simulations and evaluates the implications of this approach with a focus on crop yields and soil organic carbon dynamics under various agricultural management practices.

Keywords

Sequential coupling · Integrated coupling · Soil dynamics prediction · Water availability prediction

Introduction

In modern agricultural systems, the ability to accurately predict soil, water, and crop dynamics is critical for optimizing productivity and sustainability (Asseng et al. 2018; Baroudy et al. 2020; Kheir et al. 2019, 2021a). Agriculture is the cornerstone of global food security, yet it faces unprecedented challenges due to climate change, land degradation, and increasing resource constraints (Ding et al. 2021; Kheir et al. 2024a). Accurate modeling of soil moisture, water availability, and crop growth dynamics is essential for guiding effective management practices, ensuring water-use efficiency, and enhancing crop yields under varying environmental conditions (Kuang et al. 2021). Moreover, soil health and water dynamics are integral to maintaining ecosystem services, mitigating the effects of extreme weather events, and supporting resilience in agricultural systems (Peethani et al. 2024). Traditional methods of predicting these dynamics have often relied on mechanistic, process-based models (Magyar et al. 2023), which simulate the physical, chemical, and biological interactions that govern agroecosystems. However, while these models are comprehensive in their representation of environmental processes, their accuracy is often limited by uncertainties related to model structure, parameterization, and data availability (Uusitalo et al. 2015). The increasing complexity of agricultural systems and the need for precise, location-specific predictions demand advancements in modeling techniques to improve predictive accuracy and robustness.

To address these challenges, the coupling of process-based models with machine learning techniques has emerged as a promising approach (Attia et al. 2022; Kheir et al. 2022b). Process-based models offer a mechanistic understanding of agroecosystems, allowing for the simulation of interactions between climate, soil, and crops. However, these models are computationally intensive and often struggle to account for all sources of variability in large-scale systems. On the other hand, machine learning excels in analyzing large datasets, identifying patterns, and making data-driven predictions without requiring explicit knowledge of the underlying processes (Kheir et al. 2024c; van Klompenburg et al. 2020). The integration of these two modeling paradigms seeks to leverage the strengths of both approaches.

By coupling the mechanistic insights of process-based models with the data-processing power of machine learning algorithms, it is possible to enhance predictive performance and reduce uncertainties. This hybrid approach enables the modeling of complex, nonlinear relationships in agricultural systems while still retaining the interpretability and scientific grounding provided by process-based models. This coupling is especially valuable in scenarios where data is sparse, and process-based models require additional flexibility to capture the full range of environmental variability. Machine learning can be used to enhance process-based models by improving model calibration, filling gaps in datasets, and providing predictions for scenarios where mechanistic models may be insufficient (Zhang et al. 2022). Conversely, process-based models can inform and constrain machine learning predictions, ensuring that they adhere to known physical laws and biological principles (Droutsas et al. 2022).

Machine learning is increasingly being applied to enhance soil health, optimize water availability, and improve crop productivity across diverse environmental conditions and management practices (Awais et al. 2023; Umutoni and Samadi 2024). By leveraging large datasets from soil sensors, weather stations, and remote sensing technologies, machine learning algorithms can model complex interactions between soil properties, water dynamics, and crop growth (Minasny et al. 2024; Ye et al. 2024). These models enable precise predictions and recommendations tailored to specific conditions, such as soil type, climate variability, and crop type (Kheir et al. 2024b). For instance, machine learning can identify patterns in soil nutrient levels and water usage, helping farmers make data-driven decisions about fertilization and irrigation (Islam et al. 2023). Additionally, machine learning approaches can analyze historical crop performance and environmental data to forecast future productivity and recommend best practices for sustainable agriculture (Prity et al. 2024). By integrating machine learning with real-time data, agricultural systems can become more resilient, adaptive, and efficient, ultimately supporting the goal of maintaining soil health and optimizing crop yields under changing conditions.

This chapter explores the novel integration of process-based models and ML to improve predictions of soil, water, and crop dynamics in agricultural systems. The primary objective of this study is to demonstrate how coupling these methodologies can enhance forecasting accuracy and robustness, particularly in the context of sustainable agriculture and climate resilience. By applying this hybrid approach, we aim to provide more reliable assessments of soil health, water availability, and crop productivity under a wide range of environmental conditions and management practices. The scope of this work includes the evaluation of case studies in which process-based models are enhanced with machine learning techniques to predict soil moisture, water availability, and crop yields. These case studies serve as proof of concept for the proposed modeling framework and highlight the advantages of combining mechanistic knowledge with data-driven approaches. The implications of this approach for decision-making in agricultural management, as well as its potential to address global challenges such as food security and resource conservation, are discussed in detail. By advancing the integration of process-based and machine learning models, this work contributes to the development of robust tools for agricultural research and practice. Ultimately, this coupling aims to support the sustainable intensification of agriculture, helping to mitigate the environmental impacts of farming while ensuring continued productivity and resilience in the face of future challenges.

The Use of Process-Based Models in Agriculture

Process-based models have been fundamental in agricultural research and practice, offering mechanistic insights into the interactions between soil, water, and crops (Ali et al. 2020, 2022). These models simulate the physical, chemical, and biological processes underlying agroecosystems, allowing for detailed assessments of crop growth, soil fertility, and water dynamics. Notable examples include the Agricultural Production Systems Simulator (APSIM), which integrates various biophysical processes to predict crop yields under different management practices and environmental conditions (McCown et al. 1996; Pasley et al. 2023). DSSAT is a widely used process-based model that simulates crop growth, soil processes, and climate interactions to aid decision-making in agricultural management (Jones et al. 2003; Thorp et al. 2008).

Developed over several decades, DSSAT integrates various models to represent different aspects of crop production, including growth, yield, and nutrient dynamics. It allows users to evaluate the effects of different management practices, such as irrigation, fertilization, and planting dates, on crop performance under varying environmental conditions. DSSAT is known for its detailed simulation of crop physiological processes, including photosynthesis, respiration, and biomass accumulation, making it a valuable tool for understanding the impacts of climate change and management practices on crop yields. The model's flexibility in incorporating various crops and management scenarios enhances its applicability across diverse agricultural systems.

Simulateur mulTIdisciplinaire pour les Cultures Standard (STICS) is another prominent process-based model designed to simulate the interactions between soil, crop, and climate variables (Brisson et al. 2003). Developed primarily for European agricultural systems, STICS focuses on crop growth, soil water balance, and nitrogen dynamics (Brisson et al. 2003). The model provides a comprehensive representation of crop development stages, including emergence, flowering, and maturity, as well as soil processes, such as water infiltration, evaporation, and nutrient cycling. STICS is particularly useful for evaluating the impacts of different agricultural practices on soil fertility and crop productivity. Its ability to simulate multiple crops and management strategies makes it a versatile tool for optimizing agricultural practices and assessing environmental impacts.

The versatile crop simulation model that helps researchers understand and predict crop growth, yield, and water use (CROPSYST) is another process-based model designed to simulate crop growth, soil processes, and water dynamics over time

(Stöckle et al. 2003). It integrates various sub-models to represent crop development, soil water balance, and nutrient cycling. The CROPSYST is known for its long-term simulation capabilities, allowing users to evaluate the effects of different management practices and environmental conditions on crop productivity and soil health. The model's flexibility in simulating multiple crops and management scenarios makes it a useful tool for exploring sustainable agricultural practices and assessing the long-term impacts of different strategies.

There are other models described in Kheir et al. (2022a, b, c) and presented in Table 16.1. Despite their robustness, these models face limitations, such as high computational demands, complex parameterization, and difficulties in scaling predictions to diverse environments (Silva and Giller 2020). As a result, enhancing these models' accuracy and applicability remains a key focus in agricultural research.

Exploration of Machine Learning Applications in Agricultural Forecasting

Machine learning techniques have gained prominence for their ability to analyze large volumes of data and identify patterns that may not be evident through traditional methods. In agriculture, machine learning has been applied to various forecasting tasks, including soil health assessment, water-use optimization, and crop yield prediction (Rani et al. 2023; Trentin et al. 2024), and climate change downscaling (Kheir et al. 2023a). For instance, machine learning algorithms have been used to predict soil properties from spectral data, providing rapid and cost-effective soil analysis (Padarian et al. 2019). In terms of water management, machine learning models can optimize irrigation scheduling by predicting soil moisture levels and crop water requirements based on historical and real-time data (Gu et al. 2021). Furthermore, machine learning techniques have been employed to forecast crop yields by analyzing weather patterns, historical yield data, and remote sensing information, enabling more precise and timely decision-making (Arshad et al. 2023). The flexibility and adaptability of machine learning models make them particularly valuable for handling the complexity and variability inherent in agricultural systems. Figure 16.1 represents the different types of machine learning and the workflow based on the aim of the study.

Studies on the Integration of Process-Based Models with Machine Learning in Related Fields

The integration of process-based models with machine learning represents a promising approach to overcoming the limitations of each method when used in isolation. Several studies have explored this hybrid approach to enhance predictive accuracy and robustness in agricultural contexts (Feng et al. 2019; Shahhosseini et al. 2021). For example, Attia et al. (2022) demonstrated how combining

Table 16.1 List of the common process-based models used in monocrop and agroforestry systems

Model name	Description/documentation	References/d/r Github
APSIM classic (APSIM7.10)	https://www.apsim.info/documentation/model-documentation/technical-documentation/	https://github.com/APSIMInitiative/APSIM710; Holzworth et al. (2014)
APSIM next generation	https://apsimnextgeneration.netlify.app/modeldocumentation/	https://github.com/APSIMInitiative/ApsimX; Kheir et al. (2023b) and Zhu et al. (2021)
AFRCWHEAT2	Information provided upon request from John Porter (jrp@plen.ku.dk)	Porter (1993)
AQUACROP (V.7.1)	https://www.fao.org/aquacrop/overview/whatisaquacrop/en/	https://github.com/KUL-RSDA/AquaCrop; Foster et al. (2017) and Panek-Chwastyk et al. (2024)
APSIM-Wheat (V.7.5)	https://www.apsim.info/documentation/model-documentation/crop-module-documentation/wheat/	Ahmed et al. (2016) and Brown et al. (2018)
CropSyst (V.3.04.08)	https://www.quantitative-plant.org/model/CropSyst	https://github.com/NSSAC/vic-cropsyst-runner.git; Malek et al. (2017) and Stöckle et al. (2003)
DSSAT-CERES-Wheat (CSCER, V.4.8.2.0)	https://dssat.net/training/getting-started/; https://dssat.net/models-overview/components/ceres-wheat/	https://github.com/DSSAT/dssat-csm-os.git; https://github.com/Attia81/optimization-algorithms-for-DSSAT-CERES-wheat-CUL-ECO.git; Shawon et al. (2024), Kheir et al. (2021b)
DSSAT-Nwheat (WHAPS, V.4.8.2.0)	Referenced in DSSAT's official resource hub	Kassie et al. (2016), Kheir et al. (2025), Pequeno et al. (2024) and Röll et al. (2020)
DSSAT-CROPSIM (CSCRP, V4.8.2)	Available on the DSSAT network page	Paff et al. (2023)
DAISY (V.5.24)	http://www.daisy-model.org	https://github.com/daisy; Le Noë et al. (2023)
EPIC-I (V1102)	https://epicapex.tamu.edu/manuals-and-publications/	https://github.com/ritviksahajpal/EPIC.git; Jones et al. (1991) and Wang et al. (2022)
EPIC-Wheat	https://epicapex.tamu.edu/manuals-and-publications/	Yue et al. (2015)
GLAM (V.2 updated)	https://www.quantitative-plant.org/model/GLAM	Challinor et al. (2004)
HERMES	https://www.zalf.de/de/forschung_lehre/mas/Seiten/modelle.aspx	https://github.com/zalf-rpm/Hermes2Go.git
INFOCROP	https://www.quantitative-plant.org/model/InfoCrop	Aggarwal et al. (2006)

LINTUL4	https://models.pps.wur.nl/lintul-4-generic-simulation-model-crop-growth-under-potential-water-limited-and-n-limited	Wang et al. (2024)
SIMPLACE	https://www.simplace.net/index.php/documentation	https://github.com/gk-crop/simplace_rpkg.git; Enders et al. (2023)
LPJmL	https://www.pik-potsdam.de/en/institute/departments/activities/biosphere-water-modelling/lpjml	https://github.com/PIK-LPJmL/LPJmL.git
MCWLA-Wheat	Request information from taofl@igsnrr.ac.cn	Li et al. (2021) and Zhang et al. (2020)
MONICA	https://soil-modeling.org/resources-links/model-portal/monica	https://github.com/zalf-rpm/monica.git; Nendel (2014)
Expert-N—CERES	Documentation at Helmholtz Munich's resource site	Johnsson et al. (1987)
Expert-N—GECROS	Accessible on Helmholtz Munich's modeling page	Johnsson et al. (1987)
Expert-N—SPASS	Details available at Helmholtz Munich's database	Johnsson et al. (1987)
Expert-N—SUCROS	Find resources on the Helmholtz Munich platform	Johnsson et al. (1987)
Sirius	https://eclipse.dev/sirius/resource-center.html	https://github.com/sirius-ms/sirius.git; Jamieson et al. (1998)
SALUS	https://www.cibotechnologies.com/salus-model/	https://github.com/jdeines/Deines_etal_LEMA_SALUS.git
SiriusQuality	https://eng-lepse.montpellier.hub.inrae.fr/modelling/process-based-model/siriusquality	https://github.com/AgriculturalModelExchangeInitiative/SQ_Wheat_Phenology.git
SSM-Wheat	Request from afshin.soltani@gmail.com	Manschadi et al. (2022)
STICS	https://stics.inrae.fr/eng/resources/documentation	https://github.com/SticsRPacks/SticsOnR.git; Brisson et al. (2003)
WheatGrow	Request details from yanzhu@njau.edu.cn	Zhang et al. (2016)
WOFOST	https://www.wur.nl/en/research-results/research-institutes/environmental-research/facilities-tools/software-models-and-databases.htm	https://github.com/ajwdewit/WOFOST_crop_parameters.git; Berghuijs et al. (2024) and de Wit et al. (2019)
Hi-sAFe	https://www1.montpellier.inra.fr/wp-inra/hi-safe/en/	https://github.com/kevinwolz/hisafer.git; Dupraz et al. (2019)
APEX	http://epicapex.tamu.edu/apex/	Pan et al. (2021)
Hydrus 2D-3D	https://www.pc-progress.com/en/Default.aspx?hydrus	https://github.com/jknappe/hydrus2r.git; Morianou et al. (2023)

Modified after Asseng et al. (2019)

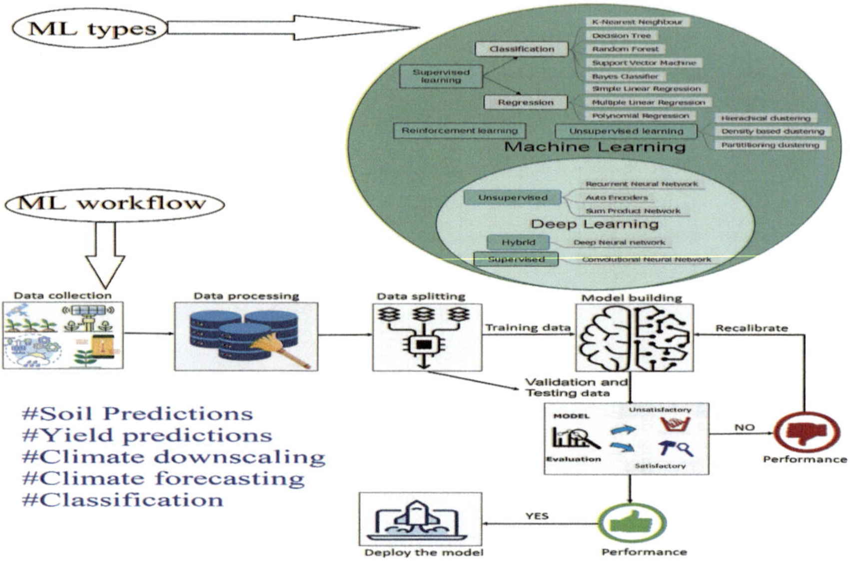

Fig. 16.1 Types and workflow of the machine learning (ML) in predictions of soil, yield, and climate

process-based models with machine learning algorithms improved the prediction of maize yield and evapotranspiration under arid conditions. By leveraging the mechanistic insights of process-based models and the data-driven capabilities of machine learning, their approach provided more accurate and region-specific forecasts.

Similarly, Kheir et al. (2022a) integrated machine learning techniques with process-based models to enhance wheat yield simulations in arid regions, showing that machine learning could effectively refine model predictions and account for variability not captured by the mechanistic models alone. These studies highlight the potential of coupling these methodologies to address complex challenges in agricultural systems, such as climate variability, soil health, and resource management, ultimately leading to more robust and actionable insights for sustainable agriculture.

The flowchart (Fig. 16.2) outlines a proposed framework for integrating machine learning with crop models to enhance predictions of crop yield (GY) and evapotranspiration (ET) under future climate change (CC) scenarios. The process begins by gathering baseline data (1981–2020) from various sources, including satellite data, climate data, soil data, and water data. This data is then processed using Google Earth Engine (GEE) and Google Colaboratory (Colab) to extract relevant features and prepare it for modeling. Crop models, such as those described in the Kheir et al. (2022b) were used to simulate baseline GY and ET under historical climate conditions. These simulated values were compared with observed baseline values to assess the model's accuracy. Machine learning algorithms, such as random forest, support vector machines, or neural networks, were then employed to predict

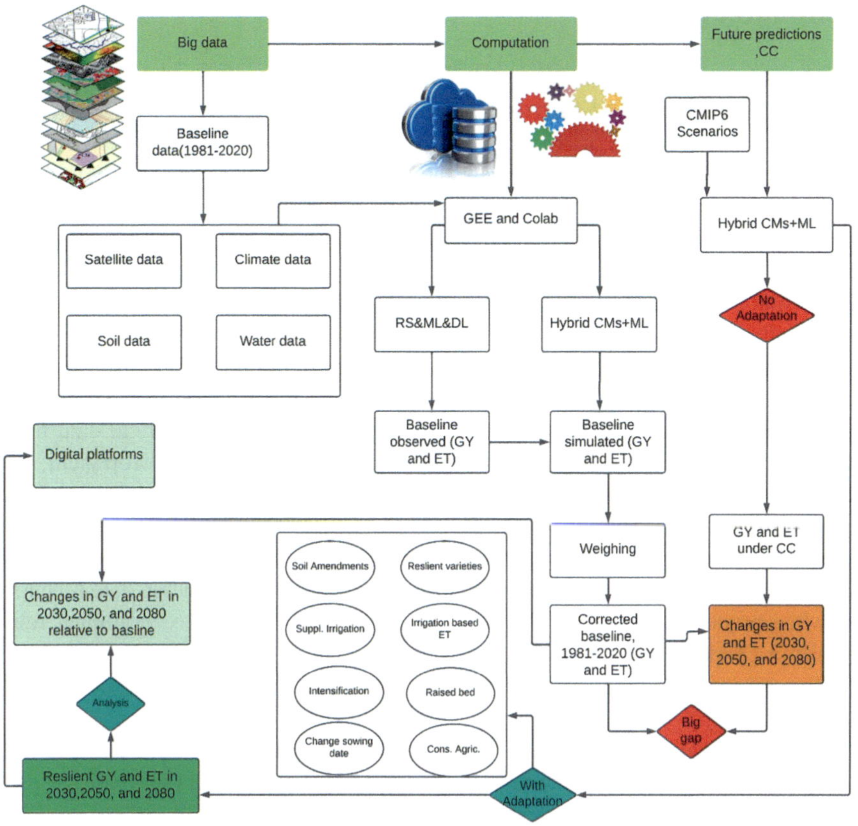

Fig. 16.2 A flowchart illustrating the integration of machine learning with crop models for predicting GY and ET under CC. The framework incorporates data collection, crop model simulations, machine learning predictions, adaptation strategy evaluation, and impact assessment

future GY and ET under various CC scenarios generated by global climate models (GCMs) from the Coupled Model Intercomparison Project 6 (CMIP6). These predictions were made using a hybrid approach that combines crop model outputs with machine learning-driven predictions. The framework also incorporates adaptation strategies, such as soil amendments, resilient varieties, supplemental irrigation, irrigation based on ET, intensification, raised beds, and changes in sowing date. These strategies were evaluated to assess their potential to mitigate the negative impacts of CC on GY and ET. Finally, the results were analyzed to identify the most effective adaptation strategies and to quantify the changes in GY and ET under different CC scenarios with and without adaptation. This information can be used to inform decision-making and develop strategies for sustainable agriculture in the face of climate change.

Methodology

Explanation of the Coupling Strategy for Process-Based Models and Machine Learning Algorithms

The coupling of process-based models and machine learning algorithms is designed to leverage the strengths of both approaches, combining the mechanistic understanding provided by process-based models with the predictive power and pattern recognition capabilities of machine learning. The strategy begins with the selection of a suitable process-based model, such as DSSAT or APSIM, that simulates the biophysical processes of crop growth, soil dynamics, and water availability. These models were chosen for their ability to incorporate environmental conditions, management practices, and physiological processes in a mechanistic framework. The process-based model was first calibrated and validated using historical data for a specific crop and region. The output data from these models, such as crop yield, soil moisture, or nutrient dynamics, were then used as input features for training machine learning algorithms. This coupling allows the machine learning model to learn from the process-based simulations and enhance its ability to predict outcomes for a wider range of environmental and management conditions, even when data is sparse or when process-based models alone struggle to capture complex, nonlinear relationships (Sarker 2021).

Two potential coupling strategies could be employed:

- *Sequential Coupling:* The output of the process-based model is fed as input into the machine learning algorithm. The machine learning model enhances the predictions by filling gaps, correcting biases, or improving accuracy where process-based models fall short due to simplifications or incomplete data.
- *Integrated Coupling:* The process-based model and machine learning algorithm work simultaneously, with the machine learning algorithm adjusting process-based model parameters in real time based on the emerging patterns from the data. This dynamic interaction allows the process-based model to be more adaptive to local conditions while ensuring that ML predictions adhere to known physical and biological constraints.

The ultimate goal of this coupling is to achieve more robust and accurate predictions across a range of spatial and temporal scales, making the approach well-suited for addressing challenges related to climate variability, resource limitations, and changing management practices (Materia et al. 2024; Zhang et al. 2024).

Description of Data Collection and Preprocessing Techniques

Data collection is a critical component of this methodology, as it provides the foundation for both process-based model simulations and machine learning training (Ammar et al. 2022). Data is gathered from a variety of sources, including:

- *Field Experiments:* Data on soil properties, crop growth, yield, and management practices are collected through field experiments. Sensors are deployed to monitor soil moisture, temperature, nutrient levels, and other key variables over the growing season.
- *Remote Sensing:* Satellite and drone imagery are used to collect data on land cover, vegetation indices (e.g., NDVI), and climate, such as precipitation and evapotranspiration. This data allows for broader spatial coverage and enhances the ability to scale predictions from field to regional levels.
- *Weather Stations:* Historical and real-time weather data (e.g., temperature, precipitation, solar radiation) are integrated to simulate climate effects on crop growth and soil dynamics. These datasets are critical for both process-based modeling and the training of machine learning algorithms (Azeem and Dev 2024).

Once collected, the data undergoes several preprocessing steps to ensure consistency and quality:

- *Data Cleaning:* Missing values, outliers, and erroneous data points are identified and addressed using statistical techniques such as interpolation, smoothing, or removal.
- *Normalization/Scaling:* The data is normalized or scaled to ensure uniformity, particularly when integrating different types of data (e.g., field measurements vs. remote sensing data). This process improves the performance and convergence of machine learning algorithms.
- *Feature Engineering:* Relevant features are extracted from the raw data, including both direct measurements (e.g., soil moisture content, crop height) and derived features (e.g., vegetation indices, degree days). These features serve as inputs for both the process-based models and the machine learning algorithms (Maestrini et al. 2022).

Details on Model Calibration, Validation, and Evaluation Procedures

The coupling methodology relies on rigorous calibration, validation, and evaluation procedures to ensure the accuracy and reliability of both the process-based models and the machine learning algorithms.

Process-based models are first calibrated using observed data from field experiments or historical datasets. Calibration involves adjusting key model parameters (e.g., crop coefficients, soil hydraulic properties) to align the model outputs with observed data. This step is essential for ensuring that the process-based model accurately simulates crop growth, soil dynamics, and water availability under specific environmental conditions. Machine learning models, on the other hand, are trained using a portion of the available dataset (typically 70–80%) to learn the relationships between input features (e.g., soil properties, climate data) and target variables (e.g., crop yield, soil moisture). Hyperparameter tuning is performed to optimize the performance of the machine learning models, with techniques such as cross-validation used to prevent overfitting (Wu et al. 2019).

Once calibrated, the models are validated using independent datasets that were not used during the training or calibration phase. For process-based models, validation typically involves comparing simulated outputs with observed field data across different seasons or locations. For machine learning models, validation is performed using a separate validation dataset, and metrics such as root mean square error (RMSE), mean absolute error (MAE), and R-squared (R^2) are used to assess the model's predictive accuracy. In the context of the coupled models, validation also examines how well the machine learning algorithm improves the process-based model's performance. This might involve testing the coupled system under new environmental conditions or management scenarios to evaluate its robustness and generalizability (Amini et al. 2024).

After validation, the final evaluation of the coupled models is carried out using a test dataset, often from a different location or time period to assess the model's ability to generalize beyond the conditions it was trained or calibrated on. The performance of the coupled model is compared against that of the individual process-based and machine learning models. Performance metrics such as accuracy, bias, precision, recall, and F1 score are used to evaluate the effectiveness of the coupling strategy. Additionally, scenario analysis is performed to test the model's performance under different environmental conditions and management practices, such as varying levels of irrigation, fertilization, or extreme weather events. The evaluation also includes sensitivity analysis to identify which parameters have the greatest influence on model outcomes, helping to refine the model further and improve its predictive capability (Rounsevell et al. 2021).

By rigorously calibrating, validating, and evaluating the coupled process-based and machine learning models, the methodology ensures that the final system provides robust, accurate, and scalable predictions of soil health, water availability, and crop productivity under diverse conditions. This approach not only enhances the decision-making process for agricultural management but also contributes to sustainable intensification and climate resilience in farming systems (Essenfelder and Giupponi 2020).

Case Studies

Application of the Coupled Approach to Soil Dynamics Prediction

Predicting soil dynamics is a critical component of sustainable agricultural management, as it directly influences crop productivity, water retention, nutrient cycling, and soil health (Awais et al. 2023). Traditional crop models, such as APSIM and DSSAT, have been extensively used to simulate the effects of various agricultural practices and environmental conditions on soil properties. However, the increasing complexity of agroecosystems and the demand for higher prediction accuracy have led researchers to explore the integration of machine learning techniques with these crop models. The coupled crop model and machine learning approach leverages the strengths of both methodologies. Crop models offer mechanistic insights into soil-crop-climate interactions, capturing the intricate relationships between weather patterns, soil properties, and crop growth (You et al. 2022). Machine learning, on the other hand, excels at handling large datasets, identifying patterns, and making accurate predictions even in scenarios where traditional models struggle due to nonlinear or complex relationships (Kheir et al. 2024a). In this integrated approach, crop models generate a baseline understanding of soil processes by simulating key variables such as soil moisture, nutrient levels, and temperature under various management practices and climate scenarios. The outputs from these models serve as features that are fed into machine learning algorithms. Machine learning techniques such as Random Forest, Gradient Boosting, or Deep Learning models are then trained on large datasets that may include observational data from sensors, remote sensing, and historical records (Li et al. 2023). These models learn from discrepancies between observed and predicted data, thus enhancing the accuracy of soil dynamics predictions.

Applications of this coupled approach include predicting soil moisture content, erosion rates, nutrient leaching, and carbon sequestration potential under different climate change scenarios (Golchin and Misaghi 2024). Furthermore, it enables farmers and policymakers to optimize irrigation strategies, fertilization schedules, and land management practices with higher precision, reducing resource waste and enhancing resilience to environmental stresses (Raza et al. 2024). The key advantage of this method is its ability to handle uncertainty and variability more effectively than traditional models alone. By continually improving through data input, machine learning can adapt to new conditions, such as changing climate patterns or the introduction of novel agricultural technologies (Parra-López et al. 2024). This provides a more dynamic and flexible system for predicting soil behavior, crucial for long-term sustainability in agriculture. As more datasets become available through advancements in sensor technology, remote sensing, and big data (Delfani et al. 2024; Padhiary et al. 2024), the coupled crop model and machine learning approach will play an increasingly vital role in advancing precision agriculture. Its capacity to simulate, predict, and optimize soil dynamics with high accuracy represents a significant step forward in addressing the global challenges of food security, climate resilience, and environmental sustainability (Mana et al. 2024).

Analysis of Yield and Water Availability Forecasts Using Integrated Models

Accurate forecasting of water availability is a critical challenge for modern agriculture, particularly in the context of climate change and increasing water scarcity (Preite and Vignali 2024). Water availability directly impacts crop growth, yield, and overall farm productivity. Traditionally, crop models like APSIM, AquaCrop, and DSSAT have been employed to predict water requirements and availability based on soil, crop, and climate parameters (Kephe et al. 2021; Kheir et al. 2021a).

These models offer mechanistic insights into how water interacts with crops and soil over different growth stages. However, the increasing complexity of water dynamics driven by climatic variability, coupled with the need for highly accurate forecasts, has encouraged the integration of machine learning techniques with these crop models (Umutoni and Samadi 2024).

The integration of crop models and machine learning for water availability forecasting merges the predictive power of process-based simulation with the adaptability and pattern recognition capabilities of ML (Chen et al. 2023; Li et al. 2024). Crop models provide baseline simulations of water balance components, including evapotranspiration, soil water content, infiltration, runoff, and percolation under various climate scenarios and management practices. Machine learning models, on the other hand, enhance these forecasts by utilizing vast datasets that crop models may not fully exploit. These datasets can include real-time meteorological data, remote sensing data, soil moisture sensors, and historical water availability records (Feng et al. 2024).

In this integrated approach, crop models generate simulations that serve as input features for machine learning algorithms. These algorithms, such as support vector machines, neural networks, or ensemble learning methods, are trained to detect complex nonlinear patterns in water dynamics that may not be captured by traditional models. The machine learning models can then adjust and refine the forecasts based on discrepancies between the observed data and model outputs, resulting in more accurate and reliable predictions (Rodrigues et al. 2024).

The integrated approach offers several key advantages. First, it can improve the accuracy of water availability predictions by accounting for local variations and environmental factors that crop models might not fully consider. Second, machine learning allows for continuous improvement of the model as new data becomes available, adapting forecasts in near real time to reflect changing conditions such as sudden droughts, extreme rainfall events, or shifts in irrigation practices. This dynamic ability to learn from data makes the approach particularly valuable for addressing the growing uncertainty in water resources due to climate variability (Albahri et al. 2024).

Practical applications of this integration include more precise irrigation scheduling, better water resource management, and improved drought preparedness. For example, farmers can use these enhanced forecasts to optimize irrigation timing and amounts, reducing water waste and enhancing water-use efficiency. On a larger scale, policymakers can benefit from these predictions to implement more effective water allocation strategies and drought mitigation measures in water-scarce regions.

As the availability of high-resolution, real-time data from sensors and remote sensing technologies increases, the potential for integrated crop models and machine learning to provide accurate and timely water availability forecasts will continue to grow. This hybrid approach represents a promising avenue for advancing sustainable water management in agriculture, ensuring food security in the face of evolving climatic and environmental pressures (He et al. 2019).

Discussion, Conclusion, and Outlook

The integration of process-based models with machine learning techniques represents a promising approach to enhancing predictions of soil, water, and crop dynamics in agricultural systems. This chapter demonstrates the efficacy of this hybrid methodology in improving forecasting accuracy and robustness, particularly in the context of sustainable agriculture and climate resilience. By combining the mechanistic understanding of process-based models with the data-driven capabilities of machine learning algorithms, the proposed approach provides a more comprehensive and efficient assessment of agroecosystems, facilitating informed decision-making (Procopio et al. 2023; Zhou et al. 2021).

One of the significant advantages of coupling process-based models with machine learning is the ability to address the limitations of each approach in isolation (Johnston et al. 2023). Process-based models, while offering mechanistic insights, often struggle with computational demands and data scarcity (Bishnu et al. 2023). Machine learning, on the other hand, excels in data-driven predictions but lacks the mechanistic interpretability of process-based models (Lu et al. 2023). By integrating these methodologies, we can leverage the strengths of both, leading to more accurate and robust predictions.

Furthermore, this approach allows for the incorporation of a wider range of environmental variability and management options. Process-based models, while flexible in specifying management practices, may be limited by computational constraints and data availability (Wynn and Clarkson 2018). Machine learning techniques can be used to mimic process-based model predictions, extending the model's coverage to a broader range of conditions.

The case studies presented in this chapter illustrate the effectiveness of the coupled approach in predicting soil moisture, water availability, and crop yields. By integrating process-based models with machine learning, we can enhance the accuracy and reliability of these predictions, supporting sustainable agricultural practices and climate resilience.

While the coupled approach offers significant benefits, it is important to acknowledge potential challenges and limitations. The success of this integration relies on the quality and availability of data (Ammar et al. 2022), as well as the appropriate selection and tuning of process-based models and machine learning algorithms (Bernardini et al. 2024). Additionally, understanding the interactions between the two methodologies is crucial for effective coupling.

The coupling of process-based models with machine learning represents a promising avenue for advancing agricultural research and practice. By addressing the limitations and capitalizing on the strengths of this hybrid methodology, we can contribute to a more sustainable and resilient future for agriculture.

The integration of process-based models with machine learning techniques represents a promising avenue for advancing agricultural research and practice. This study has demonstrated the efficacy of this hybrid methodology in improving predictions of soil, water, and crop dynamics, leading to significant implications for agricultural practices and decision-making. The coupled approach offers several advantages, including enhanced prediction accuracy, increased robustness, improved flexibility, and enhanced decision-making. By combining the mechanistic understanding of process-based models with the data-driven insights from machine learning algorithms, the coupled approach can provide more accurate and reliable forecasts, particularly in the face of uncertainty and variability in agricultural systems. Moreover, the approach allows for the incorporation of a wider range of environmental conditions and management options, making it more adaptable to changing circumstances.

The findings of this study have significant implications for agricultural practices and decision-making. Accurate predictions of soil moisture, water availability, and crop yields can help farmers optimize the use of resources such as irrigation water and fertilizers. Sustainable agricultural practices can be promoted by using the coupled approach to assess the impacts of different management practices on soil health, water quality, and biodiversity. Additionally, the coupled approach can help farmers adapt to climate change by providing insights into the effects of changing environmental conditions on crop productivity and resource availability.

Future research should focus on improving the quality and availability of data, further developing process-based models and machine learning algorithms, integrating other factors into the coupled approach, enhancing scalability, and promoting interdisciplinary collaboration. By addressing these recommendations, future research can contribute to the further development and application of the coupled approach for sustainable and resilient agriculture. In conclusion, the coupling of process-based models with machine learning offers a valuable approach for enhancing predictions in agricultural systems. By addressing the limitations and capitalizing on the strengths of this hybrid methodology, we can contribute to a more sustainable and resilient future for agriculture.

References

Aggarwal PK, Banerjee B, Daryaei MG, Bhatia A, Bala A, Rani S, Chander S, Pathak H, Kalra N (2006) InfoCrop: a dynamic simulation model for the assessment of crop yields, losses due to pests, and environmental impact of agro-ecosystems in tropical environments. II. Performance of the model. Agric Syst 89(1):47–67. https://doi.org/10.1016/j.agsy.2005.08.003

Ahmed M, Akram MN, Asim M, Aslam M, Hassan F-U, Higgins S, Stöckle CO, Hoogenboom G (2016) Calibration and validation of APSIM-Wheat and CERES-Wheat for spring wheat under rainfed conditions: models evaluation and application. Comput Electron Agric 123:384–401. https://doi.org/10.1016/j.compag.2016.03.015

Albahri AS, Khaleel YL, Habeeb MA, Ismael RD, Hameed QA, Deveci M, Homod RZ, Albahri OS, Alamoodi AH, Alzubaidi L (2024) A systematic review of trustworthy artificial intelligence applications in natural disasters. Comput Electr Eng 118:109409. https://doi.org/10.1016/j.compeleceng.2024.109409

Ali MGM, Ibrahim MM, El Baroudy A, Fullen M, Omar E-SH, Ding Z, Kheir AMS (2020) Climate change impact and adaptation on wheat yield, water use and water use efficiency at North Nile Delta. Front Earth Sci 14(3):522–536. https://doi.org/10.1007/s11707-019-0806-4

Ali MGM, Ahmed M, Ibrahim MM, El Baroudy AA, Ali EF, Shokr MS, Aldosari AA, Majrashi A, Kheir AMS (2022) Optimizing sowing window, cultivar choice, and plant density to boost maize yield under RCP8.5 climate scenario of CMIP5. Int J Biometeorol 66(5):971–985. https://doi.org/10.1007/s00484-022-02253-x

Amini H, Alanne K, Kosonen R (2024) Building simulation in adaptive training of machine learning models. Autom Constr 165:105564. https://doi.org/10.1016/j.autcon.2024.105564

Ammar KA, Kheir AMS, Manikas I (2022) Agricultural big data and methods and models for food security analysis—a mini-review. PeerJ 10:e13674. https://doi.org/10.7717/peerj.13674

Arshad S, Kazmi JH, Javed MG, Mohammed S (2023) Applicability of machine learning techniques in predicting wheat yield based on remote sensing and climate data in Pakistan, South Asia. Eur J Agron 147:126837. https://doi.org/10.1016/j.eja.2023.126837

Asseng S, Kheir AMS, Kassie BT, Hoogenboom G, Abdelaal AIN, Haman DZ, Ruane AC (2018) Can Egypt become self-sufficient in wheat? Environ Res Lett 13(9):094012. https://doi.org/10.1088/1748-9326/aada50

Asseng S, Martre P, Maiorano A, Rötter RP, O'Leary GJ, Fitzgerald GJ, Girousse C, Motzo R, Giunta F, Babar MA, Reynolds MP, Kheir AMS, Thorburn PJ, Waha K, Ruane AC, Aggarwal PK, Ahmed M, Balkovič J, Basso B et al (2019) Climate change impact and adaptation for wheat protein. Glob Chang Biol 25(1):155–173. https://doi.org/10.1111/gcb.14481

Attia A, Govind A, Qureshi AS, Feike T, Rizk MS, Shabana MMA, Kheir AMS (2022) Coupling process-based models and machine learning algorithms for predicting yield and evapotranspiration of maize in arid environments. Water 14(22):3647. https://www.mdpi.com/2073-4441/14/22/3647

Awais M, Naqvi S, Zhang H, Li L, Zhang W, Awwad FA, Ismail EAA, Khan MI, Raghavan V, Hu J (2023) AI and machine learning for soil analysis: an assessment of sustainable agricultural practices. Bioresour Bioprocess 10(1):90. https://doi.org/10.1186/s40643-023-00710-y

Azeem MA, Dev S (2024) A performance and interpretability assessment of machine learning models for rainfall prediction in the Republic of Ireland. Decis Anal J 12:100515. https://doi.org/10.1016/j.dajour.2024.100515

Baroudy AAE, Ali AM, Mohamed ES, Moghanm FS, Shokr MS, Savin I, Poddubsky A, Ding Z, Kheir AMS, Aldosari AA, Elfadaly A, Dokukin P, Lasaponara R (2020) Modeling land suitability for rice crop using remote sensing and soil quality indicators: the case study of the Nile Delta. Sustain For 12(22):9653. https://www.mdpi.com/2071-1050/12/22/9653

Berghuijs HNC, Silva JV, Reidsma P, de Wit AJW (2024) Expanding the WOFOST crop model to explore options for sustainable nitrogen management: a study for winter wheat in the Netherlands. Eur J Agron 154:127099. https://doi.org/10.1016/j.eja.2024.127099

Bernardini LG, Rosinger C, Bodner G, Keiblinger KM, Izquierdo-Verdiguier E, Spiegel H, Retzlaff CO, Holzinger A (2024) Learning vs. understanding: when does artificial intelligence outperform process-based modeling in soil organic carbon prediction? New Biotechnol 81:20–31. https://doi.org/10.1016/j.nbt.2024.03.001

Bishnu SK, Alnouri SY, Al-Mohannadi DM (2023) Computational applications using data driven modeling in process systems: a review. Digit Chem Eng 8:100111. https://doi.org/10.1016/j.dche.2023.100111

Brisson N, Gary C, Justes E, Roche R, Mary B, Ripoche D, Zimmer D, Sierra J, Bertuzzi P, Burger P, Bussière F, Cabidoche YM, Cellier P, Debaeke P, Gaudillère JP, Hénault C, Maraux F, Seguin B, Sinoquet H (2003) An overview of the crop model STICS. Eur J Agron 18(3):309–332. https://doi.org/10.1016/S1161-0301(02)00110-7

Brown H, Huth N, Holzworth D (2018) Crop model improvement in APSIM: using wheat as a case study. Eur J Agron 100:141–150. https://doi.org/10.1016/j.eja.2018.02.002

Challinor AJ, Wheeler TR, Craufurd PQ, Slingo JM, Grimes DIF (2004) Design and optimisation of a large-area process-based model for annual crops. Agric For Meteorol 124(1):99–120. https://doi.org/10.1016/j.agrformet.2004.01.002

Chen C, Zhang H, Shi W, Zhang W, Xue Y (2023) A novel paradigm for integrating physics-based numerical and machine learning models: a case study of eco-hydrological model. Environ Model Softw 163:105669. https://doi.org/10.1016/j.envsoft.2023.105669

de Wit A, Boogaard H, Fumagalli D, Janssen S, Knapen R, van Kraalingen D, Supit I, van der Wijngaart R, van Diepen K (2019) 25 years of the WOFOST cropping systems model. Agric Syst 168:154–167. https://doi.org/10.1016/j.agsy.2018.06.018

Delfani P, Thuraga V, Banerjee B, Chawade A (2024) Integrative approaches in modern agriculture: IoT, ML and AI for disease forecasting amidst climate change. Precis Agric. https://doi.org/10.1007/s11119-024-10164-7

Ding Z, Ali EF, Elmahdy AM, Ragab KE, Seleiman MF, Kheir AMS (2021) Modeling the combined impacts of deficit irrigation, rising temperature and compost application on wheat yield and water productivity. Agric Water Manag 244:106626. https://doi.org/10.1016/j.agwat.2020.106626

Droutsas I, Challinor AJ, Deva CR, Wang E (2022) Integration of machine learning into process-based modelling to improve simulation of complex crop responses. in silico Plants 4(2):diac017. https://doi.org/10.1093/insilicoplants/diac017

Dupraz C, Wolz KJ, Lecomte I, Talbot G, Vincent G, Mulia R, Bussière F, Ozier-Lafontaine H, Andrianarisoa S, Jackson N, Lawson G, Dones N, Sinoquet H, Lusiana B, Harja D, Domenicano S, Reyes F, Gosme M, Van Noordwijk M (2019) Hi-sAFe: a 3D agroforestry model for integrating dynamic tree–crop interactions. Sustain For 11(8):2293. https://www.mdpi.com/2071-1050/11/8/2293

Enders A, Vianna M, Gaiser T, Krauss G, Webber H, Srivastava AK, Seidel SJ, Tewes A, Rezaei EE, Ewert F (2023) SIMPLACE—a versatile modelling and simulation framework for sustainable crops and agroecosystems. in silico Plants 5(1):diad006. https://doi.org/10.1093/insilicoplants/diad006

Essenfelder AH, Giupponi C (2020) A coupled hydrologic-machine learning modelling framework to support hydrologic modelling in river basins under Interbasin Water Transfer regimes. Environ Model Softw 131:104779. https://doi.org/10.1016/j.envsoft.2020.104779

Feng P, Wang B, Li Liu D, Waters C, Yu Q (2019) Incorporating machine learning with biophysical model can improve the evaluation of climate extremes impacts on wheat yield in South-Eastern Australia. Agric For Meteorol 275:100–113

Feng X, Bi S, Li H, Qi Y, Chen S, Shao L (2024) Soil moisture forecasting for precision irrigation management using real-time electricity consumption records. Agric Water Manag 291:108656. https://doi.org/10.1016/j.agwat.2023.108656

Foster T, Brozović N, Butler AP, Neale CMU, Raes D, Steduto P, Fereres E, Hsiao TC (2017) AquaCrop-OS: an open source version of FAO's crop water productivity model. Agric Water Manag 181:18–22. https://doi.org/10.1016/j.agwat.2016.11.015

Golchin A, Misaghi M (2024) Investigating the effects of climate change and anthropogenic activities on SOC storage and cumulative CO2 emissions in forest soils across altitudinal gradients using the century model. Sci Total Environ 943:173758. https://doi.org/10.1016/j.scitotenv.2024.173758

Gu Z, Zhu T, Jiao X, Xu J, Qi Z (2021) Neural network soil moisture model for irrigation scheduling. Comput Electron Agric 180:105801. https://doi.org/10.1016/j.compag.2020.105801

He X, Estes L, Konar M, Tian D, Anghileri D, Baylis K, Evans TP, Sheffield J (2019) Integrated approaches to understanding and reducing drought impact on food security across scales. Curr Opin Environ Sustain 40:43–54. https://doi.org/10.1016/j.cosust.2019.09.006

Holzworth DP, Huth NI, deVoil PG, Zurcher EJ, Herrmann NI, McLean G, Chenu K, van Oosterom EJ, Snow V, Murphy C (2014) APSIM–evolution towards a new generation of agricultural systems simulation. Environ Model Softw 62:327–350

Islam MR, Oliullah K, Kabir MM, Alom M, Mridha MF (2023) Machine learning enabled IoT system for soil nutrients monitoring and crop recommendation. J Agric Food Res 14:100880. https://doi.org/10.1016/j.jafr.2023.100880

Jamieson PD, Semenov MA, Brooking IR, Francis GS (1998) Sirius: a mechanistic model of wheat response to environmental variation. Eur J Agron 8(3):161–179. https://doi.org/10.1016/S1161-0301(98)00020-3

Johnsson H, Bergstrom L, Jansson P-E, Paustian K (1987) Simulated nitrogen dynamics and losses in a layered agricultural soil. Agric Ecosyst Environ 18(4):333–356

Johnston DB, Pembleton KG, Huth NI, Deo RC (2023) Comparison of machine learning methods emulating process driven crop models. Environ Model Softw 162:105634. https://doi.org/10.1016/j.envsoft.2023.105634
Jones CA, Dyke PT, Williams JR, Kiniry JR, Benson VW, Griggs RH (1991) EPIC: an operational model for evaluation of agricultural sustainability. Agric Syst 37(4):341–350. https://doi.org/10.1016/0308-521X(91)90057-H
Jones JW, Hoogenboom G, Porter CH, Boote KJ, Batchelor WD, Hunt LA, Wilkens PW, Singh U, Gijsman AJ, Ritchie JT (2003) The DSSAT cropping system model. Eur J Agron 18(3):235–265. https://doi.org/10.1016/S1161-0301(02)00107-7
Kassie BT, Asseng S, Porter CH, Royce FS (2016) Performance of DSSAT-Nwheat across a wide range of current and future growing conditions. Eur J Agron 81:27–36. https://doi.org/10.1016/j.eja.2016.08.012
Kephe PN, Ayisi KK, Petja BM (2021) Challenges and opportunities in crop simulation modelling under seasonal and projected climate change scenarios for crop production in South Africa. Agric Food Secur 10(1):10. https://doi.org/10.1186/s40066-020-00283-5
Kheir AMS, El Baroudy A, Aiad MA, Zoghdan MG, Abd El-Aziz MA, Ali MGM, Fullen MA (2019) Impacts of rising temperature, carbon dioxide concentration and sea level on wheat production in North Nile delta. Sci Total Environ 651:3161–3173. https://doi.org/10.1016/j.scitotenv.2018.10.209
Kheir AMS, Alkharabsheh HM, Seleiman MF, Al-Saif AM, Ammar KA, Attia A, Zoghdan MG, Shabana MMA, Aboelsoud H, Schillaci C (2021a) Calibration and validation of AQUACROP and APSIM models to optimize wheat yield and water saving in arid regions. Land 10(12):1375. https://www.mdpi.com/2073-445X/10/12/1375
Kheir AMS, Alrajhi AA, Ghoneim AM, Ali EF, Magrashi A, Zoghdan MG, Abdelkhalik SAM, Fahmy AE, Elnashar A (2021b) Modeling deficit irrigation-based evapotranspiration optimizes wheat yield and water productivity in arid regions. Agric Water Manag 256:107122. https://doi.org/10.1016/j.agwat.2021.107122
Kheir, A. M. S., Ammar, K. A., Attia, A., Elnashar, A., Ahmad, S., El-Gioushy, S.F., & Ahmed, M. (2022a). Cereal crop modeling for food and nutrition security. In M. Ahmed (Ed.), Global agricultural production: resilience to climate change (pp. 183–195). Springer. doi:https://doi.org/10.1007/978-3-031-14973-3_6
Kheir AMS, Ammar KA, Amer A, Ali MGM, Ding Z, Elnashar A (2022b) Machine learning-based cloud computing improved wheat yield simulation in arid regions. Comput Electron Agric 203:107457. https://doi.org/10.1016/j.compag.2022.107457
Kheir AMS, Elnashar A, Mosad A, Govind A (2023a) An improved deep learning procedure for statistical downscaling of climate data. Heliyon 9(7). https://doi.org/10.1016/j.heliyon.2023.e18200
Kheir AMS, Mkuhlani S, Mugo JW, Elnashar A, Nangia V, Devare M, Govind A (2023b) Integrating APSIM model with machine learning to predict wheat yield spatial distribution. Agron J 115(6):3188–3196. https://doi.org/10.1002/agj2.21470
Kheir A, Ali O, Shawon AR, Elrys A, Ali M, Darwish M, Elmahdy A, Abou-Hadid A, Nóia Júnior RDS, Feike T (2024a) Impacts of climate change on spatial wheat yield and nutritional values using hybrid machine learning. Environ Res Lett 19:104049. http://iopscience.iop.org/article/10.1088/1748-9326/ad75ab
Kheir AMS, Govind A, Nangia V, Devkota M, Elnashar A, Omar MED, Feike T (2024b) Developing automated machine learning approach for fast and robust crop yield prediction using a fusion of remote sensing, soil, and weather dataset. Environ Res Commun 6(4):041005. https://doi.org/10.1088/2515-7620/ad2d02
Kheir A, Gosme M, Bakhtiary N, Kephe P, Assou J, Langhof M, Strassemeyer J, Feike T (2024c) Developing Hi-sAFe-machine learning hybrid approach as a field-specific decision support system for agroforestry systems. In: International research symposium on agricultural greenhouse gas mitigation from research to implementation, Berlin, Germany
Kheir AMS, Govind A, Nangia V, El-Maghraby MA, Elnashar A, Ahmed M, Aboelsoud H, Gamal R, Feike T (2025) Hybridization of process-based models, remote sensing, and machine

learning for enhanced spatial predictions of wheat yield and quality. Comput Electron Agric 234:110317. https://doi.org/10.1016/j.compag.2025.110317
Kuang N, Ma Y, Hong S, Jiao F, Liu C, Li Q, Han H (2021) Simulation of soil moisture dynamics, evapotranspiration, and water drainage of summer maize in response to different depths of subsoiling with RZWQM2. Agric Water Manag 249:106794. https://doi.org/10.1016/j.agwat.2021.106794
Le Noë J, Manzoni S, Abramoff R, Bölscher T, Bruni E, Cardinael R, Ciais P, Chenu C, Clivot H, Derrien D, Ferchaud F, Garnier P, Goll D, Lashermes G, Martin M, Rasse D, Rees F, Sainte-Marie J, Salmon E et al (2023) Soil organic carbon models need independent time-series validation for reliable prediction. Commun Earth Environ 4(1):158. https://doi.org/10.1038/s43247-023-00830-5
Li Z, Zhang Z, Zhang L (2021) Improving regional wheat drought risk assessment for insurance application by integrating scenario-driven crop model, machine learning, and satellite data. Agric Syst 191:103141. https://doi.org/10.1016/j.agsy.2021.103141
Li F, Yigitcanlar T, Nepal M, Nguyen K, Dur F (2023) Machine learning and remote sensing integration for leveraging urban sustainability: a review and framework. Sustain Cities Soc 96:104653. https://doi.org/10.1016/j.scs.2023.104653
Li Z, Nie Z, Li G (2024) Integrating crop modeling and machine learning for the improved prediction of dryland wheat yield. Agronomy 14(4):777. https://www.mdpi.com/2073-4395/14/4/777
Lu SC, Swisher CL, Chung C, Jaffray D, Sidey-Gibbons C (2023) On the importance of interpretable machine learning predictions to inform clinical decision making in oncology. Front Oncol 13:1129380. https://doi.org/10.3389/fonc.2023.1129380
Maestrini B, Mimić G, van Oort PAJ, Jindo K, Brdar S, Athanasiadis IN, van Evert FK (2022) Mixing process-based and data-driven approaches in yield prediction. Eur J Agron 139:126569. https://doi.org/10.1016/j.eja.2022.126569
Magyar T, Fehér Z, Buday-Bódi E, Tamás J, Nagy A (2023) Modeling of soil moisture and water fluxes in a maize field for the optimization of irrigation. Comput Electron Agric 213:108159. https://doi.org/10.1016/j.compag.2023.108159
Malek K, Stöckle C, Chinnayakanahalli K, Nelson R, Liu M, Rajagopalan K, Barik M, Adam JC (2017) VIC–CropSyst-v2: a regional-scale modeling platform to simulate the nexus of climate, hydrology, cropping systems, and human decisions. Geosci Model Dev 10(8):3059–3084. https://doi.org/10.5194/gmd-10-3059-2017
Mana AA, Allouhi A, Hamrani A, Rehman S, el Jamaoui I, Jayachandran K (2024) Sustainable AI-based production agriculture: exploring AI applications and implications in agricultural practices. Smart Agric Technol 7:100416. https://doi.org/10.1016/j.atech.2024.100416
Manschadi AM, Palka M, Fuchs W, Neubauer T, Eitzinger J, Oberforster M, Soltani A (2022) Performance of the SSM-iCrop model for predicting growth and nitrogen dynamics in winter wheat. Eur J Agron 135:126487. https://doi.org/10.1016/j.eja.2022.126487
Materia S, García LP, van Straaten C, Sungmin O, Mamalakis A, Cavicchia L, Coumou D, de Luca P, Kretschmer M, Donat M (2024) Artificial intelligence for climate prediction of extremes: state of the art, challenges, and future perspectives. WIREs Clim Change 15(6):e914. https://doi.org/10.1002/wcc.914
McCown RL, Hammer GL, Hargreaves JNG, Holzworth DP, Freebairn DM (1996) APSIM: a novel software system for model development, model testing and simulation in agricultural systems research. Agric Syst 50(3):255–271. https://doi.org/10.1016/0308-521X(94)00055-V
Minasny B, Bandai T, Ghezzehei TA, Huang Y-C, Ma Y, McBratney AB, Ng W, Norouzi S, Padarian J, Rudiyanto, Sharififar A, Styc Q, Widyastuti M (2024) Soil science-informed machine learning. Geoderma 452:117094. https://doi.org/10.1016/j.geoderma.2024.117094
Morianou G, Kourgialas NN, Karatzas GP (2023) A review of HYDRUS 2D/3D applications for simulations of water dynamics, root uptake and solute transport in tree crops under drip irrigation. Water 15(4):741. https://www.mdpi.com/2073-4441/15/4/741
Nendel C (2014) MONICA: a simulation model for nitrogen and carbon dynamics in agro-ecosystems. In: Mueller L, Saparov A, Lischeid G (eds) Novel measurement and assessment tools for

monitoring and management of land and water resources in agricultural landscapes of Central Asia. Springer, pp 389–405. https://doi.org/10.1007/978-3-319-01017-5_23
Padarian J, Minasny B, McBratney AB (2019) Using deep learning to predict soil properties from regional spectral data. Geoderma Reg 16:e00198. https://doi.org/10.1016/j.geodrs.2018.e00198
Padhiary M, Saha D, Kumar R, Sethi LN, Kumar A (2024) Enhancing precision agriculture: a comprehensive review of machine learning and AI vision applications in all-terrain vehicle for farm automation. Smart Agric Technol 8:100483. https://doi.org/10.1016/j.atech.2024.100483
Paff K, Timlin D, Fleisher DH (2023) A comparison of wheat leaf-appearance rate submodules for DSSAT CROPSIM-CERES (CSCER). Ecol Model 482:110406. https://doi.org/10.1016/j.ecolmodel.2023.110406
Pan F, Feng Q, McGehee R, Engel BA, Flanagan DC, Chen J (2021) A framework for automated and spatially-distributed modeling with the Agricultural Policy Environmental eXtender (APEX) model. Environ Model Softw 144:105147. https://doi.org/10.1016/j.envsoft.2021.105147
Panek-Chwastyk E, Ozbilge CN, Dąbrowska-Zielińska K, Gurdak R (2024) Advancing crop yield predictions: AQUACROP model application in Poland's JECAM fields. Agronomy 14(4):854. https://www.mdpi.com/2073-4395/14/4/854
Parra-López C, Ben Abdallah S, Garcia-Garcia G, Hassoun A, Sánchez-Zamora P, Trollman H, Jagtap S, Carmona-Torres C (2024) Integrating digital technologies in agriculture for climate change adaptation and mitigation: state of the art and future perspectives. Comput Electron Agric 226:109412. https://doi.org/10.1016/j.compag.2024.109412
Pasley H, Wenham K, Bell L, Huth N, Holzworth D, Chaki A, Gaydon D, Collins M (2023) APSIM next generation mungbean model: a tool for advancing mungbean production. Field Crop Res 298:108955. https://doi.org/10.1016/j.fcr.2023.108955
Peethani S, Pathakoti M, Mahalakshmi DV, Kheir AMS, Singh A, Govind A (2024) Variability of ENSO teleconnections indices, and its impacts on Moroccan agriculture. J Atmos Sol Terr Phys 265:106386. https://doi.org/10.1016/j.jastp.2024.106386
Pequeno DNL, Ferreira TB, Fernandes JMC, Singh PK, Pavan W, Sonder K, Robertson R, Krupnik TJ, Erenstein O, Asseng S (2024) Production vulnerability to wheat blast disease under climate change. Nat Clim Chang 14(2):178–183. https://doi.org/10.1038/s41558-023-01902-2
Porter JR (1993) AFRCWHEAT2: a model of the growth and development of wheat incorporating responses to water and nitrogen. Eur J Agron 2(2):69–82. https://doi.org/10.1016/S1161-0301(14)80136-6
Preite L, Vignali G (2024) Artificial intelligence to optimize water consumption in agriculture: a predictive algorithm-based irrigation management system. Comput Electron Agric 223:109126. https://doi.org/10.1016/j.compag.2024.109126
Prity FS, Hasan MDM, Saif SH, Hossain MM, Bhuiyan SH, Islam MA, Lavlu MTH (2024) Enhancing agricultural productivity: a machine learning approach to crop recommendations. Hum-Centric Intell Syst 4(4):497–510. https://doi.org/10.1007/s44230-024-00081-3
Procopio A, Cesarelli G, Donisi L, Merola A, Amato F, Cosentino C (2023) Combined mechanistic modeling and machine-learning approaches in systems biology—a systematic literature review. Comput Methods Prog Biomed 240:107681. https://doi.org/10.1016/j.cmpb.2023.107681
Rani S, Mishra AK, Kataria A, Mallik S, Qin H (2023) Machine learning-based optimal crop selection system in smart agriculture. Sci Rep 13(1):15997. https://doi.org/10.1038/s41598-023-42356-y
Raza A, Vianna MDS, Ahmadi SH, Habib-ur-Rahman M, Gaiser T (2024) Comparison of predictive modeling approaches to estimate soil erosion under spatially heterogeneous field conditions. Environ Model Softw 180:106145. https://doi.org/10.1016/j.envsoft.2024.106145
Rodrigues M, Miguéis V, Freitas S, Machado T (2024) Machine learning models for short-term demand forecasting in food catering services: a solution to reduce food waste. J Clean Prod 435:140265. https://doi.org/10.1016/j.jclepro.2023.140265
Röll G, Memic E, Graeff-Hönninger S (2020) Implementation of an automatic time-series calibration method for the DSSAT wheat models to enhance multi-model approaches. Agron J 112(5):3891–3912. https://doi.org/10.1002/agj2.20328

Rounsevell MDA, Arneth A, Brown C, Cheung WWL, Gimenez O, Holman I, Leadley P, Luján C, Mahevas S, Maréchaux I, Pélissier R, Verburg PH, Vieilledent G, Wintle BA, Shin Y-J (2021) Identifying uncertainties in scenarios and models of socio-ecological systems in support of decision-making. One Earth 4(7):967–985. https://doi.org/10.1016/j.oneear.2021.06.003

Sarker IH (2021) Machine learning: algorithms, real-world applications and research directions. SN Comput Sci 2(3):160. https://doi.org/10.1007/s42979-021-00592-x

Shahhosseini M, Hu G, Huber I, Archontoulis SV (2021) Coupling machine learning and crop modeling improves crop yield prediction in the US Corn Belt. Sci Rep 11(1):1606. https://doi.org/10.1038/s41598-020-80820-1

Shawon AR, Memic E, Kottmann L, Uptmoor R, Hackauf B, Feike T (2024) Comprehensive evaluation of the DSSAT-CSM-CERES-Wheat for simulating winter rye against multi-environment data in Germany. Agron J 116(4):1844–1868. https://doi.org/10.1002/agj2.21590

Silva JV, Giller KE (2020) Grand challenges for the 21st century: what crop models can and can't (yet) do. J Agric Sci 158(10):794–805. https://doi.org/10.1017/S0021859621000150

Stöckle CO, Donatelli M, Nelson R (2003) CropSyst, a cropping systems simulation model. Eur J Agron 18(3):289–307. https://doi.org/10.1016/S1161-0301(02)00109-0

Thorp KR, DeJonge KC, Kaleita AL, Batchelor WD, Paz JO (2008) Methodology for the use of DSSAT models for precision agriculture decision support. Comput Electron Agric 64(2):276–285. https://doi.org/10.1016/j.compag.2008.05.022

Trentin C, Ampatzidis Y, Lacerda C, Shiratsuchi L (2024) Tree crop yield estimation and prediction using remote sensing and machine learning: a systematic review. Smart Agric Technol 9:100556. https://doi.org/10.1016/j.atech.2024.100556

Umutoni L, Samadi V (2024) Application of machine learning approaches in supporting irrigation decision making: a review. Agric Water Manag 294:108710. https://doi.org/10.1016/j.agwat.2024.108710

Uusitalo L, Lehikoinen A, Helle I, Myrberg K (2015) An overview of methods to evaluate uncertainty of deterministic models in decision support. Environ Model Softw 63:24–31. https://doi.org/10.1016/j.envsoft.2014.09.017

van Klompenburg T, Kassahun A, Catal C (2020) Crop yield prediction using machine learning: a systematic literature review. Comput Electron Agric 177:105709. https://doi.org/10.1016/j.compag.2020.105709

Wang Z, Ye L, Jiang J, Fan Y, Zhang X (2022) Review of application of EPIC crop growth model. Ecol Model 467:109952. https://doi.org/10.1016/j.ecolmodel.2022.109952

Wang D, Guo M, Li J, Wu S, Cheng Y, Shi L, Liu S, Ge J, Dong Q, Li Y, Wu F, Jiang T (2024) Impact of climate change on the winter wheat productivity under varying climate scenarios in the loess plateau: an APSIM analysis (1961–2100). Agronomy 14(11):2609. https://www.mdpi.com/2073-4395/14/11/2609

Wu J, Chen X-Y, Zhang H, Xiong L-D, Lei H, Deng S-H (2019) Hyperparameter optimization for machine learning models based on Bayesian optimization b. J Electron Sci Technol 17(1):26–40. https://doi.org/10.11989/JEST.1674-862X.80904120

Wynn DC, Clarkson PJ (2018) Process models in design and development. Res Eng Des 29(2):161–202. https://doi.org/10.1007/s00163-017-0262-7

Ye Z, Yin S, Cao Y, Wang Y (2024) AI-driven optimization of agricultural water management for enhanced sustainability. Sci Rep 14(1):25721. https://doi.org/10.1038/s41598-024-76915-8

You Y, Tian H, Pan S, Shi H, Bian Z, Gurgel A, Huang Y, Kicklighter D, Liang X-Z, Lu C, Melillo J, Miao R, Pan N, Reilly J, Ren W, Xu R, Yang J, Yu Q, Zhang J (2022) Incorporating dynamic crop growth processes and management practices into a terrestrial biosphere model for simulating crop production in the United States: toward a unified modeling framework. Agric For Meteorol 325:109144. https://doi.org/10.1016/j.agrformet.2022.109144

Yue Y, Li J, Ye X, Wang Z, Zhu AX, Wang J-A (2015) An EPIC model-based vulnerability assessment of wheat subject to drought. Nat Hazards 78(3):1629–1652. https://doi.org/10.1007/s11069-015-1793-8

Zhang L, Guo CL, Zhao LY, Zhu Y, Cao WX, Tian YC, Cheng T, Wang X (2016) Estimating wheat yield by integrating the WheatGrow and PROSAIL models. Field Crop Res 192:55–66. https://doi.org/10.1016/j.fcr.2016.04.014

Zhang Z, Li Z, Chen Y, Zhang L, Tao F (2020) Improving regional wheat yields estimations by multi-step-assimilating of a crop model with multi-source data. Agric For Meteorol 290:107993. https://doi.org/10.1016/j.agrformet.2020.107993

Zhang N, Zhou X, Kang M, Hu B-G, Heuvelink E, Marcelis LFM (2022) Machine learning versus crop growth models: an ally, not a rival. AoB Plants 15(2):plac061. https://doi.org/10.1093/aobpla/plac061

Zhang J-L, Huang X-M, Sun Y-Z (2024) Multiscale spatiotemporal meteorological drought prediction: a deep learning approach. Adv Clim Chang Res 15(2):211–221. https://doi.org/10.1016/j.accre.2024.04.003

Zhou T, Gani R, Sundmacher K (2021) Hybrid data-driven and mechanistic modeling approaches for multiscale material and process design. Engineering 7(9):1231–1238. https://doi.org/10.1016/j.eng.2020.12.022

Zhu J, Parker A, Gou F, Agnew R, Yang L, Greven M, Raw V, Neal S, Martin D, Trought MCT, Huth N, Brown HE (2021) Developing perennial fruit crop models in APSIM Next Generation using grapevine as an example. in silico Plants 3(2):diab021. https://doi.org/10.1093/insilicoplants/diab021

Fusion of Process-Based Models, Machine Learning, and Remote Sensing for Yield Gap Assessment

17

Ahmed M. S. Kheir, Mahmoud M. A. Shabana, Marwa G. M. Ali, Ahmed Attia, Mohamed A. Abd El-Aziz, and Til Feike

Abstract

This chapter delves into the innovative integration of process-based models, machine learning techniques, and remote sensing to significantly improve the precision and reliability of yield gap assessments. Yield gaps—the disparities between potential and actual crop yields—serve as critical indicators for understanding the underlying limitations in agricultural productivity and for formulating strategies to close these gaps. Dynamic models, such as APSIM and DSSAT, ensure a comprehensive mechanistic framework that simulates crop growth, development, and environmental interactions under varying management practices and climatic conditions. These models are invaluable for estimating potential yields across diverse agroecosystems. On the other hand, machine learning excels in processing large datasets, uncovering complex patterns, and offering predictive insights that traditional methods may overlook. When paired with remote sensing, which provides high-resolution and robust data on crop health, growth stages, and environmental conditions, this integrated approach addresses

A. M. S. Kheir (✉)
Julius Kühn Institute (JKI)—Federal Research Centre for Cultivated Plants, Institute for Strategies and Technology Assessment, Kleinmachnow, Germany

Soils, Water and Environment Research Institute, Agricultural Research Center, Giza, 12112, Egypt
e-mail: Ahmed.Kheir@julius-kuehn.de

M. M. A. Shabana · M. A. Abd El-Aziz
Soils, Water and Environment Research Institute, Agricultural Research Center, Giza, 12112, Egypt

M. G. M. Ali · A. Attia · T. Feike
Julius Kühn Institute (JKI)—Federal Research Centre for Cultivated Plants, Institute for Strategies and Technology Assessment, 14532, Kleinmachnow, Germany
e-mail: til.feike@julius-kuehn.de

A. E.-H. Zohry, S. Ouda (eds.), *Resilient Agroecosystems*, Sustainability Sciences in Asia and Africa, https://doi.org/10.1007/978-981-96-9630-7_17

the spatial and temporal variability inherent in agricultural systems. By combining the strengths of these technologies, this chapter proposes a robust methodology for yield gap assessments, offering improved predictive accuracy and scalability. This fusion of advanced techniques not only enhances the precision of yield forecasts but also provides actionable insights for policymakers and agricultural stakeholders, enabling the development of targeted, sustainable interventions that are essential for future food security in the face of resource constraints and climate change.

Keywords

APSIM · DSSAT · Regression models · Support vector machine · Random forest · Deep learning

Introduction

The accurate assessment of yield gaps—the difference between potential and actual yields—has long been a crucial pursuit in agricultural research, aimed at identifying opportunities to improve productivity and ensure food security (Neumann et al. 2010; Zhi et al. 2022). Over the decades, various approaches have been employed to assess yield gaps, each with its own set of strengths and limitations. Early approaches relied heavily on empirical regression methods, which used simple linear relationships between environmental variables and crop yields (Miti et al. 2024; Rattalino Edreira et al. 2017; van Ittersum et al. 2013). These methods were easier to apply but often failed to capture the complex, nonlinear dynamics of crop-environment interactions, especially in the presence of multiple stressors like climate change, pest outbreaks, and variable soil conditions.

In response to the limitations of empirical models, process-based crop models such as APSIM (Keating et al. 2003; Kheir et al. 2023a; McCown et al. 1996) and DSSAT (Jones et al. 2003; Kheir et al. 2022a) were developed, offering a more mechanistic approach to simulating crop growth. These models take into account detailed biophysical processes governing plant development, such as photosynthesis, evapotranspiration, and nutrient cycling. Although process-based models have significantly improved yield predictions by incorporating environmental interactions, they remain limited by their reliance on extensive, high-quality input data and their inability to scale easily across large, heterogeneous landscapes (Antle et al. 2017).

Machine learning have recently expanded novel ideas for yield gap assessment. Machine learning models, particularly deep learning algorithms, excel at identifying patterns in vast, complex datasets (Kheir et al. 2024a, b; Liu et al. 2024) Furthermore, machine learning can process multidimensional data driven include remote sensing and environmental sensors, to predict crop yields with high accuracy. Unlike traditional methods, machine learning models are less constrained by the need for detailed crop parameters, explaining the complexed relationships between inputs and yields (Paudel et al. 2021; Priyatikanto et al. 2023). However, these models are often criticized for their "black box" nature and dependency on large amounts of training data, which limits their interpretability and reliability when applied to regions with sparse data (Priyatikanto et al. 2023).

The integration of remote sensing with crop models and machine learning has further advanced yield gap assessments by providing continuous, high-resolution data on crop conditions through wide area (Karmakar et al. 2024; Lobell 2013; Wang et al. 2023; Wu et al. 2022). Remote sensing, through satellites and drones, captures valuable spatial and temporal information on vegetation indices, land surface temperature, and soil moisture (Krishnan and Indu 2023; Neinavaz et al. 2021). This information, when fed into crop models or machine learning algorithms, allows for the precise monitoring of crop health and environmental conditions across diverse agroecosystems.

This chapter introduces a synergistic approach, combining process-based models, machine learning techniques, and remote sensing to address the limitations of previous methods. By leveraging the detailed simulations of process-based models, the pattern-recognition capabilities of machine learning, and the real-time spatial data from remote sensing, this integrated approach offers a more robust and scalable framework for yield gap assessments. This fusion enables the accurate prediction of yields across various spatial scales, from individual fields to entire regions, while also accounting for the dynamic interactions between crops and their environments. The methodology presented in this chapter not only enhances the accuracy of yield gap assessments but also provides actionable insights for policymakers and farmers, aiding in the development of targeted strategies for sustainable agricultural intensification under changing climatic conditions.

Process-Based Models in Agriculture

Overview

Process-based models have become fundamental tools in agricultural research and decision-making, offering a mechanistic framework for simulating crop growth, development, and interactions with environmental factors such as weather, soil properties, and management practices (Kheir et al. 2019, 2022a). These models are built upon scientific principles, such as photosynthesis, nutrient cycling, and water balance, which dictate crop performance under various conditions (Fodor et al. 2017; Pasley et al. 2022). By simulating these processes, process-based models provide insights into how crops respond to different agronomic and climatic scenarios, helping researchers, policymakers, and farmers optimize management practices for improved productivity and resource use efficiency.

Two of the most widely used process-based models in agriculture are the Agricultural Production Systems sIMulator (APSIM) and the Decision Support System for Agrotechnology Transfer (DSSAT). APSIM was developed to predict biophysical processes in agricultural systems, focusing on crop production, soil dynamics, and the impact of management interventions. APSIM is highly flexible, allowing users to configure different crop types, environmental settings, and management practices. It has been applied extensively across diverse agroecosystems to explore issues such as crop rotations, climate variability, and soil fertility management (Bosi et al. 2020). DSSAT, on the other hand, is a suite of crop models that simulates the growth and yield of over 42 crops. It integrates data on crop genetics, soil properties, weather conditions, and management practices to simulate development and yield of crops and assess the effects of different factors such as climate and biotic and abiotic stresses. DSSAT is widely used for simulating potential yields, analyzing the impacts of climate change, and evaluating different cropping systems at field, regional, and global scales (Thorp et al. 2008).

These models are grounded in fundamental agronomic processes, making them valuable for conducting virtual experiments (Jones et al. 2017), testing hypotheses (Liman Harou et al. 2021), and optimizing cropping systems under varying conditions (Asseng et al. 2018). As climate change continues to affect agricultural systems worldwide, process-based models like APSIM and DSSAT are increasingly important for understanding how crops will respond to future environmental stresses.

Strengths and Limitations of Process-Based Models

Process-based models offer mechanistic insights into crop growth and environmental interactions, distinguishing them from empirical or statistical models that merely describe relationships between inputs and outputs (Ammar and Davies 2019; Hajjarpoor et al. 2022). Process-based models explicitly simulate the physiological and biophysical processes driving crop development, such as photosynthesis, respiration, evapotranspiration, nutrient uptake, and carbon allocation. This allows for a more detailed representation of crop-environment interactions, providing greater accuracy in predicting crop growth under different management scenarios and environmental conditions (Hajjarpoor et al. 2022).

Another strength of process-based models is their flexibility and extensibility. These models can simulate a wide range of crops, soils, climates, and management practices, making them highly adaptable to diverse agricultural systems around the world (van Keulen 2013). APSIM, for instance, allows researchers to simulate complex cropping systems, including crop rotations and intercropping (Holzworth et al. 2014), while DSSAT supports detailed simulations of management interventions, such as irrigation, fertilization, and pest control (Ali et al. 2022).

Despite their strengths, process-based models face several challenges. One of the primary limitations is their high computational demand. Process-based models require extensive data inputs, including detailed soil properties, climate data, and management histories, to accurately simulate crop growth (Droutsas et al. 2019; Gavasso-Rita et al. 2024; Kephe et al. 2021). Collecting and processing these data can be time-consuming and costly, particularly for large-scale or regional applications. Furthermore, these models often require significant computational power, especially when applied to large datasets or when simulations are conducted over long periods or across multiple regions (Gavasso-Rita et al. 2024). Another limitation is the sensitivity to input data quality. Process-based models are highly reliant on accurate and precise data inputs, such as soil texture, initial soil moisture content, and weather conditions. Any inaccuracies or gaps in this data can lead to significant errors in model outputs (Kephe et al. 2021). Additionally, process-based models often require detailed crop-specific parameters (e.g., growth stages, leaf area index, and rooting depth), which are not always readily available for all crops or regions.

Moreover, while process-based models offer mechanistic depth, they may struggle to scale up efficiently. Applying these models at larger scales, such as for regional or national yield assessments, can be challenging due to the complexity of the interactions they simulate and the variability of environmental factors across large landscapes (Droutsas et al. 2019; Gavasso-Rita et al. 2024; Kephe et al. 2021).

Consequently, while process-based models like APSIM and DSSAT are powerful tools for simulating crop growth and environmental interactions, their effectiveness is often limited by their data requirements and computational demands. However, when these challenges are addressed, process-based models provide highly detailed and insightful analyses that are crucial for improving agricultural productivity, especially in the face of climate variability and resource constraints. Accordingly, these models may struggle to capture spatial heterogeneity, limiting their usefulness in certain large-scale applications without significant adjustments or integration with other models, such as machine learning.

Machine Learning Techniques

Overview

Machine learning signifies a substantial breakthrough in data analysis, providing techniques that enable computers to "learn" from data and enhance their performance progressively without direct programming (El Bouchefry and de Souza 2020; Kheir et al. 2023b). Machine learning can be categorized into three primary types: supervised learning, unsupervised learning, semi-supervised learning, and re-enforcement learning (Table 17.1).

Artificial intelligence and machine learning algorithms are transforming image recognition, offering significant benefits to communities (Soori et al. 2023). These technologies can automate complex methods, adapt them to individual situations, and minimize errors, ultimately boosting efficiency and accuracy in imaging (Fig. 17.1). This helps standardize reporting procedures and boosts confidence in imaging results. Recent developments demonstrate artificial intelligence's promise for predictive management and intervention skills (Olawade et al. 2024). Digital twins that mimic interventional therapies can be produced by combining artificial intelligence with virtual planning. Personalized therapy options are made possible by this breakthrough, which enhances patient outcomes and care (Puccetti et al. 2024). In agriculture, machine learning has become particularly valuable for predicting crop yields (Kheir et al. 2023a), diagnosing crop health (Ahmed and Yadav

Table 17.1 Types and examples of machine learning techniques

Types of machine learning	Function	Examples
Supervised learning (Seetharam et al. 2019)	The dataset includes labels, results, and inferences drawn from the data to make predictions	Includes Bayesian networks, artificial neural networks, logistic regression, ridge regression, and elastic net regression
Unsupervised learning (Seetharam et al. 2019)	The dataset finds important associations without labels	This includes principal component analysis, k-means clustering, and hierarchical clustering
Semi-supervised learning (Seetharam et al. 2019)	A mix of supervised and unsupervised learning that uses the incentive system to carry out tasks	Widely employed in speech and picture recognition
Re-enforcement learning (Seetharam et al. 2019)	The learning process typically involves trial and error, where the agent balances exploration (trying new actions) and exploitation (using known strategies that give good rewards). Algorithms like Q-learning, Deep Q-Networks (DQN), and policy gradient methods are popular approaches in RL	Robotics, gaming (e.g., AlphaGo), self-driving cars, and recommendation systems

2023), optimizing resource use (Xun et al. 2020), and improving decision-making processes. Machine learning models rely on algorithms to uncover complex, nonlinear relationships between input data (e.g., weather patterns, soil conditions, remote sensing data) and outputs, such as crop yields. Unlike traditional statistical methods that rely on predefined assumptions about data distribution, machine learning can process large, complex datasets, identifying patterns and relationships that may not be immediately evident.

The common machine learning algorithms applied in yield prediction include:

- *Regression models*, which use statistical relationships between input features (e.g., soil moisture, temperature) and yield. These models are simple and interpretable but often struggle with nonlinearity in complex agricultural systems (Heilemann et al. 2024).
- *Support vector machine (SVM)*, which classify data points by finding a hyperplane that best separates different classes. SVMs have been widely used for classifying crop types and predicting yield based on remote sensing data (Adugna et al. 2022).
- *Random forest (RF)*, a popular ensemble learning method that builds multiple decision trees and merges their predictions for more accurate results. Random forests can effectively handle large datasets and account for variable interactions, making them highly suitable for agricultural applications (Adugna et al. 2022).

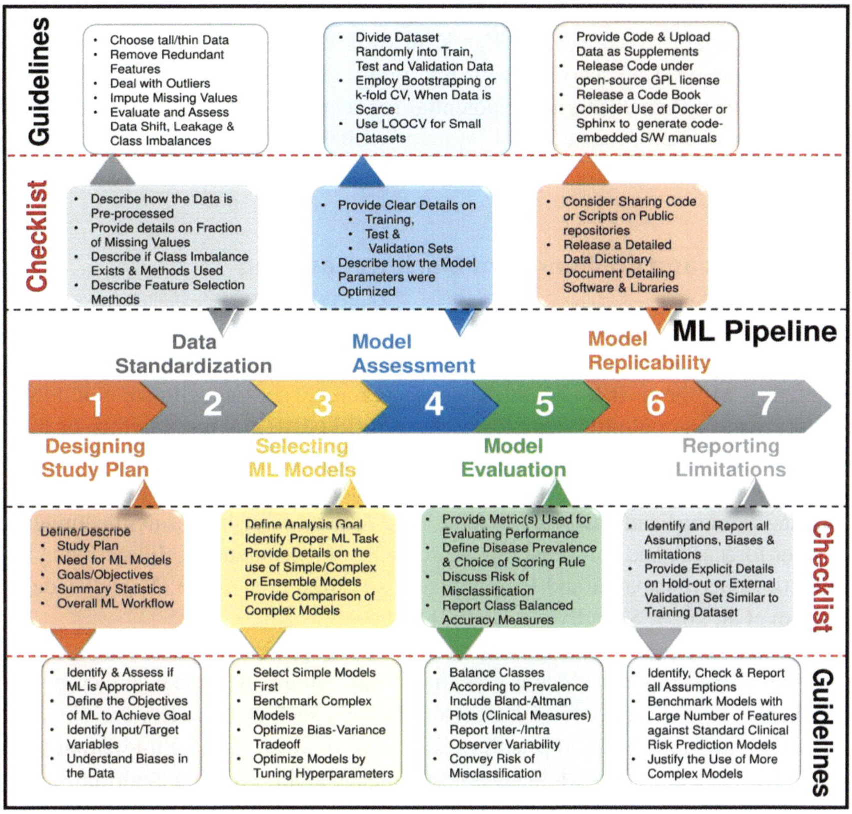

Fig. 17.1 The items in a checklist for reporting and the procedures for creating a machine learning (ML) pipeline. (Seetharam et al. 2020; Sengupta et al. 2020)

- *Artificial neural networks (ANNs)* are composed of layers of interconnected "neurons" that process input data and iteratively adjust their internal weights to reduce prediction errors. They are particularly effective at handling nonlinear relationships between variables but can be prone to overfitting if not properly managed (Moghaddam and Cao 2024).
- *Deep learning (DL)*, a subset of machine learning, involves multilayered neural networks that automatically extract high-level features from raw data. Deep learning has gained traction in agriculture, particularly for applications involving large, complex datasets, such as satellite imagery and sensor data. Models such as convolutional neural networks (CNNs) and long short-term memory (LSTM) networks have been applied to yield forecasting, identifying stress conditions, and monitoring crop phenology.

Recently, DL represented as a leading technique for crop yield prediction. For example, LSTM networks have been successfully applied to predict maize and rice yields months ahead of harvest by capturing long-term dependencies between environmental variables (Jiang et al. 2020; Kang et al. 2020). Furthermore, hybrid models that combine deep learning with traditional crop models, such as APSIM or DSSAT, have shown even greater potential for accurate, early-season yield predictions across diverse agricultural systems (Shahhosseini et al. 2021; Zhang et al. 2022).

Advantages and Challenges of Machine Learning

A key advantage of machine learning is its capability to manage large and complex datasets, which are becoming more prevalent in contemporary agriculture (Ennaji et al. 2023; Meshram et al. 2021). Advances in remote sensing, climate monitoring, and precision agriculture technologies have resulted in a massive influx of high-dimensional data. Machine learning models are adept at processing these data, identifying patterns that may not be obvious through conventional methods. For instance, machine learning can integrate diverse data sources—such as satellite imagery, weather forecasts, and soil sensor measurements—to generate highly accurate yield predictions across spatial and temporal scales (Elbasi et al. 2023; Naghdyzadegan Jahromi et al. 2023; Shendryk et al. 2021).

Machine learning models also excel in uncovering complex, nonlinear relationships between variables, which is essential in agricultural systems where crop yield is influenced by a multitude of interacting factors, such as climate variability, soil conditions, and management practices (Elbasi et al. 2023). Unlike traditional models, which often require predefined assumptions, machine learning algorithms adapt to the data, learning relationships directly from observations. Additionally, machine learning models can continuously improve over time as more data become available. This capability is particularly useful in dynamic agricultural environments where new technologies and practices are regularly introduced (Shendryk et al. 2021).

Remote Sensing for Agricultural Monitoring

Remote sensing has become a crucial technology for agricultural monitoring, offering unique capabilities to collect and analyze data from large areas of land (Rogan and Chen 2004). This section delves into the various types of remote sensing technologies, their applications, and the key data products used in agricultural monitoring and yield gap assessments.

Types of Remote Sensing

- *Satellite-Based Remote Sensing:* Satellites such as MODIS, Landsat, and Sentinel provide large-scale, repetitive coverage of agricultural landscapes

(Radeloff et al. 2024). They capture data on vegetation indices, land surface temperature, and soil moisture, which are essential for monitoring crop growth, detecting stress conditions, and assessing yield potentials.
- *Unmanned Aerial Vehicles (UAVs):* UAVs, or drones, offer higher resolution data than satellites and are ideal for monitoring smaller plots of land (Alvarez-Vanhard et al. 2021). They can capture detailed imagery of crop health, growth stages, and soil properties. UAVs are particularly useful for precision agriculture, where decisions on irrigation, fertilization, and pesticide applications are made at the field level.

Key Data Products for Yield Gap Assessment

- *Normalized Difference Vegetation Index (NDVI):* NDVI is a commonly utilized metric for assessing plant health and vitality by analyzing the contrast between near-infrared and red light reflectance (Farias et al. 2023). It provides insights into the photosynthetic activity of crops, allowing for early detection of stress conditions.
- *Soil Moisture:* Satellite-derived soil moisture data, such as those from SMAP or Sentinel-1, are crucial for assessing water availability and detecting drought conditions (Nanda et al. 2024). Soil moisture is a key driver of crop yield, especially in rainfed agriculture.
- *Leaf Area Index (LAI):* LAI quantifies the total leaf area relative to a unit of ground area and is closely associated with crop biomass and potential yield (Pérez et al. 2022). It is derived from satellite imagery and helps in estimating crop productivity.

Applications in Agriculture

Remote sensing technologies have transformed agricultural monitoring by delivering real-time, spatially detailed information on crop conditions and environmental variables. These technologies have diverse applications that enhance agricultural productivity, sustainability, and resilience to climate change as follows:

Monitoring Crop Health and Growth Stages

- Remote sensing allows for continuous monitoring of crop health throughout the growing season. Vegetation indices, such as NDVI and enhanced vegetation index (EVI), help detect areas of low vigor that may indicate nutrient deficiencies, water stress, or pest infestations (Radočaj et al. 2023). By analyzing temporal trends in vegetation indices, farmers can optimize input use and improve overall crop management.
- Growth stages, including emergence, flowering, and maturity, can be tracked using time-series satellite data. Early detection of anomalies in crop development enables timely interventions to mitigate yield loss (Duan et al. 2024).

Monitoring Environmental Conditions

- Remote sensing provides crucial data on environmental factors such as temperature, precipitation, soil moisture, and solar radiation (Zhao et al. 2023). These variables influence crop growth and productivity and can be integrated into crop models to improve yield predictions. For instance, thermal imagery from satellites helps monitor canopy temperature, an indicator of crop water stress.
- Water indices like the Land Surface Water Index (LSWI) assess crop and soil moisture status, particularly in rice paddy systems, where water management is critical (Qiu et al. 2015; Tornos et al. 2015).

Integration with Other Data Sources

- Remote sensing data can be integrated with ground-based observations, weather data, and process-based crop models to reduce uncertainty of predictions. By combining satellite imagery with data from on-the-ground sensors, such as those measuring soil moisture or temperature, farmers can gain realistic understanding of crop conditions (Jin et al. 2018).
- In recent years, advances in machine learning and deep learning have further improved the ability to analyze remote sensing data. These techniques can uncover complex patterns in the data, allowing for more precise yield forecasts and the identification of factors contributing to yield gaps (Shawon et al. 2025).

The remote sensing's ability to provide high-resolution, continuous data across large areas makes it indispensable for modern agriculture. Its integration with machine learning, process-based models, and other datasets is paving the way for more accurate and scalable solutions to yield gap assessment and agricultural monitoring.

Integrating Process-Based Models, Machine Learning, and Remote Sensing for Yield Gap Assessment

The integration of process-based models, machine learning, and remote sensing technologies offers a powerful and comprehensive approach to agricultural monitoring, yield gap assessments, and policy development. This framework combines crop models, machine learning algorithms, and remote sensing data within a cloud computing environment to provide accurate yield predictions, close yield gaps, and inform agricultural decision-making (Benti et al. 2024; Jeong et al. 2024).

The key components of the hybrid approach (Fig. 17.2) include:

- *Process-Based Crop Models:* Crop models, such as DSSAT and APSIM, simulate crop growth, development, and yield under varying conditions of water, nutrients, and climate. These models are calibrated using field data to simulate potential yields under optimal management practices (full irrigation and fertilization), serving as a baseline for identifying yield gaps—the difference between potential and actual farmer yields. By simulating different management strate-

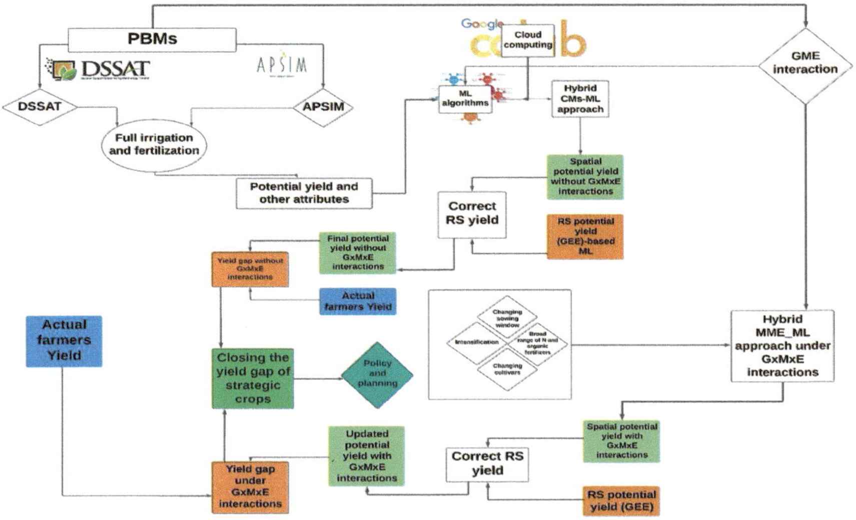

Fig. 17.2 Integrated framework for closing agricultural yield gaps. This flowchart illustrates the integration of process-based models (PBMs), machine learning (ML), and remote sensing (RS) data within a cloud-computing environment. The framework operates in the following sequence: PBMs simulate potential yields under optimal management practices, RS data provides spatially continuous and real-time updates, and ML refines predictions by correcting discrepancies between modeled and observed yields. The framework further accounts for Genotype × Management × Environment (G × M × E) interactions to offer dynamic, region-specific agricultural solutions. This hybrid approach enhances yield gap analysis, facilitates informed policy decisions, and promotes sustainable agricultural practices under changing environmental conditions

gies, such as adjusting sowing dates, irrigation, fertilization, and cultivar choices, the models explore strategies for improving productivity under specific Genotyp e × Management × Environment (G × M × E) interactions (Xin and Tao 2019).

- *Remote Sensing Data:* Remote sensing provides spatially continuous data on variables like vegetation indices (e.g., NDVI), soil moisture, and land surface temperature. Remote sensing data, especially from platforms such as Google Earth Engine (GEE), is used to estimate spatial yield potential and update crop model predictions with real-time field conditions that may not be fully captured in process-based models. Remote sensing data is particularly valuable for scaling predictions across large regions and correcting model estimates through regression analysis, thereby enhancing the accuracy of yield predictions (Woldemariam et al. 2024).
- *Machine Learning Algorithms:* Machine learning algorithms, run on cloud computing platforms like Google Colab, process vast amounts of remote sensing data and crop model outputs to predict yields. By learning patterns from historical data, machine learning techniques refine yield predictions and correct discrepancies in potential yield estimates provided by crop models. The hybrid crop model-machine learning approach leverages the strengths of both process-based and data-driven methods for robust yield prediction. The multi-model ensemble-

machine learning approach integrates multiple crop models simulating G × M × E interactions with machine learning algorithms identifying the best-performing strategies (Sharma et al. 2022).

- *Hybrid Models Under G × M × E Interactions:* This hybrid approach incorporates G × M × E interactions, modeling how different genotypes respond to varying management strategies and environmental conditions. The machine learning algorithms refine these models by integrating outputs from multiple crop models and RS data to optimize predictions under complex agricultural systems. This approach provides crucial insights for policy and planning, offering recommendations for resource allocation, cultivar selection, and management practices to close yield gaps (Crossa et al. 2024).

Steps of the Integrated Approach

- *Simulating Potential Yield:* Process-based models simulate crop growth under optimal management (full irrigation and fertilization) to estimate potential yields without considering G × M × E interactions.
- *Correcting Remote Sensing Yields:* Remote sensing-derived yield estimates are corrected using machine learning outputs, generating more accurate yield predictions across large spatial areas.
- *Yield Gap Without G × M × E Interactions:* The difference between simulated potential yield and actual farmer yield represents the initial yield gap without G × M × E interactions.
- *Incorporating G × M × E Interactions:* Crop models are refined to simulate yields under different G × M × E scenarios, representing the complexity of real-world systems. These updates provide a more accurate potential yield estimate.
- *Machine Learning Refinement with G × M × E Data:* Machine learning models refine yield predictions by integrating updated remote sensing data with crop model simulations under G × M × E interactions.
- *Closing the Yield Gap:* The hybrid approach provides policy recommendations based on optimized management strategies, identified through both crop models and machine learning predictions, to close the yield gap (von Bloh et al. 2024).

Advantages of the Integrated Approach

- *Scalability:* Cloud computing platforms (e.g., Google Colab) enable the scalability of this approach, allowing for the processing of large datasets across different regions.
- *Enhanced Accuracy:* By combining biophysical models, data-driven machine learning predictions, and remote sensing data, the approach provides more accurate yield estimates for agricultural decision-making.

- *Dynamic Adaptation:* The ability to incorporate G × M × E interactions allow for dynamic adaptation to changing environmental conditions, promoting sustainable agricultural practices and climate resilience (Bhatnagar et al. 2024).

Implications of Integrating Process-Based Models, Machine Learning, and Remote Sensing

The integration of process-based models, machine learning, and remote sensing technologies represents a significant advancement in the field of agricultural monitoring and decision-making. It offers several key benefits, chief among them being the enhanced accuracy and scalability of yield gap assessments. Process-based crop models like DSSAT and APSIM provide detailed simulations of biophysical processes (Kim et al. 2018), while remote sensing adds spatially continuous data across large areas, capturing real-time crop conditions (Brook et al. 2020). Machine learning further refines the outputs of these models by learning complex patterns from vast datasets, improving the overall accuracy of predictions. The use of cloud computing platforms like Google Earth Engine (GEE) and Google Colab ensures that this approach is scalable, allowing researchers and policymakers to assess yield gaps across different regions and under various management and environmental conditions (Kheir et al. 2022b). This combined approach is especially valuable in closing yield gaps under G × M × E interactions. By considering how different crop genotypes respond to varying management strategies and environmental stresses, the framework dynamically adapts to real-world conditions (Gao et al. 2023; Webber et al. 2022; Kheir et al. 2025). This flexibility makes the integrated approach highly relevant for policy development and strategic planning, offering robust data-driven insights for resource allocation, cultivar selection, and management optimization.

Improved Accuracy and Scalability of Yield Gap Assessments

Traditional yield gap assessments often rely on generalized models or single data sources, which can limit their accuracy and scalability (Lobell 2013). The described hybrid framework overcomes these limitations by blending process-based models with data-driven machine learning and spatially comprehensive remote sensing data. This integration is crucial for scaling agricultural assessments from small, local studies to broader regional or even global scales. For instance, cloud computing platforms enable the simultaneous processing of massive datasets, making it possible to monitor yield gaps across various ecosystems and management scenarios without the computational bottlenecks of traditional methods (Crossa et al. 2022). Moreover, the ability to incorporate G × M × E interactions offers enhanced precision in predicting how different genotypes and management practices perform under varying environmental conditions (Crossa et al. 2022; de Leon et al. 2016). This allows for yield gap assessments that are not only more accurate but also tailored to specific regions and crop varieties. Such detailed assessments have far-reaching implications for global food security, especially in regions where yield gaps remain significant.

Enhanced Decision-Making and Policy Development

The integration of process-based models, machine learning, and remote sensing into a unified framework enhances decision-making at multiple levels of agricultural management. By providing more accurate predictions of yield potentials and identifying specific yield gaps, this approach informs better resource allocation, management strategies, and policy interventions. For example, policymakers can use updated yield estimates to develop programs that target regions with significant yield gaps, focusing on improved management practices, crop variety adoption, and infrastructure investments that address local challenges. At the farm level, this framework offers actionable insights into optimizing practices such as sowing windows, fertilizer application, and irrigation scheduling. Farmers, advisors, and policymakers alike can benefit from the detailed predictions generated by the hybrid models, enabling more efficient and sustainable agricultural practices that reduce input costs and environmental impacts while increasing productivity. Furthermore, the enhanced scalability of this approach allows for more effective long-term planning. By continuously updating yield estimates with real-time remote sensing data and refining them through machine learning, decision-makers can dynamically adjust policies and practices in response to changing environmental conditions, such as climate variability, drought, or pest outbreaks. This adaptability promotes agricultural resilience and sustainability, crucial in the face of climate change and other global challenges (Delfani et al. 2024).

Challenges and Future Directions

Despite the benefits of integrating process-based models, machine learning, and remote sensing, several technical and logistical challenges remain. One major challenge is the harmonization of diverse data sources (Abiri et al. 2023; Gebresenbet et al. 2023). Process-based models, remote sensing data, and machine learning algorithms often have different data formats, resolutions, and temporal scales, making integration complex and time-consuming. For example, while crop models simulate yields based on daily weather inputs, remote sensing data is often available only on a weekly or biweekly basis, requiring interpolation and careful data fusion. Moreover, inconsistencies in remote sensing data due to cloud cover, atmospheric conditions, or sensor calibration issues can introduce errors that affect the accuracy of yield predictions. Another logistical challenge is the need for computational resources to process the vast amounts of data generated by remote sensing and machine learning. While cloud computing platforms mitigate some of these issues, access to high-quality computational infrastructure may still be limited in some regions, particularly in developing countries where yield gaps are most pronounced. Capacity building and investment in technology infrastructure are therefore critical to realizing the full potential of this integrated approach globally.

Future research should focus on addressing these challenges by developing more robust methods for harmonizing and integrating diverse data sources. Advances in data fusion techniques, such as deep learning algorithms designed to handle multimodal inputs (e.g., combining remote sensing and model data), could improve the accuracy and reliability of yield predictions (Li et al. 2022). Additionally, more work is needed to refine the handling of G × M × E interactions within this framework. While current models capture some of the complexity of these interactions, further research is needed to simulate more nuanced responses to environmental stresses, such as multiyear droughts or extreme weather events. There is also great potential for expanding the framework's applicability by incorporating new data sources, such as farmer-reported yield data or social-economic factors, to further contextualize yield gap assessments (Xin and Tao 2019). By integrating socioeconomic data, models could better reflect the on-ground realities of farm management practices, labor availability, and market access, providing even more actionable insights for both policymakers and farmers. Additionally, as remote sensing technologies continue to evolve, with new satellite constellations offering higher resolution and more frequent coverage, the precision and timeliness of remote sensing data will improve, further enhancing yield predictions and agricultural monitoring capabilities.

Conclusions

In conclusion, the integration of process-based crop models, machine learning, and remote sensing offers a transformative approach to agricultural monitoring and decision-making. This hybrid framework not only improves the accuracy and scalability of yield gap assessments but also provides actionable insights for enhancing agricultural productivity and sustainability. Although challenges persist in integrating data sources and expanding computational capacity, continuous research and technological progress offer the potential to address these issues, enabling more efficient agricultural management and policy formulation in the future. This chapter has demonstrated the potential of combining these approaches to more accurately assess yield gaps and inform sustainable agricultural practices, particularly in the face of environmental and management challenges.

References

Abiri R, Rizan N, Balasundram SK, Shahbazi AB, Abdul-Hamid H (2023) Application of digital technologies for ensuring agricultural productivity. Heliyon 9(12):e22601. https://doi.org/10.1016/j.heliyon.2023.e22601

Adugna T, Xu W, Fan J (2022) Comparison of random forest and support vector machine classifiers for regional land cover mapping using coarse resolution FY-3C images. Remote Sens 14(3):574. https://www.mdpi.com/2072-4292/14/3/574

Ahmed I, Yadav PK (2023) A systematic analysis of machine learning and deep learning based approaches for identifying and diagnosing plant diseases. Sustain Oper Comput 4:96–104. https://doi.org/10.1016/j.susoc.2023.03.001

Ali MGM, Ahmed M, Ibrahim MM, El Baroudy AA, Ali EF, Shokr MS, Aldosari AA, Majrashi A, Kheir AMS (2022) Optimizing sowing window, cultivar choice, and plant density to boost maize yield under RCP8.5 climate scenario of CMIP5. Int J Biometeorol 66(5):971–985. https://doi.org/10.1007/s00484-022-02253-x

Alvarez-Vanhard E, Corpetti T, Houet T (2021) UAV & satellite synergies for optical remote sensing applications: a literature review. Sci Remote Sens 3:100019. https://doi.org/10.1016/j.srs.2021.100019

Ammar ME, Davies EGR (2019) On the accuracy of crop production and water requirement calculations: process-based crop modeling at daily, semi-weekly, and weekly time steps for integrated assessments. J Environ Manag 238:460–472. https://doi.org/10.1016/j.jenvman.2019.03.030

Antle JM, Basso B, Conant RT, Godfray HCJ, Jones JW, Herrero M, Howitt RE, Keating BA, Munoz-Carpena R, Rosenzweig C, Tittonell P, Wheeler TR (2017) Towards a new generation of agricultural system data, models and knowledge products: design and improvement. Agric Syst 155:255–268. https://doi.org/10.1016/j.agsy.2016.10.002

Asseng S, Kheir AMS, Kassie BT, Hoogenboom G, Abdelaal AIN, Haman DZ, Ruane AC (2018) Can Egypt become self-sufficient in wheat? Environ Res Lett 13(9):094012. https://doi.org/10.1088/1748-9326/aada50

Benti NE, Chaka MD, Semie AG, Warkineh B, Soromessa T (2024) Transforming agriculture with Machine Learning, Deep Learning, and IoT: perspectives from Ethiopia—challenges and opportunities. Discov Agric 2(1):63. https://doi.org/10.1007/s44279-024-00066-7

Bhatnagar S, Chaudhary R, Sharma S, Janjhua Y, Thakur P, Sharma P, Keprate A (2024) Exploring the dynamics of climate-smart agricultural practices for sustainable resilience in a changing climate. Environ Sustain Indic 24:100535. https://doi.org/10.1016/j.indic.2024.100535

Bosi C, Sentelhas PC, Huth NI, Pezzopane JRM, Andreucci MP, Santos PM (2020) APSIM-tropical pasture: a model for simulating perennial tropical grass growth and its parameterisation for palisade grass (Brachiaria brizantha). Agric Syst 184:102917. https://doi.org/10.1016/j.agsy.2020.102917

Brook A, De Micco V, Battipaglia G, Erbaggio A, Ludeno G, Catapano I, Bonfante A (2020) A smart multiple spatial and temporal resolution system to support precision agriculture from satellite images: proof of concept on Aglianico vineyard. Remote Sens Environ 240:111679. https://doi.org/10.1016/j.rse.2020.111679

Crossa J, Montesinos-López OA, Pérez-Rodríguez P, Costa-Neto G, Fritsche-Neto R, Ortiz R, Martini JWR, Lillemo M, Montesinos-López A, Jarquin D, Breseghello F, Cuevas J, Rincent R (2022) Genome and environment based prediction models prediction models and methods of complex traits complex traits incorporating genotype × environment interaction. In: Ahmadi N, Bartholomé J (eds) Genomic prediction of complex traits: methods and protocols. Springer US, pp 245–283. https://doi.org/10.1007/978-1-0716-2205-6_9

Crossa J, Montesinos-Lopez OA, Costa-Neto G, Vitale P, Martini JWR, Runcie D, Fritsche-Neto R, Montesinos-Lopez A, Pérez-Rodríguez P, Gerard G, Dreisigacker S, Crespo-Herrera L, Pierre CS, Lillemo M, Cuevas J, Bentley A, Ortiz R (2024) Machine learning algorithms translate big data into predictive breeding accuracy. Trends Plant Sci 30(2):167–184. https://doi.org/10.1016/j.tplants.2024.09.011

de Leon N, Jannink J-L, Edwards JW, Kaeppler SM (2016) Introduction to a special issue on genotype by environment interaction. Crop Sci 56(5):2081–2089. https://doi.org/10.2135/cropsci2016.07.0002in

Delfani P, Thuraga V, Banerjee B, Chawade A (2024) Integrative approaches in modern agriculture: IoT, ML and AI for disease forecasting amidst climate change. Precis Agric 25(5):2589–2613. https://doi.org/10.1007/s11119-024-10164-7

Droutsas I, Challinor AJ, Swiderski M, Semenov MA (2019) New modelling technique for improving crop model performance—application to the GLAM model. Environ Model Softw 118:187–200. https://doi.org/10.1016/j.envsoft.2019.05.005

Duan K, Vrieling A, Schlund M, Bhaskar Nidumolu U, Ratcliff C, Collings S, Nelson A (2024) Detection and attribution of cereal yield losses using Sentinel-2 and weather data: a case study in South Australia. ISPRS J Photogramm Remote Sens 213:33–52. https://doi.org/10.1016/j.isprsjprs.2024.05.021

El Bouchefry K, de Souza RS (2020) Chapter 12—learning in big data: introduction to machine learning. In: Škoda P, Adam F (eds) Knowledge discovery in big data from astronomy and earth observation. Elsevier, pp 225–249. https://doi.org/10.1016/B978-0-12-819154-5.00023-0

Elbasi E, Zaki C, Topcu AE, Abdelbaki W, Zreikat AI, Cina E, Shdefat A, Saker L (2023) Crop prediction model using machine learning algorithms. Appl Sci 13(16):9288. https://www.mdpi.com/2076-3417/13/16/9288

Ennaji O, Vergütz L, El Allali A (2023) Machine learning in nutrient management: a review. Artif Intell Agric 9:1–11. https://doi.org/10.1016/j.aiia.2023.06.001

Farias GD, Bremm C, Bredemeier C, de Lima Menezes J, Alves LA, Tiecher T, Martins AP, Fioravanço GP, da Silva GP, de Faccio Carvalho PC (2023) Normalized Difference Vegetation Index (NDVI) for soybean biomass and nutrient uptake estimation in response to production systems and fertilization strategies [original research]. Front Sustain Food Syst 6:959681. https://doi.org/10.3389/fsufs.2022.959681

Fodor N, Challinor A, Droutsas I, Ramirez-Villegas J, Zabel F, Koehler A-K, Foyer CH (2017) Integrating plant science and crop modeling: assessment of the impact of climate change on soybean and maize production. Plant Cell Physiol 58(11):1833–1847. https://doi.org/10.1093/pcp/pcx141

Gao L, Kantar MB, Moxley D, Ortiz-Barrientos D, Rieseberg LH (2023) Crop adaptation to climate change: an evolutionary perspective. Mol Plant 16(10):1518–1546. https://doi.org/10.1016/j.molp.2023.07.011

Gavasso-Rita YL, Papalexiou SM, Li Y, Elshorbagy A, Li Z, Schuster-Wallace C (2024) Crop models and their use in assessing crop production and food security: a review. Food Energy Secur 13(1):e503. https://doi.org/10.1002/fes3.503

Gebresenbet G, Bosona T, Patterson D, Persson H, Fischer B, Mandaluniz N, Chirici G, Zacepins A, Komasilovs V, Pitulac T, Nasirahmadi A (2023) A concept for application of integrated digital technologies to enhance future smart agricultural systems. Smart Agric Technol 5:100255. https://doi.org/10.1016/j.atech.2023.100255

Hajjarpoor A, Nelson WCD, Vadez V (2022) How process-based modeling can help plant breeding deal with G x E x M interactions. Field Crop Res 283:108554. https://doi.org/10.1016/j.fcr.2022.108554

Heilemann J, Klassert C, Samaniego L, Thober S, Marx A, Boeing F, Klauer B, Gawel E (2024) Projecting impacts of extreme weather events on crop yields using LASSO regression. Weather Clim Extrem 46:100738. https://doi.org/10.1016/j.wace.2024.100738

Holzworth DP, Huth NI, deVoil PG, Zurcher EJ, Herrmann NI, McLean G, Chenu K, van Oosterom EJ, Snow V, Murphy C, Moore AD, Brown H, Whish JPM, Verrall S, Fainges J, Bell LW, Peake AS, Poulton PL, Hochman Z et al (2014) APSIM—evolution towards a new generation of agricultural systems simulation. Environ Model Softw 62:327–350. https://doi.org/10.1016/j.envsoft.2014.07.009

Jeong S, Ko J, Ban J-O, Shin T, Yeom J-M (2024) Deep learning-enhanced remote sensing-integrated crop modeling for rice yield prediction. Eco Inform 84:102886. https://doi.org/10.1016/j.ecoinf.2024.102886

Jiang Z, Liu C, Ganapathysubramanian B, Hayes DJ, Sarkar S (2020) Predicting county-scale maize yields with publicly available data. Sci Rep 10(1):14957. https://doi.org/10.1038/s41598-020-71898-8

Jin X, Kumar L, Li Z, Feng H, Xu X, Yang G, Wang J (2018) A review of data assimilation of remote sensing and crop models. Eur J Agron 92:141–152. https://doi.org/10.1016/j.eja.2017.11.002

Jones JW, Hoogenboom G, Porter CH, Boote KJ, Batchelor WD, Hunt LA, Wilkens PW, Singh U, Gijsman AJ, Ritchie JT (2003) The DSSAT cropping system model. Eur J Agron 18(3):235–265. https://doi.org/10.1016/S1161-0301(02)00107-7

Jones JW, Antle JM, Basso B, Boote KJ, Conant RT, Foster I, Godfray HCJ, Herrero M, Howitt RE, Janssen S, Keating BA, Munoz-Carpena R, Porter CH, Rosenzweig C, Wheeler TR (2017) Brief history of agricultural systems modeling. Agric Syst 155:240–254. https://doi.org/10.1016/j.agsy.2016.05.014

Kang Y, Ozdogan M, Zhu X, Ye Z, Hain C, Anderson M (2020) Comparative assessment of environmental variables and machine learning algorithms for maize yield prediction in the US Midwest. Environ Res Lett 15(6):064005. https://doi.org/10.1088/1748-9326/ab7df9

Karmakar P, Teng SW, Murshed M, Pang S, Li Y, Lin H (2024) Crop monitoring by multimodal remote sensing: a review. Remote Sens Appl Soc Environ 33:101093. https://doi.org/10.1016/j.rsase.2023.101093

Keating BA, Carberry PS, Hammer GL, Probert ME, Robertson MJ, Holzworth D, Huth NI, Hargreaves JNG, Meinke H, Hochman Z, McLean G, Verburg K, Snow V, Dimes JP, Silburn M, Wang E, Brown S, Bristow KL, Asseng S et al (2003) An overview of APSIM, a model designed for farming systems simulation. Eur J Agron 18(3):267–288. https://doi.org/10.1016/S1161-0301(02)00108-9

Kephe PN, Ayisi KK, Petja BM (2021) Challenges and opportunities in crop simulation modelling under seasonal and projected climate change scenarios for crop production in South Africa. Agric Food Secur 10(1):10. https://doi.org/10.1186/s40066-020-00283-5

Kheir AMS, El Baroudy A, Aiad MA, Zoghdan MG, Abd El-Aziz MA, Ali MGM, Fullen MA (2019) Impacts of rising temperature, carbon dioxide concentration and sea level on wheat production in North Nile delta. Sci Total Environ 651:3161–3173. https://doi.org/10.1016/j.scitotenv.2018.10.209

Kheir AMS, Hoogenboom G, Ammar KA, Ahmed M, Feike T, Elnashar A, Liu B, Ding Z, Asseng S (2022a) Minimizing trade-offs between wheat yield and resource-use efficiency in the Nile Delta—a multi-model analysis. Field Crop Res 287:108638. https://doi.org/10.1016/j.fcr.2022.108638

Kheir AMS, Ammar KA, Amer A, Ali MGM, Ding Z, Elnashar A (2022b) Machine learning-based cloud computing improved wheat yield simulation in arid regions. Comput Electron Agric 203:107457. https://doi.org/10.1016/j.compag.2022.107457

Kheir AMS, Mkuhlani S, Mugo JW, Elnashar A, Nangia V, Devare M, Govind A (2023a) Integrating APSIM model with machine learning to predict wheat yield spatial distribution. Agron J 115(6):3188–3196. https://doi.org/10.1002/agj2.21470

Kheir AMS, Elnashar A, Mosad A, Govind A (2023b) An improved deep learning procedure for statistical downscaling of climate data. Heliyon 9(7):e18200. https://doi.org/10.1016/j.heliyon.2023.e18200

Kheir A, Ali O, Shawon AR, Elrys A, Ali M, Darwish M, Elmahdy A, Abou-Hadid A, Nóia Júnior RDS, Feike T (2024a) Impacts of climate change on spatial wheat yield and nutritional values using hybrid machine learning. Environ Res Lett 19:104049. http://iopscience.iop.org/article/10.1088/1748-9326/ad75ab

Kheir AMS, Govind A, Nangia V, Devkota M, Elnashar A, Omar MED, Feike T (2024b) Developing automated machine learning approach for fast and robust crop yield prediction using a fusion of remote sensing, soil, and weather dataset. Environ Res Commun 6(4):041005. https://doi.org/10.1088/2515-7620/ad2d02

Kheir el al. 2025

Kim KS, Yoo BH, Shelia V, Porter CH, Hoogenboom G (2018) START: a data preparation tool for crop simulation models using web-based soil databases. Comput Electron Agric 154:256–264. https://doi.org/10.1016/j.compag.2018.08.023

Krishnan S, Indu J (2023) Assessing the potential of temperature/vegetation index space to infer soil moisture over Ganga Basin. J Hydrol 621:129611. https://doi.org/10.1016/j.jhydrol.2023.129611

Li J, Hong D, Gao L, Yao J, Zheng K, Zhang B, Chanussot J (2022) Deep learning in multimodal remote sensing data fusion: a comprehensive review. Int J Appl Earth Obs Geoinf 112:102926. https://doi.org/10.1016/j.jag.2022.102926

Liman Harou I, Whitney C, Kung'u J, Luedeling E (2021) Crop modelling in data-poor environments—a knowledge-informed probabilistic approach to appreciate risks and uncertainties in flood-based farming systems. Agric Syst 187:103014. https://doi.org/10.1016/j.agsy.2020.103014

Liu L, Zhou W, Guan K, Peng B, Xu S, Tang J, Zhu Q, Till J, Jia X, Jiang C, Wang S, Qin Z, Kong H, Grant R, Mezbahuddin S, Kumar V, Jin Z (2024) Knowledge-guided machine learning can improve carbon cycle quantification in agroecosystems. Nat Commun 15(1):357. https://doi.org/10.1038/s41467-023-43860-5

Lobell DB (2013) The use of satellite data for crop yield gap analysis. Field Crop Res 143:56–64. https://doi.org/10.1016/j.fcr.2012.08.008

McCown RL, Hammer GL, Hargreaves JNG, Holzworth DP, Freebairn DM (1996) APSIM: a novel software system for model development, model testing and simulation in agricultural systems research. Agric Syst 50(3):255–271. https://doi.org/10.1016/0308-521X(94)00055-V

Meshram V, Patil K, Meshram V, Hanchate D, Ramkteke SD (2021) Machine learning in agriculture domain: a state-of-art survey. Artif Intell Life Sci 1:100010. https://doi.org/10.1016/j.ailsci.2021.100010

Miti C, Milne AE, Giller KE, Lark RM (2024) The concepts and quantification of yield gap using boundary lines. A review. Field Crop Res 311:109365. https://doi.org/10.1016/j.fcr.2024.109365

Moghaddam SNM, Cao H (2024) Artificial intelligence. In: Artificial intelligence-driven geographies: revolutionizing urban studies. Springer Nature Singapore, pp 11–69. https://doi.org/10.1007/978-981-97-5116-7_2

Naghdyzadegan Jahromi M, Zand-Parsa S, Razzaghi F, Jamshidi S, Didari S, Doosthosseini A, Pourghasemi HR (2023) Developing machine learning models for wheat yield prediction using ground-based data, satellite-based actual evapotranspiration and vegetation indices. Eur J Agron 146:126820. https://doi.org/10.1016/j.eja.2023.126820

Nanda A, Das N, Singh G, Bindlish R, Andreadis KM, Jayasinghe S (2024) Harnessing SMAP satellite soil moisture product to optimize soil properties to improve water resource management for agriculture. Agric Water Manag 300:108918. https://doi.org/10.1016/j.agwat.2024.108918

Neinavaz E, Schlerf M, Darvishzadeh R, Gerhards M, Skidmore AK (2021) Thermal infrared remote sensing of vegetation: current status and perspectives. Int J Appl Earth Obs Geoinf 102:102415. https://doi.org/10.1016/j.jag.2021.102415

Neumann K, Verburg PH, Stehfest E, Müller C (2010) The yield gap of global grain production: a spatial analysis. Agric Syst 103(5):316–326. https://doi.org/10.1016/j.agsy.2010.02.004

Olawade DB, Aderinto N, Olatunji G, Kokori E, David-Olawade AC, Hadi M (2024) Advancements and applications of artificial intelligence in cardiology: current trends and future prospects. J Med Surg Public Health 3:100109. https://doi.org/10.1016/j.glmedi.2024.100109

Pasley H, Brown H, Holzworth D, Whish J, Bell L, Huth N (2022) How to build a crop model. A review. Agron Sustain Dev 43(1):2. https://doi.org/10.1007/s13593-022-00854-9

Paudel D, Boogaard H, de Wit A, Janssen S, Osinga S, Pylianidis C, Athanasiadis IN (2021) Machine learning for large-scale crop yield forecasting. Agric Syst 187:103016. https://doi.org/10.1016/j.agsy.2020.103016

Pérez G, Coma J, Chàfer M, Cabeza LF (2022) Seasonal influence of leaf area index (LAI) on the energy performance of a green facade. Build Environ 207:108497. https://doi.org/10.1016/j.buildenv.2021.108497

Priyatikanto R, Lu Y, Dash J, Sheffield J (2023) Improving generalisability and transferability of machine-learning-based maize yield prediction model through domain adaptation. Agric For Meteorol 341:109652. https://doi.org/10.1016/j.agrformet.2023.109652

Puccetti M, Pariano M, Schoubben A, Giovagnoli S, Ricci M (2024) Biologics, theranostics, and personalized medicine in drug delivery systems. Pharmacol Res 201:107086. https://doi.org/10.1016/j.phrs.2024.107086

Qiu B, Li W, Tang Z, Chen C, Qi W (2015) Mapping paddy rice areas based on vegetation phenology and surface moisture conditions. Ecol Indic 56:79–86. https://doi.org/10.1016/j.ecolind.2015.03.039

Radeloff VC, Roy DP, Wulder MA, Anderson M, Cook B, Crawford CJ, Friedl M, Gao F, Gorelick N, Hansen M, Healey S, Hostert P, Hulley G, Huntington JL, Johnson DM, Neigh C, Lyapustin A, Lymburner L, Pahlevan N et al (2024) Need and vision for global medium-resolution Landsat and Sentinel-2 data products. Remote Sens Environ 300:113918. https://doi.org/10.1016/j.rse.2023.113918

Radočaj D, Šiljeg A, Marinović R, Jurišić M (2023) State of major vegetation indices in precision agriculture studies indexed in web of science: a review. Agriculture 13(3):707. https://www.mdpi.com/2077-0472/13/3/707

Rattalino Edreira JI, Mourtzinis S, Conley SP, Roth AC, Ciampitti IA, Licht MA, Kandel H, Kyveryga PM, Lindsey LE, Mueller DS, Naeve SL, Nafziger E, Specht JE, Stanley J, Staton MJ, Grassini P (2017) Assessing causes of yield gaps in agricultural areas with diversity in climate and soils. Agric For Meteorol 247:170–180. https://doi.org/10.1016/j.agrformet.2017.07.010

Rogan J, Chen D (2004) Remote sensing technology for mapping and monitoring land-cover and land-use change. Prog Plan 61(4):301–325. https://doi.org/10.1016/S0305-9006(03)00066-7

Seetharam K, Shrestha S, Sengupta PP (2019) Artificial intelligence in cardiovascular medicine. Curr Treat Options Cardiovasc Med 21(6):25. https://doi.org/10.1007/s11936-019-0728-1

Seetharam K, Brito D, Farjo PD, Sengupta PP (2020) The role of artificial intelligence in cardiovascular imaging: state of the art review [review]. Front Cardiovasc Med 7:618849. https://doi.org/10.3389/fcvm.2020.618849

Sengupta PP, Shrestha S, Berthon B, Messas E, Donal E, Tison GH, Min JK, D'Hooge J, Voigt JU, Dudley J, Verjans JW, Shameer K, Johnson K, Lovstakken L, Tabassian M, Piccirilli M, Pernot M, Yanamala N, Duchateau N et al (2020) Proposed requirements for cardiovascular imaging-related machine learning evaluation (PRIME): a checklist: reviewed by the American College of Cardiology Healthcare Innovation Council. JACC Cardiovasc Imaging 13(9):2017–2035. https://doi.org/10.1016/j.jcmg.2020.07.015

Shahhosseini M, Hu G, Huber I, Archontoulis SV (2021) Coupling machine learning and crop modeling improves crop yield prediction in the US Corn Belt. Sci Rep 11(1):1606. https://doi.org/10.1038/s41598-020-80820-1

Sharma V, Tripathi AK, Mittal H (2022) Technological revolutions in smart farming: current trends, challenges & future directions. Comput Electron Agric 201:107217. https://doi.org/10.1016/j.compag.2022.107217

Shawon SM, Ema FB, Mahi AK, Niha FL, Zubair HT (2025) Crop yield prediction using machine learning: an extensive and systematic literature review. Smart Agric Technol 10:100718. https://doi.org/10.1016/j.atech.2024.100718

Shendryk Y, Davy R, Thorburn P (2021) Integrating satellite imagery and environmental data to predict field-level cane and sugar yields in Australia using machine learning. Field Crop Res 260:107984. https://doi.org/10.1016/j.fcr.2020.107984

Soori M, Arezoo B, Dastres R (2023) Artificial intelligence, machine learning and deep learning in advanced robotics, a review. Cogn Robotics 3:54–70. https://doi.org/10.1016/j.cogr.2023.04.001

Thorp KR, DeJonge KC, Kaleita AL, Batchelor WD, Paz JO (2008) Methodology for the use of DSSAT models for precision agriculture decision support. Comput Electron Agric 64(2):276–285. https://doi.org/10.1016/j.compag.2008.05.022

Tornos L, Huesca M, Dominguez JA, Moyano MC, Cicuendez V, Recuero L, Palacios-Orueta A (2015) Assessment of MODIS spectral indices for determining rice paddy agricultural practices and hydroperiod. ISPRS J Photogramm Remote Sens 101:110–124. https://doi.org/10.1016/j.isprsjprs.2014.12.006

van Ittersum MK, Cassman KG, Grassini P, Wolf J, Tittonell P, Hochman Z (2013) Yield gap analysis with local to global relevance—a review. Field Crop Res 143:4–17. https://doi.org/10.1016/j.fcr.2012.09.009

van Keulen H (2013) Simulation models as tools for crop management. In: Christou P, Savin R, Costa-Pierce BA, Misztal I, Whitelaw CBA (eds) Sustainable food production. Springer, New York, pp 1459–1476. https://doi.org/10.1007/978-1-4614-5797-8_300

von Bloh M, Lobell D, Asseng S (2024) Knowledge informed hybrid machine learning in agricultural yield prediction. Comput Electron Agric 227:109606. https://doi.org/10.1016/j.compag.2024.109606

Wang J, Chen J, Zhang J, Yang S, Zhang S, Bai Y, Xu R (2023) Consistency and uncertainty of remote sensing-based approaches for regional yield gap estimation: a comprehensive assessment of process-based and data-driven models. Field Crop Res 302:109088. https://doi.org/10.1016/j.fcr.2023.109088

Webber H, Rezaei EE, Ryo M, Ewert F (2022) Framework to guide modeling single and multiple abiotic stresses in arable crops. Agric Ecosyst Environ 340:108179. https://doi.org/10.1016/j.agee.2022.108179

Woldemariam W, Gessesse Awoke B, Vargas Maretto R (2024) Remote sensing vegetation Indices-Driven models for sugarcane evapotranspiration estimation in the semiarid Ethiopian Rift Valley. ISPRS J Photogramm Remote Sens 215:136–156. https://doi.org/10.1016/j.isprsjprs.2024.07.004

Wu B, Zhang M, Zeng H, Tian F, Potgieter AB, Qin X, Yan N, Chang S, Zhao Y, Dong Q, Boken V, Plotnikov D, Guo H, Wu F, Zhao H, Deronde B, Tits L, Loupian E (2022) Challenges and opportunities in remote sensing-based crop monitoring: a review. Natl Sci Rev 10(4):nwac290. https://doi.org/10.1093/nsr/nwac290

Xin Y, Tao F (2019) Optimizing genotype-environment-management interactions to enhance productivity and eco-efficiency for wheat-maize rotation in the North China Plain. Sci Total Environ 654:480–492. https://doi.org/10.1016/j.scitotenv.2018.11.126

Xun L, Tran-Thanh L, Al-Hashimi BM, Merrett GV (2020, March 9–13) Optimising resource management for embedded machine learning. In: 2020 design, automation & test in Europe conference & exhibition (DATE)

Zhang N, Zhou X, Kang M, Hu B-G, Heuvelink E, Marcelis LFM (2022) Machine learning versus crop growth models: an ally, not a rival. AoB PLANTS 15(2):plac061. https://doi.org/10.1093/aobpla/plac061

Zhao S, Liu M, Tao M, Zhou W, Lu X, Xiong Y, Li F, Wang Q (2023) The role of satellite remote sensing in mitigating and adapting to global climate change. Sci Total Environ 904:166820. https://doi.org/10.1016/j.scitotenv.2023.166820

Zhi J, Cao X, Zhang Z, Qin T, Qu LA, Qi L, Ge L, Guo A, Wang X, Da C, Sun Y, Liu W, Zhang H, Fu X (2022) Identifying the determinants of crop yields in China since 1952 and its policy implications. Agric For Meteorol 327:109216. https://doi.org/10.1016/j.agrformet.2022.109216

18 Conclusions and Recommendations to "Resilient Agroecosystems: Innovations in Cropping Systems and Climate Change Mitigation"

Samiha Ouda and Abd El-Hafeez Zohry

Abstract

This chapter consolidates the fundamental challenges regarding conclusions and suggestions of agroecosystems and offers insights derived from the cases presented in the volume. This chapter concentrates on the concepts recorded in this book. We delineate five primary areas of contribution: innovative cropping systems for resilience, augmenting the soybean intercropping system with maize, untraditional approaches to maximize agricultural production, climate change impacts and mitigation, and machine learning use in crop production; in addition the last chapter presents the conclusions and recommendations highlighted by the authors. The conclusions will be based on the authors' insights into their findings and limitations. This chapter also presents a series of recommendations to guide future research towards innovation aimed at sustaining the agroecosystem for enhanced resilience.

Keywords

Agroecological intensification · Greenhouses and soilless agriculture · Fishpond wastewater · Climate change · Water and crops · Climate change and soil in the Nile Delta of Egypt · Machine learning in agricultural production

S. Ouda (✉)
Water Requirements and Field Irrigation Research Department, Soils, Water and Environment Research Institute, Agricultural Research Center, Giza, Egypt
e-mail: souda1958@arc.sci.eg

A. E.-H. Zohry
Crop Intensification Research Department, Field Crops Research Institute, Agricultural Research Center, Giza, Egypt
e-mail: azohry1954@arc.sci.eg

A. E.-H. Zohry, S. Ouda (eds.), *Resilient Agroecosystems*, Sustainability Sciences in Asia and Africa, https://doi.org/10.1007/978-981-96-9630-7_18

Introduction

An agroecosystem is an ecological system comprising one or more annual or perennial agricultural crops, related species, and several abiotic elements. The principal aim of agricultural research ought to be the enhancement of these systems. Agroecosystems are anticipated to supply food and feed amid shifting climatic circumstances and demographic demands. The increasing need for food and the necessity for climate change mitigation have exerted pressure on natural resources, such as land, water, and biodiversity. Inadequately managed agriculture has led to resource pollution and depletion, land degradation and desertification, as well as heightened greenhouse gas emissions.

This chapter delineates the overarching results about advancements in cropping systems, climate change mitigation, and their effects on the resilience of agroecosystems. The results obtained encompass significant improvements aimed at enhancing sustainable food production. Moreover, these findings are significant for other scholars in the discipline to advance their subsequent investigations. The following topics are addressed in no specific sequence.

Conclusions

Innovative Cropping Systems for Resilience

The swift rise in global food demand has resulted in the advent of agroecological intensification, a strategy of crop diversification designed to enhance agricultural efficiency by decreasing external inputs and mitigating environmental effects. Altering the conventional cropping sequence (a summer crop followed by a winter crop) to a novel cropping sequence (a summer crop, an early legume winter crop, and a winter crop), along with the adoption of intercropping systems that incorporate legume crops, represents innovative practices that can expedite microbial activity and nitrogen fixation, improve soil health, and diminish chemical inputs. An illustration of a novel cropping sequence is the integration of fahl clover (a short-season variety of Egyptian clover) between summer peanuts and winter wheat, which led to an enhancement in wheat output. The addition of fahl clover increased the nitrogen concentration of the soil, facilitating greater assimilate partitioning.

Intercropping is an essential approach for agroecological intensification, involving the simultaneous cultivation of two or more crops in the same field for a portion of the growing season. Legume-cereal intercropping systems confer various advantageous agroecological benefits, such as crop diversity, nitrogen self-sufficiency for farms, enhanced soil health, insect management, and reduction of yield losses. In intercropping, interspecific plant interactions can be categorized into four principles: competition, complementarity, collaboration, and compensation, collectively referred to as the "4C" method. Enhancing agronomic characteristics in intercropping systems can alleviate the negative effects of competition while fostering complementarity, collaboration, and compensation. Complementarity, characterized by the more complete utilization of niche space with the addition of species, is affected by variables like planting density and both intra- and interspecific competition. Intercropping promotes synergistic nitrogen use between legumes and grains, improving nitrogen fixation and soil nitrogen absorption. Intercropping systems can enhance resource acquisition by improving the complementarity and selection impacts of intercrops.

Collaboration improves soil organic carbon, humus levels, and bacterial proliferation. Intercropping with legumes can augment soil organic matter, promote usage efficiency, and improve biological pest control. Cooperation in intercropping systems is facilitated by modifications to the microenvironment surrounding intercrops, which alleviates high temperatures and enhances relative humidity. Compensation sometimes transpires in intercropping, wherein one crop produces lower yields than expected while another exceeds projections. The vulnerability of one intercrop to abiotic and biotic stress triggers compensation, wherein the failure of the initial intercrop is mitigated by the subsequent one.

The 4C approach evaluates intercrop performance throughout the growing season, emphasizing land equivalent ratio, which assesses land utilization efficiency within the intercropping system, and water equivalent ratio, which correlates intercrop yield with water utilization. The intercropping system of cowpea and maize exhibits the 4C methodology. Although competition for light, soil nutrients, and water is modest in this system, a synergistic impact is evident, leading to increased soil nitrogen and organic matter. The cooperative effect occurs through the release of insoluble soil nutrients and the reduction of maize predators. The compensating impact is achieved by the alleviation of water stress and the extremely efficient utilization of water by intercrops.

Augment the Soybean Intercropping System with Maize

Intercropping soybeans with maize is a highly beneficial system that enhances soil health and biodiversity, yielding a net nitrogen advantage for the agroecosystem. Intercropping soybeans with maize results in diminished light intensity within the canopy's interior. Variations in light exposure can impact the growth and development of crops, potentially affecting soil nutrient absorption, nutrient uptake, and overall plant health and production. Factors influencing light interception and production encompass the height of soybean plants, which impacts their capacity to absorb sunlight, and the spacing between maize hills. Effectively balancing maize plant densities can alleviate the adverse effects of shading on soybean yield while optimizing the advantages of intercropping.

The arrangement of plants and crop ratio affect light interference and utilization efficiency in intercropping systems of soybean and maize. A reduced distance between maize and soybean plants might exacerbate light interference, thereby diminishing crop yields owing to resource competition. An increased soybean ratio in the intercropping system can improve light interference and usage efficiency, advantageous for both crops. The effectiveness of light use in soybean and maize intercropping systems is affected by canopy architecture and photosynthetic ability. Choosing the suitable soybean cultivars and plant density can optimize these potentials, resulting in enhanced yields of both crops.

The Egyptian soybean varieties cultivated in various intercropping systems with maize may generate differing ultimate seed outputs. Intercropping Giza 21, Giza 22, or Giza 35 with maize in a 2:2 maize:soybean intercropping system yielded greater seed yields per plant, along with the highest land equivalent ratio and system production index values. Intercropping Giza 111 with maize in a 2:2 maize:soybean intercropping system or a mixed intercropping system led to reduced seed yields per plant. This results from the shading provided by intercropped maize plants, which diminishes dry matter accumulation and negatively affects seed yield. Moreover, intercropping Giza 21 with maize in mixed and 2:2 maize:soybean intercropping system resulted in a reduced seed yield per hectare compared to the 2:4 maize:soybean intercropping system.

Reports indicate that the Giza 111 and Giza 82 genotypes exhibited enhanced shading intensity, likely attributable to differences in solar radiation penetration. Intercropping Giza 111 with maize exhibited greater shading intensity in the middle and lower regions of the canopy, whereas sole cultivation of Giza 82 has shown reduced shading intensity. Furthermore, intercropping Giza 35 with maize exhibited elevated shading intensity levels, potentially impacting plant growth and development.

Additionally, soybean varieties Giza 21 and Giza 111 exhibited reduced sensitivity to soybean mosaic virus (SMV) infection, attributed to the abundant pubescence on their leaves, which deterred aphids from alighting and feeding on them. Moreover, the expression of genes associated with SMV tolerance was elevated due to intercropping. Protein synthesis is crucial for plant defense mechanisms and response to pathogenic threats. The Pd1 and Pd2 genes are crucial in breeding initiatives designed to improve soybean yield and environmental adaptability. Pyramiding genes, which integrate numerous resistance genes, has shown effective in controlling insect pests in soybeans. Giza 22, a pyramided cultivar, has shown a notable reduction in infestations of aphids, whiteflies, thrips, and leaf miners through intercropping, resulting in increased seed output.

Untraditional Approaches to Enhance Agricultural Production

Greenhouses and soilless agricultural systems serve as alternatives to conventional farming methods, offering advantages such as enhanced crop yield, improved food accessibility, and more resource efficiency. Greenhouses control temperature, humidity, and light, facilitating continuous cultivation and enhanced agricultural production. While they tackle issues like climatic fluctuations, resource limitations, and pest control, they also present disadvantages including substantial initial expenditures, considerable energy usage, maintenance demands, and possible ecological repercussions.

Moreover, soilless agriculture, including hydroponics and aeroponics, can substantially enhance crop yield and resource efficiency, rendering it optimal for regions with constrained water supplies and cultivable land. These systems can address issues such as soil deterioration and water scarcity, rendering them viable alternatives to traditional soil-based agriculture. Soilless agricultural techniques enhance resource utilization efficiency, adaptability, and sustainability, therefore minimizing water consumption and waste. They provide economical alternatives for small-scale farmers, as they can adjust to evolving agricultural requirements and market variances. Nonetheless, they encounter obstacles, including substantial initial expenditures, technical intricacies, nutrient management, environmental repercussions, and reliance on technology. Substantial initial expenditures encompass equipment and infrastructure, whereas technological intricacy needs comprehensive maintenance and oversight. Effective nutrient management is essential, and environmental consequences encompass synthetic nutrient solutions and artificial illumination. Aquaponics integrates hydroponics and aquaculture, uniting fish and plants to create a sustainable ecosystem. Nonetheless, obstacles encompass the control of fish and plant requirements, intricate system administration, possible nutrient level imbalances, and substantial initial expenditure.

Utilizing fishpond wastewater in agriculture sustainably enhances crop production by recycling nutrients and reducing waste. The nutrient-rich wastewater of fishponds acts as a fertilizer, improving soil fertility and encouraging plant growth while promoting water conservation, especially in drought-prone regions. Fish release nutrients that crops assimilate via the decomposition of organic waste, while aquaculture water recycling generates additional irrigation resources, diminishing fertilizer requirements and supplying advantageous nutrients such as nitrogen and phosphorus for plant development. This method enhances crop attributes by supplying vital nutrients and improving soil quality.

Aquaculture can adversely affect the environment through pollution, contamination by chemicals and pathogens, soil degradation, and the proliferation of diseases. The swift growth of aquaculture can provide both beneficial and detrimental outcomes, contingent upon variables such as the species cultivated, management methodologies, production sites, and the quality and administration of feed. Consequently, it is essential to equilibrate the application of fishpond effluent with other agricultural inputs to mitigate its adverse environmental impacts. Issues like substandard water quality management, nutrient discrepancies, and insufficient treatment facilities have been recognized. Integrated aquaculture encounters obstacles including resource diversification, variable nutrient composition, disease's introduction, environmental contamination, and compliance with water quality and environmental standards.

Climate Change Impacts and Mitigation

Climate change is an urgent worldwide issue impacting the environment and human civilization. It encompasses alterations in air conditions, temperature, precipitation, and other climatic variables, resulting in global warming and potentially irreversible transformations to Earth's biological and geological processes. Climate change is impacting water resources by altering temperature and precipitation patterns, hence influencing the hydrological cycle, as well as the quality and quantity of water supplies. Elevated temperatures and diminished precipitation are resulting in heightened evaporation, impacting surface water and soil moisture levels. This results in heightened demand for water, especially in agriculture. Climate variability and change impact groundwater replenishment and exacerbate saltwater intrusion issues in coastal aquifers. It directly affects groundwater and surface water resources via recharge and evaporation, and indirectly through alterations in water usage. These alterations are impacting water accessibility and quality, requiring adaptive agricultural methods in at-risk regions.

The substantial problems posed by climate change on water resources require the implementation of mitigation and adaptation strategies. Mitigation measures encompass the reduction of greenhouse gas emissions, the promotion of sustainable farming practices, the restoration of ecosystems, and the enhancement of community engagement. Adaptation solutions encompass infrastructure improvements, including reservoirs and water storage, as well as enhanced water management methodologies such as integrated water resource management. Strategies for water conservation, modifications in policy and legislation, and technological advancements such as desalination and wastewater reclamation can mitigate water scarcity. Technologies that improve water use efficiency are crucial for sustainable agricultural advancement, especially in arid and semiarid areas. Contemporary irrigation methods, including drip irrigation and soil moisture sensing, can markedly enhance water use efficiency and agricultural productivity. Adaptation strategies must account for the environmental consequences of climate change and the necessity for sustainable water management methods.

Modifying agricultural techniques can mitigate water scarcity induced by climate change. Tillage can enhance water retention and diminish water loss. Conservation tillage techniques are crucial for enhancing water use efficiency and reducing water losses. Organic substances, including compost and organic fertilizers, are essential for soil carbon sequestration and water retention. The stability of soil organic carbon is affected by feedstock and decomposition rate, whereas the long-term application of organic amendments benefits agricultural productivity and sustainability goals.

Furthermore, the modification in agricultural techniques to align with the changing environment can enhance crop productivity and decrease water consumption. The application of biochar enhances water use efficiency in irrigated agriculture by fostering stable soil aggregate formation and increasing microbial activity. Precision irrigation and smart irrigation water-saving technologies are crucial for the management of water resources, enhancement of crop output, and mitigation of environmental impacts through water conservation and decreased reliance on fertilizers and pesticides. It also decreases operational expenses and minimizes waste. Moreover, precision farming enhances a sustainable agricultural future by increasing productivity and yields while minimizing overall inputs.

Climate change substantially affects food systems, resulting in diminished production, biodiversity, and intensified drought and saline soils. Climate change adversely impacts crop development, potentially obstructed by elevated CO_2 concentrations. Crops, especially C3 plants, could benefit from elevated CO_2 concentrations; however, this may necessitate improved sink capacity. Despite advancements in Egypt's breeding of drought-tolerant wheat and maize varieties, their yields are anticipated to decline due to alterations in precipitation and heightened greenhouse gas emissions. Climate change requires the creation of climate-resilient cultivars that can reduce methane and nitrous oxide emissions while enhancing carbon sequestration. Selective breeding to improve water and nitrogen use efficiency is essential. Creating heat-resistant genotypes and enhancing CO_2 fertilization may offer remedies to forthcoming climatic challenges.

Strategies for adapting to climate change encompass crop rotation and the cultivation of climate-resilient crop varieties. Adaptation initiatives encounter obstacles such as insufficient knowledge and policy, inadequate adaptive potential, a lack of scientific data, limited financial resources, and fragile institutional frameworks. Employing advanced breeding techniques such as in vitro culture, genetic engineering, marker technology, and gene editing can enable the creation of new resilient genotypes. Genetically modified technologies can improve sustainable agriculture by increasing crop yields and elevating farmers' incomes. Strategies to augment seed yield resilience, agricultural diversity, and ecological robustness in agriculture encompass genetic engineering, crop varietal enhancement, optimized management practices, and the integration of traditional approaches alongside marginal habitats.

Maize production in Egypt fluctuates due to weather circumstances, irrigation practices, and agricultural methodologies. Egypt experiences considerable discrepancies between maize production and consumption as a result of fast population growth. Furthermore, climate change poses a substantial risk to maize output, with projection indicating a 7% decline in productivity for every degree rise in world mean temperature. Employing simulation models to project the effects of climate change on maize yield is an effective approach. The AquaCrop simulation model was employed to assess the effects of climate change on maize productivity, considering different planting dates and irrigation levels following its calibration and validation. The AquaCrop model was verified for maize grain and biological yield, demonstrating great predictive accuracy. The simulation results for maize grain and biological yields were closely aligned with the observed values, indicating a strong correlation between the predicted and actual data. The research indicated that different maize planting dates and soil moisture levels markedly affected grain and biological yields. Delaying maize planting restricted the vegetative and reproductive growth stages, leading to reduced dry matter buildup in plant organs and lower yield. The application of the future climatic scenario SSP5-8.5 for the periods 2020–2039 and 2040–2059 indicates substantial reductions in maize and biological yields under the third planting date and the lowest irrigation level. The expected reductions in maize sown on the optimal planting date and irrigation level were the lowest, whereas the highest reduction is expected on late maize sowing with the lowest irrigation level. Increased temperatures from the year of 2040 to the year of

2059 are anticipated to decrease moisture in the root zone, resulting in diminished yields. The actual evapotranspiration of maize is anticipated to increase due to rising temperatures. Water productivity metrics derived from grain yield or biological yield will diminish, with negligible reductions noted for maize sown at the optimal planting date and irrigation level.

Legume-based intercropping systems can enhance agricultural productivity, nutritional quality of food, and soil health, thereby contributing to the attainment of sustainable development goals of the United Nations. Legume-based intercropping systems can improve resistance to climate change by maximizing the utilization of natural resources in food production.

Climate change is anticipated to affect soil microorganisms and their communities, which are essential for facilitating carbon and nitrogen cycles. Legume-based intercropping systems can improve nutrient uptake efficiency by increasing the density of absorptive roots and root hairs, hence improving nutrient absorption and enhancing soil quality and resource acquisition. Climate change also impacts soil enzyme activity, which is crucial for sustaining soil quality and fertility. Legume-based intercropping systems can boost soil fertility by mobilizing soil organic matter, phosphate, iron, manganese, and zinc.

Climate change is anticipated to elevate soil evaporation losses, thereby diminishing the moisture accessible to crops. Legume-based intercropping systems can diminish runoff, enhance soil moisture utilization, and augment yield per unit of irrigation water utilized. Legumes can augment soil micro- and macropores, hence enhancing water infiltration to deeper root zone depths. Moreover, legume-based intercropping systems can improve soil physical-hydraulic properties and optimize soil structure, hence increasing soil water retention and availability.

Climate change is anticipated to exacerbate extreme weather phenomena. Intercropping tall plants with shorter ones can alleviate drought and heat stress by offering partial shade and diminishing wind speed, so lowering soil water loss. Climate change may exacerbate soil degradation and erosion. Legume-based intercropping systems can improve soil nutrient circulation, increase soil nitrogen content, and alter soil pH. Legumes can also extract calcium and magnesium from soil, lowering soil pH and inhibiting N_2O gas emissions. Legume-based intercropping systems can facilitate carbon sequestration, as increased biomass production in these systems boosts carbon input to the soil, hence augmenting soil organic matter.

Climate change is anticipated to elevate the probability of plant infections by diseases, hence diminishing agricultural output. Legume-based intercropping systems can interrupt insect and disease cycles, hence promoting biodiversity and diminishing pest populations. The augmented canopy cover from legume-based intercropping can also decrease weed biomass, hence reducing insect prevalence and constraining the reproductive potential of diseases. Moreover, legume-based intercropping systems can markedly diminish greenhouse gas emissions, especially N_2O and CO_2, by enhancing carbon and nitrogen management within agricultural ecosystems.

A prime illustration of a legume-based intercropping system is the cultivation of cowpeas at 50% of their planting density alongside maize at 100% of its planting density, which facilitates a compensatory soil moisture effect, enabling cowpea plants to access water from upper soil layers while maize plants draw moisture from deeper strata. This results in a cowpea yield equivalent to its monoculture and a maize yield above its monoculture. Intercropping cowpea with maize functions as a climate change adaptation method, bolstering the resilience of nonlegume crops to climate-induced stress.

A case study was provided on the impact of climate change on Egypt's Nile Delta, highlighting how climate change and unsustainable human behaviors exacerbated soil degradation. The Nile Delta of Egypt is a region with rich and fertile soils and has been significantly impacted by climate change, leading to reduced agricultural production and increase nutrients loss. The principal degradation processes encompass sea-level rise, land subsidence, salinization, waterlogging, erosion, soil compaction, depletion of soil organic matter, and desertification. Climate change intensifies land subsidence via sea-level rise, modified sediment deposition, saline water intrusion, and heightened groundwater extraction. The lands of the Nile Delta's are vulnerable to sinking due to increased salinity in groundwater, which weakens the structural integrity of the land.

Salinization and waterlogging are significant soil degradation issues in the Nile Delta, affecting plant development and agricultural productivity. Salts remain after water evaporates, exacerbated by insufficient drainage and excessive irrigation, which is intensified by climate change. Waterlogging, a condition where soil is saturated with water, is a major issue due to level topography, insufficient drainage systems, and flood irrigation. Elevated temperatures and diminished precipitation exacerbate drought conditions, adversely affecting plant growth and heightening disease vulnerability. The loss of soil organic matter resulting from climate change accelerates decomposition, diminishes soil fertility, and restricts moisture retention. Soil compaction is intensified by increasing temperatures and modified precipitation patterns, resulting in denser soil formations and diminished water infiltration and drainage. Desertification, induced by diminished water flow, jeopardizes the sustainability of long-term agriculture.

Rising temperatures in the Nile Delta are expected to lead to shifts in precipitation patterns, affecting agriculture, water resources, and environmental health. Elevated temperatures augment evaporation rates, exacerbating the risk of soil desiccation and salinization. Inadequate water management techniques can intensify water scarcity, rendering soils more vulnerable to deterioration.

The average annual temperature in the Nile Delta governorates has consistently increased over the past 23 years. This consistent trend of warming is particularly noticeable in the coastal governorates like Alexandria and Port Said. Furthermore, coastal governorates are already experiencing saltwater intrusion and erosion, while inland governorates are experiencing accelerated loss of organic matter, depletion of nutrients, and erosion.

Furthermore, climate change is causing unpredictable rainfall patterns in the Nile Delta, leading to droughts and intense storms. This instability can compromise soil stability, making it more susceptible to erosion. Shifts in rainfall patterns within the Nile Basin could impact the flow of the Nile River. Variations in annual precipitation totals in eight governorates over 23 years (2000–2023) show significant shifts in precipitation patterns. These patterns could impact water scarcity, soil erosion, and agricultural land degradation. Moreover, Sharkia governorate in the Nile Delta of Egypt has experienced a significant increase in rainfall over the past two decades, indicating a shift in local climate trends.

Furthermore, soil microbial biodiversity, particularly in the Nile Delta of Egypt, can play a crucial role in mitigating climate change. Diverse microbial communities enhance soil health, nutrient cycling, and carbon storage, which are essential for resilience against climate change impacts. Soil microbial diversity contributes to soil stability and functionality, enhancing soil nitrogen cycling and plant productivity. Soils with higher microbial diversity are more resistant to stressors, including those induced by climate change.

The interaction between plant diversity and soil microbial communities is crucial for soil health and resilience. Diverse plant communities support different microbial populations, which is essential in the Nile Delta, where agricultural practices often lead to monocultures. Enhanced plant diversity through crop rotation and cover cropping can foster a more resilient soil microbial community, improving soil carbon storage and mitigating greenhouse gas emissions. Climate variability, particularly changes in precipitation and temperature, can significantly impact soil microbial diversity. Maintaining microbial diversity is essential for sustaining agricultural productivity and ecosystem resilience, especially in the context of climate change. Strategies such as reducing tillage, minimizing chemical inputs, and implementing organic farming practices can enhance soil microbial communities, improve soil structure and fertility, and contribute to carbon sequestration. Promoting soil microbial biodiversity through sustainable agricultural practices can improve ecosystem functions, support agricultural productivity, and contribute to climate change mitigation efforts.

Higher temperatures promote root elongation and plant-microbe interactions, while lower temperatures result in less colonization. Frost, a major wintertime weather phenomenon in Northern Egypt, negatively impacts plant development and soil microbial activity. Frost-related cycles of freezing and thawing physically disturb the soil structure, impacting microbial communities and nutrient availability. Nematodes, root rot fungi, and microbial populations are also affected by frost.

Controlling microbial processes, such as carbon fixation and methane oxidation, can affect greenhouse gas emissions and sequestration. Bacteria play a role in the carbon cycle, promoting plant development and decomposing organic materials. Climate variables, such as temperature and precipitation, impact microbial metabolic processes, leading to increased CO_2 emissions. Microbial enzyme systems help lower CO_2 levels through respiration, sequestering carbon, and controlling atmospheric carbon concentrations. The nitrogen cycle in soil is driven by microbial activities such as nitrogen fixation, nitrification, assimilation, ammonification, and denitrification.

Soil moisture availability affects microbial activity, with ideal conditions promoting respiration, enzyme activity, and biomass generation. Variations in soil moisture can affect crop yield and soil fertility, especially in agricultural systems that rely on consistent microbial contributions to carbon and nitrogen cycling. Climate change has increased the unpredictability of soil moisture, leading to changes in microbial activities, greenhouse gas emissions, and nutrient leaching. Adaptive strategies like osmoregulation, dormancy, and enzyme synthesis are essential for microbial communities to cope with moisture stress.

Soil microbiomes, controlled by temperature and moisture, are affected by drought, with some species preferring Gram-positive species over Gram-negative ones. Soil microorganisms use tactics like increased production of extracellular polymeric substances, osmolytes, and organic chemicals to survive. Drought also affects soil enzyme activities, photosynthetic output, water usage efficiency, and nutrient availability.

Moreover, climate change and irrigation methods contribute to increased salinity, which affects rhizosphere bacteria and soil microbial activities, including enzyme functions. Salinity also affects the bacteria/fungi ratio, as fungi are more susceptible to salt stress. Excessive salinity affects soil health, community makeup, and microbial activity. Pathogenic nematodes and root rot pathogens may adapt to saline conditions, allowing salt-tolerant pathogenic organisms to thrive.

Machine Learning Use in Crop Production

Machine learning approaches have gained interest in agriculture, including smart irrigation, crop management, disease detection, and soil management. Crop yield prediction using machine learning while considering climate change impacts is a useful approach. Classification models, regression models, and deep learning models can be adopted for crop yield prediction under climate change. Classification methods involve categorizing data into distinct classes, such as semiparametric neural networks, bagging, and Bayesian hyperparameter optimization. Alternatively, regression methods aim to understand the relationship between a dependent variable and one or more independent variables. Crop yield prediction algorithms and boosted regression trees (BRTs) were used to predict crop yield. Other classifications, namely, DecisionTreeRegressor, RandomForestRegressor, and ExtraTreesRegressor, were also used. Furthermore, ensemble learning models were also developed for crop yield prediction. Furthermore, deep learning approaches have been increasingly used in smart agriculture, including plant disease detection, crop identification, water management, and fruit counting. Examples of deep learning approaches are long short-term memory and gated recurrent unit architectures.

Additionally, the integration of process-based models with machine learning techniques is a promising approach for improving predictions of soil, water, and crop dynamics in agricultural systems. This approach addresses the limitations of each approach in isolation, enabling more accurate and robust predictions. It also allows for a wider range of environmental variability and management options. The success of this integration depends on the quality and availability of data, the appropriate selection and tuning of process-based models, machine learning algorithms, and understanding the interactions between the two methodologies.

Furthermore, the integration of process-based crop models, machine learning, and remote sensing offers another transformative approach to agricultural monitoring and decision-making. This hybrid framework not only improves the accuracy and scalability of yield gap assessments but also provides actionable insights for enhancing agricultural productivity and sustainability. Although challenges persist in integrating data sources and expanding computational capacity, continuous research and technological progress offer the potential to address these issues, enabling more efficient agricultural management and policy formulation in the future.

To conclude, the resiliency of the agroecosystem in Egypt can be attained through changing traditional practices with using more innovated practices, such as changing crop sequence from two crops per year to three crops per year, and implementing intercropping system including legume crops. Using greenhouses and soilless agricultural as well as integrated farming systems, in which livestock, aquatic species, fruit-bearing trees, and vegetables constitute one production system, provides farmers with efficient farm production with optimized resource use and ecological preservation. Using fishpond wastewater in agriculture enhances crop productivity by recycling nutrients, which reduces waste and environmental impact, as well as it promotes water conservation and reduces the need for freshwater irrigation. Furthermore, to increase the resiliency of the agroecosystem to climate change, mitigation and adaptation measures should be implemented for water resources, crop production, and soil. Additionally, using simulation models and integrating process-based crop models, machine learning, and remote sensing offer transformative approaches to agricultural monitoring and decision-making.

Recommendations

Throughout this book project, the editorial teams noted some areas that could be explored to further improvement. Based on the authors' findings and conclusions, this section offers a set of recommendations providing suggestions for future researchers, stakeholders, and decision-makers as follows:

Innovative Cropping Systems for Resilience

- Changing cropping sequence and implementing intercropping system can help farmers improve yields, maximize agricultural resources, and increase food security.
- Including a legume crop in the cropping sequence and as an intercrop accelerates microbial activity and nitrogen fixation, which enhance soil health and reduce chemical inputs.
- Agroecological intensification, particularly legume-cereal intercropping systems, has the potential to enhance crop diversification, farm nitrogen self-sufficiency, soil health, pest control, and yield stability.

- The advantageous agroecological effects of legume-cereal intercropping systems assist in addressing the challenge of creating agroecosystems that attain high crop yields with minimal input levels.
- The 4C approach enables a thorough examination of how its distinct principles augment the beneficial effects of intercropping by classifying interspecific plant-plant interactions into four categories: competition, complementarity, collaboration, and compensation.
- When implementing intercropping, farmers should consider crop-specific responses to environmental factors to enhance overall productivity.
- In intercropping, farmers can modify planting density of the intercrops to enhance light utilization, reduce competition, and attain optimal crop growth, maximizing biomass and grain yield.
- When intercropping, farmers should cultivate tall and short species as intercrops to enhance light interception through expanded ground cover.
- Farmers should improve agronomic practices in intercropping systems to mitigate the adverse consequences of competition while promoting complementarity, cooperation, and compensation.

Augment the Soybean Intercropping System with Maize

- The productivity of intercropping soybeans with maize depends on the spatial arrangement of the two crops. By strategically planting soybeans between rows of maize, farmers can maximize sunlight exposure for both crops, leading to increased overall productivity.
- Selecting complementary varieties of soybeans and maize with similar growth rates can further enhance light interception and productivity in an intercropping system.
- Plant density and variety are essential for optimizing light usage efficiency in soybean and maize intercropping systems, which provides valuable insights into sustainable farming practices and optimizes crop yield while minimizing environmental impacts.
- Utilizing more efficient and sustainable agricultural practices, farmers can benefit from the production of maize and soybeans, ensuring a stable and productive ecosystem.
- Using insect-tolerant soybean varieties can reduce pesticide usage and promote sustainable agriculture.
- Soybean varieties Giza 21 and Giza 111 exhibited tolerance to soybean mosaic virus infection. Using these varieties can reduce the reliance on chemical pesticides, leading to increased agricultural output and improved sustainable farming systems.
- Soybean variety Giza 22 is a multi-pest tolerance that can be used to reduce the reliance on chemical insecticides, promoting sustainable agricultural practices.

Untraditional Approaches to Maximize Agricultural Production

- Greenhouses and soilless agriculture systems are considered viable alternatives for enhancing resource efficiency, crop yields, and food accessibility.
- Greenhouse systems offer various benefits that increase agricultural output and sustainability as well as contribute to more efficient farming techniques.
- Soilless agriculture techniques, such as hydroponics and aeroponics, provide novel solutions to many of the problems associated with traditional soil-based farming.
- Despite initial high costs for establishment and energy demands, overcoming these barriers is critical for wider acceptance and long-term sustainability.
- Continued innovation in them, investment, and collaboration are essential for creating resilient agroecosystems that will support future generations.
- Using fishpond wastewater in agriculture enhances crop productivity by recycling nutrients, which reduces waste and environmental impact.
- It can enhance farm profitability by lowering input costs and increasing crop yields, as it eliminates the need for chemical fertilizers and freshwater irrigation.
- The fishpond wastewater promotes water conservation and reduces the need for freshwater irrigation, benefiting drought-prone areas.
- To ensure successful implementation, it is essential to address potential issues such as nutrient imbalance, pathogen risks, and regulatory compliance.

Climate Change Impacts and Mitigation

- Mitigating the effects of climate change on water resources necessitates collaborative efforts from governments, companies, and communities to formulate effective adaptation strategies and minimize further environmental damage.
- Mitigating irrigation water shortages caused by climate change necessitates a comprehensive strategy that includes technology advancements, strategic agricultural practices, robust policy frameworks, and community education.
- Modifying irrigation techniques and implementing sophisticated technology are essential for alleviating the effects of climate change on agriculture.
- Policymakers must prioritize investments in these domains to facilitate sustainable agricultural development amid persistent environmental changes.
- Precision irrigation methods provide a comprehensive strategy to mitigate the negative impacts of climate change on agriculture.
- Farmers can save water resources and sustain or augment agricultural yield by embracing contemporary technologies, optimizing management techniques, and improving soil health.
- Advanced irrigation technologies and precision agriculture should be implemented to enhance water use efficiency to overcome the negative effect of climate change.
- Biotechnology, encompassing genetic engineering and various molecular breeding techniques in conjunction with traditional breeding, is essential for develop-

ing resilient crops to climate change capable of enduring biotic and abiotic challenges.

- Expanding crop diversity beyond mainstays such as wheat and maize can enhance resilience, drought tolerance, and insect resistance, such as quinoa, forage and grain millets, canola, and barley, which has high resilience and capacity to thrive in marginal soils and stressful conditions.
- Enhanced integrated management strategies should be conducted, including optimum planting schedules, cropping systems, and crop rotation to defeat the negative impacts of the expected climate change.
- It is essential to create methods that alleviate climate change risks affecting maize productivity by determining ideal planting dates to endure the abiotic stressors associated with climate change.
- The AquaCrop model was exceptionally effective in precisely predicting maize production under the different planting dates and irrigation levels.
- AquaCrop projected a decline in maize grain output, biological yield, and water productivity under the future scenario SSP5-8.5.
- It is advisable to sow maize in early May and irrigate it with 20% available soil moisture deficit to alleviate the effects of climate change.
- In instances of water scarcity, employing deficit irrigation of maize at 40% available soil moisture deficit can conserve water while minimizing yield losses.
- Legume-based intercropping systems should be adapted to mitigate GHG emissions and enhancing resilience to climate change and mitigating the aforementioned adverse effects.
- Legume-based intercropping systems promote soil nutrient cycling, augmenting soil organic matter which is strongly associated with enhanced water retention capacity, which makes it suitable to be used to mitigate the effect of climate change.
- Legume-based intercropping systems mitigate yield losses due to climate change, stabilizing grain production and enhancing crop resilience.
- Legume-based intercropping systems can enhance soil carbon sequestration and reduce runoff and soil erosion as well as interrupt the life cycles of pests, diseases, and weeds; all are expected to increase under climate change.
- To counter the problems imposed by climate change on the Nile Delta of Egypt, a variety of adaptive agricultural techniques should be used, namely, improved irrigation methods, soil moisture conservation strategies, the installation of appropriate drainage systems, and the introduction of crops resistant to salt and drought.
- A comprehensive strategy that incorporates the combination of physical and bioremediation technologies should be conducted to successfully address the problem of soil degradation in the Nile Delta.
- Soil fertility, structure, and crop resilience to climate stressors can all be improved by increasing soil microbial diversity through the use of microbial inoculants, biochar additions, and sustainable water management practices.

- Advanced soil monitoring techniques and predictive models should be used to aid in the early detection of environmental changes, allowing for swift intervention to mitigate their detrimental effects on agriculture.
- Future research should focus on optimizing microbial-based treatments to enhance plant-microbe interactions, promote soil carbon sequestration, and develop climate-resilient crop types.
- The use of microbial technologies to conventional farming practices and sustainable soil management strategies should potentially be encouraged by policy frameworks.
- Cooperation between researchers, legislators, and farmers will be essential to developing all-encompassing plans to mitigate the effects of climate change on soil ecosystems.
- By encouraging microbial biodiversity and using innovative soil management strategies, agricultural systems can become more resilient to the ongoing effects of climate change, ensuring both environmental sustainability and long-term output.

Machine Learning Use in Crop Production

- Probabilistic machine learning models could be explored as they provide uncertainty measures which could be beneficial for decision-making in smart agriculture.
- More research could be devoted to interpretable machine learning models for crop yield prediction so that useful insights could be gained.
- By combining the mechanistic understanding of process-based models with the data-driven capabilities of machine learning algorithms, a more comprehensive and efficient assessment of agroecosystems facilitates informed decision-making as well as leverages the strengths of both, leading to more accurate and robust predictions.
- By integrating process-based models with machine learning, we can enhance the accuracy and reliability of the predictions, supporting sustainable agricultural practices and climate resilience.
- The coupling of process-based models with machine learning represents a promising avenue for advancing agricultural research and practice.
- By addressing the limitations and capitalizing on the strengths of this hybrid methodology, we can contribute to a more sustainable and resilient future for agriculture.
- Future research should focus on improving the quality and availability of data, further developing process-based models and machine learning algorithms, integrating other factors into the coupled approach, enhancing scalability, and promoting interdisciplinary collaboration.
- The integration of process-based crop models, machine learning, and remote sensing offers a transformative approach to agricultural monitoring and decision-making.

- This hybrid framework not only improves the accuracy and scalability of yield gap assessments but also provides actionable insights for enhancing agricultural productivity and sustainability.
- Continuous research and technological progress offer the potential to address these issues, enabling more efficient agricultural management and policy formulation in the future.

Last But Not Least

- It is critical to execute integrated policies that address the underlying causes of food insecurity, with continual study and collaboration among stakeholders required to develop successful solutions.
- Researchers, policymakers, and practitioners will need to work together to shape resilient and sustainable agricultural practices.
- Capacity building programs and training for farmers and stakeholders are essential for ensuring long-term viability and sustainable development.
- Continued technological breakthroughs, sustainability, and supportive legislation will be the key to realizing these systems' full potential, ultimately addressing global food security and environmental sustainability.

GPSR Compliance

The European Union's (EU) General Product Safety Regulation (GPSR) is a set of rules that requires consumer products to be safe and our obligations to ensure this.

If you have any concerns about our products, you can contact us on ProductSafety@springernature.com

In case Publisher is established outside the EU, the EU authorized representative is:

Springer Nature Customer Service Center GmbH
Europaplatz 3
69115 Heidelberg, Germany

Batch number: 10704270

Printed by Printforce, the Netherlands